Lecture Notes in Physics

Volume 961

The series Lecture Notes in Physics (LNP), founded in 1969, reports new developments in physics research and teaching - quickly and informally, but with a high quality and the explicit aim to summarize and communicate current knowledge in an accessible way. Books published in this series are conceived as bridging material between advanced graduate textbooks and the forefront of research and to serve three purposes:

- to be a compact and modern up-to-date source of reference on a well-defined topic;
- to serve as an accessible introduction to the field to postgraduate students and non-specialist researchers from related areas;
- to be a source of advanced teaching material for specialized seminars, courses and schools.

Both monographs and multi-author volumes will be considered for publication. Edited volumes should however consist of a very limited number of contributions only. Proceedings will not be considered for LNP.

Volumes published in LNP are disseminated both in print and in electronic formats, the electronic archive being available at springerlink.com. The series content is indexed, abstracted and referenced by many abstracting and information services, bibliographic networks, subscription agencies, library networks, and consortia.

Proposals should be sent to a member of the Editorial Board, or directly to the responsible editor at Springer:

Dr Lisa Scalone
lisa.scalone@springernature.com

Alex Amato • Elvezio Morenzoni

Introduction to Muon Spin Spectroscopy

Applications to Solid State and Material Sciences

 Springer

Alex Amato
Laboratory for Muon Spin Spectroscopy
Paul Scherrer Institute
Villigen PSI, Switzerland

Elvezio Morenzoni
Laboratory for Muon Spin Spectroscopy
Paul Scherrer Institute
Villigen PSI, Switzerland

ISSN 0075-8450   ISSN 1616-6361   (electronic)
Lecture Notes in Physics
ISBN 978-3-031-44958-1   ISBN 978-3-031-44959-8   (eBook)
https://doi.org/10.1007/978-3-031-44959-8

This Springer imprint is published by the registered company Springer Nature Switzerland AG
The registered company address is: Gewerbestrasse 11, 6330 Cham, Switzerland

Paper in this product is recyclable.

To our wives Rebecca and Giovanna

Preface

This book describes the use of polarized positive muons as local spin probes of matter. The method is called μSR, which stands for muon spin rotation, relaxation, or resonance. Implanted and stopped in matter, the polarized muons provide information at atomic scale about the physical and chemical properties of the material under investigation. The seeds of the μSR method were planted in 1957 in the seminal paper by Garwin et al. (Physical Review (1957) 105, 1415), who discovered that the decays of the muon and the pion (the muon's parent particle) violate parity conservation. With great foresight, they predicted the possibility "that polarized positive and negative muons will become a powerful tool for exploring magnetic fields in nuclei, atoms, and interatomic regions".

This prediction has come true especially for the positive muon. The μSR technique is now applied to the study of virtually all types of materials in the solid, liquid, and gas phases. Systems studied include metals, insulators, magnetic and the expanding new classes of superconductors and topological materials, semiconductors, organic materials, and even biological samples. The development of a beam of tunable very low-energy muons has made it possible to investigate, in addition to the traditional bulk samples, thin films, multilayers, and heterostructures, so that the specific questions arising from reduced dimensions or the juxtaposition of different materials can be addressed.

In the last decades since the 1970s and 1980s, when the μSR method gained importance and its use increased enormously, there has been a not too large number of books exclusively covering the μSR technique. For many years, the last in the three-volume series edited by W. Hughes and C.S. Wu (*Muon Physics*, Academic Press, 1975) and especially the book by A. Schenck (*Muon Spin Rotation Spectroscopy*, Adam Hilger Ltd, 1985) have served as reference books. These books were followed by the monograph by V.P. Smilga and Yu. M. Belousov (*The Muon Method in Science*, Nova Science Publishers, 1994), which focuses more on theoretical aspects, and by the book of K. Nagamine (*Introductory Muon Science*, Cambridge University Press, 2003), which includes sections on applied muon science as well as an introduction to μSR. The book by A. Yaouanc and P. Dalmas de Réotier (*Muon Spin Rotation, Relaxation and Resonance*, Oxford University Press, 2011) is most recent monograph about μSR, but also more than 10 years old. Although rich in experimental information, it emphasizes the theoretical aspects, giving a detailed and coherent description of the theoretical tools used to understand

and analyze the μSR spectra and extract the physical information. Complementary to the above textbooks, there are graduate school proceedings on μSR, in which chapters, written by a large team of specialists, have been compiled into a book. Such books have appeared regularly: *Muons and Pions in Materials Research*, J. Chappert and R.I. Grynszpan eds., North-Holland, 1983; *Muon Science: Muons in Physics, Chemistry, and Materials*, S. Lee, S. H. Kilcoyne and R. Cywinski eds., Institute of Physics Publishing, 1999; and most recently *Muon Spectroscopy. An Introduction*, S. J. Blundell, R. De Renzi, T. Lancaster and F. L. Pratt eds., Oxford University Press, 2022. They reflect very well the evolution of μSR and its use over time. This evolution is also well represented in the Proceedings of the International μSR Conference, held every 3 years or so after the first one in 1975 in Rorschach, Switzerland. In addition, there are books dealing with specific applications of μSR such as in chemistry (which we do not cover in this book), and several review articles on specific topics. These are mentioned in the book where appropriate.

Parts of the book have evolved from a one-semester course for last year's undergraduate and graduate students with the generic title "Physics with Muons: from Atomic Physics to Condensed Matter Physics" (but with a focus on μSR), taught at the ETH and the University of Zurich by EM until 2017 and then continued at the University by AA. While roughly maintaining the structure of the lectures, the book represents a major expansion and improvement. It contains much more discussion, details, new examples, and figures (mostly newly drawn), so that its material goes beyond a one-semester course.

In the first part of the book, Chaps. 1–4 follow the logical sequence of steps in a μSR experiment, i.e., description of the muon properties relevant to μSR, generation of polarized muon beams at proton accelerators (Chap. 1, which also contains a brief historical look), muon implantation and thermalization in the sample (Chap. 2), experimental setup of a μSR measurement (Chap. 3), and analysis of the muon polarization data (Chap. 4).

In the second part of the book, Chaps. 5–8 discuss the application of the μSR technique to extract key physical information in the major fields: magnetism (Chap. 5), superconductivity (Chap. 6), study of semiconductors with muonium (the bound μ^+-electron atom) as hydrogen isotope (Chap. 7), thin-film physics with low-energy muons (Chap. 8). In the final part of the book, we cover two aspects that go beyond μSR but are still closely related. First, in Chap. 9, the use of negative muons for μ^-SR experiments and, more important and topical, the spectroscopy of muonic atoms and its use in applied sciences are presented. In the last chapter, the role of the muon as an elementary particle in particle physics is briefly described. In particular, the theoretical understanding of the decay properties of the muon, on which the μSR technique is based, is elucidated. Seven appendices provide information that we feel is either complementary or too lengthy for the main text. At the end of the book, problems (with solutions) have been added to test comprehension and to acquire practical familiarity with the material.

Our principal aim for this book was to provide a tutorial self-contained introduction to μSR by showing how and what kind of physical information can be extracted through a μSR experiment, by presenting illustrative examples of experiments.

We emphasize the connection to the experiments rather than giving an overview of the μSR research in a particular field or discussing in depth specific physical questions. We are well aware that there are many other interesting and topical μSR experiments, which would have deserved to be selected as examples. We apologize to the researchers whose contributions we have not mentioned. The literature on μSR is now very extensive and broad and, to avoid overloading the book, we could include only a few examples.

Prerequisites for a profitable reading of the book are basic knowledge of condensed matter physics, quantum mechanics, and familiarity with mechanics and electromagnetism (and for Chap. 10 special relativity) at a level which is achieved in the final year of an undergraduate physics program. To facilitate understanding, we have made extensive use of cross-reference links, which are particularly efficient in the electronic version. The book is intended not only for students, but also for researchers in other fields who want to use the μSR technique or understand μSR results for their research, and last but not least, for practitioners.

The book has benefited indirectly from discussions with current and former colleagues at the Laboratory for Muon Spin Spectroscopy at the Paul Scherrer Institute, Villigen, Switzerland, and with many colleagues from other institutions with whom we have collaborated. There are too many to mention here; instead, we refer to the author lists of the publications that we have co-authored and that are cited in the book. We are grateful to Toni Shiroka and Zaher Salman for their comments and suggestions after visioning the book or parts of it. We would also like to thank Ms. Lisa Scalone for her editorial assistance and for her patience in waiting for this work to be completed.

Villigen PSI, Switzerland
August 2023

Elvezio Morenzoni
Alex Amato

Contents

Physical Constants, Symbols and Abbreviations

The following tables summarize the values of fundamental physical constants and quantities, the definition of the most important symbols, and the acronyms/abbreviations used in this book.

For the fundamental physical constants, we have used the current (2018) values recommended by the Committee on Data for Science and Technology (CODATA), E. Tiesinga et al., Rev. Mod. Phys. (2021), 93, 025010, which can also be found on the website of the National Institute of Standards and Technology USA (NIST) https://physics.nist.gov/cuu/Constants/index.html. These values are also used to calculate quantities not listed by CODATA (e.g., γ_μ, γ_{Mu}^T ...). Frequency of vacuum muonium ν_0 is taken from W. Liu et al., Phys. Rev. Lett. (1999), 82, 711. The muon lifetime is from the Particle Data Group (2022) R. L. Workman et al., Progress of Theoretical and Experimental Physics (2022), 2022, 083C01.

We define the g-factors g_e, g_μ, ..., gyromagnetic ratios γ_e, γ_μ ... and the Larmor angular frequencies ω_μ, ω_e ... as positive quantities and take into account the opposite sign of the positive muon and electron values with a plus or minus sign when necessary.

The units are expressed in SI, except for atomic or particle properties such as energy, momentum, or mass, which are expressed in eV, eV/c, and eV/c^2, respectively. Spin is always meant in units of $\hbar$, i.e., dimensionless. In Chap. 10 and related Appendix G dealing with aspects of particle physics, we use natural units ($\hbar = c = \varepsilon_0 = 1$).

Some symbols have different meanings depending on the context. For example, (i) A is used for the atomic mass number/atomic weight and for a generic hyperfine coupling constant; (ii) α is used for the fine structure constant, for the ratio $\Delta(0)/k_B T_c$ characterizing a superconductor, and as a fitting parameter of μSR spectra to account for different solid angles and efficiencies of positron detectors of a μSR spectrometer; and (iii) λ is the exponential relaxation rate and the magnetic penetration depth. We have retained these multiple definitions to conform to the most common use of these symbols in the literature. It should always be clear from the context which definition applies.

We use the term magnetic field instead of magnetic induction (or magnetic flux density) and this for macroscopic averaged values as well as for values at microscopic scale (e.g., for $\mathbf{B}_\mu$).

Values of Important Physical Constants (Grouped by Subject)

Symbol	Definition
	Value (SI or eV-units)
$a_0 = \dfrac{4\pi\varepsilon_0\hbar^2}{m_e e^2} = \dfrac{\hbar}{m_e c\alpha}$	Bohr radius (nucleus mass = ∞)
	$5.29177210903(80) \times 10^{-11}$ m
c	Speed of light (in vacuum)
	299,792,458 m s^{-1}
e	Elementary charge (electron charge magnitude)
	$1.602176634 \times 10^{-19}$ C
$\hbar$	Reduced Planck constant
	$1.054571817\ldots \times 10^{-34}$ J s
k_B	Boltzmann constant
	1.380649×10^{-23} J K^{-1}
$k_e = \dfrac{1}{4\pi\varepsilon_0}$	Coulomb constant
	$8.9875517923(14) \times 10^{9}$ kg m^3 s^{-2} C^{-2}
N_A	Avogadro constant
	$6.02214076 \times 10^{23}$ mol^{-1}
$R_y = \dfrac{m_e e^4}{(4\pi\varepsilon_0)^2 2\hbar^2} = \dfrac{m_e c^2 \alpha^2}{2}$	Rydberg energy (H atom with nucleus mass = ∞)
	$13.605693122994(26)$ eV
$\alpha = \dfrac{1}{4\pi\varepsilon_0}\dfrac{e^2}{\hbar c}$	Fine structure constant
	$7.2973525693(11) \times 10^{-3}$
	$1/137.035999084(21)$
ε_0	Vacuum electric permittivity
	$8.8541878128(13) \times 10^{-12}$ F m^{-1}
μ_0	Vacuum magnetic permeability
	$1.25663706212(19) \times 10^{-6}$ N A^{-2}
$\nu_0 = \dfrac{\omega_0}{2\pi}$	Hyperfine frequency (splitting) of the free muonium ground state
	$4,463,302,776\,(51)$ Hz
Φ_0	Magnetic flux quantum
	$2.067833848\ldots \times 10^{-15}$ Wb
τ_μ	Muon lifetime
	$2.1969811(22) \times 10^{-6}$ s
m_μ	Muon mass
	$105.6583755(23)$ MeV c^{-2}
	$1.883531627(42) \times 10^{-28}$ kg
$\dfrac{m_\mu}{m_e}$	Muon-electron mass ratio
	$206.7682830(46)$
$\dfrac{m_\mu}{m_p}$	Muon-proton mass ratio
	$0.1126095264(25)$
$\mu_\mu = g_\mu \mu_B^\mu I_\mu$	Muon magnetic moment
	$4.49044830(10) \times 10^{-26}$ J T^{-1}

Symbol	Definition Value (SI or eV-units)
$\mu_{\mathrm{B}}^{\mu} = \dfrac{e\hbar}{2m_{\mu}}$	Muon Bohr magneton $4.48521889(10) \times 10^{-26}$ J T^{-1}
g_{μ}	Free muon g-factor 2.0023318418(13)
$\gamma_{\mu} = \dfrac{\mu_{\mu}}{\hbar I_{\mu}} = g_{\mu}\dfrac{e}{2m_{\mu}} = \dfrac{g_{\mu}\mu_{\mathrm{B}}^{\mu}}{\hbar}$	Muon gyromagnetic ratio $8.51615457(19) \times 10^{8}$ rad s^{-1} T^{-1} $2\pi \times 135.538810(3)$ MHz T^{-1}
$\dfrac{\mu_{\mu}}{\mu_{p}}$	Muon-proton magnetic moment ratio 3.183345142(71)
$\dfrac{\mu_{\mu}}{\mu_{e}}$	Muon-electron magnetic moment ratio $4.83636198\,(11) \times 10^{-3}$
$\gamma_{\mathrm{Mu}}^{\mathrm{T}} = \dfrac{\gamma_{e} - \gamma_{\mu}}{2}$	Low-field gyromagnetic ratio of triplet Mu m$_{F} = \pm 1$ $2\pi \times 13{,}944.706$ MHz T^{-1}
m_{e}	Electron mass $0.51099895000(15)$ MeV c^{-2} $9.1093837015(28) \times 10^{-31}$ kg
$\mu_{e} = g_{e}\mu_{\mathrm{B}}S$	Electron magnetic moment $9.2847647043(28) \times 10^{-24}$ J T^{-1}
$\mu_{\mathrm{B}} = \dfrac{e\hbar}{2m_{e}}$	(Electron) Bohr magneton $9.2740100783(28) \times 10^{-24}$ J T^{-1}
g_{e}	Free electron g-factor 2.00231930436256(35)
$\gamma_{e} = \dfrac{\mu_{e}}{\hbar S} = g_{e}\dfrac{e}{2m_{e}} = \dfrac{g_{e}\mu_{\mathrm{B}}}{\hbar}$	Electron gyromagnetic ratio $1.76085963023(53) \times 10^{11}$ rad s^{-1} T^{-1} $2\pi \times 28{,}024.9514242(85)$ MHz T^{-1}
m_{p}	Proton mass $938.27208816(29)$ MeV c^{-2} $1.67262192369(51) \times 10^{-27}$ kg
$\dfrac{m_{p}}{m_{e}}$	Proton-electron mass ratio 1836.15267343(11)
μ_{p}	Proton magnetic moment $1.41060679736(60) \times 10^{-26}$ J T^{-1}
$\mu_{\mathrm{NM}} = \dfrac{e\hbar}{2m_{p}}$	Nuclear magneton $5.0507837461(15) \times 10^{-27}$ J T^{-1}
$\dfrac{\mu_{p}}{\mu_{\mathrm{NM}}}$	Proton magnetic moment to nuclear magneton ratio 2.79284734463(82)
m_{n}	Neutron mass $939.56542052(54)$ MeV c^{-2} $1.67492749804(95) \times 10^{-27}$ kg
μ_{n}	Neutron magnetic moment $-9.6623651(23) \times 10^{-27}$ J T^{-1}
$\dfrac{\mu_{n}}{\mu_{\mathrm{NM}}}$	Neutron magnetic moment to nuclear magneton ratio $-1.91304273(45)$

Important Symbols

Symbol	Definition
$A\ (\tilde{A})$	Generic symbol for hyperfine coupling constant (or tensor) (energy unit)
A_0	Initial experimental asymmetry
$A(t)$	Asymmetry spectrum or function
$A_{\mathrm{Mu}} = \hbar\omega_0 = h\nu_0$	Hyperfine coupling in free muonium
$\mathbf{B}$	(Local) magnetic field vector (generic)
B_{c}	Thermodynamic critical field
B_{c1}	Lower critical field
B_{c2}	Upper critical field
$\mathbf{B}_{\mathrm{cont},fd}$	RKKY enhanced magnetic contact field vector
$\mathbf{B}_{\mathrm{dem}}$	Demagnetizing field vector
$\mathbf{B}_{\mathrm{dip}}(\mathbf{r}_\mu)$	Dipolar field vector at muon site $\mathbf{r}_\mu$
$\mathbf{B}_{\mathrm{ext}}$	Externally applied magnetic field vector (by convention $\hat{\mathbf{z}}$-axis)
B_{ext}	Absolute value of $\mathbf{B}_{\mathrm{ext}}$
$\mathbf{B}_{\mathrm{F.\ cont}}$	Fermi magnetic contact field vector
$\mathbf{B}_{\mathrm{Lor}}$	Lorentz field vector
$\mathbf{B}_\mu$	Local magnetic field vector probed by the muon
B_μ	Absolute value of $\mathbf{B}_\mu$
E_{e^+}	Energy of a muon decay positron
$E_{e^+}^{\mathrm{max}} \cong \dfrac{m_\mu c^2}{2}$	Maximum energy of a decay positron
$E_{\mathrm{kin}},\ E_\mu$	Kinetic and total energy of a muon
E_{F}	Fermi energy
$E_{n,\mu} = -\dfrac{\overline{m}_\mu\, Z^2 e^4}{(4\pi\varepsilon_0)^2 2n^2\hbar^2}$	Energy of the level n of a muonic atom with atomic number Z
$f(\mathbf{B}_\mu)$ or $f(\mathbf{B})$	Distribution of vector $\mathbf{B}_\mu$ or $\mathbf{B}$
$f_m(B_\mu)$ or $f_m(B)$	Distribution of the field modulus
$\mathbf{I}_\mu$	Muon spin vector and operator (in units of $\hbar$)
I_μ	Muon spin (in units of $\hbar$)
$\hat{\jmath}^\alpha$	Muon spin 4-vector
$\mathbf{I}_{\mathrm{N}}$	Nuclear spin vector and operator (in units of $\hbar$)
I_{N}	Spin quantum number of a nucleus
$\mathbf{j}_s$	Supercurrent vector
$\mathbf{k}$	wave vector in reciprocal space
K_{exp}	Experimental relative frequency shift (Knight shift)
K_μ	Muon Knight shift
ℓ	Mean free path
$\mathbf{M}$	Magnetization vector
$\mathbf{M}_{\mathrm{S}}$	Saturation magnetization vector

Symbol	Definition
m_{I_μ}	Muon spin quantum number
m_S	Electron spin quantum number
m^*	Effective carrier mass
$\overline{m}_\mu = \dfrac{m_\mu m_A}{m_\mu + m_A}$	Reduced mass in a muonic atom with mass number A
$n^\uparrow / n^\downarrow$	Electron density with spin parallel/antiparallel to the quantization direction (e.g., applied magnetic field direction)
$n_e = n^\uparrow + n^\downarrow$	Density of conduction electrons
n_{sc}	Density of superconducting carriers, superfluid density
$\mathbf{P}(t)$	Polarization vector of the muon ensemble
$P(t) = \mathbf{P}(t) \cdot \hat{\mathbf{n}}$	Projection of the polarization vector along the direction of observation $\hat{\mathbf{n}}$
$P_x(t)$	Transverse polarization function
$P_z(t)$	Longitudinal polarization function
q	Charge of a particle (with sign)
R	Total range of a particle in matter $[\mathrm{g\,cm}^{-2}]$
R_p	Projected range: projection of R into direction of incoming particle $[\mathrm{g\,cm}^{-2}]$
$r_{n,\mu} = \dfrac{4\pi \varepsilon_0 \hbar^2 \, n^2}{\overline{m}_\mu Z e^2}$	Radius of an orbit with quantum number n in a muonic atom with atomic number Z
$r_s = \dfrac{1}{a_0}(\dfrac{3}{4\pi n_e})^{\frac{1}{3}}$	Electron density parameter: radius occupied by one conduction electron in units of a_0
$\mathbf{S}$	Electron spin vector and operator (in units of $\hbar$)
S	Electron spin
$S = -\dfrac{dE}{dx}$	Energy loss (or mass stopping power) $[\mathrm{MeV\,cm^2\,g^{-1}}]$, $x\,[\mathrm{g\,cm}^{-2}]$ mass per unit area
T_c	Critical temperature
T_C	Curie temperature
T_N	Néel temperature
T_f or T_g	Freezing or glass temperature
v	Velocity of a particle or relative velocity between two inertial frames of reference
v_F	Fermi velocity
$\hat{\mathbf{z}}$	Unit vector along z direction
$\alpha = \dfrac{\Delta(0)}{k_B T_c}$	Parameter of the α-model
α	Fit parameter of the experimental asymmetry
β	Fit parameter of the experimental asymmetry (generally $\beta \equiv 1$)
β	Stretch parameter of the stretched exponential function
$\beta = \dfrac{v}{c}$	Ratio of velocity over the speed of light
$\sqrt{\langle \Delta B_\mu^2 \rangle}$	Standard deviation of B_μ, field width
$\Delta(T,\mathbf{k})$ or $\Delta(T,\phi)$ or $\Delta(T)$	Superconducting gap function
ΔR_p	Projected range straggling (FWHM)
$\varepsilon = \dfrac{E_{e^+}}{E_{e^+}^{\max}}$	Scaled decay positron energy

Symbol	Definition
ϵ	Detector efficiency
$\gamma = \dfrac{1}{\sqrt{1-(v^2/c^2)}}$	Lorentz factor
$\gamma^\alpha\ (\alpha = 0, 1, 2, 3),\ \gamma^5$	γ-matrices
$\kappa = \dfrac{\lambda}{\xi_{GL}}$	Ginzburg-Landau parameter
λ	Exponential depolarization rate
λ_L	Longitudinal depolarization rate
λ_T	Transverse depolarization rate
λ	Magnetic penetration depth
λ_L	London magnetic penetration depth
$\boldsymbol{\mu}_\mu = g_\mu\, \mu_B^\mu\, \mathbf{I}_\mu$	Muon magnetic moment vector and operator
$\boldsymbol{\mu}_e = g_e \mu_B \mathbf{S}$	Electron magnetic moment vector and operator
$\boldsymbol{\mu}_j$	Electronic magnetic moment vector and operator at site j
$\boldsymbol{\mu}_{N,j}$	Nuclear magnetic moment vector and operator at site j
$\nu = \dfrac{1}{\tau_c}$	Fluctuation rate
$\nu_\mu = \dfrac{\omega_\mu}{2\pi}$	Muon spin precession frequency [Hz]
ξ_{BCS}	BCS coherence length
ξ_{GL}	GL coherence length
$\rho_{sc} \equiv \dfrac{n_{sc}}{m^*}$	Superfluid density
$\boldsymbol{\sigma}$	Pauli matrices $\sigma_x, \sigma_y, \sigma_z$
σ	Gaussian depolarization rate
σ_N	Gaussian depolarization rate by nuclear moments
τ_c	Fluctuation/correlation time
φ	Initial muon spin phase
$\omega_\mu = \gamma_\mu B_\mu$ or $\gamma_\mu B_{ext}$	Angular frequency of the muon spin precession (Larmor precession, $[\text{rad s}^{-1}]$) in the local field B_μ or in the applied field B_{ext}

Acronyms and Abbreviations

Symbol	Definition
AF	Antiferromagnetism
at%	Atomic percent
BCS	Bardeen-Cooper-Schrieffer (theory)
BNL	Brookhaven National Laboratory
BSCCO	Bismuth Strontium Calcium Copper Oxide, generic formula $Bi_2Sr_2Ca_{n-1}Cu_nO_{2n+4+x}$, mostly $n = 2$
CERN	European Organization for Nuclear Research
CODATA	Committee on Data for Science and Technology
DFT	Density Functional Theory
DMS	Dilute Magnetic Semiconductors
FFT	Fast Fourier Transform
FLL	Flux-Line Lattice
FNAL	Fermi National Accelerator Laboratory (Fermilab)
FWHM	Full Width at Half Maximum
FM	Ferromagnetism
GKT	Gaussian Kubo-Toyabe (function)
GL	Ginzburg-Landau (theory)
HWHM	Half Width at Half Maximum
IUPAC	International Union of Pure and Applied Chemistry
J-PARC	Japan Proton Accelerator Research Complex
KEK	High Energy Accelerator Research Organization
KT	Kubo-Toyabe (function)
LEM	Low-Energy Muon (beam)
LE-μSR	Low-Energy Muon Spin Rotation/Relaxation/Resonance
LF-μSR	Longitudinal Field Muon Spin Rotation/Relaxation/Resonance
LKT	Lorentzian Kubo-Toyabe (function)
LSCO	Lanthanum Strontium Copper Oxide, $La_{2-x}Sr_xCuO_4$ (214)
Mev	Million events (recorded in an experiment)
MIXE	Muon Induced X-ray Emission
ML	Machine Learning
μSR	Muon Spin Rotation/Relaxation/Resonance
NC	Nuclear Capture
NMR	Nuclear Magnetic Resonance
PIXE	Proton or Particle Induced X-ray Emission
ppb	Parts per billion
ppm	Parts per million
PS	Polystyrene
PSI	Paul Scherrer Institute
QCP	Quantum Critical Point

Symbol	Definition
QED	Quantum Electrodynamics
QFT	Quantum Field Theory
QPT	Quantum Phase Transition
QSL	Quantum Spin Liquid
rms	Root mean square
RAL	Rutherford Appleton Laboratory
RE	Rare Earth
RKKY	Ruderman-Kittel-Kasuya-Yosida (interaction)
SCA	Strong Collision Approximation
SCR	Self-Consistent Renormalization (theory)
SRO	Strontium Ruthenate, Sr_2RuO_4
SM	Standard Model of particle physics
SMM	Single Molecule Magnet
$S\mu S$	Swiss Muon Source
STP	Standard Temperature and Pressure
TF-μSR	Transverse Field Muon Spin Rotation/Relaxation/Resonance
TRIM.SP	Transport and Range of Ions in Matter (Monte Carlo Program)
TRIUMF	TRI-University Meson Facility
TRSB	Time-Reversal Symmetry Breaking
UHV	Ultra-High Vacuum
V-A	Vector minus Axial (interaction)
wTF	Weak Transverse Field
wt%	Weight percent
XRF	X-Rays Fluorescence
YBCO	Yttrium Barium Copper Oxide, $YBa_2Cu_3O_{7-\delta}$ or $YBa_2Cu_3O_{6+x}$ (also known as Y123)
ZF, LF, TF	Zero, Longitudinal, Transverse applied Field $\mathbf{B}_{ext}$
ZF-μSR	Zero Field Muon Spin Rotation/Relaxation/Resonance

Fundamentals

1

In the Muon Spin Rotation/Relaxation technique (μSR), positive polarized muons produced at an accelerator, are implanted in a sample where they thermalize and act as a local spin or magnetic probe of the compound under study.

Since its discovery in the 1930's, see Sect. 1.2, the muon has played an important role in particle physics, see Chap. 10. On the other hand, condensed matter or materials scientists are generally less familiar with the unstable[1] muon than with the electron, neutron, and proton, which are the constituents of an atom. However, the specific properties of the muon make it an excellent spin probe and, although unstable, the only elementary particle other than the electron / positron, that is used as a probe of matter.

In this chapter, after a brief historical overview, we introduce the properties of the muon and describe how these particles produced at an accelerator can be manipulated to form a polarized beam that can be implanted in a sample.

1.1   The Muon as Elementary Particle

The muon (μ) is one of the few elementary particles. It has unit electric charge and spin $1/2$ (fermion). It is a charged lepton, that does not participate in the strong interaction and is one of the fundamental fermions of the Standard Model[2] (SM) of particle physics, which describes the electromagnetic, weak and strong interactions and is very successful (although not complete) in predicting a wide range of experimental results. Note that the negative muon (μ^-) is the particle, while the positively charged muon (μ^+) is by definition the antiparticle (Fig. 1.1).

---

[1] The muon is present in nature in the cosmic rays and is therefore less exotic than other unstable particles.

[2] For a comprehensive description of the Standard Model, see, e.g., Thomson (2013).

© Springer Nature Switzerland AG 2024
A. Amato, E. Morenzoni, *Introduction to Muon Spin Spectroscopy*,
Lecture Notes in Physics 961, https://doi.org/10.1007/978-3-031-44959-8_1

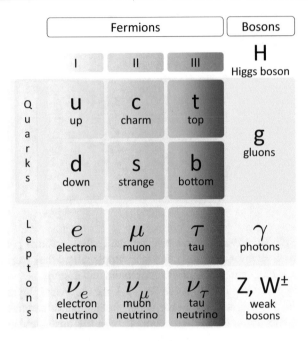

Fig. 1.1 The 12 fundamental spin $1/2$ fermions and the 4 fundamental bosons (integer spin) of the Standard Model of particle physics. Yellow shaded regions indicate which bosons couple to which fermions

1.2   A Brief History of the Muon

The history of the muon discovery is interesting because it exemplifies how science progresses step by step, small or big, sometimes in the wrong direction, but always testing accumulated knowledge against new observations and hypotheses. Here we summarize the timeline of milestones associated with the discovery and identification of the muon leading to its eventual use as a magnetic microprobe.

1785 Charles Augustin Coulomb finds that electroscopes[3] can discharge spontaneously by the action of the air.

1879 William Crookes shows that the rate of spontaneous discharge decreases as the pressure in a gas discharge tube decreases. This indicates that the discharge is due to the ionization of the air.

1896 Henri Becquerel discovers natural radioactivity. This triggers interest in the origin of the spontaneous electrical discharge previously observed in the air. The obvious hypothesis is that the discharge is

---

[3] The electroscope is an instrument used to detect the presence and sign of static electric charges via the Coulomb force.

caused by the radioactive materials on Earth, although this is difficult to prove.

1899   Julius Elster and Hans Geitel find that surrounding a (gold leaf) electroscope with a thick metal box reduces its spontaneous discharge. From this observation, they conclude that the discharge is due to highly penetrating ionizing agents outside of the container.

1909–1910   Theodor Wulf performs experiments with precise electroscopes that detect natural sources of radiation on the ground. Wulf takes his electroscope to the top of the Eiffel Tower to test whether the radiation is coming from the Earth or from above. He finds that the radiation intensity is slightly lower at the top of the Eiffel Tower. Because of absorption, if the radiation is coming from the Earth, the intensity should be much lower at the top. However, the results are not conclusive.

1910   Albert Gockel arranges the first balloon flights with the purpose of studying the properties of penetrating radiation. He measures the levels of ionizing radiation up to an altitude of 3000 meters and concludes that the ionization does not decrease with altitude and therefore cannot have a purely terrestrial origin. He also introduces the term "cosmic radiation" (kosmische Strahlung).

1907–1911   Domenico Pacini observes that underwater the ionization is significantly lower than at the sea surface. For these measurements, he immerses an electroscope about three meters deep into the sea near Livorno, Italy. This demonstrates that part of the ionization itself must be due to sources other than terrestrial radioactivity. Pacini concludes that "... *sizable cause of ionization exists in the atmosphere, originating from penetrating radiation, independent of the direct action of radioactive substances in the ground.*" (De Angelis 2010).

1911–1912   Victor Hess makes charge measurements from balloons up to an altitude of 5.2 km. He measures an increasing charge with increasing altitude (Fig. 1.2).

1913   The results by Hess are confirmed by the young Werner Kolhörster in subsequent flights up to 9.2 km.

1924–1926   Millikan first questions the existence of cosmic rays after a flight over Texas up to 15 km (the results were obscured by the latitude geomagnetic effect[4]). "*We conclude, therefore, that there exists no such penetrating radiation as we have assumed*" (Russel et al. 1924).

---

[4] Since the interstellar charged particles approaching at the level of the equator must travel in a direction perpendicular to the Earth's magnetic field, they are deflected by the Lorentz force, and only very energetic particles reach the Earth. The low-energy particles will have a spiral trajectory around the field line and will be directed toward the Earth's poles. Also, near the poles, incoming particles have a higher probability that their trajectory will be along the magnetic field lines and therefore not affected by the Lorentz force.

Fig. 1.2 Victor F. Hess, center, departing from Vienna, circa 1911. Picture: V. F. Hess Society, Echophysics, Schloss Pöllau, Austria

Using unmanned balloons to perform experiments at even higher altitudes, Millikan completely changes his mind and coins the term 'cosmic rays'. "*. . . all this constitutes pretty unambiguous evidence that the high-altitude rays do not originate in our atmosphere, very certainly not in the lower nine-tenths of it, and justify the designation 'cosmic rays'*" (Millikan & Cameron 1928).

Millikan's move to take the credit for the discovery provokes an angry response by Hess (1926): "*. . . The recent determination by Millikan and his colleagues of the high-penetrating power of high-altitude radiation has been an occasion for American scientific journals such as 'Science' and 'Scientific Monthly' to propose the name 'Millikan Rays'. Since his work is merely a confirmation and extension of the results obtained by Gockel, myself and Kolhörster from 1910 to 1913 using balloon borne measurements of the rays, this appellation should be rejected as it is misleading and unjustified. . .*".

1927 The geomagnetic effect on the cosmic rays is discovered by Clay et al. (1934) (Fig. 1.3).

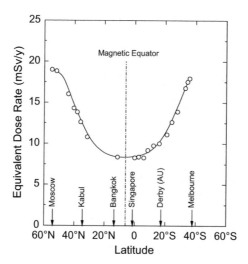

Fig. 1.3 Change of cosmic radiation as a function of the latitude. The experiment has been performed during an aircraft flight in 1988 from Melbourne to Frankfurt. The data have been obtained at an altitude of about 10 km; the neutron component is removed. Note that the radiation does not disappear near the magnetic equator, because the geomagnetic effect has a larger effect on the part of the cosmic ray with relatively low energy, i.e., the plasma coming from the Sun. Modified from Bonka (1990), under CC-BY license. ©The Authors

1933 First picture of a muon track (originally misidentified as a byproduct of a nuclear explosion) in a Wilson cloud chamber[5] by Kunze (1933) (Fig. 1.4).

1936 V. Hess receives the Nobel Prize for the discovery of the cosmic radiation.

1936 Discovery of the muon by C. Anderson and S. Neddermeyer using a cloud chamber at an altitude of 4300 m on Pikes Peak, Colorado (Anderson & Neddermeyer 1936; Neddermeyer & Anderson 1938). They name it "mesotron" (Anderson & Neddermeyer 1938) (i.e., intermediate particle, with mass between electron and proton). The muon, however, is initially misinterpreted as the so-called Yukawa's particle, proposed to mediate the nuclear force (Yukawa 1935). The muon is so unexpected that its discovery led I. I. Rabi[6] to say "who ordered that?".

---

[5] A cloud chamber is a device for detecting the path of ionizing particles. In a container a volume of saturated water or alcohol vapor is made supersaturated by sudden adiabatic expansion. An energetic charged particle ionizes the gas molecules leaving a trail of ionized particles. The resulting ions act as condensation centers, forming small droplets that are visible as a "cloud" track persisting for several seconds. The tracks are characteristic for different particles.

[6] I. I. Rabi received in 1944 the Nobel Prize for his discovery of nuclear magnetic resonance in molecular beams.

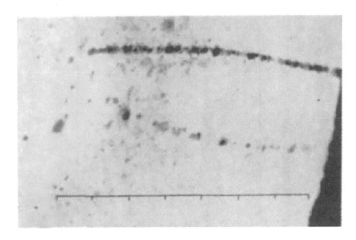

Fig. 1.4 First observation of a muon track (top track) but not correctly identified. In his paper, Kunze correctly noticed that a particle producing this track would have a mass between that of a positron and a proton, but concluded the track to be a product of a nuclear explosion. Reprinted from Kunze (1933), © Springer Nature. Reproduced with permission. All rights reserved

1941   B. Rossi and D. B. Hall determine the muon lifetime to be $\tau_\mu = 2.4(3) \times 10^{-6}$ s (Rossi & Hall 1941) .

1945–1947   Conversi et al. (1947) measure the lifetimes of positive and negative muons. The lifetimes are too long for strongly interacting particles. It turns out that the Yukawa's particle is actually the pion.

1946   Discovery of the pion by C. F. Powell et al. (Lattes et al. 1947) studying the cosmic rays with special photographic emulsions at high-altitude. It decays primarily into a muon and a muon neutrino, see Sect. 1.4.1.

1956   Lee and Yang (1956) predict that any process governed by the weak interaction should lead to a violation of the parity. The 1957 Nobel Prize in Physics is jointly awarded to them.

     Between Christmas of 1956 and New Year, National Bureau of Standards (NBS) scientists led by Columbia University Prof. C. S. Wu (Wu et al. 1957) confirm that the emission of beta particles from cobalt-60 nuclei oriented with a strong magnetic field is asymmetric.

1957   R. Garwin et al. observe the parity violation in weak decay of the muon (Garwin et al. 1957).[7] This work is followed a few months later by one of Friedman and Telegdi (1957), who also discuss the possible formation of muonium (μ^+e^- bound state) as a plausible cause of muon spin depolarization. The paper of Garwin et al. already lists the main muon properties, which are also essential for its use as a

---

[7] The muons were produced at the Nevis Columbia Cyclotron, Columbia University, New York, USA.

magnetic probe: the parity violation of pion and muon decays, the asymmetric distribution of the decay positrons, the muon spin $1/2$, and the first experimental determination of the muon g-factor $g_\mu = 2.00 \pm 0.10$.

Furthermore, one finds the remarkable prediction: *It seems possible that polarized positive and negative muons will become a powerful tool for exploring magnetic fields in nuclei ..., atoms, and interatomic regions*, thus predicting the later use of muons by solid state physicists. A spectrum showing the muon spin precession in a magnetic field and hence a prototype of a modern μSR spectrum, can also be found in this seminal paper.

1957–~1970 Between 1957 and the early '70s, a few groups use the parity violating μ^+ decay for condensed matter studies (Swanson 1958) and theoretical studies are pursued to understand muon spin depolarization processes in matter (Ivanter & Smilga 1969). First experiments are conducted at the CERN Synchro-Cyclotron, Geneva, Switzerland, at the Joint Institute for Nuclear Research, Dubna, former USSR, and at the Lawrence Berkeley Laboratory, Berkeley, USA. A short bibliography of the early period can be found in Brewer et al. (1975) and Brewer and Crowe (1978).

1970– Higher beam intensities at the so-called "meson factories"[8] enable the development and versatile use of μSR.

1.3 Atmospheric Muons

Muons are the main component of the cosmic radiation at sea level, with a flux of about one muon per cm^2 per minute. They originate from the decay of pions (see Sect. 1.4.3) which are produced by nuclear reactions of the primary cosmic rays in the upper atmosphere, Fig. 1.5. They have a broad spectrum of energies with a mean energy of the muons reaching sea level of about 4 GeV. They lose energy when interacting with atmospheric molecules, see Chap. 2. The energy loss is about 2 MeV per g/cm^2. Since the interaction depth of the atmosphere is about $1000 \, g/cm^2$, muons lose about 2 GeV in passing through the atmosphere, thus suggesting an initial energy of about 6 GeV, see Chap. 2 and Exercise 2.1.

Most muons are created at altitudes of about 15 km and travel to Earth with other particles in conical showers within about $1°$ of the trajectory of the primary particle that created them. The measurement of the muon flux at different altitudes, Fig. 1.6, is an instructive example of relativistic time dilation. Without this effect the cosmic muons would not reach Earth, see Exercise 1.1.

Atmospheric muons are not very useful for μSR experiments, but they can be used to make a radiography ("muography") of large massive objects, such as the

---

[8] 1972: LAMPF at Los Alamos, USA, 1974: SIN (now PSI) in Villigen, Switzerland and TRIUMF in Vancouver, Canada. Pulsed muon beams are provided at KEK, Tsukuba, Japan in 1980 (since 2008 at J-PARC, Tokai) and since 1987 at RAL, Didcot, UK. See also footnote 11.

Fig. 1.5 Schematic view of the particle shower produced by cosmic rays

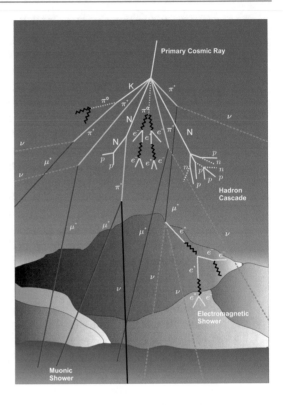

Fig. 1.6 Total fluxes of muons, pions, protons, neutrons and helium nuclei (alpha particles) as a function of atmospheric depth. Modified from Hansen et al. (2003), © American Physical Society. Reproduced with permission. All rights reserved

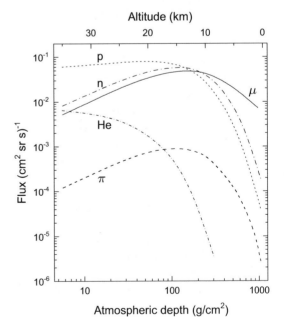

well-known examples of pyramids (Alvarez et al. 1970). Recently, a large cavity in the pyramid of Cheops (Khufu) has been discovered by using atmospheric muons (Morishima et al. 2017; Procureur et al. 2023). In this experiment, very large muon detectors were installed at the base of the pyramid and absorption measurements of atmospheric muons were performed. A variation of the muon counts as a function of spherical angles reflects a change in the density of the pyramid with an excess of muons revealing the presence of a cavity in the pyramid, Fig. 1.7. Muography is also used to study the interior of the Earth (e.g., a volcano) (Tanaka 2019, 2014). Figure 1.8 shows a large, shallow-depth, low-density region beneath a crater of the Satsuma-Iwojima volcano, detected by this technique.

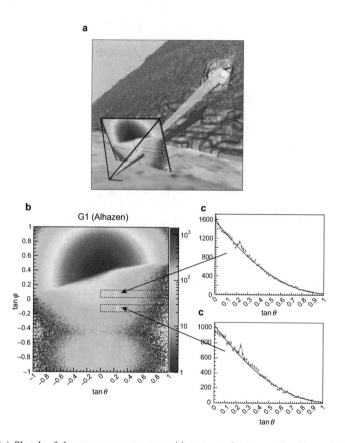

Fig. 1.7 (**a**) Sketch of the measurement setup with a muon detector at the Cheops's (Khufu's) pyramid. Muons are collected as a function of azimuth (ϕ) and elevation (θ) angles after they have passed through the pyramid. Two cones (yellow and red) with excess of muons are observed, see also panel (**b**). (**c**) The two regions where excess of muons are observed. The lower one represents the known Grand Gallery. The upper one represents an unknown cavity. Modified from Morishima et al. (2017), © Springer Nature. Reproduced with permission. All rights reserved

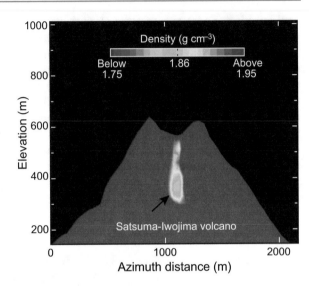

Fig. 1.8 Muography of the Satsuma-Iwojima volcano shows a low density region interpreted as degassing magma, with a high proportion of bubbles. Modified from Tanaka (2019), © Royal Society. Reproduced with permission. All rights reserved

1.4   Pion: The Parent Particle

1.4.1   Pion Properties

Positive and negative muons are obtained from the decay of positive and negative pions (π). The pion, which exists in three charge states π^-, π^+ and π^0 is classified as a meson because it consists of a quark and an antiquark (u and $\bar{d}$ for π^+). The exchange of virtual pions was proposed by Yukawa (1935) as an explanation for the (residual) strong force between nucleons. Table 1.1 gives the most important properties of the pion.

Table 1.1 Main properties of the pion, data from Workman et al. (2022)

	π^+	π^-	π^0
Lifetime (s)	$26.033 \pm 0.005 \times 10^{-9}$	$26.033 \pm 0.005 \times 10^{-9}$	$8.52 \pm 0.18 \times 10^{-17}$
Spin	0	0	0
Mass (MeV/c^2)	139.57039 ± 0.00018	139.57039 ± 0.00018	134.9768 ± 0.0005
Main decay	$\rightarrow \mu^+ + \nu_\mu$	$\rightarrow \mu^- + \bar{\nu}_\mu$	$\rightarrow \gamma + \gamma$

1.4.2   Pion Production Reactions

Pions are produced through high-energy collisions between hadrons. For a nucleon collision in an accelerator, the available energy in the center of mass must exceed the pion rest mass of $\sim 140\,\mathrm{MeV/c^2}$. For a proton this implies a threshold kinetic energy to produce a charged pion of about 290 MeV, see Exercise 1.2. Typical reactions to produce a single positive pion are[9]

$$p + p \rightarrow \ p + n + \pi^+$$

$$p + n \rightarrow \ n + n + \pi^+ \ .$$

The cross sections increase rapidly with energy. As shown in Fig. 1.9 the optimum energy for pion production, corresponding to the maximum of the cross section,[10] is above 500 MeV. This defines the energy of an accelerator needed to generate pion (and muon) beams.

One often measures double differential scattering cross sections, which are determined by observing the particles produced at a given energy under a defined angular position (Fig. 1.10). From the double differential scattering cross section, one obtains the total cross section by integrating over the energy of the particles produced and over the solid angle.

Fig. 1.9 Energy dependence of the cross section for pion production in some nucleon-proton reactions. Modified from Jones (2008), © AIP Publishing. Reproduced with permission. All rights reserved

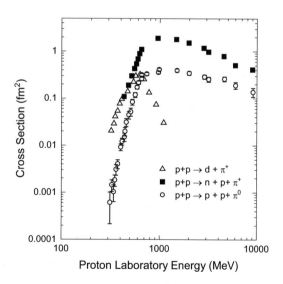

---

[9] As we are primarily interested in the μSR applications when writing about pions and muons we generally mean the positively charged particles.

[10] The cross section is used to express the probability that two particles will interact and can be considered as the effective area for the collision. The natural unit of cross section is an area unit, but it is often given in barn, with $1\,\mathrm{b} = 10^{-28}\,\mathrm{m^2} = 100\,\mathrm{fm^2}$.

Fig. 1.10 Double differential π^+ production cross section in proton carbon collisions as a function of the positive pion energy. Data extracted from Meshkovskii et al. (1958). For the incoming protons of 600 MeV, the pions have a broad spectrum of energies around a few hundred MeV. The different curves show the production in a differential solid angle $d\Omega$ for different values of the polar angle θ

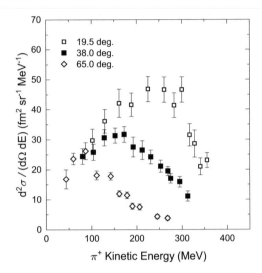

For a maximum number of single pion reactions the incident proton beams should have energies in the range 500–1000 MeV. At higher energies, it is possible to have reactions producing a pair of pions (such as $p + p \rightarrow p + p + \pi^- + \pi^+$). However, one needs more energetic (and more expensive) accelerators since the threshold energy is of the order of 600 MeV with the cross sections reaching saturation values of ≈ 10 mb above 1.5 GeV.[11]

1.4.3   The Pion Decay

As shown in Table 1.1, the pion lifetime is about 26 ns and the primary decay mode, with a branching fraction of 0.9998770(4), is a leptonic decay into a muon and a muon neutrino

$$\pi^+ \rightarrow \mu^+ + \nu_\mu \ . \tag{1.1}$$

Two characteristics of this decay process are important for our scope:

- It is a two-body decay. The conservation of momentum and energy implies that, in the pion reference frame, the muon and the neutrino are emitted in

---

[11] The J-PARC muon facility MUSE in Tokai, Japan makes use for the muon production of a 3 GeV proton synchrotron ring (see Sect. 1.8.1) (Miyake et al. 2010). The ISIS (Hillier et al. 2019) and the RIKEN-RAL facilities (Matsuzaki et al. 2001) at the Rutherford Appleton Laboratory (RAL), Didcot UK, make use of a 800 MeV synchrotron ring. The continuous beam facilities, CMMS at TRIUMF, Vancouver Canada (Baartman et al. 2003) and SμS at PSI, Villigen Switzerland (Grillenberger et al. 2021) use cyclotrons (see Sect. 1.8.1) to accelerate protons at $\sim$ 500 and 590 MeV, respectively.

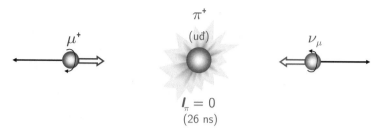

π^+

$(u\bar{d})$

μ^+

ν_μ

$I_\pi = 0$
(26 ns)

Fig. 1.11 Schematic of the π^+ decay. The black arrows represent the momentum and the colored arrows represent the spins of the particles

opposite directions with equal momentum and well-defined energies. Neglecting the neutrino mass, the muon has a momentum of 29.79 MeV/c (corresponding to a kinetic energy of 4.12 MeV, see Exercise 1.3).

- The pion has spin $I_\pi = 0$ and the total spin is conserved during the decay. This means that the muon spin $I_\mu = 1/2$ is pointing opposite to the neutrino spin. The decay is governed by the weak interaction, which does not conserve parity.[12] In nature, only so-called left-handed neutrinos are produced (spin direction opposite to the momentum or helicity -1). Therefore, the mirror image of Fig. 1.11 (corresponding to performing the parity operation on the pion decay) is not realized in nature, and all decay positive muons, in the pion rest frame, have their spin antiparallel to the momentum, see Fig. 1.12. The parity violation of the π decay allows therefore the production of 100% polarized muon beams. The π^- decay is obtained by performing the charge conjugation and parity operations (CP) on the π^+ decay, Fig. 1.12c.

1.5   Muon Properties

Table 1.2 summarizes the most important muon properties relevant for muon spin spectroscopy experiments.[13]

Some properties are crucial for the use of polarized μ^+'s as spin and magnetic probes:

- As a spin $1/2$ particle the muon has no electric quadrupole moment and is therefore a purely magnetic probe. Classically, the electric quadrupole moment

---

[12] The weak interaction involves the exchange of an intermediate vector boson, W and Z (in this case W$^+$, see Fig. 10.1). It has a range of about 10^{-18} m which is about 0.1% of the diameter of a proton. A discussion of the weak interaction can be found in Chap. 10.

[13] More properties can be found in the list of physical constants and symbols on page xvii.

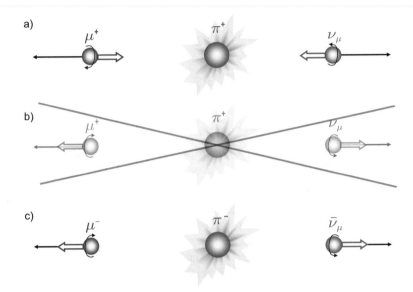

Fig. 1.12 Parity violation of the π^+ decay. The black arrows represent the momenta and the colored arrows the spins of the particles. (**a**) Natural decay. (**b**) Mirrored decay (corresponding to applying the parity operation on (**a**)). The process of Fig. (**b**) does not exist, since right-handed ν_μ's do not exist. (**c**) Charge conjugated process of (**b**), representing the π^- decay

Table 1.2 Main properties of the positive muon. μ_e, μ_p, μ_n and μ_{NM} are the magnetic moments of the electron, proton, neutron, and nucleon, respectively

Lifetime τ_μ	$2.1969811(22) \times 10^{-6}$ s
Charge q	e ($-e$ for μ^-)
Spin quantum number I_μ	$^1/_2$
Mass m_μ	$105.6583745(24)$ MeV c^{-2}
	$206.768\ m_e$
	$0.1126\ m_p$
Magnetic Moment $\mu_\mu = g_\mu \dfrac{e\hbar}{2m_\mu} I_\mu$	$4.49044830(10) \times 10^{-26}$ J T^{-1}
	$4.836 \times 10^{-3}\ \mu_B$
	$4.836 \times 10^{-3}\ \mu_e$
	$3.183\ \mu_p$
	$-4.647\ \mu_n$
	$8.891\ \mu_{NM}$
Gyromagnetic ratio $\gamma_\mu = \dfrac{\mu_\mu}{\hbar I_\mu}$	$2\pi \times 135.53881$ MHz T^{-1}
Decay (branching ratio $\approx 100\%$)	$\mu^+ \rightarrow e^+ + \nu_e + \bar{\nu}_\mu$

of a charged object with charge density distribution $\rho(\mathbf{r})$ is given by

$$Q = \int (3z^2 - r^2)\rho(r)\, d^3r \ . \tag{1.2}$$

For a spherical charge distribution, $\langle x^2 \rangle = \langle y^2 \rangle = \langle z^2 \rangle = 1/3 \langle r^2 \rangle$, so the quadrupole moment vanishes.[14]

In quantum mechanics the quadrupole moment is defined as the expectation value of the quadrupole tensor Q_{20} in the substate $|I, M = I\rangle$, so-called spectroscopic quadrupole moment $Q_s(I)$, which is zero for a spin 1/2 particle.[15]

- The muon magnetic moment (or equivalently the gyromagnetic ratio) is large, making this particle a very sensitive magnetic probe.
- The muon lifetime is long compared to the time scale of many dynamical phenomena in solids and is easily accessible with modern detectors and timing techniques.
- In some materials, the positive muon can capture an electron and form a bound state (muonium), which can be considered a light hydrogen isotope. Muon spin

---

[14] In classical mechanics, we note that

$$Q = \int (3z^2 - r^2)\,\rho(r)\, d^3r = \int r^2 (3\cos^2\theta - 1)\,\rho(r)\, d^3r = \sqrt{\frac{16\pi}{5}} \int r^2\, Y_2^0\, \rho(r)\, d^3r = Q_{20}\ , \tag{1.3}$$

where Y_2^0 is the spherical function with $l = 2$ and $m = 0$.

[15] We recall the Wigner-Eckart theorem

$$\langle J\, M | T_{(k)}^q | J'\, M' \rangle = \frac{\langle J'\, M'\, k\, q | J\, M \rangle \langle J \| T_{(k)} \| J' \rangle}{\sqrt{2J+1}}\ . \tag{1.4}$$

Here, $T_{(k)}^q$ is the q-th component of the spherical tensor operator $T_{(k)}$ of rank k (in our case $k = 2$), $\langle J'M'kq|JM \rangle$ is the Clebsch-Gordan coefficient for coupling J' with k to get J, and $\langle J \| T_{(k)} \| J' \rangle$ called the "reduced matrix element". This means that the matrix element of a tensor operator can be factored into a part that is independent of the tensor itself, but involves the projection quantum numbers (the Clebsch-Gordan coefficient), and a part that does not involve the projection quantum numbers (the reduced matrix element). We therefore can write

$$Q_s(I) = \langle I, M = I | Q_{20} | I, M = I \rangle$$

$$= \langle I, M = I\, 2\, 0 | I, M = I \rangle \frac{\langle I \| Q_2 \| I \rangle}{\sqrt{2J+1}}\ , \tag{1.5}$$

where

$$\langle I, M = I\, 2\, 0 | I, M = I \rangle = \sqrt{\frac{I(2I-1)}{(2I+3)(I+1)}} \tag{1.6}$$

is the appropriate Clebsch-Gordan coefficient. For $I = 0$ and $I = 1/2$ the quadrupolar interaction vanishes.

spectroscopy on muonium provides spectroscopic information that can be used to investigate the behavior of hydrogen-like states in matter, see Chap. 7.

1.6   The Muon Decay

1.6.1   Kinematics

The muon is an unstable particle and decays with a mean lifetime of $\tau_\mu \cong 2.197\,\mu s$ as follows

$$\mu^+ \rightarrow e^+ + v_e + \bar{v}_\mu$$

$$\mu^- \rightarrow e^- + \bar{v}_e + v_\mu \ . \tag{1.7}$$

Since this is a three-body decay, the kinetic energy of the emerging positron takes values ranging from zero up to a maximum value $E_{e^+}^{\max}$. The case of zero kinetic energy represents the situation where the neutrino and the antineutrino emerge in opposite directions and carry away all the available kinetic energy. On the other hand, the kinetic energy of the positron is maximum when the neutrino and antineutrino travel together in the opposite direction of the positron (Fig. 1.13). The maximum and mean positron energies resulting from the muon decay are given by[16] (see Exercise 1.4)

$$E_{e^+}^{\max} = \frac{m_\mu^2 + m_e^2}{2m_\mu} c^2 = 52.830\,\text{MeV} \ \ \text{and} \tag{1.8}$$

$$\overline{E}_{e^+} = 36.980\,\text{MeV} \ . \tag{1.9}$$

1.6.2   Differential Positron Emission

The theoretical treatment of the muon decay, which is due to the weak interaction, is described in Chap. 10. The differential positron emission probability per unit of time as a function of energy and solid angle for a muon with spin pointing in the $\theta = 0$ direction is given by Michel (1950)

$$d\Gamma = W(\varepsilon, \theta)\, d\varepsilon\, d\Omega = \frac{1}{4\pi\,\tau_\mu} 2\varepsilon^2(3 - 2\varepsilon)\left[1 + \frac{2\varepsilon - 1}{3 - 2\varepsilon}\cos\theta\right] d\varepsilon\, d\Omega \ , \tag{1.10}$$

---

[16] We concentrate the discussion on μ^+, but the considerations are analogous for μ^-, where the electron is the decay product.

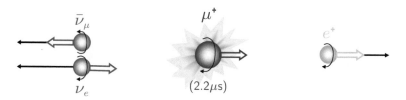

Fig. 1.13 Schematic of a positive muon decay. The black arrows represent the directions of flight and the colored arrows represent the spins of the particles. The sketch shows a decay producing a positron of maximum energy

where $\varepsilon \equiv E_{e+}/E_{e+}^{\max}$, and $d\Omega = \sin\theta \; d\theta \; d\phi$ is the solid angle (with θ the polar and ϕ the azimuthal angle). Note that $d\Gamma$ is independent of ϕ.

The important term in Eq. 1.10 is the energy dependent asymmetry term

$$a(\varepsilon) = \frac{2\varepsilon - 1}{3 - 2\varepsilon} \; , \tag{1.11}$$

which is a direct consequence of the muon decay being governed by the weak interaction; the positron is emitted asymmetrically with respect to the muon spin direction, Fig. 1.14.[17] As we will see in Chap. 3, the asymmetric positron emission is at the base of the μSR technique.

Using Eq. 1.11 and $E(\varepsilon) \equiv 2\varepsilon^2 (3 - 2\varepsilon)$ we can rewrite Eq. 1.10[18]

$$d\Gamma = W(\varepsilon, \theta) \, d\varepsilon \, d\Omega = \frac{1}{4\pi \tau_\mu} 2\varepsilon^2 (3 - 2\varepsilon) \left[1 + a(\varepsilon) \cos\theta \right] d\varepsilon \, d\Omega$$

$$= \frac{1}{4\pi \tau_\mu} E(\varepsilon) \left[1 + a(\varepsilon) \cos\theta \right] d\varepsilon \, d\Omega \; . \tag{1.12}$$

We can understand the asymmetric positron emission by considering the situation for positrons emitted with kinetic energies of the order of $E_{e+}^{\max}$. These antiparticles are ultrarelativistic ($E_{e+}^{\max} \gg m_{e+}c^2$) and the Dirac theory tells us that they behave as antineutrinos with helicity $+1$, i.e., with spin pointing in the propagation direction (see Fig. 1.13). Therefore, in order to conserve the spin during the muon decay, the positron emission direction is asymmetric, preferentially along the muon spin direction at the moment of the decay. For $\varepsilon \to 1$, $a(\varepsilon) \to 1$, i.e., the asymmetry is maximum. On the other hand, the asymmetry disappears for $\varepsilon = 1/2$ and even becomes even negative for lower positron energies.

---

[17] Note that for negative muons $a(\varepsilon)$ has the opposite sign.

[18] A curve with polar parametrization $r = a + b \cos\theta$ is called limaçon of Pascal (Pascal's snail). The curve for the special case $b = \pm 1$ is heart-shaped and named cardio.

Fig. 1.14 Anisotropic
distribution of the positron
emission with respect to the
direction of the muon spin at
the time of decay (cardioid
curves). Energies between
$\varepsilon = 0.5$ and $\varepsilon = 1.0$ are
shown. For lower positron
energies, the
forward-backward asymmetry
becomes negative (see
Fig. 1.15)

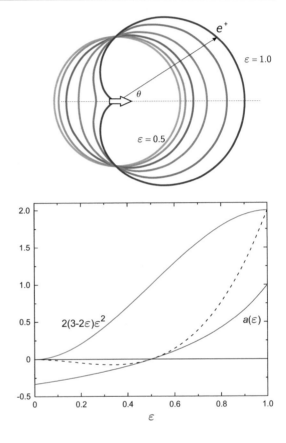

Fig. 1.15 Red curve:
Normalized energy spectrum
of the emitted positrons $E(\varepsilon)$.
Blue curve: Energy
dependence of the asymmetry
term $a(\varepsilon)$. Dashed line:
Weighted positron emission
asymmetry $a(\varepsilon)E(\varepsilon)$

Integrating Eq. 1.10 over the energy and the solid angle we obtain the total decay
rate

$$\Gamma = \int_0^{2\pi} \int_0^{\pi} \int_0^1 W(\varepsilon, \theta)\, d\varepsilon\, \sin(\theta)\, d\theta\, d\phi = \frac{1}{\tau_\mu} \ , \qquad (1.13)$$

as expected.

The decay rate as a function of the positron energy, i.e., the energy spectrum
of the positrons independent of the emission angle (also called Michel spectrum
(Michel 1950)) is obtained from Eq. 1.10 by integrating over the angles θ and ϕ
(see Fig. 1.15)

$$d\Gamma = W(\varepsilon)\, d\varepsilon = \frac{1}{\tau_\mu} 2\varepsilon^2 (3 - 2\varepsilon) d\varepsilon = \frac{1}{\tau_\mu} E(\varepsilon) d\varepsilon \ . \qquad (1.14)$$

Since both the emission probability of positrons and their decay asymmetry increase with energy, the asymmetry of the total angular distribution is mainly due to the high-energy positrons. If we integrate over all energies (experimentally this corresponds to the ideal case where all the positrons are detected with the same efficiency), the average asymmetry $\overline{A}$ is given by

$$\overline{A} = \int_0^1 a(\varepsilon) E(\varepsilon) \, d\varepsilon = \frac{1}{3} \, , \qquad (1.15)$$

and we can write the angular distribution of the positrons originating from the muon decay as

$$d\Gamma = W(\theta) \, d\Omega = \frac{1}{4\pi \tau_\mu} \left(1 + \frac{1}{3} \cos\theta \right) d\Omega \, . \qquad (1.16)$$

1.6.3   Decay of a Muon Ensemble

The μSR experiments are based on the observation of the decay of an ensemble of $N_{\mu,0}$ muons, which have been implanted in a sample, see Chap. 3. The decay is monitored by observing of the emitted positrons. The number of positrons $N_{e^+}(t)$ emitted at time t is given by the number of muons decaying in the interval dt at time t. By looking at the full solid angle around the muons (assumed to be all within a small spot with respect to the positron detectors), we have[19]

$$N_{e^+}(t) = -\frac{dN_\mu}{dt} = \Gamma N_\mu(t) = \frac{1}{\tau_\mu} N_\mu(t) = \frac{1}{\tau_\mu} N_{\mu,0} \, e^{-\frac{t}{\tau_\mu}} \, . \qquad (1.17)$$

If we now restrict the observation to a direction in space, defined by a positron detector subtending a solid angle $d\Omega$, we have (see Eq. 1.12)

$$dN_{e^+}(t) = N_\mu(t) d\Gamma = \frac{N_{\mu,0}}{4\pi \tau_\mu} \, e^{-\frac{t}{\tau_\mu}} \, E(\varepsilon) \, [1 + a(\varepsilon) \cos\theta] \, d\varepsilon \, d\Omega \, . \qquad (1.18)$$

In a μSR experiment the positron energy is not measured. In the ideal case that all positrons are detected with the same efficiency, that they are not scattered or absorbed and that the solid angle subtended by the detector is small, this corresponds

---

[19] $N_\mu(t)$ is the number of muons remaining at time t.

to averaging Eq. 1.18 over ε

$$\frac{dN_{e^+}(t)}{d\Omega} = \frac{N_{\mu,0}}{4\pi\tau_\mu} e^{-\frac{t}{\tau_\mu}} \left[1 + \overline{A}\cos\theta\right] . \tag{1.19}$$

This expression assumes that the muon ensemble is and remains fully polarized, i.e., $|\mathbf{P}(t)| = 1$ and pointing in the $\theta = 0$ direction. Of course, this is not the case when the muon spin interacts with its environment. The time evolution of $\mathbf{P}(t)$ will modify the detected asymmetry in positron counts and make it time dependent, which is exactly what the μSR technique measures, see Chap. 3 and Eq. 3.3.

1.7 Muon Magnetic Moment and Spin Precession

1.7.1 Muon Magnetic Moment

The magnetic moment of a muon is simply related to its spin

$$\boldsymbol{\mu}_\mu = \gamma_\mu \hbar \mathbf{I}_\mu = g_\mu \frac{q}{2m_\mu} \mathbf{I}_\mu = g_\mu \frac{\pm e}{2m_\mu} \mathbf{I}_\mu , \tag{1.20}$$

where the $+$ and $-$ signs stand for μ^+ and μ^-, respectively. The value of the magnetic moment μ_μ is given by[20]

$$\mu_\mu = \gamma_\mu \hbar I_\mu = g_\mu \frac{e\hbar}{2m_\mu} I_\mu = \frac{\gamma_\mu \hbar}{2} , \tag{1.21}$$

where $I_\mu = 1/2$ is the muon spin quantum number and the gyromagnetic ratio γ_μ is the ratio between the value of the magnetic moment and the spin

$$\gamma_\mu = g_\mu \frac{e}{2m_\mu} . \tag{1.22}$$

The g-factor of the muon is predicted to be $g_\mu = 2$ by the Dirac equation, which gives a relativistic quantum mechanical description of massive elementary spin $1/2$ particles and antiparticles, see Appendix G.[21] The presently accepted value is very slightly larger $g_\mu = 2.0023318418(13)$.[22]

---

[20] Note that the proton and neutron magnetic moments can also be written as $\mu_p = g_p \frac{e}{2m_p}\hbar I_p = \frac{g_p\mu_{NM}}{2}$ and $\mu_n = \frac{g_n\mu_{NM}}{2}$ with $g_p = 5.5857$ and $g_n = -3.826$, hence with g-factors very different from 2, which reflects the composite character of both nucleons.

[21] The g=2 result can also be obtained within the nonrelativistic Schrödinger equation (Feynman 2018; Alder & Martin 1992).

[22] The very small difference between the real value of g_μ and 2 (of the order of 0.1%) is called the anomalous magnetic dipole moment and is due to interactions with virtual particles, accounted for by quantum electrodynamics (QED), weak and hadronic interactions. Its precise determination is an important test of the Standard Model of particle physics in the search for so-called "new" physics beyond the Standard Model, see Sect. 10.5.

The numerical value of μ_μ and its relation to the magnetic moments of other particles are summarized in Table 1.2. Note that the value of the muon magnetic moment is large compared to that of proton and neutron, and therefore larger than the nuclear magnetic moments that are relevant for defining the sensitivity of a related spin technique such as Nuclear Magnetic Resonance (NMR). This makes the interaction between the muon moment and the local field large (see Eq. 1.29) and is the basis of the μSR sensitivity.

The gyromagnetic ratio is usually given as $\gamma_\mu = 2\pi \times 135.538810(3)\,\text{MHz T}^{-1}$ and $\gamma_\mu/2\pi$ is the proportionality constant between the muon spin precession frequency and the local magnetic field

$$\nu_\mu = \frac{\gamma_\mu}{2\pi} B_\mu \ . \tag{1.23}$$

1.7.2   Muon Spin Precession

When the muon experiences a magnetic field, its spin (or its magnetic moment) will precess (Larmor precession). We can understand the Larmor precession of the muon spin either classically or from a quantum mechanical point of view.

Classical View   Classically, a magnetic field $\mathbf{B}_\mu$[23] creates a torque on the magnetic moment of the muon

$$\boldsymbol{\tau} = \boldsymbol{\mu}_\mu \times \mathbf{B}_\mu = \gamma_\mu\, \hbar \mathbf{I}_\mu \times \mathbf{B}_\mu \ . \tag{1.24}$$

The torque can be expressed as the rate of change of the muon spin

$$\boldsymbol{\tau} = \hbar\, \frac{d\mathbf{I}_\mu}{dt} \ , \tag{1.25}$$

and Eq. 1.24 becomes (see Fig. 1.16)

$$\frac{d\mathbf{I}_\mu}{dt} = \gamma_\mu\, \mathbf{I}_\mu \times \mathbf{B}_\mu \ . \tag{1.26}$$

With

$$\boldsymbol{\tau} = \hbar \frac{\Delta I_\mu}{\Delta t} = \frac{\hbar I_\mu \sin\theta\, \Delta\phi}{\Delta t} = \gamma_\mu \hbar I_\mu B_\mu \sin\theta \tag{1.27}$$

---

[23] We call $\mathbf{B}_\mu$ the total magnetic field sensed by the muon.

Fig. 1.16 Classical view of
the Larmor precession

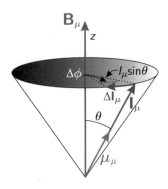

and $\frac{\Delta\phi}{\Delta t}$ giving the angular velocity, we obtain a spin (or magnetic moment or polarization[24]) precession with angular velocity (Larmor precession)

$$\frac{\Delta\phi}{\Delta t} = \omega_\mu = \gamma_\mu B_\mu \; . \tag{1.28}$$

Quantum Mechanical View For the quantum mechanical treatment we start from the Hamiltonian describing the interaction of the magnetic moment with the local field

$$\mathcal{H} = -\boldsymbol{\mu}_\mu \cdot \mathbf{B}_\mu = -\gamma_\mu \, \hbar \mathbf{I}_\mu \cdot \mathbf{B}_\mu = -\gamma_\mu \, \hbar (I_{\mu,x} B_x + I_{\mu,y} B_y + I_{\mu,z} B_z) \; . \tag{1.29}$$

Without loss of generality we can take the field along the z-axis of quantization

$$\mathcal{H} = -\gamma_\mu \, B_z \hbar I_{\mu,z} \; . \tag{1.30}$$

The time evolution of the spin state, described by the nonrelativistic time dependent Schrödinger equation, can be expressed with the unitary operator

$$\mathcal{U}(t) = \exp\left(-\frac{i\mathcal{H}t}{\hbar}\right) \; . \tag{1.31}$$

Assuming that at $t = 0$ the spin is pointing in the direction defined by the angles (θ, ϕ), the muon spin is in the state (see Eq. A.22)

$$|\chi\rangle(0) = e^{-i\frac{\phi}{2}} \cos\frac{\theta}{2} |\uparrow\rangle + e^{i\frac{\phi}{2}} \sin\frac{\theta}{2} |\downarrow\rangle \; . \tag{1.32}$$

---

[24] Note that Eq. 1.26 is also valid for time-varying $\mathbf{B}_\mu$. It holds not only for the spin and magnetic moment but also for the polarization $\mathbf{P}(t)$, and quantum mechanically as well.

The eigenvalues of the Hamiltonian acting on the eigenstates $|\uparrow\rangle$ (spin up) and $|\downarrow\rangle$ (spin down) are $-\gamma_\mu B_\mu \hbar/2$ and $\gamma_\mu B_\mu \hbar/2$, respectively. Therefore

$$
\begin{aligned}
|\chi\rangle(t) &= \mathcal{U}(t,0)|\chi\rangle(0) \\
&= e^{-i\mathcal{H}t/\hbar}\left(e^{-i\frac{\phi}{2}}\cos\frac{\theta}{2}|\uparrow\rangle + e^{i\frac{\phi}{2}}\sin\frac{\theta}{2}|\downarrow\rangle\right) \\
&= e^{+i(-\phi+\gamma_\mu B_\mu t)/2}\cos\frac{\theta}{2}|\uparrow\rangle + e^{+i(\phi-\gamma_\mu B_\mu t)/2}\sin\frac{\theta}{2}|\downarrow\rangle \ .
\end{aligned}
\tag{1.33}
$$

Comparing with the state at $t = 0$, we see that the spin state at time t has rotated by an azimuthal angle of $\gamma_\mu B_\mu t$, corresponding to a spin precession with Larmor angular velocity (or angular frequency)

$$
\omega_\mu = \gamma_\mu B_\mu \ ,
\tag{1.34}
$$

as in the classical treatment.[25]

1.8 Muon Beams

1.8.1 Proton Accelerators

Muon spin spectroscopy experiments require medium- or high-energy proton accelerators. The extracted proton beam impinging on a target (production target) produces among other particles pions, see Sect. 1.4.2. The muons from the pion decays are collected usually into one or more muon beams, which are transported and focused to the μSR instrument. Essentially, three basic types of accelerators can be used to accelerate protons: linear accelerators (linacs), cyclotrons, and synchrotrons.

Linac A linac accelerates particles in a straight line. The particles pass through electrodes in a tube-shaped vacuum chamber. A driving radio frequency (RF) is timed to positively and negatively charge the electrodes thus creating an oscillating electric field in the gap between them that accelerates the particles, Fig. 1.17.

Cyclotron A cyclotron accelerates particles along a spiral trajectory. The particles are held on the trajectory by a static magnetic field perpendicular to it. Schematically, protons from a source are injected into the center of the cyclotron in a vacuum chamber between two hollow metal electrodes (because of the similarity of this semicircular structure to the capital D, the electrodes are called "dees"). An alternating RF voltage is alternately applied to one dee and the other, accelerating

---

[25] Note that the muon spin rotates in clockwise direction with respect to the axis defined by the magnetic field. see Fig. 1.16.

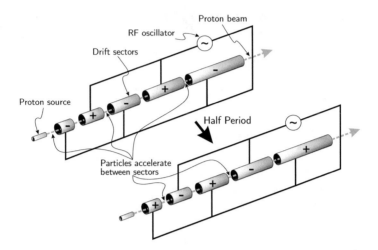

Fig. 1.17 Schematic of a linear accelerator

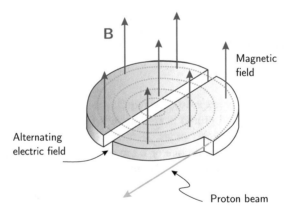

Fig. 1.18 Classical cyclotron: operation principle. The magnetic field bends the trajectory of the protons and the square-wave alternating electric field accelerates the protons as they pass through the gap, twice per turn

the particles and increasing the diameter of their circular orbit with each revolution and turning it into a spiral (see Fig. 1.18). The RF frequency (and hence the proton repetition frequency) is on the order of some tens of MHz and therefore leads to the production of quasi-continuous muon beams, see Sect. 3.3.1.

Such a classical cyclotron does not take into account the relativistic increase of the proton mass with velocity. As the mass increases, the orbital frequency decreases, and the particles may cross the gap at times when the electric field slows them down. To overcome this, one method is to strengthen the magnetic field near the periphery of the dees, thus keeping the cyclotron angular velocity $\omega_c = eB/\gamma m$

constant ($\gamma = (\sqrt{1 - v^2/c^2})^{-1}$, Lorentz factor).[26] Cyclotrons operated in this way are called isochronous and can produce beams with energies up to ~ 1 GeV. Another method is to vary the frequency of the RF voltage, i.e., reducing the frequency as the mass of the proton increases with energy at large radii. Cyclotrons operated in this way are called synchrocyclotrons. A disadvantage of a synchrocyclotron is that it produces a pulsed beam of relatively low intensity.

Synchrotron A synchrotron, like a cyclotron, is a cyclic accelerator that sends particles into a closed-loop orbit. Acceleration is achieved by applying high-frequency electric fields to cavities along the circumference of the ring. Unlike cyclotrons, the synchrotron's orbit is not a spiral, so the magnetic fields bending the proton trajectory must be increased synchronously with the acceleration in order to keep the particles on a constant-radius trajectory.[27] Synchrotrons produce pulsed proton beams, with a typical pulse frequency of 50 Hz, and can reach very high energies (up to TeV). Muon beams produced at synchrotrons are therefore pulsed beams (see Sect. 3.3.3). Due to the closed path shape, a synchrotron needs an injection accelerator (usually a linac system).

1.8.2 Example of a Proton Accelerator for μSR

As an example, we present the proton accelerator complex used at the Paul Scherrer Institute to produce pion and muon beams (Fig. 1.19). By ionizing hydrogen atoms, protons are obtained which are then accelerated in three stages.

The first stage is a Cockcroft-Walton accelerator, Fig. 1.20, which is a device that acts as a voltage multiplier, generating a high DC voltage. Since the proton source is located at the high-voltage end, the extracted protons are accelerated to 870 keV and then fed into the Injector II.

The Injector II, Fig. 1.21, is a small ring cyclotron. Acting as a pre-accelerator, it accelerates protons to 72 MeV before injecting them into the center of the main ring cyclotron, Fig. 1.22.

The large ring isochronous cyclotron with a diameter of approximately 15 meters is the heart of the PSI proton accelerator complex. The protons are accelerated in about 200 revolutions to a final kinetic energy of 590 MeV, corresponding to $v/c \approx 0.80$.

After reaching the final energy of 590 MeV, the protons are extracted and transported in the so-called proton-channel to impinge on two graphite pion-production targets (Kiselev et al. 2021) where nuclear reactions producing pions take place, see Sect. 1.4.2. Pions and muons are collected from the targets, transported and focused to form several so-called "secondary" beamlines. After traversing the two targets,

---

[26] In the classical limit the principle of a cyclotron is based on equating the centrifugal force to the Lorentz force, i.e., $evB = mv^2/r$. In this limit, the cyclotron angular velocity $\omega_c = v/r = eB/m$ does not depend on the particle velocity or the radius of the trajectory.

[27] Besides a circle, the path can be an oval, or even a polygon with rounded corners.

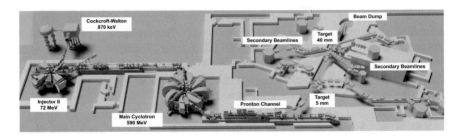

Fig. 1.19 Sketch of the High-energy Proton Accelerator (HIPA) at PSI with the primary proton beamline, the two targets and the secondary (μ and π) beamlines. The proton beam continues to the neutron spallation source (not shown). See text for details. © Paul Scherrer Institute. Reprinted with permission. All rights reserved

Fig. 1.20 The PSI Cockcroft-Walton accelerator. © Paul Scherrer Institute, Markus Fischer. Reprinted with permission. All rights reserved

the proton beam has about 60%[28] of the initial current of presently 2.4 mA. These protons are transported to a neutron spallation source (Blau et al. 2009) to produce

---

[28] Note that the proton losses are mainly due to scattering and not to absorption by the pion production processes, see Exercise 1.5.

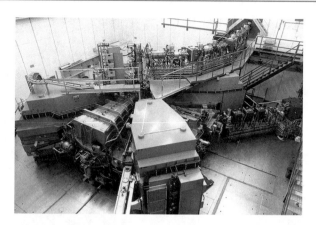

Fig. 1.21 The first stage of PSI's cyclotron, Injector II, which pre-accelerates the protons to 72 MeV. The green components are bending magnets. An RF cavity (silver) between two magnets is also visible. © Paul Scherrer Institute. Reprinted with permission. All rights reserved

Fig. 1.22 Top view of the main ring cyclotron at PSI, which accelerates the protons to 590 MeV for the production of pion, muon and neutron beams. Several RF cavities (silver) are positioned between the magnets (green). The main components are eight sector magnets and four accelerator cavities. © Paul Scherrer Institute. Reprinted with permission. All rights reserved

Fig. 1.23 Picture of the thick graphite target at PSI (40 mm target in Fig. 1.19). The target has a conical shape, oriented so that the protons traverse a section of the conical surface, which can be 40 or 60 mm thick. To reduce the heat load, the target rotates at one turn per second so that the protons do not dwell on the same spot for too long time. Still, the operating temperature reaches 1700 K. © Paul Scherrer Institute. Reprinted with permission. All rights reserved

neutron beams for condensed matter and material science studies, see Fig. 1.19. Figure 1.23 shows the second (thick) production target at PSI.

We leave it as an exercise (Exercise 1.5) to estimate the rate of pion production, from the information available in this chapter.

1.8.3   Surface and Decay Muon Beams

For μSR investigations of bulk materials,[29] two types of beams are used: the so-called "surface" muon beams and the decay muon beams. A low-energy muon beam with tunable energy in the keV range has been developed and is in operation at PSI for thin film studies and depth dependent investigations on the nanometer scale. Another low-energy beam, based on a different principle, is under development at the J-PARC accelerator complex in Tokai, Japan. See Chap. 8 for details and references about low-energy muon beams.

Surface Muons   The majority of available muon beamlines in the various facilities around the world are surface muon beams.[30] In a surface muon beam the muons are extracted directly from the production target, see Figs. 1.24 and 1.25. They originate from pions decaying at rest inside and near the surface of the production target, hence the name "surface" muons.[31]

As shown in Sect. 1.4.3 and Exercise 1.3, these muons are nominally 100% polarized and monochromatic with a relatively low momentum of 29.8 MeV/c.

---

[29] Sample thickness from a fraction of a mm to several cm.

[30] Especially in the early days, in memory of the pioneering work of the University of Arizona group at the Lawrence Berkeley Laboratory, USA, they were also called "Arizona" beams (Pifer et al. 1976).

[31] Muons from pions decaying deep inside the production target do not have enough energy to escape it, see Sect. 2.2.1 about range of muons in matter.

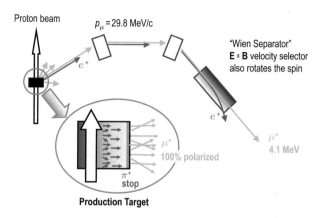

Proton beam p_μ = 29.8 MeV/c

"Wien Separator"
E × **B** velocity selector
also rotates the spin

e^+

e^+

μ^+
4.1 MeV

μ^+
100% polarized

π^+
stop

Production Target

Fig. 1.24 Principle and main components of a surface muon beamline. Adapted from a drawing of J. Brewer, UBC (Canada)

Fig. 1.25 Realization of a surface muon beamline (π M3 beamline at PSI). On the bottom left the beam comes from the production target. A Wien filter (lower left), which acts also as a spin rotator, quadrupole magnets (red) and a dipole magnet (blue) are visible. The beam is split and sent to two instruments. On the top right, the second μSR instrument (FLAME) is visible. © Paul Scherrer Institute. Reprinted with permission. All rights reserved

They have a mean penetration depth in matter of about 140 mg/cm^2 (see Chap. 2), allowing one to study relatively thin samples with typical thickness of a few hundreds μm. In addition, the beam can be focused to a relatively small spot of a few mm diameter, since they originate from a spatially well defined source that can be imaged at the final focus.[32] In a plot of muon intensity versus momentum, Fig. 1.26, the surface muons appear as an edge at $\sim$ 30 MeV/c.[33]

---

[32] Another advantage is that they can be easily manipulated, for example their spin polarization can be rotated, see Sect. 1.8.4.

[33] Momenta slightly below this value are mainly from pions decaying in the $\sim$ 1 mm thick "skin" of the target.

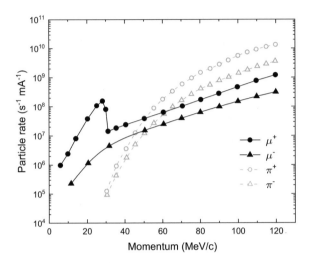

Fig. 1.26 Intensities of positive and negative muons and pions as a function of momentum (normalized to 1 mA proton current) achieved at the πE5 beamline at PSI. High-momenta are dominated by pions, which are more copiously produced. Low-momenta pions are suppressed as a consequence of the decay in flight due to their short lifetime of 26 ns. Clearly visible is the pronounced momentum edge of the surface muons, which can be optimally transported to the experiments. There are no negative surface muons, see Sect. 9.1. Negative muon intensities are lower because nuclear reaction with protons produce more positive than negative pions

High-Energy or Decay Muons Some experiments require muons with energies higher than the 4.1 MeV of the surface muons. This is the case, for instance, when the sample is placed in a pressure cell or for liquid targets in a vessel. The principle of a high-energy (or "decay") muon beamline is shown in Fig. 1.27. First, high-energy pions are extracted from the production target. Pions of a given momentum are selected by a bending magnet and are transported to a long superconducting solenoid ("decay muon channel") where they decay in flight. Additional dipole magnets select the muon momentum (see Figs. 1.27 and 1.28), which can be up to a few hundred MeV/c for the muons decaying in the same direction as the pion momentum ("forward" muons). In practice, "backward" muons, which are less energetic but much less contaminated by other particles, are more commonly used. For high-pressure experiments, one generally utilizes momenta in the range of 60–125 MeV/c (Khasanov et al. 2016; Khasanov 2022).

Muon Spin Polarization of Decay Muons Depending on the direction of the muon decay, muon beams with different polarizations can be obtained.

In practice, two extreme cases are used, Exercise 1.6:

- The muon is emitted in the direction of the pion momentum, i.e., forward direction. The momenta $\mathbf{p}_\mu^0 = 29.79$ MeV/c (corresponding to the muon momentum in the pion reference frame) and $\mathbf{p}_\pi$ add to a final muon momentum $\mathbf{p}_\mu$ greater than

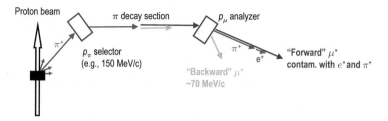

Fig. 1.27 Principle and main components of a high-energy (decay) muon beamline. See text for details. Adapted from a drawing of J. Brewer, UBC (Canada)

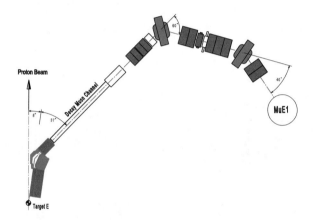

Fig. 1.28 Realization of a decay muon beamline (μE1 at PSI) with the solenoidal decay channel. The dipole magnets are shown in blue, the quadrupole magnets in red. © Paul Scherrer Institute. Reprinted with permission. All rights reserved

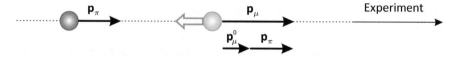

Fig. 1.29 Schematic view of the creation of forward muons in the muon decay channel

that of the pion. The muon spin points in the opposite direction of its propagation (as for a surface muon), Fig. 1.29. These muons lie on the dark blue line of the kinematics diagram of Fig. 1.31.

- The muon is emitted opposite to the direction of the pion momentum, i.e., backward direction; $\mathbf{p}_\mu^0$ and $\mathbf{p}_\pi$ are antiparallel and the final muon momentum $\mathbf{p}_\mu$ is less than that of the pion. The muon spin is parallel to its momentum, Fig. 1.30. These muons lie on the green line of the kinematics diagram of Fig. 1.31.

The choice between these two extreme cases can be made by selecting the muon momentum with the first bending magnet after the decay channel, see Fig. 1.28. The polarization of these muons is about 80%.

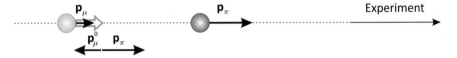

Fig. 1.30 Schematic view of the creation of backward muons in the muon decay channel

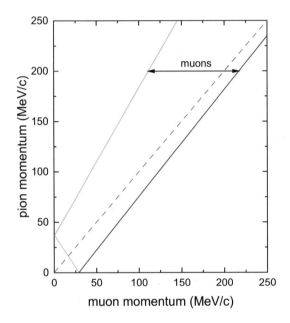

Fig. 1.31 Kinematics of the pion decay. The kinematically allowed region lies between the forward (dark blue curve) and the backward muons (green curve). The magenta dashed line corresponds to muons with momentum equal to that of the pions

The calculations of the forward and backward muon momenta as a function of the pion momentum and of the decay length of pions in a decay beam are the object of Exercise 1.6.

1.8.4   Beam Optics and Beamline Elements

The optics of a muon beam is the process of transporting a charged particle such as the muon from the source to the experiment. In simple terms, this is essentially done with bending (dipole) and focusing (quadrupole) magnets, see for example Fig. 1.25.

Fig. 1.32 Schematic view of
a dipole magnet with field
lines. The Lorentz force
acting on a positive muon
entering the paper plane is
shown

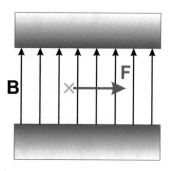

Dipole Magnet A dipole magnet provides a constant field **B** with field lines
running from the north to the south pole (see Fig. 1.32).

Considering the Lorentz force acting on a μ^+ moving in a circular orbit

$$|\mathbf{F}| = |\frac{d\mathbf{p}}{dt}| = e\,|(\mathbf{v} \times \mathbf{B})| = \frac{m_\mu\,\gamma\,v^2}{r} \quad, \tag{1.35}$$

where γ is the Lorentz factor and r is the bending radius and taking into account
that **B** and **v** are perpendicular (Fig. 1.32), we obtain

$$\frac{1}{r} = \frac{e\,B}{p} \quad. \tag{1.36}$$

Numerically

$$\frac{1}{r}\,[\text{m}^{-1}] = 299.8\,\frac{B\,[\text{T}]}{p\,[\text{MeV/c}]} \quad. \tag{1.37}$$

Given Eq. 1.36, a dipole magnet performs momentum selection; for a given bending
radius, determined by the geometry of the beamline, one selects the momentum of
the particle moving along the correct trajectory by tuning the field.[34]

Quadrupole Magnet In general, quadrupole magnets (Fig. 1.33) are used to focus
the beam. A quadrupole consists of a set of four magnetic poles arranged so that
north and south poles of hyperbolic shape alternate in order to produce a magnetic
field corresponding to the $n = 2$ term of the magnetic multipole series, see Eq. B.5
and Appendix B. They produce a magnetic field whose strength grows as a function
of the distance from the beam center axis.

---

[34] Note that other particles with different mass but equal momentum are transported in the same
way. To "clean" the beam, a velocity selection must be performed by the combined action of
perpendicular electric and magnetic fields (Wien filter), see Sect. 1.8.4. Slit systems are used to
control beam intensity, shape and momentum width. Horizontal slits are generally installed after
one of the first dipole magnets to adjust the momentum width accepted (also called momentum
bite) $\Delta p/p$ (FWHM), which typically varies between 1% and 10%.

Fig. 1.33 Quadrupole magnet, showing the four coils and the pole shoes. © Paul Scherrer Institute. Reprinted with permission. All rights reserved

The field can be determined by taking into account that there are no electric currents between the pole shoes, so that **B** must satisfy the Maxwell equations

$$\nabla \cdot \mathbf{B} = 0$$

$$\nabla \times \mathbf{B} = 0 \ . \tag{1.38}$$

In Appendix B the general solution is given in terms of a multipole expansion. Taking from Eq. B.9 the $n = 2$ term of a normal quadrupole,[35] we have for the field components[36]

$$B_x = \frac{B_0}{R_0} b_2 \, y \tag{1.39}$$

$$B_y = \frac{B_0}{R_0} b_2 \, x \quad \text{with} \tag{1.40}$$

$$g \equiv \frac{\partial B_y}{\partial x} = \frac{\partial B_x}{\partial y} = \frac{B_0}{R_0} b_2 \ . \tag{1.41}$$

---

[35] See Appendix B for the definition of a normal quadrupole.

[36] With the choice of B_0 to be the field at the pole center and R_0 the radial distance of the pole center, $b_2 = 1$, see Appendix B.

Fig. 1.34 Magnetic field lines in a quadrupole magnet. The pole shoes have the shape of hyperbolas. The vector forces acting on a positive muon flying into the paper are shown

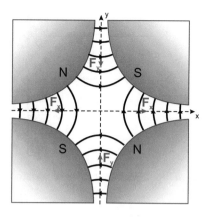

We see that, in a normal quadrupole for $y = 0$ the field is vertical and it increases linearly with x, while for $x = 0$ the field is horizontal and increasing linearly with y.

From $\mathbf{B} = -\nabla \Phi$ it follows that $\Phi(x, y) = -(B_0/R_0) b_2 xy$, hence the equipotential lines are hyperbolas $xy = \text{const.}$, the pole shoes have hyperbolic shape, and the field lines are perpendicular to them (Fig. 1.34).

As it can be inferred from the force vectors plotted in Fig. 1.34, a single quadrupole focuses the beam in one direction (in this case the y direction) and defocuses it in the other. In the focusing and defocusing planes a quadrupole creates a field, which is proportional to the lateral deviation of the trajectory. For example in Fig. 1.34 for $x = 0$ $B_x = gy$ and $F_y = -ev_z gy$. An overall focusing effect can be achieved by so-called FODO cells, consisting of a focusing (e.g., in the y direction) F quadrupole, a drift space O, a defocusing (in the y direction) D quadrupole and again a drift space O. Such a device is a quadrupole doublet. In the beamline layout of Fig. 1.28 quadrupole doublets and triplets ave visible.[37]

We can calculate the focal length of a quadrupole (see Fig. 1.35). If L denotes the length of the quadrupole, the deflection angle for a particle traveling in the beam direction at a distance $y = R$ from the central axis is given by (using Eqs. 1.39 and 1.36 and considering the thin lens approximation)

$$\alpha \simeq \frac{L}{r} = \frac{e B_x}{p} L \simeq \frac{e g R}{p} L \ . \tag{1.42}$$

---

[37] The name FODO is used as a general name for a focusing lattice. The magnet arrangement may be more complicated.

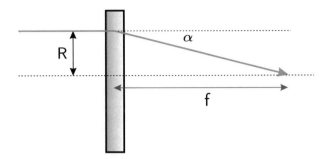

Fig. 1.35 Schematic of the focal length of a focusing quadrupole

The focal length f is therefore

$$\alpha \simeq \frac{R}{f} \rightarrow \frac{1}{f} = \frac{e\,g}{p} L = k\,L \ , \tag{1.43}$$

where we have defined the quadrupole strength $k = eg/p$, which normalizes the field gradient to the momentum of the particle (analogous to the bending strength $1/r$ defined in Eq. 1.36). Numerically we have

$$k\,[\mathrm{m}^{-2}] = 299.8\,\frac{g\,[\mathrm{T/m}]}{p\,[\mathrm{MeV/c}]} \ . \tag{1.44}$$

Motion of the Muon Spin What happens to the muon spin during transport through magnetic elements? i.e., is an initial full polarization of a surface muon beam, where ideally all muons have spin antiparallel to their momentum, maintained?

In Sect. 1.7.2 we have derived the Larmor expression for the motion of a (resting or nonrelativistic) muon spin in a magnetic field. A full relativistic treatment[38] includes the Thomas precession and modifies Eq. 1.34 to[39]

$$\omega_\mu = \gamma_\mu B = g_\mu \frac{e}{2m_\mu} B - (\gamma - 1)\frac{eB}{m_\mu \gamma} \ . \tag{1.45}$$

---

[38] Note that surface muons have $\beta = 0.28$.

[39] For the general relativistic equation of motion for a spin in uniform or slowly varying external **B** and **E** fields see Jackson (1998).

The change of trajectory (or momentum direction) is given relativistically by the muon cyclotron frequency

$$\omega_c = \frac{eB}{\gamma m_\mu} \ .$$ (1.46)

Since $g_\mu \cong 2.002$ the Larmor frequency is to a very good approximation practically identical to the cyclotron frequency.[40] As a consequence, the bending of the muon trajectory by a magnetic field is accompanied by an equal change of the spin direction. This means that the initial polarization of a muon beam is practically preserved during the transport of the muons from their source (production target or decay channel) to the experiment via magnetic elements.

Wien Filter: Separator and Spin Rotator The magnetic elements of a beamline are purely momentum selective, Eqs. 1.37 and 1.44, i.e., they transport all particles, also the unwanted ones like the positrons,[41] with the same momentum as the one selected for the μ^+. To separate and suppress unwanted particles one needs a so-called "Wien filter" or "separator", which can also be operated as a spin rotator.

A Wien filter[42] is a device with homogeneous magnetic and electric fields that are perpendicular to each other[43] and transverse to the velocity of the incoming particles. It works as a velocity filter when the electric and magnetic forces are equal

$$\mathbf{F} = q(\mathbf{E} + \mathbf{v} \times \mathbf{B}) = 0 \ .$$ (1.47)

The central undeflected trajectory is followed by particles with velocity[44]

$$v = \frac{E}{B} \ .$$ (1.48)

---

[40] For our purposes, the small deviation from $g_\mu = 2$ can be neglected during beam transport. For particle physics, however, the exact determination of $g_\mu - 2$ is of fundamental importance, see Chap. 10. The so-called $g_\mu - 2$ experiment (see Sect. 10.5) measures this deviation with great accuracy, by exploiting the fact that the cyclotron frequency and the Larmor frequency are not exactly equal.

[41] Positrons come from the μ^+ decay in flight, but also from the pair conversion of the γ's from the π_0 decay, Table 1.1.

[42] Named after the German physicist Wilhelm Wien, Nobel Prize recipient in 1911.

[43] Hence the name $\mathbf{E} \times \mathbf{B}$ filter.

[44] Note that we are considering here the classical case. To high-momenta muons, a static electric field appears as a magnetic field and corrections to the formula are necessary.

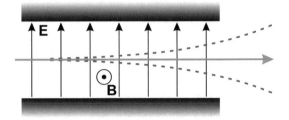

Fig. 1.36 Schematic of a Wien filter. Particles following the green trajectory have the selected velocity. Those with a lower (higher) velocity are deflected to the top (bottom) of the figure

With the geometry of Fig. 1.36 particles with lower (higher) velocity will be deflected upwards (downwards). Numerically the total deflection is given by

$$\Delta\phi = \frac{L[\text{m}]}{p[\text{MeV/c}]} \left(\frac{E\,[\text{MV/m}]}{\beta} - 299.8\ B\,[\text{T}] \right) \ . \tag{1.49}$$

Since only the magnetic part of the force acts on the spin vector, the spin will be rotated by a corresponding amount. For a surface muon beam, the angle introduced by a Wien filter used as a separator is typically $5°$–$10°$.

A Wien filter is a practical device for surface muon beams, Sect. 1.8.3, and low-energy muon beams, Sect. 8.2. Because of the high velocity and large lateral spread of the decay muon beams, a Wien filter cannot be used for them, as an effective separation between muons and positrons cannot be achieved with reasonable fields. For a surface muon beam, the electrodes creating the electrical field have a gap of typically 0.2 m and a voltage difference of about 80 kV is applied (in separator mode).

For μSR experiments, it is often necessary to apply a magnetic field perpendicular to the muon spin direction at the sample position; this is called a transverse field configuration, see Sect. 3.4.2. For high-energy muon beams, a reasonable magnetic field can be safely applied without special considerations. For beams with lower momentum, such as the surface muon beams, only a small transverse magnetic field can be applied ($\lesssim 20$ mT), since it notably deflects the trajectory.[45] The solution is to markedly rotate the muon spin with respect to the momentum direction prior to muon implantation into the sample. This can also be achieved by a Wien filter, in spin rotator mode, by applying much higher magnetic fields (and consequently a much higher electric fields, see Eq. 1.48) than in the separator mode.

---

[45] Note that even for high-energy muon beams, depending on momentum and applied field strength, the influence on the muon trajectory at the sample position is not completely negligible and sample and detector positions must be adjusted.

Fig. 1.37 Double spin rotator unit installed in the πE3 beamline of the Paul Scherrer Institute with a refocusing quadrupole element (red) in between. © Paul Scherrer Institute. Reprinted with permission. All rights reserved

A typical spin rotator has a length of about 3 m and rotates the spin by about 45° to 65°. For a complete 90° rotation, two $\mathbf{E} \times \mathbf{B}$ units are used, with a refocusing quadrupole doublet in between to achieve a good overall transmission, Fig. 1.37, (Vrankovic et al. 2012). The typical voltage difference necessary in this case is of the order of 500 kV and requires careful design (e.g., of electrodes and feedthroughs) to avoid voltage breakdown and achieve stable operation.

Exercises

1.1 Cosmic muons and relativity
The mean energy of the muons reaching sea level is about 4 GeV. Assume that $N_0 = 10^6$ muons are created at a height of 15 km. Calculate classically and relativistically the number of muons that have survived at sea level. Assume that the muons lose no energy during their flight.

1.2 Threshold proton energy for pion production
Calculate the minimum kinetic energy in a proton-proton collision to create a positive pion.

1.3 Pion decay kinematics
Calculate the momentum, energy, and kinetic energy of a muon created from a pion decaying at rest.

1.4 Muon decay kinematics
Determine the maximum and average energy of the positrons emitted in muon decays.

1.5 Pion production in a graphite target

Using a typical cross section for pion production estimate the rate of positive pions produced in the thick (60 mm) production target of the PSI cyclotron operating at 2.4 mA proton current.

1.6 Momentum of forward and backward decay muons

Determine the momentum of the forward and backward emitted decay muons as a function of the pion momentum. Write the expression for the pion decay length.

References

Alder, R. J., & Martin, R. A. (1992). *American Journal of Physics, 60*, 837–839.

Alvarez, L. W., Anderson, J. A., Bedwei, F. E., et al. (1970). *Science, 167*, 832.

Anderson, C. D., & Neddermeyer, S. H. (1936). *Physical Review, 50*, 263–271.

Anderson, C. D., & Neddermeyer, S. H. (1938). *Nature, 142*, 878–878.

Baartman, R., Bricault, P., Bylinsky, I., et al. (2003). *Proceedings of the 2003 particle accelerator conference* (Vol. 3, p. 1584).

Blau, B., Clausen, K. N., Gvasaliya, S., et al. (2009). *Neutron News, 20*, 5.

Bonka, H. (1990). *Physikalische Blätter, 46*, 126.

Brewer, J. H., & Crowe, K. M. (1978). *Annual Review of Nuclear and Particle Science, 28*, 239.

Brewer, J. H., Crowe, K. M., Gygax, F. N. et al. (1975). In V. W. Hughes, & C. S. Wu (Eds.), *Muon Physics: Chemistry and Solids* (Vol. III). Academic Press. ISBN: 978-0123606037.

Clay, J., van Alphen, P., & Hooft, C. (1934). *Physica, 1*, 829.

Conversi, M., Pancini, E., & Piccioni, O. (1947). *Physical Review, 71*, 209.

De Angelis, A. (2010). *La Rivista del Nuovo Cimento, 33*, 713–756.

Feynman, R. P. (2018). *Quantum Electrodynamics* (1st ed.). Advanced Books Classics. CRC Press. ISBN: 0-429-96179-0.

Friedman, J. I., & Telegdi, V. L. (1957). *Physical Review, 106*, 1290.

Garwin, R. L., Lederman, L. M., & Weinrich, M. (1957). *Physical Review, 105*, 1415.

Grillenberger, J., Baumgarten, C., & Seidel, M. (2021). *SciPost Physics Proceedings, 5*, 002.

Hansen, P., Carlson, P., Mocchiutti, E., et al. (2003). *Physical Review D, 68*, 103001.

Hess, V. (1926). *Physikalische Zeitschrift, 27*, 159.

Hillier, A. D., Lord, J. S., Ishida, K. et al. (2019). *Philosophical Transactions of the Royal Society A: Mathematical, Physical and Engineering Sciences, 377*, 20180064.

Ivanter, I. G., & Smilga, V. P. (1969). *Soviet Physics JETP, 28*, 286.

Jackson, J. D. (1998). *Classical Electrodynamics*. Wiley. ISBN: 978-0471309321.

Jones, G. (2008). *AIP Conference Proceedings, 79*, 15.

Khasanov, R. (2022). *Journal of Applied Physics, 132*, 190903.

Khasanov, R., Guguchia, Z., Maisuradze, A., et al. (2016). *High Pressure Research, 36*, 140.

Kiselev, D., Duperrex, P. A., Jollet, S., et al. (2021). *SciPost Physics Proceedings, 5*, 003.

Kunze, P. (1933). *Zeitschrift für Physik, 83*, 1.

Lattes, C. M. G., Muirhead, H., Occhialini, G. P. S., et al. (1947). *Nature, 159*, 694.

Lee, T. D., & Yang, C. N. (1956). *Physical Review, 104*, 254.

Matsuzaki, T., Ishida, K., Nagamine, K., et al. (2001). *Nuclear Instruments and Methods in Physics Research - Section A, 465*, 365.

Meshkovskii, A. G., Shalamov, I. I., & Shebarov, V. A. (1958). *Soviet Physics JETP, 7*, 987.

Michel, L. (1950). *Proceedings of the Physical Society. - Section A, 63*, 514.

Millikan, R., & Cameron, G. (1928). *Nature, 121*, 19.

Miyake, Y., Shimomura, K., Kawamura, N., et al. (2010). *Journal of Physics: Conference Series, 225*, 012036.

Morishima, K., et al. (2017). *Nature, 552*, 386.

Neddermeyer, S. H., & Anderson, C. D. (1938). *Physical Review, 54*, 88–89.

Pifer, A., Bowen, T., & Kendall, K. (1976). *Nuclear Instruments and Methods, 135*, 39.

Procureur, S., Morishima, K., Kuno, M. et al. (2023). *Nature Communications, 14*, 1144.

Rossi, B., & Hall, D. B. (1941). *Physical Review, 59*, 223.

Russel, M., Otis, R., & Millikan, R. (1924). *Physical Review, 23*, 760.

Swanson, R. A. (1958). *Physical Review, 112*, 580.

Tanaka, H. K. M. (2014). *Annual Review of Earth and Planetary Sciences, 42*, 535.

Tanaka, H. K. M. (2019). *Philosophical Transactions of the Royal Society A: Mathematical, Physical and Engineering Sciences, 377*, 20180142.

Thomson, M.. (2013). *Modern Particle Physics*. Cambridge University Press. ISBN: 978-1107034266.

Vrankovic, V., Gabard, A., Meier, I., et al. (2012). *IEEE Transactions on Applied Superconductivity, 22*, 4101204.

Workman, R. L., et al. (2022). *Progress of Theoretical and Experimental Physics, 2022*, 083C01. Particle Data Group.

Wu, C. S., Ambler, E., Hayward, R. W., et al. (1957). *Physical Review, 105*, 1413.

Yukawa, H. (1935). *Proceedings of the Physico-Mathematical Society of Japan. 3rd Series, 17*, 48.

Muon Implantation and Thermalization in Matter

2

2.1 Energy Loss of Particles in Matter

Particles passing through matter undergo a gradual loss of energy and scattering. The slowing down occurs by collision processes with atoms and electrons and by radiative processes with probabilities given by the interaction cross sections. Energy loss and scattering processes occur not only in the muon production target, but also by passing material (e.g., cryostat walls, windows), counters (e.g., muon counter) and in the sample. So, when performing a μSR experiment the sample must be thick enough to allow the muons to thermalize inside it.

We briefly review here the most important concepts relevant for the use of muons as spin probes in matter. Energy loss and scattering also play a role in the detection of the decay positrons, which have to traverse material (sample, sample holder, vacuum chamber's walls,...) before detection.

For charged particles, one defines the "energy loss" (or stopping power) as the mean energy loss per path length l in the material

$$-\frac{dE}{dl} \quad, \tag{2.1}$$

where E is the charged particle energy and the negative sign expresses the loss of energy. The higher the stopping power, the shorter the penetration into the material. The SI units of the stopping power are J/m, but generally, by scaling with the density ρ, it is expressed as a quantity almost independent of density: the (mass) stopping power

$$S = -\frac{1}{\rho}\frac{dE}{dl} = -\frac{dE}{dx} \quad, \tag{2.2}$$

© Springer Nature Switzerland AG 2024
A. Amato, E. Morenzoni, *Introduction to Muon Spin Spectroscopy*,
Lecture Notes in Physics 961, https://doi.org/10.1007/978-3-031-44959-8_2

where $x = l\rho$ is generally given in g/cm^2 units and the energy loss is in MeV cm^2/g.

After a relatively well-defined distance, called the "range" R,[1] the particles come at rest having lost all their kinetic energy. Since slowing down and scattering are statistical processes, path length variations for identical monoenergetic particles are observed. The statistical variation of the path length (FWHM) is called the "straggling".

For practical purposes, the so-called projected range and straggling R_p and ΔR_p (FWHM) obtained by projecting the stopping trajectory to the direction of the incoming particle, are more relevant since they define the effective depth of penetration of the muon into the sample.

There are several possible types of interactions; the dominant processes depend on the particle momentum, see, e.g., Fig. 2.4. For the muons relevant for μSR experiments with momenta $|\mathbf{p}_\mu| \lesssim 200$ MeV/c (see Sect. 1.8.3) the main channels of energy loss are electronic processes.

Depending on the velocity, we can roughly distinguish three energy loss regimes: (i) the so-called Bethe-Bloch regime, where the dominant process is the transfer of energy to target electrons leading to ionization or excitation of the target atoms or molecules; (ii) the rapid charge cycles regime at lower velocity, where, because of their positive electric charge, the muons repeatedly capture electrons (formation of muonium) and lose electrons (ionization of muonium); (iii) at even lower velocities the charge exchange cycles stop and the muons thermalize as either a positive (diamagnetic μ^+) or neutral species (paramagnetic muonium), see Sect. 2.3.[2]

As long as the energy of the particles is large compared to the ionization energies, the energy loss in each collision with an electron is only a small fraction of the energy of the particle and the energy loss can be treated as continuous. The history of the energy loss calculations is marked by the work of Bohr in 1913 and 1915 (Bohr 1913, 1915). Quantum mechanical versions based on perturbation theory,

---

[1] Note that, following the scaling of the energy loss, the range is often given in the units of g/cm^2, which corresponds to the mean penetration length L multiplied with the material density $L\rho$.

[2] In general, the elastic collision between a muon and an atom or ion as a whole (also called nuclear energy loss, but it has nothing to do with nuclear forces), which may play a role only at very low momenta, (see Fig. 2.4) can be safely neglected. In the classical limit, the maximum energy transfer in an elastic collision of a particle with mass m with an atom of mass m_A is given by

$$\Delta E_{max} = \frac{1}{2}mv^2 \left(\frac{4mm_A}{(m+m_A)^2} \right)$$

$$= \frac{2m_A v^2}{1 + 2\frac{m_A}{m} + \frac{m_A^2}{m^2}}, \tag{2.3}$$

which is negligible for $m_A \gg m_\mu$, $\Delta E_{max}/E_{kin} \cong 4m_\mu/m_A$. However, this process is important in the generation of low-energy muons, which are obtained by moderating a surface muon beam down to ~ 10 eV, see Chap. 8.

Fig. 2.1 Model for
calculating the energy lost by
a charged particle of mass m
velocity v and charge ze in a
collision with an electron.
The parameter b is the impact
parameter

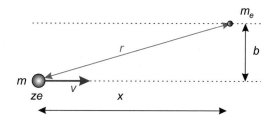

first nonrelativistic then relativistic, were later presented by Bethe (1930) and Bloch
(1933). For a historical overview of the various calculations, see in Ziegler (1999).

2.1.1 Energy Loss by Ionization: Classical Approach

Following Bohr (1913, 1915), we first derive the classical formula for the interaction
of a particle of charge $q = ze$ and mass m with an electron in matter (Fig. 2.1).

The momentum transferred to a stationary unbound electron is equal to

$$\Delta p = \int_{-\infty}^{\infty} F_{\text{Coul}} dt \ . \tag{2.4}$$

For the Coulomb force we only need to consider the transverse component, since
the contribution from the longitudinal component cancels out

$$F_{\text{Coul}}^{\perp} = F_{\text{Coul}} \frac{b}{r} = F_{\text{Coul}} \frac{b}{\sqrt{b^2 + x^2}} = \frac{k_e ze^2}{b^2 + x^2} \frac{b}{\sqrt{b^2 + x^2}} \ , \tag{2.5}$$

where k_e is the Coulomb constant and z the charge number of the incident particle.
Therefore

$$\Delta p = \int_{-\infty}^{\infty} \frac{k_e ze^2}{b^2 + x^2} \frac{b}{\sqrt{b^2 + x^2}} \frac{dx}{v} \ , \tag{2.6}$$

where v is the velocity and we obtain

$$\Delta p = \frac{2 k_e ze^2}{v b} \ . \tag{2.7}$$

The energy transfer in one collision is

$$\Delta E(b) = \frac{\Delta p^2}{2m_e} = \frac{2 k_e^2 z^2 e^4}{m_e v^2 b^2} \ . \tag{2.8}$$

Fig. 2.2 In the path length
dl and with impact parameter
between b and $b + db$, the
charged particle collides with
the electrons contained in the
cylindrical ring

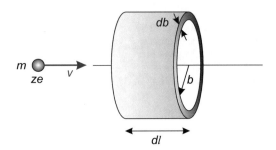

To determine the number of collisions, we consider a cylindrical volume of thickness db and length dl, Fig. 2.2. There are

$$\frac{ZN_A}{A}\rho \, 2\pi \, db \, dl \tag{2.9}$$

electrons in this volume, where Z is the atomic number, A the atomic weight, N_A the Avogadro number, and ρ the mass density. The total energy lost to these electrons for a volume dV of thickness db is

$$\Delta E = \frac{2\,k_e^2\,z^2 e^4 Z N_A}{m_e v^2 A b^2}\rho \, 2\pi b \, db \, dl \quad . \tag{2.10}$$

To find the total number of collisions we have to integrate over the possible impact parameters in the range between b_{min} and b_{max}.

The maximum energy transfer in a collision ("hard" collision) is $\Delta E_{max} = 2m_e v^2$ (consider Eq. 2.3 with $m_A = m_e \ll m$). We have

$$\Delta E_{max} = 2m_e v^2 = \Delta E(b_{min}) = \frac{2\,k_e^2\,z^2 e^4}{m_e v^2 b_{min}^2} \tag{2.11}$$

$$b_{min} = \frac{k\,z e^2}{m_e v^2} \quad . \tag{2.12}$$

The minimum value of the energy transfer corresponds to the so-called "mean excitation energy" I, $\Delta E_{min} = I$ ("soft" collision). The value of I is not easily calculated and has to be determined experimentally. For $Z > 1$ it can be approximated as follows

$$I = 16\,\mathrm{eV} \times Z^{0.9} \quad , \tag{2.13}$$

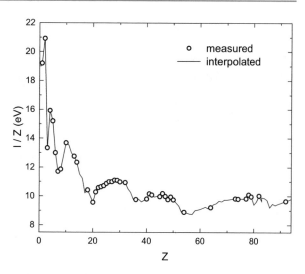

Fig. 2.3 Mean excitation energy I versus atomic number determined experimentally as adopted by the International Commission on Radiation Units and Measurements (ICRU). Modified from Tanabashi et al. (2018), © American Physical Society. Reproduced with permission. All rights reserved

and for $Z > 20$ by

$$I = 10\,\text{eV} \times Z \ . \tag{2.14}$$

Fig. 2.3 displays I against the atomic number.

Therefore

$$\Delta E_{\min} = I = \frac{2\,k_e^2\,z^2 e^4}{m_e v^2 b_{\max}^2}$$

$$b_{\max} = \frac{k_e\, z e^2}{v} \sqrt{\frac{2}{m_e I}} \ . \tag{2.15}$$

Remembering that $dx = \rho\, dl$ and introducing a negative sign to take into account the loss of energy we have

$$-\frac{dE}{dx} = \int_{b_{\min}}^{b_{\max}} \frac{2\,k_e^2\,z^2 e^4 Z N_A}{m_e v^2 A b}\, 2\pi\; db = \frac{4\pi\,k_e^2\,z^2 e^4 Z N_A}{m_e v^2 A}\frac{1}{2}\ln\left(\frac{2 m_e v^2}{I}\right)$$

$$= \frac{4\pi\,k_e^2\,z^2 e^4 Z N_A}{m_e \beta^2 c^2 A}\frac{1}{2}\ln\left(\frac{2 m_e \beta^2 c^2}{I}\right) \ , \tag{2.16}$$

which is Bohr's classical derivation of the energy loss of a particle in matter.

2.1.2 Energy Loss: Bethe Formula

The full quantum mechanical derivation including relativistic effects gives the Bethe equation[3] (Workman et al. 2022)

$$
\begin{aligned}
-\frac{dE}{dx} &= \frac{4\pi\, k_e^2\, z^2 e^4 Z N_A}{m_e \beta^2 c^2 A}\left(\frac{1}{2}\ln\frac{2m_e \beta^2 c^2 \gamma^2 T_{\max}}{I^2} - \beta^2 - \frac{\delta(\beta\gamma)}{2}\right)\\
&= K\, z^2 \frac{Z}{A}\frac{1}{\beta^2}\left(\frac{1}{2}\ln\frac{2m_e \beta^2 c^2 \gamma^2 T_{\max}}{I^2} - \beta^2 - \frac{\delta(\beta\gamma)}{2}\right) .
\end{aligned}
\tag{2.17}
$$

$T_{\max}$ is the maximum kinetic energy transfer to an electron calculated relativistically

$$
T_{\max} = \frac{2 m_e c^2 \beta^2 \gamma^2}{1 + 2\gamma\dfrac{m_e}{m} + \dfrac{m_e^2}{m^2}} ,
\tag{2.18}
$$

$\delta(\beta\gamma)$ is the so-called density effect correction to the energy loss by ionization, and the coefficient

$$
K = \frac{4\pi\, k_e^2\, e^4 N_A}{m_e c^2} = 0.307075\ \text{MeV cm}^2\ \text{mol}^{-1} .
\tag{2.19}
$$

The Bethe formula describes the mean rate of energy loss in the range $0.1 \lesssim \beta\gamma \lesssim 1000$, see Fig. 2.4. A few remarks on this formula:

- Since $-dE/dx$ is proportional to the square of the charge of the incoming projectile, it predicts the same stopping power for positive ($z = 1$) and negative ($z = -1$) muons .
- For $\gamma m_e \ll m$ the stopping power dE/dx is practically independent of the mass of the incoming particle. The Bethe equation depends essentially only on the velocity of the incoming particle. This means that, for particles with the same charge, it is a universal curve that depends on $\beta\gamma$.
- It is also only slightly dependent on the target material; only the factors Z/A, which is almost constant over a wide range of materials and $\ln I$ are (weakly) material dependent, see Fig. 2.5 (we can use this information to solve Exercise 2.1).
- For $\beta\gamma \lesssim 3$ (corresponding to $\lesssim 300$ MeV/c muon momentum) $-dE/dx \propto 1/\beta^2$. This is the relevant range for muon beams for μSR.

---

[3] There are different versions of the relativistic stopping power formula, sometimes also called Bethe-Bloch formula.

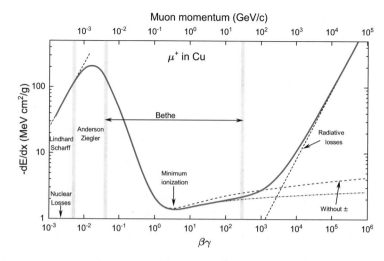

Fig. 2.4 Stopping power $-dE/dx$ for positive muons in copper as a function of momentum p_μ and $\beta\gamma = p_\mu/m_\mu c$. The validity range of the Bethe formula is indicated by gray bars. There are two regions where the Bethe formula is no longer valid: (i) at very high momenta ($\beta\gamma \gtrsim 1000$), where radiative energy processes such as bremsstrahlung dominate (these momenta are not relevant for our scope), (ii) around the stopping power maximum and below it at very low momenta, where one observes a linear increase of the stopping power with velocity. Modified from Tanabashi et al. (2018), © American Physical Society. Reproduced with permission. All rights reserved. See also (Workman et al. 2022)

- When the kinematic factor $1/\beta^2$ decreases, a minimum at $\beta\gamma \sim 3-4$ is reached. Particles at this point are called minimum ionising particles (MIP). Inspection of Fig. 2.4 shows that a muon loses about 13 MeV/cm at the MIP point in copper (density 8.96 g/cm^3).
- At low energies and when the incoming particle mass is larger than that of the electron, $m \gg \gamma m_e$, Eq. 2.18 simplifies to

$$T_{\max} \simeq 2m_e c^2 \beta^2 \gamma^2 \ . \tag{2.20}$$

Introducing this into 2.17 and taking $\gamma \simeq 1$, we see that the Bethe result is identical to the classical one, except by a factor of 2.[4]

- At higher energies there is a logarithmic increase due to relativistic effects. It reflects the relativistic increase of the transverse component of the electric field (Lorentz transformation of the field). This leads to collisions at larger distances and to more collisions.

---

[4] This factor arises from the fact that the classical limit does not correctly describe the very distant collisions and that the atomic binding of the electrons cannot be neglected.

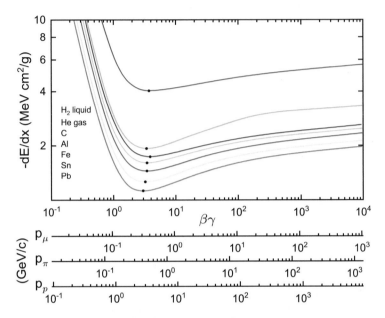

Fig. 2.5 Stopping power $-dE/dx$ of μ, π and p in various materials: liquid hydrogen, helium gas, carbon, aluminum, iron, tin and lead. Except for hydrogen, the energy loss in different materials is similar. Particles of the same charge and velocity have the same energy loss. The dots mark the minimum ionizing point, which is essentially determined only by the velocity. Modified from Tanabashi et al. (2018), © American Physical Society. Reproduced with permission. All rights reserved

Bethe's expression fails to explain the region around $\beta\gamma \sim 0.01 - 0.1$, where the stopping power has a broad maximum, and below, because it assumes that the projectile has a velocity much greater than the typical velocity of the electrons with which it collides $v \gg v_e$, and that its effective charge does not change with velocity.

At lower energies, $v \approx v_e$, a positively charged particle can pick up an electron, thus reducing its effective charge and therefore the stopping power (Lindhard et al. 1963). Moreover, repeated charge changing cycles of electron-pickup and electron-loss in successive collisions become possible. Quantum mechanical calculations (e.g., energy loss in an electron gas) and also experiments with light ion projectiles give an energy loss proportional to velocity $dE/dx \propto v$ (Lindhard & Scharff 1961), which is well fulfilled in metals.

There is no satisfactory theory for the broad maximum between $0.01 < \beta\gamma < 0.05$. To obtain values of the stopping power S also in this intermediate energy range an interpolation

$$\frac{1}{S} = \frac{1}{S_{\text{low}}} + \frac{1}{S_{\text{high}}} \tag{2.21}$$

is often applied, in which tabulated values for the region below S_{low} and above S_{high} are used. Extensive tables are compiled in Ziegler et al. (1985) and Andersen and Ziegler (1977) and in a report of the International Commission on Radiation Units and Measurement (ICRU) (Berger et al. 1993), which also reviews the stopping power theory for protons and heavier ions. For the muons one can use the proton formulas and data velocity scaled.[5]

2.2 Range and Slowing Down Time

2.2.1 Range of Muons

The total range R (in g/cm^2) of a particle with initial kinetic energy E_{kin} (or total energy E_{tot}) can be formally obtained from the integral

$$R = \int_{E_{\text{kin}}}^{0} \frac{1}{dE/dx} dE$$

$$= \int_{E_{\text{tot}}}^{mc^2} \frac{1}{dE/dx} dE \quad . \tag{2.22}$$

Within its validity range, the integration can be performed using the Bethe formula. For the different ranges, approximations are often used. For example for a muon having an initial momentum $p_\mu = \gamma m_\mu v = \beta \gamma m_\mu c$, one can use

$$-\frac{dE}{dx} = \begin{cases} a\, \dfrac{\ln(\beta)}{\beta^2}, & \text{for } \beta\gamma = \dfrac{p_\mu}{m_\mu c} \ll 1 \\[2ex] b\, \dfrac{1}{\beta^2}, & \text{for } \beta\gamma = \dfrac{p_\mu}{m_\mu c} \lesssim 1 \\[2ex] c, & \text{for } \beta\gamma = \dfrac{p_\mu}{m_\mu c} \simeq 3-4 \\[2ex] c + d\ln(\beta), & \text{for } \beta\gamma = \dfrac{p_\mu}{m_\mu c} \gg 1 \end{cases} \tag{2.23}$$

---

[5] This velocity scaling and the linear proportionality $\frac{dE}{dx} \propto v$ fail when the muon energy is of the order of the ionization threshold, which in insulators can be as high as ~ 10–$20\,\text{eV}$. This is of no importance when dealing with surface muons, but it is of significance in the moderation process of low-energy muons, Sect. 8.2.3.

where a, b, c, and d are taken as constants. For example, for backward muons from a decay beamline, Fig. 1.31, the second approximation is valid

$$-\frac{dE}{dx} = b\,\frac{1}{\beta^2}\;. \tag{2.24}$$

We have (see Exercise 2.2 for details)

$$R = \int_{E_\mu}^{m_\mu c^2} \frac{1}{dE/dx}dE = \frac{1}{b}\frac{E_{\rm kin}^2}{E_\mu}\;. \tag{2.25}$$

For a decay muon in copper with a momentum $p_\mu = 110\,\text{MeV/c}$[6] (total energy $E_\mu \simeq 152.5\,\text{MeV}$ and $\gamma \simeq 1.444$) and with $b \simeq 1.3\,\text{MeV}\,\text{cm}^2/\text{g}$) one obtains a total range $R \simeq 11.1\,\text{g/cm}^2$, which represents a path length $R_l = R/\rho_{\rm Cu} \simeq 1.24\,\text{cm}$ (as $\rho_{\rm Cu} = 8.96\,\text{g/cm}^3$). From Eq. 2.25 we note that a muon with the same kinetic energy as a proton has a stopping range $m_p/m_\mu \simeq 9$ times longer.

For smaller momenta (such as those of a surface muon beam, $\beta = 0.28$), the above formula underestimates the range by quite a bit. This is a consequence of neglecting the logarithmic term in the stopping power and of the limited range of validity of the Bethe formula.

As mentioned before, for practical use, the relevant quantities are the projection of the range $R_{\rm p}$ and corresponding straggling $\Delta R_{\rm p}$ along the incoming trajectory. For muons with an initial momentum corresponding to the region where $-dE/dx \propto (1/\beta)^2$ the following semi-empirical formula is found to be applicable (Pifer et al. 1976)

$$R_{\rm p} = C\,p_\mu^{3.5}\;, \tag{2.26}$$

where C is a constant that depends on the material. The typical intrinsic range straggling is about 10% of $R_{\rm p}$, see Fig. 2.6, so that the total straggling is given by

$$\frac{\Delta R_{\rm p}}{R_{\rm p}} = \sqrt{0.1^2 + \left(3.5\frac{\Delta p_\mu}{p_\mu}\right)^2}\;. \tag{2.27}$$

For surface muon beams with $p_\mu \simeq 30\,\text{MeV/c}$ and a typical momentum bite $\Delta p_\mu/p_\mu \simeq 0.03$, the relative straggling $\Delta R_{\rm p}/R_{\rm p} \simeq 14\%$. $R_{\rm p}$ expressed in mg/cm^2 is almost independent of the material and is about 140 mg/cm^2. This

---

[6] This momentum is typically used for μSR studies under pressure.

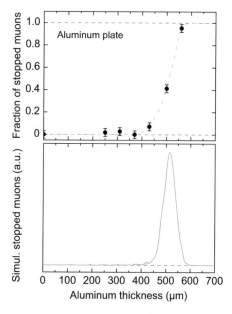

Fig. 2.6 Top: Fraction of a surface muon beam stopped in an aluminum foil as a function of thickness. The measurements were performed with the GPS instrument at PSI. The beam (after traversing some additional material such as vacuum window) is stopped by an aluminum foil with thickness of the order of $L = 0.50 \pm 0.05$ mm, which corresponds to a range $R_p = L\rho \simeq$ 135 mg/cm^2. Bottom: Stopping distribution of the surface muons. No momentum uncertainty, $\Delta p_\mu / p_\mu = 0$, is assumed in the calculation, i.e., the width of the stopping distribution represents the intrinsic range straggling of about 10% of R_p

number[7] defines the typical minimum sample thickness necessary to perform a μSR experiment on bulk samples.[8] Figure 2.7 shows the energy dependence of the projected range in Cu.

How the energy is lost along the path of the particle is called the Bragg curve. According to the Bethe formula, the energy of the particle decreases as it passes through matter, and consequently the specific energy loss (mainly by ionization) increases. The Bragg curve shows a flat plateau region and a broad peak near the end of the particle path (Bragg peak).[9] The shape of the Bragg curve can be roughly determined starting from Eq. 2.24, see Exercise 2.3.

---

[7] It corresponds to a penetration depth of about a few hundred μm in solids, a few mm in liquids, or a few m in gases at STP.

[8] If a sample cannot be grown to the required thickness, one can still perform a μSR experiment by placing a degrader (typically a Ti, Al, or Kapton foil) in front of the sample. The exact thickness of the degrader should be carefully calculated to avoid stopping muons in it.

[9] Note that this Bragg peak should not be confused with the nuclear or magnetic Bragg peak in neutron scattering.

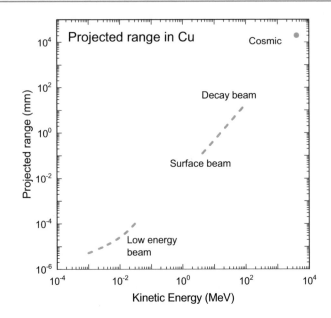

Fig. 2.7 Mean projected range of positive muons implanted in copper as a function of kinetic energy. Surface or decay muon beams are used to study the bulk properties of matter; low-energy muons (LEM) are used to perform depth dependent investigation at the nanometer scale and to study thin films. Cosmic muons are used for muon radiography

2.2.2 Thermalization Time

The thermalization time[10] can be estimated by

$$t_{\text{th}} = \int_{E_{\text{kin}}}^{0} dt = \int_{E_{\text{kin}}}^{0} \frac{dl}{v} = \int_{E_{\text{kin}}}^{0} \frac{1}{v \dfrac{dE}{dl}} \, dE$$

$$= \int_{E_{\text{kin}}}^{0} \frac{1}{v \rho \dfrac{dE}{dx}} \, dE \quad . \tag{2.28}$$

It is inversely proportional to the density of the material and of the order of $\simeq 10$ ps for usual solids. Since the slowing down processes are only electronic and very fast, the muon polarization of the implanted beam is not affected and the thermalization time can be considered instantaneous. This is not the case in gases, which have much

---

[10] Possible and putative delayed processes in the thermalization regime, see Sect. 2.3, can be discarded for most of the experiments.

lower density and where the thermalization time can reach tens of nanoseconds (Fleming et al. 1982; Senba et al. 2006).

2.2.3 Multiple Scattering

A muon passing through a material experiences many small scattering events due to the Coulomb interaction (Fig. 2.8). This effect has to be taken into account when designing a μSR spectrometer, where the muon beam may have to pass through a muon detector and various types of windows and cryogenic shields before reaching the sample, see Sect. 3.3.1. The resulting projected (plane) and nonprojected (space) angular distributions can be approximated by a Gaussian distribution (Tanabashi et al. 2018)

$$F(\theta_{\text{plane}})d\theta_{\text{plane}} = \frac{1}{2\pi\theta_0^2}\exp\left(-\frac{\theta_{\text{plane}}^2}{2\theta_0^2}\right)d\theta_{\text{plane}} \tag{2.29}$$

$$F(\theta_{\text{space}})d\Omega = \frac{1}{2\pi\theta_0^2}\exp\left(-\frac{\theta_{\text{space}}^2}{2\theta_0^2}\right)d\Omega \ , \tag{2.30}$$

where $\theta_0 = \theta_{\text{plane}}^{\text{rms}} = \frac{1}{\sqrt{2}}\theta_{\text{space}}^{\text{rms}}$ and

$$\theta_0 = \frac{13.6\,\text{MeV}}{\beta\,cp}z\sqrt{\frac{x}{X_0}}\left(1+0.038\ln\frac{x\,z^2}{X_0\,\beta^2}\right) \ . \tag{2.31}$$

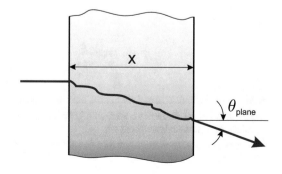

Fig. 2.8 A muon incident from the left and passing through a material of thickness x undergoes multiple scattering

x/X_0 is the thickness of the scattering material, expressed in so-called radiation lengths X_0.[11]

2.3   Muon States in Matter

The state reached by the muon after deceleration and thermalization in the sample depends on the chemical and physical properties of the sample material, which can be a solid, a liquid, or a gas. A fraction of the thermalized muons end up in a diamagnetic environment, in the so-called "diamagnetic" state. This includes the "bare" or "free" muon state,[12] e.g., at an interstitial state of a lattice, but also if it plays the role of a proton in a diamagnetic molecule such as $C_6H_{11}\mu^+$ or in bonded states such as $O^-\mu^+$. Another fraction can capture an electron during (or to a lesser extent after) the thermalization process and form muonium, Mu (μ^+e^-), a neutral "paramagnetic" state in matter. Muonium, as we will see, can assume different electronic configurations, which can be investigated by μSR spectroscopy, see Chap. 7. In principle the negative muonium state, Mu^-, is also possible in materials such as semiconductors. Like "free" μ^+, it is also a diamagnetic state, where the muon spin precesses with the same Larmor frequency as a "bare" μ^+; spectroscopically, it is not possible to distinguish between different diamagnetic environments.

The "bare" muon state is the most important state for the use of a polarized muon as a local magnetic probe. It represents a charged impurity in the sample material. Due to its positive charge it is repelled by the positively charged ions and generally occupies an interstitial position in the crystal lattice, which relaxes with a slight local lattice expansion, see Fig. 2.9 for examples of muon sites in some materials. In a metal, the "bare" muon is the stable state; its charge modifies the local charge density of the conduction electrons.

The relative enhancements of charge densities as a function of the bulk electron density have been calculated self-consistently for a range of metallic densities using a spin density functional formalism (Jena & Singwi 1978).[13] Figure 2.10 depicts

---

[11] The radiation length for a material with atomic number Z and mass number A can be approximated by

$$X_0 = 716.4 \frac{g}{cm^2} \frac{A}{Z(Z+1)\ln\left(\frac{287}{\sqrt{Z}}\right)} ,$$

see Workman et al. (2022) and references therein for details.

[12] The quotation marks indicate that, strictly speaking, the muon is not bare or free, because it is embedded in a chemical environment.

[13] There might be also an influence on a possibly preexisting electronic spin polarization, e.g., in a magnetic material.

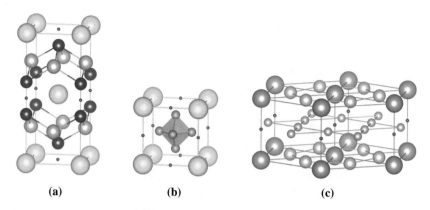

Fig. 2.9 Examples of experimentally determined muon stopping sites (red spheres) for different systems. For all systems, the sites are crystallographically equivalent, i.e. they satisfy the same symmetry operations that leave these points of the crystal unchanged. Structures drawn with VESTA, © 2006–2022 Koichi Momma and Fujio Izumi (Momma & Izumi 2011). (a) $CeRu_2Si_2$. (b) CeB_6. (c) UPd_2Al_3

the enhancement at the muon site and the oscillatory behavior with distance. We use this Figure for a simplified qualitative discussion of the effect of the muon charge.[14]

The unperturbed electron density is expressed as $3[4\pi(r_s a_0)^3]^{-1}$ where a_0 is the Bohr radius and r_s the so-called electron density parameter, which is generally used to express conduction electron densities. So, $r_s a_0$ represents the radius of a sphere containing one electron. In Fig. 2.10 a typical value $r_s = 2$ is chosen, which gives an unperturbed electron density

$$n_{e,0}(r_s = 2) = \frac{1}{\frac{4}{3}\pi(2a_0)^3} \ . \tag{2.32}$$

The Figure shows that in the presence of the muon a significant increase of electron density at the muon site ($r = 0$) takes place with

$$n_e(r = 0) \simeq 16\, n_{e,0}(r_s = 2) = \frac{16}{\frac{4}{3}\pi(2a_0)^3} \ . \tag{2.33}$$

Note that the electron density of muonium corresponds to Eq. 2.32 with $r_s = 1$. Therefore, the electron density at the muon site in a metal is about twice that in muonium

$$n_{e,Mu} = \frac{1}{\frac{4}{3}\pi a_0^3} \simeq \frac{1}{2}\, n_e(r = 0) \ . \tag{2.34}$$

---

[14] State-of-the-art density functional theory calculations (see footnote 18, Chap. 5) of a muon stopped in typical magnetic materials (Onuorah et al. 2018) are able to provide accurate quantitative estimates of the muon coupling with its surrounding, including electronic spin degrees of freedom and contact field (see Sect. 5.1.2).

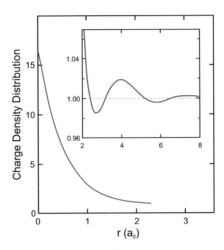

Fig. 2.10 Representative electron charge density distribution around the positive muon. The curves represent the normalized charge density $n_e(r)/n_{e,0}$ for $r_s = 2$ versus distance in units of Bohr radius, see text for details. The exact form depends on the density of the conduction electrons. The ratio $n_e(r = 0)/n_{e,0}$ strongly increases with r_s. The inset shows the large distance behavior. Modified from Jena and Singwi (1978), © American Physical Society. Reproduced with permission. All rights reserved

However, muonium is not observed in metals. The collective screening of the muon's Coulomb potential and the scattering processes with the electrons prevent the formation of a bound muon-electron state. Even if the state is formed, it is extremely short lived and destroyed by the scattering of the bound electron with the conduction electrons.

Therefore, the state of a positive muon implanted in a metal is the diamagnetic state of a "free" or "bare" muon.

The situation is different in materials where the free electron density is low, such as in insulators, semiconductors, molecular systems, liquids or gases. In these systems stable muonium formation is possible. As detailed in Chap. 7 spectroscopy of muonium in semiconductors and insulators is an important tool to study the electronic states of hydrogen in these materials.[15]

An instructive representation of the processes involved in muonium formation can be obtained by considering muon-atom (or -molecule) collision processes in a low density gas (gas model) (Senba et al. 2006). The main electronic processes responsible for energy loss and muonium formation are summarized in Table 2.1, together with the typical energy loss per collision. Excitation processes are implic-

---

[15] A rough upper limit on the electron density to observe muonium in a material was given in Cox et al. (2001) with $n_e \sim 3 \times 10^{22}$ cm^{-3}.

Table 2.1 Electronic processes contributing to muon thermalization and muonium formation (gas model). E_A ionization energy of atom A. $E_{Mu} = 13.6$ eV ionization energy of muonium

Process		ΔE: typical energy loss ($\Delta E > 0$) or gain ($\Delta E < 0$) per collision
Ionization	$\mu^+ + A \longrightarrow \mu^+ + A^+ + e$	E_A
	$Mu + A \longrightarrow Mu + A^+ + e$	E_A
Electron capture (Mu formation)	$\mu^+ + A \longrightarrow Mu + A^+$	$E_A - E_{Mu}$
Electron loss (Mu breakup)	$Mu + A \longrightarrow \mu^+ + A + e$	E_{Mu}

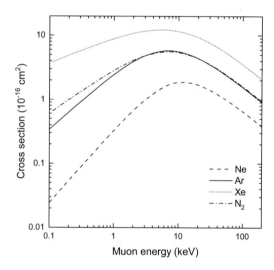

Fig. 2.11 Ionization cross section of muons in different gases. Scaled from proton data (Rudd et al. 1985). The cross section scales with velocity

itly included in the ionization.[16] The cross sections for these processes (displayed in Figs. 2.11 and 2.12) depend essentially on the velocity of the impinging particle and can be obtained by scaling the known corresponding cross sections of proton-gas collisions (Rudd et al. 1985; Nakai et al. 1987).

As already mentioned before, after the kinetic energy of the implanted muons has dropped to a few tens of keV (essentially by ionization/excitation processes) and the muon velocity becomes comparable to the orbital velocity of the electrons of the medium, the positive muon can capture an electron to form a muonium or a similar neutral muon-electron complex.[17] Generally, the muon, in a successive collision, loses this electron but it can again capture another one in a successive collision, this way undergoing a rapid series of several hundred electron pickup (capture) and stripping (loss) cycles, losing energy in each cycle. The final charge

---

[16] In principle, the formation of a negative charge state (Mu⁻) is also possible. It can be ignored here because its relative probability is low and the binding of the second electron is weak.

[17] Initially, the velocity of the muon is much higher than the typical velocity of a bound electron. While a muon from a surface beam has a velocity of 0.28 c, an electron in level n of an atom with nuclear charge Z has a velocity of $\alpha c Z/n$, α fine structure constant (from the Bohr atom model).

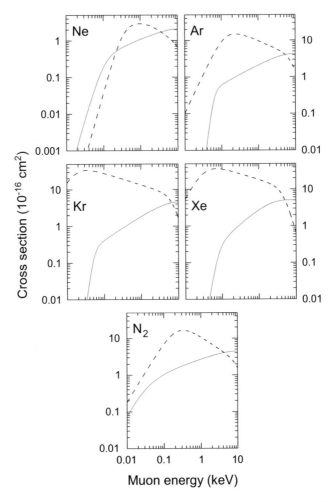

Fig. 2.12 Cross sections for electron loss (solid line) and electron capture (dashed line) of muons in different gases. Data velocity scaled from Nakai et al. (1987)

state fractions are essentially determined by the relative magnitude of the capture and loss cross sections at the end of the cyclic charge exchange regime. This process ends when the effective threshold for electron capture or loss is reached. A muonium fraction, present at these epithermal[18] energies, can finally thermalize by elastic collisions (or in solids by phonon excitation) and determines the observed muonium fraction in the sample ("prompt" or epithermal muonium formation). From Fig. 2.12 it is clear that this is the case for all gases (except Ne), where at low energies the capture cross section is much larger than the loss cross section, making it unlikely

---

[18] i.e., having an energy greater than the thermal energy (at 300 K $k_B T \cong 26$ meV).

that a muonium atom formed at low energies will lose an electron in a successive collision. In contrast, in Ne before thermalization, the loss cross section is much larger than the capture cross section, so that the probability of μ^+ picking up an electron in a successive collision is small. The gas model with velocity scaling explains satisfactorily well the thermalized charge fractions found in noble and other gases (e.g., N_2) (Fleming et al. 1982).

The applicability of "atomic" processes in solids requires that the collision processes of the projectile with the target atoms can be considered as single collision processes, which is reasonably well satisfied for muons with a kinetic energy $\gtrsim$ 1 keV. In fact, the measured energy dependent neutral fractions[19] of a muon beam exiting thin layers of solid Xe, Ar and N_2 down to energies of $\sim$1 keV are well explained by velocity scaling of the proton cross sections for electron capture and loss (Prokscha et al. 1998).

In solids, at the end of the slowing down process, i.e., just before or after thermalization, additional processes may come into play. In addition to the epithermal or prompt muonium formation process described above, a delayed muonium formation mechanism has been proposed. In this mechanism the muon thermalizes as a "bare" particle and subsequently attracts by Coulomb interaction an electron from the spur generated during slowing down forming muonium, as first observed in superfluid helium (Eshchenko et al. 1988; Krasnoperov et al. 1992). In rare gas liquids and solids the electron muon convergence and hence the delayed muonium fraction can be modified by applying an electric field parallel or antiparallel to muon's Coulomb field (Storchak et al. 1996; Eshchenko et al. 2002a). This mechanism has also been invoked to explain part of the muonium fraction and the fast muon spin depolarization observed at early times in various oxide insulators and prototypical semiconductors (Storchak et al. 1997; Eshchenko et al. 2002b). However, this has been questioned in an alternative model that attributes the fast relaxing signal to an intermediate state configuration formed epithermally and consisting of a transient muonium complex with the electron slightly separated from the muon, but weakly bound to it (Vilão et al. 2017; Brewer et al. 2020; Vilão et al. 2020).

As we will see in the next chapters, details of the thermalization mechanisms do not affect the information obtained from μSR and muonium spectroscopy experiments in solids.

Exercises

2.1 Energy loss of cosmic muons in the atmosphere

Calculate the energy loss of cosmic muons traversing the atmosphere. Hint: Use the information contained in Figs. 1.6 and 2.5.

---

[19] From the known energy dependent cross sections $\sigma_L(E)$ and $\sigma_C(E)$ the neutral equilibrium fraction is easily obtained as $F(E_{kin}) = \sigma_C(E_{kin})/(\sigma_L(E_{kin}) + \sigma_C(E_{kin}))$.

2.2 Energy loss of decay muons

Derive the approximate expression, Eq. 2.24, which describes the energy loss of decay muons.

2.3 Bragg Curve

Determine the shape of the Bragg curve starting from Eq. 2.24.

References

Andersen, H. H., & Ziegler, J. F. (1977). *The Stopping and Ranges of Ions in Matter* (Vol. 3). Pergamon Press. ISBN: 978-0080216058.

Berger, M. J., et al. (1993). *Stopping Powers and Ranges for Protons and Alpha Particles.* International Commission on Radiation Units and Measurements, Bethesda, MD (United States). ICRU Report 49.

Bethe, H. (1930). *Annalen der Physik, 5*, 325.

Bloch, F. (1933). *Annalen der Physik, 408*, 285.

Bohr, N. (1913). *The London, Edinburgh, and Dublin Philosophical Magazine and Journal of Science, 25*, 10.

Bohr, N. (1915). *The London, Edinburgh, and Dublin Philosophical Magazine and Journal of Science, 30*, 581.

Brewer, J. H., Storchak, V. G., Morris, G. D., et al. (2020). *Physical Review B, 101*, 077201.

Cox, S. F. J., Cottrell, S. P., Charlton, M., et al. (2001). *Journal of Physics: Condensed Matter, 13*, 2169.

Eshchenko, D. G., Storchak, V. G., Brewer, J. H., et al. (2002a). *Physical Review B, 66*, 035105.

Eshchenko, D. G., Storchak, V. G., Brewer, J. H., et al. (2002b). *Physical Review Letters, 89*, 226601.

Eshchenko, D., Krasnoperov, E., Barsov, S., et al. (1988). *JETP Letters, 48*, 616.

Fleming, D. G., Mikula, R. J., & Garner, D. M. (1982). *Physical Review A, 26*, 2527.

Jena, P., & Singwi, K. S. (1978). *Physical Review B, 17*, 3518.

Krasnoperov, E., Meilikhov, E., Abela, R., et al. (1992). *Physical Review Letters, 69*, 1560.

Lindhard, J., & Scharff, M. (1961). *Physical Review, 124*, 128.

Lindhard, J., Scharff, M., & Schiott, H. (1963). *Matematisk-fysiske Meddelelser udviget af Det Kongelige Danske Videnskabernes Selskab, 33*, 1.

Momma, K., & Izumi, F. (2011). *Journal of Applied Crystallography, 44*, 1272–1276.

Nakai, Y., Shirai, T., Tabata, T., et al. (1987). *Atomic Data and Nuclear Data Tables, 37*, 69.

Onuorah, I. J., Bonfà, P., & De Renzi, R. (2018). *Physical Review B, 97*, 174414.

Pifer, A., Bowen, T., & Kendall, K. (1976). *Nuclear Instruments and Methods, 135*, 39.

Prokscha, T., Morenzoni, E., Meyberg, M., et al. (1998). *Physical Review A, 58*, 3739.

Rudd, M. E., Kim, Y. K., Madison, D. H., et al. (1985). *Reviews of Modern Physics, 57*, 965.

Senba, M., Arseneau, D. J., Pan, J. J., et al. (2006). *Physical Review A, 74*, 042708.

Storchak, V., Brewer, J. H., Morris, G. D. (1996). *Physical Review Letters, 76*, 2969.

Storchak, V., Brewer, J. H., Morris, G. D. (1997). *Physical Review B, 56*, 55.

Tanabashi, M., et al. (2018). *Physical Review D, 98*, 030001. Particle Data Group.

Vilão, R. C., Vieira, R. B. L., Alberto, H. V., et al. (2017). *Physical Review B, 96*, 195205.

Vilão, R. C., Alberto, H. V., Vieira, R. B. L., et al. (2020). *Physical Review B, 101*, 077202.

Workman, R. L., et al. (2022). *Progress of Theoretical and Experimental Physics, 2022*, 083C01. Particle Data Group.

Ziegler, J. F. (1999). *Journal of Applied Physics, 85*, 1249.

Ziegler, J. F., Littmark, U., & Biersack, J. P. (1985). *The Stopping and Ranges of Ions in Matter* (Vol. 321). Pergamon. ISBN: 978-0080216034.

The μSR Technique

3

The term μSR is an acronym for Muon Spin Rotation/Relaxation/Resonance and emphasizes the analogy to NMR (Nuclear Magnetic Resonance) and ESR (Electron Spin Resonance) or EPR (Electron Paramagnetic Resonance).[1] There are, however, important differences. The μSR technique does not require specific target nuclei like NMR and is universally applicable since muons can be implanted in any kind of material. Also, while NMR and ESR rely on creating a spin polarization in thermal equilibrium (usually requiring both high magnetic fields and low temperatures), μSR starts with a fully polarized muon ensemble. It is therefore possible to perform measurements without applying a magnetic field (so-called zero field μSR, ZF-μSR) allowing to investigate magnetic systems without perturbation.[2]

The method is based upon the observation of the time evolution of the muon spin polarization $\mathbf{P}(t)$ of an ensemble of muons implanted in a sample. In a typical experiment, an ensemble of a few million of muons is recorded. The average of the spin vectors of the muon ensemble $\langle \mathbf{I}_\mu(t) \rangle$ is the relevant quantity. The polarization vector of the ensemble can be written as

$$\mathbf{P}(t) = \frac{\langle \mathbf{I}_\mu(t) \rangle}{I_\mu} \quad , \tag{3.1}$$

which can also be expressed with the Pauli matrices Eqs. A.17 as

$$\mathbf{P}(t) = \langle \boldsymbol{\sigma}(t) \rangle \quad . \tag{3.2}$$

---

[1] This acronym was coined in one of the first reviews about the use of muons in physics and chemistry (Brewer et al. 1975).

[2] Note that NQR (Nuclear Quadrupole Resonance) is also a zero field technique, but for magnetic studies less direct than zero field μSR.

© Springer Nature Switzerland AG 2024
A. Amato, E. Morenzoni, *Introduction to Muon Spin Spectroscopy*,
Lecture Notes in Physics 961, https://doi.org/10.1007/978-3-031-44959-8_3

The time dependence of the muon spin polarization contains the physical information about the interaction of the muon magnetic moment (or spin) with its local
environment. Since the positron emission is anisotropic (asymmetric) with a higher
emission probability along the spin direction of the muon at the moment of the
decay, Eq. 1.16, $\mathbf{P}(t)$ can be directly derived from the positron intensity measured
by a set of detectors as a function of time after muon implantation. The polarized
muon acts as a local very sensitive magnetic probe and is able to determine the
value, direction, distribution and/or dynamics of the internal magnetic fields present
at its localization site inside the material. Such fields can be produced, for instance,
by electronic or nuclear moments or by local currents such as those flowing in
superconductors.

3.1   Key Features of the μSR Technique

We summarize here the most important characteristics of the μSR technique. They
will be discussed in more detail in the following chapters, which show applications
in various fields.

- **Local probe** μSR provides local (mainly magnetic) information about the
  environment of its stopping site, which in a solid is mostly interstitial. In
  this sense it is complementary to other local probe techniques such as NMR
  or Mössbauer spectroscopy, where the probes are specific ion isotopes of the
  crystallographic lattice. As a spin $1/2$ particle, the muon is not directly affected by
  quadrupolar interactions, which sometimes make the interpretation of NMR and
  Mössbauer spectra difficult. The local probe character makes μSR sensitive to
  short-range or disordered magnetism and is a useful complementary technique to
  scattering techniques such as neutron diffraction, where the wave character of the
  neutron is used to determine crystallographic and magnetic structures. Neutron
  diffraction is often used to determine magnetic structures and various properties
  of a magnet as the value of the ordered magnetic moments (order parameter).
  However, the magnetic intensity of the Bragg diffraction peaks measured by
  neutron scattering is proportional to $V_{\mathrm{ord}}\,\mu_{\mathrm{ord}}^2$, so the ordered volume V_{ord} and
  the strength of the ordered phase μ_{ord} cannot be disentangled. μSR is not
  a scattering technique, and therefore magnetic structures cannot be directly
  obtained. It does, however, provide an independent estimate of the size of the
  magnetic moment[3] and a determination of the magnetic fraction of the sample,
  see Chap. 5. Similarly, superconducting and normal fractions can be determined,
  Chap. 6. μSR can test whether the superconducting state is bulk or, for example,
  filamentary, a distinction difficult to make with thermodynamic or transport
  measurements.

---

[3] For an accurate determination, the muon site and magnetic structure must be known.

- **Full initial polarization** The surface (Sect. 1.8.3), and low-energy (Chap. 8) muon beams provide 100% initial polarization[4] in zero field, independent of temperature. This allows measurements without perturbing the system. NMR requires high magnetic fields and low temperatures to achieve small polarization values of the order of 10^{-3}–10^{-4} of well-defined isotopes (see Exercise 3.1), which must be present or possibly enriched in the compound to be investigated. Muons can be implanted in any material in solid, liquid and gaseous form, providing a high **versatility** in the choice of material to be studied.
- **High sensitivity** The large magnetic moment, the full polarization, and the effective single spin detection make the polarized muons a very sensitive probe. μSR is able to detect electronic static moments of the order of 0.001 μ_B and is therefore even sensitive to nuclear magnetic moments, see Exercise 3.2 and Chap. 5.
- **Light hydrogen isotope** In insulators and semiconductors, the bound state $\mu^+ e^-$ (muonium, Mu) is generally formed with high probability. It can be used as a light hydrogen isotope for spectroscopy, impurity studies, radical chemistry, reaction kinetics, and diffusion studies, see Chap. 7.
- **Dynamical sensitivity range** As we will see in Sect. 4.3, magnetic fluctuations lead to a relaxation of the muon spin polarization. With an observation time window of typically 10–20 μs, μSR is able to detect magnetic fluctuations and relaxation phenomena with fluctuation times over a large dynamical range of 10^{-5}–10^{-11} s.

3.2 The μSR Signal

As mentioned above, in a μSR experiment, by detecting the anisotropically emitted positrons, one measures the time evolution of the spin polarization of an ensemble of muons implanted in the material under study, this way gaining access to various physical properties.

The muon ensemble is not implanted all at once into the sample, but either one by one (at continuous beams, Sect. 3.3.1) or in bunches (at pulsed beams, Sect. 3.3.3). Measuring the decay of these muons is equivalent to measuring the muon ensemble implanted at one time.[5] The overwhelming majority of experiments are performed in so-called time-differential mode,[6] see Fig. 3.1, where the muon implantation time t_0 must be known.[7]

---

[4] In decay beams, see Sect. 1.8.3 the polarization is also very high $\sim 80\%$.

[5] The density of muons in the sample, even with the most intense pulse, is negligibly small, so that muon-muon interactions can be completely ignored.

[6] Sometimes called Δt-μSR.

[7] Measurements are also performed in the so-called integral mode, where simply the number of the decay positrons in different detectors is recorded. This mode is not rate limited. It is often used in the integral avoided level crossing (ALC) or level crossing resonance (LCR) techniques (Roduner 1988), where it has been applied to determine muon and nuclei hyperfine coupling

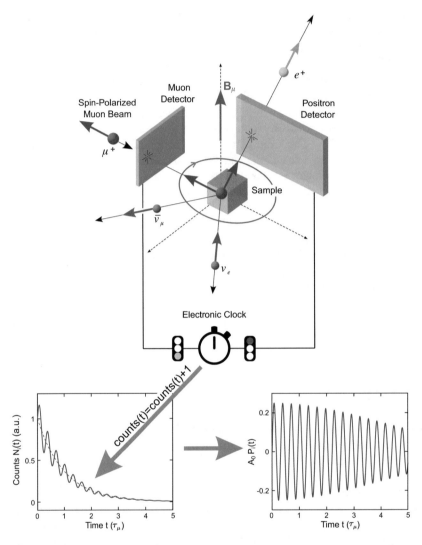

Fig. 3.1 Sketch of a time-differential μSR measurement at a continuous beam (colored arrows indicate the spin of the particles). The polarized muon comes from the upper-left part of the figure and is first detected by a (thin) muon detector. This signal starts at t_0 a very precise electronic clock (remember that the thermalization time in solids is extremely short, see Sect. 2.2.2). In this example B_μ represents the magnetic field at the muon stopping site (which is generally a vector sum of internal and applied fields). At t_0 the muon spin begins to precess with the corresponding Larmor frequency. When the muon decays, a positron is emitted preferentially along the direction of the muon spin at the time of decay. The decay is recorded by a (thick) positron detector, which also defines the direction of observation (note that in reality there are several positron detectors around the sample). The signal from the positron detector stops the clock, which measures the time t between the muon implantation and its decay. This event is added at time t to a histogram versus time. From the histogram the asymmetry spectrum is obtained, see Eq. 3.3. In this example, a clear precession signal is observed reflecting the fact that all muons in the sample sense a well-defined local field B_μ with small uncertainty. Adapted from a drawing by J. Sonier (UBC)/A. Suter (PSI)

While with a pulsed beam the implantation time of the muon bunch is obtained from the accelerator timing, with a continuous beam one determines the exact implantation time of each incoming muon with a muon detector placed in front of the sample. Figure 3.1 shows the principle of a time-differential measurement.

In the following, the discussion is limited to the case of a continuous beam, but it basically applies as well to a pulsed beam. The number of positrons detected as a function of time by a positron detector with efficiency ϵ and subtending a solid angle $\Delta\Omega$ under the direction $\hat{\mathbf{n}}$ is[8]

$$N(t) = \frac{N_{\mu,0}\,\Delta\Omega\,\epsilon}{4\pi\,\tau_\mu}\,e^{-\frac{t}{\tau_\mu}}\left[1 + A_0\,\mathbf{P}(t)\cdot\hat{\mathbf{n}}\right] + B_g$$

$$= N_0\,e^{-\frac{t}{\tau_\mu}}\left[1 + A_0\,\mathbf{P}(t)\cdot\hat{\mathbf{n}}\right] + B_g\ . \tag{3.3}$$

Compared to Eq. 1.18, we have introduced some new terms and/or notations:

The Initial Value N_0 The constant $N_0 = N_{\mu,0}\,\Delta\Omega\,\epsilon/(4\pi\,\tau_\mu)$, where $N_{\mu,0}$ is the number of implanted muons forming the ensemble.

The Background Value B_g This parameter accounts for the practically time independent background due to uncorrelated muon-positron events (see Sect. 3.3.1). In a continuous beam, it is usually less than $0.01 \times N_0$ and even smaller in a pulsed beam, Table 3.1.

The Asymmetry Term A_0 This term is called the effective or experimental asymmetry and takes into account all the factors in a real spectrometer, that modify the theoretically achievable asymmetry $\bar{A} = 1/3$, Eq. 1.15. In general, $A_0 < \bar{A}$. There are two main reasons for this:

- An energy dependent detector efficiency modifies the term $a(\varepsilon)$ in Eq. 1.15. For instance, if high energy positrons are not detected, this reduces the value of A_0 since these positrons contribute to high asymmetry values, see Fig. 1.15. Conversely, low-energy positrons may go undetected because they can be absorbed in the target, surrounding material, or vacuum chamber walls. This effect increases the effective asymmetry.

---

constants in molecules such as muoniated radicals or in muonium defect centers, thus allowing the characterization of the system and the assessment of its structure (Heming et al. 1986; Kiefl 1986), see also Chap. 7.

[8] The energy-averaged efficiency is $\epsilon = \int_0^1 E(\varepsilon)f(\varepsilon)\,d\varepsilon$, where $f(\varepsilon)$ is the efficiency of the detector as a function of the scaled energy ε.

- The solid angle of the detectors $\Delta\Omega$ is finite. Averaging over the solid angle reduces the measured asymmetry, but does not change the $\cos\theta$ form of the angular dependence.[9]

In μSR setups the experimental asymmetry is typically 0.2–0.3, see Fig. 3.1 and other examples in the book.

The Time Dependent Muon Spin Polarization P(t) This is the fundamental quantity measured in a μSR experiment. In Eq. 1.18 the polarization was assumed to be time independent and equal to one. The interaction of the muon spin with its environment makes **P** and the angle θ time dependent, since $\cos\theta(t) \propto \mathbf{P}(t) \cdot \hat{\mathbf{n}}$, where $\hat{\mathbf{n}}$ is defined by the position of the positron detector.[10] Equation 3.3 contains only parameters determined by the geometry of the setup. The μSR signal or spectrum[11] for a given detector i is defined as the measured quantity

$$A(t) = A_0\, \mathbf{P}(t) \cdot \hat{\mathbf{n}}_i$$
$$= A_0\, P_i(t) \ , \tag{3.5}$$

with $P_i(t)$ the projection of the muon spin polarization along the direction of the detector i

$$\mathbf{P}(t) \cdot \hat{\mathbf{n}}_i = P_i(t) = \frac{\langle \mathbf{I}_\mu(t) \cdot \hat{\mathbf{n}}_i \rangle}{I_\mu} \ , \tag{3.6}$$

where $\mathbf{I}_\mu$ is the muon spin and the brackets $\langle\ldots\rangle$ represent the average over the muon ensemble. Often the direction of the initial polarization is along a detector direction $\hat{\mathbf{n}}_i$ so that Eq. 3.6 can be written as

$$\mathbf{P}(t) \cdot \hat{\mathbf{n}}_i = P_i(t) = \frac{\langle \mathbf{I}_\mu(t) \cdot \mathbf{I}_\mu(0) \rangle}{I_\mu^2} \ , \tag{3.7}$$

which shows that $P_i(t)$ can also be seen as the muon spin autocorrelation function.

---

[9] The effect of a finite solid angle on the asymmetry can be written as

$$A_{0,\Delta\Omega}\,\cos\theta = \frac{\bar{A} \displaystyle\int_{\theta-\Delta\theta/2}^{\theta+\Delta\theta/2} \cos\theta'\, d\theta'}{\Delta\theta} = \bar{A}\cos\theta\,\frac{\sin\frac{\Delta\theta}{2}}{\frac{\Delta\theta}{2}} \ , \tag{3.4}$$

where θ is the observation angle between **P** and $\hat{\mathbf{n}}$ and $\Delta\theta$ is the polar angle subtended by the detector.

[10] Note that a possible reduction from the ideal value $|\mathbf{P}(0)| = 1$ is generally integrated in the asymmetry A_0.

[11] In the literature, depending on the context, the μSR signal (sometimes normalized to A_0) is also called asymmetry, polarization, depolarization, or relaxation signal.

The time evolution of the polarization is a combination of precession (spin rotation at one or more Larmor frequencies) and depolarization or relaxation (reduction of the polarization with time). It contains all the physics of the interaction between the muon and its spin or magnetic moment with the material investigated. The μSR signal is obtained by first building a histogram of the detected positron as a function of time by observing typically some 10^6 muon decays, then removing the background signal, and correcting for the radioactive exponential decay and the normalization.

The task of the scientist using the μSR technique is to extract the relevant information from the observed $P_i(t)$. As we will see in Chap. 4, this often begins with modeling $\mathbf{P}(t)$ according to specific physical phenomena. Figure 3.1 shows schematically the data processing at a continuous muon beam source with single muon and decay positron detection, and Fig. 3.2 gives a pictorial representation of the signal formation.

3.3 Experimental Setup

The experimental setup for a μSR measurement consists of the beamline transporting the muons from the production target to the experiment (see Sect. 1.8.4 for details) and the μSR instrument, where the sample is placed. The beamline may incorporate a Wien filter used as a mass filter or spin rotator. The μSR instrument itself can be divided into components such as:

- The detector system (or spectrometer), see Fig. 3.3, and associated electronics. This is the heart of the setup. Plastic scintillator detectors are used for the muon detector (in a continuous beam, Sect. 3.3.1) and to detect the decay positrons. Both particles traversing the scintillator produce light which is converted into an electronic signal by photomultipliers or, nowadays more frequently, by solid-state photodetectors such as avalanche photodiodes (APD) or silicon photomultipliers (SiPM), a single-photon avalanche diode implemented in a Si substrate (Dinu 2016; Stoykov et al. 2009). Fast electronics discriminate and process the signals. The detector system can also accommodate so-called "veto" detectors, which have the task of identifying and rejecting bad events, thus reducing the random muon-positron background events, B_g term in Eq. 3.3.
- For many experiments, it is necessary to apply external fields, mostly a magnetic field. This is the case, for example, to investigate the magnetic response of materials, to determine phase diagrams, to establish a flux-line lattice in a superconductor, to perform Knight-shift experiments for local susceptibility measurements, or to study impurity states by muonium spectroscopy. Examples are given in Chaps. 5, 6, 7, and 8. A small field transverse to the initial polarization direction is also applied to calibrate the spectrometer, Sect. 3.4.2. The magnets used are either conventional Helmholtz copper coils producing fields up to $\sim 1\,\mathrm{T}$ or

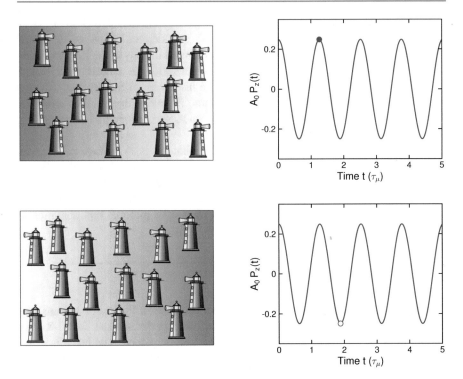

Fig. 3.2 A simple illustration of the formation of a μSR precession signal. Imagine that the muons are lighthouses with light sources (representing the muon spin), rotating in phase at a well-defined frequency (corresponding to the Larmor frequency). The positron from the muon decay corresponds to the flash of light from the lighthouse. A ship located to the right of the lighthouses represents a positron detector. When the light sources face the detector (top drawings) and some muons decay, the ship will see the light flash directed at it because some lighthouses send light in the direction of the detector; this corresponds to the μSR signal with a maximum amplitude (solid dot). If some muons decay later, after a time corresponding to 1/2 of the oscillation period, the light sources will point in the opposite direction with respect to the ship (bottom drawings) and the ship will see no or little light, i.e., the μSR signal will have a minimum amplitude (open dot). Adapted from a drawing by A. Schenck/H. Luetkens

superconducting for fields up to $\sim 10\,\text{T}$. Other external fields, such as electric fields (Storchak et al. 1996), or stimuli, such has photon irradiation (Prokscha et al. 2013) or laser excitation (Yokoyama et al. 2016) are also utilized.

- Sample environment. The sample is usually mounted on a cryostat to cool it to typically $1.5\,\text{K}$ (using helium flow cryostats) or lower temperatures, down to ~ 0.008 K, using dilution refrigerators (Pobell 1996). For some experiments, ovens with temperatures up to $1000\,\text{K}$ are used. Note that due to the rather low kinetic energy of the muon (at least for surface or low-energy muon beams), the cryostat requires a special design. For example, very thin windows (usually made of titanium with a thickness of $\sim 10\,\mu\text{m}$) must be used for vacuum insulation. For low-energy muons the cryogenic system has to be realized without windows, with the sample directly exposed to the beam. In this case, special care has to be taken to limit the thermal load on the sample, see Chap. 8.

For pressure studies the sample is mounted inside a pressure cell (typical wall thickness ∼ 6 mm), where the pressure is applied before the sample is mounted on the cryostat (Khasanov et al. 2016; Khasanov 2022). Experiments with sample under uniaxial pressure have also been performed, see Sect. 6.6 (Grinenko et al. 2021; Ghosh et al. 2020).

The following sections briefly describe typical experimental and spectrometer arrangements for continuous and pulsed beams.

3.3.1 Continuous Beam

As shown in Sect. 1.8.1 a proton cyclotron delivers a continuous muon beam.[12] The proton beam has a microstructure determined by the radiofrequency (RF) system of the accelerator. At PSI, for example, this frequency is 50 MHz, which means that the production target is bombarded every 20 ns by very short (≲ 1 ns) bunches of protons producing pions with the same pulse structure. The finite lifetime of the pion (26 ns, i.e., of the order of the repetition rate) smears out the time distribution of the muons produced by the pion decay at the production target and the muon beam appears as practically continuous.[13]

This means that the arrival time of the muon at the experiment is uncorrelated with the proton (or pion) pulse and follows a Poisson distribution. A thin muon detector placed just in front of the sample is necessary to define the implantation time t_0. After the muon signal ($M_{counter}$ in Eq. 3.8) the corresponding decay positron is expected in one of the positron detectors and the time difference is measured.

Figure 3.3 shows a typical detector setup. We remark a few points:

- The positron detectors cover a large fraction of the total solid angle around the sample. Since the instantaneous positron rate is rather low, a detector and its associated electronics can easily handle it.
- The granularity of positron detector system, compared to a spectrometer at a pulsed source, see Fig. 3.8, is low.

---

[12] Also named continuous-wave or cw-beam.

[13] High-energy decay beams retain a signature of the proton pulse because the transported muons originate mainly from a pion cloud volume with a more or less well-defined decay time.

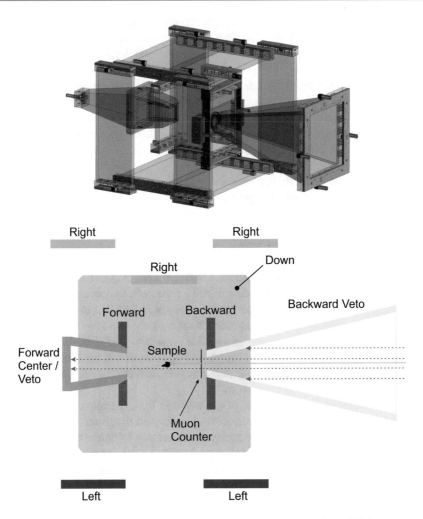

Fig. 3.3 Example of a typical μSR spectrometer at a continuous beamline, (GPS instrument at PSI (Amato et al. 2017), © The Authors. All rights reserved.). Top: 3D-view of the muon and positron detectors showing the SiPM-readout modules in the green frame. Bottom: View at the horizontal midplane (sample center) level. The muons enter the spectrometer from the right-hand side, i.e., parallel to the axis of the Backward detectors. The lateral dimension of the Down detector is 10 cm. An Up detector (not shown) is located opposite to it

- The presence of several veto detectors to reduce background events (Backward Veto and Forward Center "pyramids"). A good muon event (M_{good}) is defined by the logic

$$M_{\mathrm{good}} = M_{\mathrm{counter}} \cap \overline{(B_{\mathrm{veto}} \cup F_{\mathrm{center}})} \; . \qquad (3.8)$$

Therefore, only the muons stopping in the sample produce a start signal for the clock (green trajectory in Fig 3.3). The red dotted events do not satisfy the logic condition. This setup allows at a continuous beam the measurement of extremely small samples, corresponding to about 10 mg material. A veto system is not possible with a pulsed beam, Sect. 3.3.3, since part of the beam would always trigger it.

In a μSR experiment with a continuous beam we have to take care that only one muon at a time is present in the sample within the observation window (the data gate is generally taken 10 μs wide, corresponding to about five muon lifetimes). Otherwise the time correlation between the muon and its decay positron may be lost. This is done (i) electronically, i.e., if a 2nd muon (or a 2nd positron) arrives within the data gate all the events concerned are discarded and (ii) by limiting the incoming muon rate.

The detection of individual muon-positron (M-P) events leads to a very high time resolution of the spectrometer, which is only limited by the performance of the detectors and electronics and is not accessible with pulsed beams. On the other hand, it limits the usable rate of incoming muons. An overall time resolution as low as 60 ps (standard deviation) has been achieved for high magnetic field instruments.[14] For more conventional spectrometers, a lower time resolution is sufficient to resolve the precession frequencies observed; μSR instruments with plastic scintillators and conventional photomultiplier readout have a resolution of ~ 0.4 ns. The GPS spectrometer of Fig. 3.3, which uses silicon photomultipliers (SiPM), has an overall instrument resolution of 160 ps (standard deviation) (Amato et al. 2017).

The optimal rate of accepted muons can be calculated by taking into account that the muon arrival times are Poisson distributed. Then

$$f_{\text{Poisson}}(k, \bar{N}) = \frac{\bar{N}^k \, e^{-\bar{N}}}{k!} \quad , \tag{3.9}$$

is the probability that, given an average number of muons $\bar{N}$ in a defined time interval, a number k of muons will be found in that time interval. In our case, given that we have detected a muon, we want to make sure that no other muon event has been recorded within a data gate Δt before our event and that also no other muon event will be recorded within a data gate Δt after our event (see Fig. 3.4). So we have $k = 0$ and $\bar{N} = 2\Delta t R_\mu$, where R_μ is the rate of incoming good muons. Therefore, the accepted muon rate as a function of the total good muon rate (plotted in Fig. 3.5) is

$$R_{\mu,\text{acc}} = R_\mu \, f_{\text{Poisson}}(0, \bar{N}) = R_\mu \, e^{-2 \, \Delta t \, R_\mu} \quad . \tag{3.10}$$

---

[14] Note that the Larmor precession frequency of the muon spin in 10 T is 1.35 GHz.

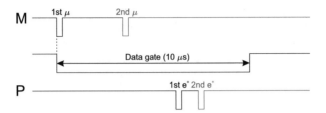

Fig. 3.4 Timing diagram of a μSR measurement at a continuous beam; M muon signal, P positron signal. A first muon, which has satisfied the logic condition defined by Eq. 3.8, starts the clock and the data gate. In the data gate only one positron has to be detected. If a 2nd muon or a 2nd positron is detected, then the complete set of events from the different detectors is rejected. Note that the 1st muon is actually a "good muon" only if there was no other muon signal in the past for at least the length of the data gate

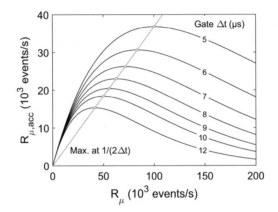

Fig. 3.5 Accepted rate as a function of the incoming rate for data gates Δt ranging from 5 to 12 μs, see also Fig. 3.4. The optimal rate lies along the line $1/(2\Delta t)$ (straight line). For a data gate of 10 μs, the theoretical optimal rate is therefore $\sim$ 50 kHz. In reality, the rate is kept slightly lower ($\sim$ 40 kHz) to avoid a too high background parameter B_g, Eq. 3.3, from uncorrelated M-P events. Such events can result from the less than 100% efficiency of the detectors. Courtesy of Toni Shiroka. All rights reserved

3.3.2   Muon-On-Request Setup

To reduce the background of uncorrelated M-P events it is possible to use an electrostatic kicker device inserted in the beamline, which allows only one muon at a time in the instrument (Beveridge 1992). Such a device, called Muon-On-Request (MORE), is implemented at PSI in a surface muon beam, Fig. 1.25 (Abela et al. 1999). An electrostatic kicker produces a gate pulse up to 5 μs wide with a fixed repetition rate (up to 40 kHz). Only when the pulse is ON, can a muon be transported to the instrument. If a muon is within this gate and is detected by the muon counter, the gate is quickly closed (with a delay of $\sim$ 400 ns) to prevent a second muon from being transported to the instrument, see Fig. 3.6.

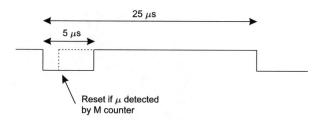

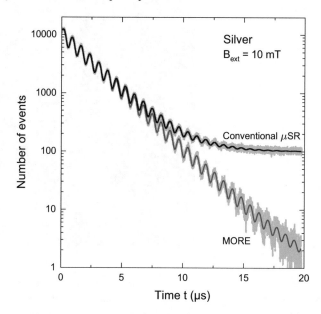

Fig. 3.6 Logic of the Muon-On-Request operation. See text for details

Fig. 3.7 μSR spectrum in silver in a magnetic field of 10 mT measured in conventional and in Muon-On-Request (MORE) mode at the GPS spectrometer at PSI. Note the remarkable random background suppression with MORE. Modified from Abela et al. (1999), © Springer Nature. Reproduced with permission. All rights reserved

This procedure reduces the random background, while keeping the excellent time resolution of the continuous beam. This allows the observation time window to be increased by a factor of about two, with a usable rate of $\sim 2 \times 10^4$ events/s, see Fig. 3.7.

3.3.3 Pulsed Beam

A synchrotron, Sect. 1.8.1, delivers a pulsed beam. It produces pulses of protons with a typical width of about 100 ns and a repetition rate of 50 Hz (sometimes 25 Hz). The muon beam originating from the production target conserves this structure

Fig. 3.8 μSR spectrometer at a pulsed beam (MUSR instrument, Rutherford Appleton Laboratory, UK). Photomultipliers are connected to 64 positron detectors. The diameter of the frame in which the photomultipliers are mounted is more than one meter. Newer instruments can have up to $\sim$ 700 detectors (Hillier et al. 2018). © Rutherford Appleton Laboratory. Reprinted with permission. All rights reserved

with the addition of an asymmetric pulse tail due to the pion lifetime (Hillier et al. 2019; Miyake et al. 2012). This means that the muons (typically 10^3–10^6) arrive at the sample grouped in $\sim$ 100 ns wide bunches making an individual detection of the muons not possible.

The pulse structure has the advantage that the implantation time is defined by the pulse and no muon detector is needed, at the cost, however, of a modest time resolution of about 40 ns (standard deviation) determined by the pulse width. All the decay positrons from a pulse are measured at once, which allows high rates. On the other hand, the large number of positrons created in a single pulse requires a high granularity (segmentation) of the detectors to minimize the loss of usable positron signals due to double hits in the same detector and also to make optimum use of the high instantaneous intensity, see Fig. 3.8. A pulsed beam has a lower background than a continuous beam since no "contamination" from the beam remains after the pulse. It is better suited for a sample environment with external stimuli (such as lasers, electric fields, radiofrequency[15]) which can be more efficiently synchronized with the muon pulse.

---

[15] The basics of RF resonance techniques for pulsed and continuous muon beams are described in Scheuermann et al. (1997), Cottrell et al. (2000), and Kreitzman (1991), respectively.

Table 3.1 Main differences between μSR instruments for continuous and pulsed beams

	Continuous beam	Pulsed beam
Muon event rate	To avoid pile-up events, limited to ~ 20 million events/h (20 Mev/h) with standard time window $10\,\mu$s	Limited by the detector granularity, up to ~ 150 Mev/h
Time resolution	Limited only by detector and electronics performance, presently down to 60 ps	Limited by the muon pulse width to about 40 ns. Can be reduced to $\sim$ 20 ns by slicing the pulse[a]
Beam size	Can be reduced to a few mm^2 by the use of active veto counters without loss of rate	A few cm^2. No use of veto detectors possible. "Flypast"[b] mode possible, but the at cost of a drastic rate reduction
Sample mass	Down to ~ 5–10 mg	About 0.2 mg or more
Background B_g/N_0	$\sim 1\%$ with standard $10\,\mu$s time window. With MORE and $20\,\mu$s time window $\sim 10^{-4}$	$\sim 10^{-5}$, essentially background free, standard time window $20\,\mu$s
Detector setup	Muon counter required. Only a few large positron detectors needed	No need of muon counter. Large number of positron detectors to handle the high instantaneous rate necessary

[a] In specific cases, the time resolution can be reduced by using pulsed magnetic fields (Shiroka et al. 1999)
[b] In the flypast mode, muons that do not stop in the sample are able to fly past it and out of the instrument without generating events in the detectors

The main characteristics of a μSR instrument at a continuous and a pulsed beam are summarized in Table 3.1.

3.4 Measurement Geometries

The magnetic field is an important physical parameter. As described in the following chapter, depending on its orientation with respect to the initial muon spin polarization vector $\mathbf{P}(0)$, information about static or dynamic properties of the investigated system can be obtained.

One speaks of longitudinal field (LF), zero field (ZF), transverse field (TF) geometry or setup.

3.4.1 Zero Field and Longitudinal Field Geometry

The zero field technique is the basic technique for studying magnetic systems.[16] Without the application of an external field the magnetic system remains undis-

---

[16] The high sensitivity of μSR requires an active compensation of the Earth's and surrounding magnetic fields to perform ZF experiments under ideal conditions. The compensation is usually better than $\sim 2\,\mu$T.

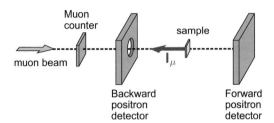

Fig. 3.9 Schematic of the ZF-μSR geometry with one muon counter and two positron detectors. In reality, the muon and positron detectors are closer to each other, the positron detector pair is located about 5 cm away from the sample, see Fig. 3.3. The designation of the detectors refers to the beam direction

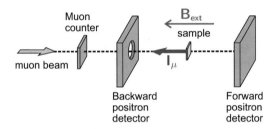

Fig. 3.10 LF-μSR geometry. The magnetic field is applied along the initial muon spin direction

turbed and $P(t)$ directly reflects the internal fields and their distribution, see Chaps. 4 and 5. ZF experiments with surface muons are usually performed with the muon spin pointing antiparallel to the muon beam direction, corresponding to the "natural" spin direction.[17]

The ZF setup, Fig. 3.9, can be considered as a special case of the LF setup, Fig. 3.10, where a magnetic field is applied along the initial muon spin direction. Longitudinal field scans allow to differentiate between static and fluctuating internal magnetic fields, both of which affect the time evolution of the muon spin polarization, see Sect. 4.2.1.

By convention and following much of the literature, we choose the z-axis as the direction of the applied external field. In the ZF and LF geometry, this is also the

---

[17] Experiments can also be performed by rotating the muon spin by 90° with a spin rotator, see Sect. 1.8.4. In this case, the Up and Down detectors, Fig. 3.13, provide the relevant information.

Fig. 3.11 Definition of the axes in the laboratory frame for ZF and LF experiments. The longitudinal magnetic field is along the z-axis

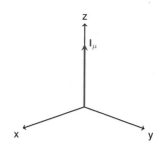

direction of the initial polarization and of the spin quantization axis, Fig. 3.11. With the Backward and Forward detectors, we therefore measure $P_z(t)$.[18]

Figure 3.12 depicts two basic examples of a μSR signal recorded in the Backward detector in a ZF experiment. When the sample is in the paramagnetic state, no effective field of electronic origin is sensed by the muons, and therefore ideally the asymmetry does not show any evolution with time.[19] On the other hand, right panel of Fig. 3.12, if the measurement is performed on a magnetic sample and the local magnetic field is perpendicular to $\mathbf{P}(0)$, the muon spins perform a Larmor precession around the local field, and the asymmetry is an oscillating signal with amplitude A_0, see Fig. 3.1.

The opposite Forward and Backward detectors contain the same information. The μSR signal can be obtained either by considering the histograms of the individual detectors or directly by combining the raw signals of two opposite detectors. Defining $\hat{\mathbf{n}}_F \parallel \hat{\mathbf{z}}$ as the direction of the Forward detector and $\hat{\mathbf{n}}_B = -\hat{\mathbf{n}}_F$ that of the Backward detector in Fig. 3.9, we have (assuming background corrected spectra)

$$N_B(t) = N_0\, e^{-\frac{t}{\tau_\mu}} \left[1 + A_0\, \mathbf{P}(t) \cdot \hat{\mathbf{n}}_B \right] \ \text{and}$$

$$N_F(t) = N_0\, e^{-\frac{t}{\tau_\mu}} \left[1 + A_0\, \mathbf{P}(t) \cdot \hat{\mathbf{n}}_F \right]$$

$$= N_0\, e^{-\frac{t}{\tau_\mu}} \left[1 - A_0\, \mathbf{P}(t) \cdot \hat{\mathbf{n}}_B \right] . \tag{3.11}$$

---

[18] As remarked earlier, we define backward and forward directions with respect to the incoming muon momentum. Sometimes backward and forward are defined with respect to the initial spin polarization. With a spin-unrotated surface muon beam, as in Fig. 3.9, this means swapping the labels of the positron detectors.

[19] We will see later that the magnetic field of the nuclear moments produces a slight change in the polarization.

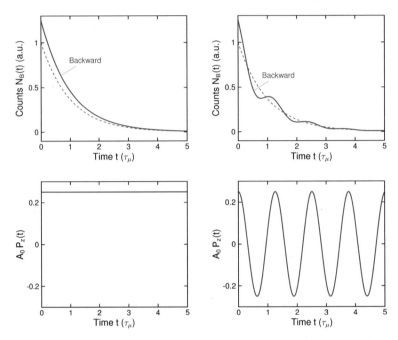

Fig. 3.12 Top panels: Ideal ZF-μSR signals in a Backward detector for paramagnetic (left) and magnetic (right) samples. In the paramagnetic sample, there is no internal field and the muon polarization remains unchanged, resulting in a constant μSR signal (bottom left). In a magnetic sample, the muon spins precess around the local field at the Larmor frequency, resulting in an oscillating asymmetry (bottom right). The μSR signal in the Forward detector corresponds to $\mathbf{P}(t) \cdot \hat{\mathbf{n}}_{\mathrm{F}} = -\mathbf{P}(t) \cdot \hat{\mathbf{n}}_{\mathrm{B}}$

From the two raw spectra, we can remove the effect of the muon decay[20] and directly extract the μSR asymmetry signal

$$A(t) \equiv A_0\, \mathbf{P}(t) \cdot \hat{\mathbf{n}}_{\mathrm{B}} = A_0\, P(t) = \frac{N_{\mathrm{B}}(t) - N_{\mathrm{F}}(t)}{N_{\mathrm{B}}(t) + N_{\mathrm{F}}(t)} \quad . \tag{3.12}$$

Equations 3.11 and 3.12 assume perfectly identical detectors, exactly placed 180° to each other and with equal solid angle and efficiency. In a real spectrometer this is not the case and in principle two additional parameters, α and β, have to be introduced

---

[20] Note that the effect of the exponential decay due to the finite muon lifetime enters in the statistical error of the histograms and of the asymmetry. Due to the Poisson statistics, the relative error of a histogram bin has an essentially exponential growth with time t, $\sim \exp(t/(2\tau_\mu))$.

to account for this. For the F and B pair of counters of Fig. 3.10, for instance

$$N_B(t) = N_{0,B}\, e^{-\frac{t}{\tau_\mu}}\left[1 + A_{0,B}\, P(t)\right] + B_{g,B}$$

$$N_F(t) = N_{0,F}\, e^{-\frac{t}{\tau_\mu}}\left[1 - A_{0,F}\, P(t)\right] + B_{g,F}\ . \tag{3.13}$$

With the definition

$$\alpha = \frac{N_{0,F}}{N_{0,B}}$$

$$\beta = \frac{A_{0,F}}{A_{0,B}}\ , \tag{3.14}$$

we can relate the raw asymmetry $A_{\text{raw}}(t)$

$$A_{\text{raw}}(t) = \frac{\left[N_B(t) - B_{g,B}\right] - \left[N_F(t) - B_{g,F}\right]}{\left[N_B(t) - B_{g,B}\right] + \left[N_F(t) - B_{g,F}\right]} \tag{3.15}$$

to the corrected asymmetry $A_{0,B}\, P(t)$ and viceversa, see Exercise 3.3,

$$A_{\text{raw}}(t) = \frac{A_{0,B}\, P(t)\big(\alpha\beta + 1\big) - (\alpha - 1)}{(\alpha + 1) - A_{0,B}\, P(t)\big(\alpha\beta - 1\big)}$$

$$A_{0,B}\, P(t) = \frac{(\alpha + 1)A_{\text{raw}}(t) + (\alpha - 1)}{(\alpha\beta + 1) + (\alpha\beta - 1)A_{\text{raw}}(t)}\ . \tag{3.16}$$

The parameter β, which takes into account the effect of possible detector differences in the asymmetry, can generally be assumed to be $\beta = 1$, see Exercise 3.4 for a justification of this assumption. With $\beta = 1$, $A_{0,B}\, P(t) = A_{0,F}\, P(t) = A_0\, P(t)$ and the corrected asymmetry simplifies to

$$A_0\, P(t) = \frac{\alpha\left[N_B(t) - B_{g,B}\right] - \left[N_F(t) - B_{g,F}\right]}{\alpha\left[N_B(t) - B_{g,B}\right] + \left[N_F(t) - B_{g,F}\right]}\ . \tag{3.17}$$

The α parameter (for each detector pair) is usually determined at the beginning of the experiment by applying a weak transverse field to the sample in a nonmagnetic or normal state and fitting the measured asymmetry to Eq. 3.17. The corrected asymmetry is centered around the zero line. In contrast, an uncorrected $\alpha \neq 1$ produces a raw asymmetry that is slightly distorted and has a nonzero asymptotic value. See Exercise 3.5 for an estimate of the effect of α variations on the asymmetry.

3.4.2   Transverse Field Geometry

The so-called transverse field (TF) geometry refers to experiments in which an external magnetic field is applied perpendicular to **P**(0). This configuration is used, for example, in the study of superconductors, magnetic materials, semiconductors, and insulators. Examples are presented in the corresponding chapters.

Also for the TF geometry, Fig. 3.13, by convention and following most of the literature, we choose the direction of the applied magnetic field as the z-axis Fig. 3.14. For the geometry depicted in the U/D detector pair replaces the B/F pair in the equations of Sect. 3.4.1 and measures $P_x(t)$.

Figure 3.15 shows an ideal μSR histogram and the corresponding asymmetry.

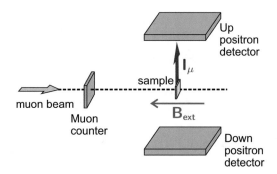

Fig. 3.13 TF-μSR geometry. The externally applied magnetic field is along the muon beam trajectory to avoid beam bending by the Lorentz force. The initial muon spin direction is ideally 90° with respect to the beam direction (rotation by a spin rotator up-stream). If only a small field is applied, one can keep the spin unrotated and apply the field perpendicular to the muon momentum (As discussed in Sect. 1.8.4, no spin rotator is used with a high-energy (decay) muon beam.)

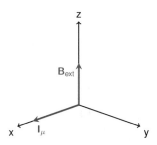

Fig. 3.14 Definition of the axes in the laboratory frame for a TF experiment with rotated spin

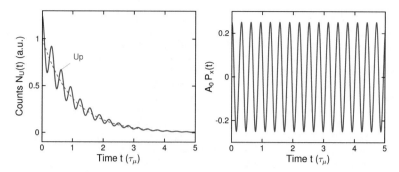

Fig. 3.15 Ideal histogram and asymmetry signals measured by the Up detector in the TF geometry of Fig. 3.13. This is the case, for example, in a paramagnetic sample, where the internal field averages to zero and the local field is essentially given by the applied field. Up and Down signals are related by $\mathbf{P}(t) \cdot \hat{\mathbf{n}}_D = -\mathbf{P}(t) \cdot \hat{\mathbf{n}}_U$

Exercises

3.1 Initial polarization in NMR
Calculate the nuclear polarization as a function of magnetic field and temperature. Estimate the value for ^{63}Cu nuclei (spin $3/2$, $\gamma_{Cu} = 2\pi \times 11.3$ MHz/T).

3.2 Sensitivity to a local field
Using what you know about the μSR technique, give an estimate of the lower limit of the local field that can be measured. What does this mean in terms of detectable electronic moments?

3.3 Asymmetry expression in a real spectrometer
Derive the modification of Eq. 3.12 to take into account that in a real spectrometer efficiency and solid angle of the forward and backward positron detectors may be different.

3.4 Evaluation of the parameter β
Justify the assumption $\beta = 1$ for the determination of the corrected asymmetry expression Eq. 3.17.

3.5 Variation of the asymmetry with the parameter α
Estimate how changing the parameter α affects the asymmetry, Eqs. 3.16 and 3.17.

References

Abela, R., Amato, A., Baines C., et al. (1999). *Hyperfine Interactions, 120,* 575.
Amato, A., Luetkens, H., Sedlak, K., et al. (2017). *Review of Scientific Instruments, 88,* 093301.
Beveridge, J. L. (1992). *Zeitschrift für Physik C Particles and Fields, 56,* S258.

Brewer, J. H., Crowe, K. M., Gygax, F. N., et al. (1975). In V. W. Hughes, & C. S. Wu (Eds.), *Muon Physics: Chemistry and Solids* (Vol. III). Academic Press. ISBN: 978-0123606037.

Cottrell, S., Cox, S., Scott, C., et al. (2000). *Physica B: Condensed Matter, 693*, 289–290.

Dinu, N. (2016). In B. Nabet (Ed.), *Photodetectors: Materials, Devices and Applications* (Vol. 255). Woodhead Publishing. ISBN: 978-1782424451.

Ghosh, S., Brückner, F., Nikitin, A., et al. (2020). *Review of Scientific Instruments, 91*, 103902.

Grinenko, V., Ghosh, S., Sarkar, R., et al. (2021). *Nature Physics, 17*, 748–754.

Heming, M., Roduner, E., Patterson, B. D., et al. (1986). *Chemical Physics Letters, 128*, 100.

Hillier, A. D., Aramini, M., Baker, P. J., et al. (2018). *JPS Conference Proceedings, 21*, 011055.

Hillier, A. D., Lord, J. S., Ishida, K., et al. (2019). *Philosophical Transactions of the Royal Society A: Mathematical, Physical and Engineering Sciences, 377*, 20180064.

Khasanov, R., (2022). *Journal of Applied Physics, 132*, 190903.

Khasanov, R., Guguchia, Z., Maisuradze, A., et al. (2016). *High Pressure Research, 36*, 140.

Kiefl, R. F. (1986). *Hyperfine Interactions, 32*, 707.

Kreitzman, S. R. (1991). *Hyperfine Interactions, 65*, 1055.

Miyake, Y., Shimomura, K., Kawamura, N., et al. (2012). *Physics Procedia, 30*, 46.

Pobell, F. (1996). *Matter and Methods at Low Temperatures*. Springer. ISBN: 978-3540463566.

Prokscha, T., Chow, K. H., Stilp, E., et al. (2013). *Scientific Reports, 3*, 2569.

Roduner, E. (1988). *The Positive Muon as a Probe in Free Radical Chemistry*. Lecture notes in chemistry. Springer. ISBN: 978-3540500216.

Scheuermann, R., Schimmele, L., Schmidl, J., et al. (1997). *Applied Magnetic Resonance, 13*, 195.

Shiroka, T., Bucci, C., De Renzi, R., et al. (1999). *Physical Review Letters, 83*, 4405.

Storchak, V., Brewer, J. H., & Morris G. D. (1996). *Physical Review Letters, 76*, 2969.

Stoykov, A., Scheuermann, R., Sedlak, K., et al. (2009). *Nuclear Instruments and Methods in Physics Research - Section A, 610*, 374.

Yokoyama, K., Lord, J. S., Murahari, P., et al. (2016). *Review of Scientific Instruments, 87*, 125111.

Polarization Functions

4

In this chapter we discuss muon polarization functions for different types of static and dynamic local fields corresponding to specific physical situations, with and without the application of external fields. We first consider the case where the internal fields seen by the muon are static, i.e., constant over the typical observation window $\Delta t \gtrsim 5 - 10 \times \tau_\mu$. In this case, the decay of the polarization of the muon ensemble is caused by a distribution of local fields and hence by a spreading of the precession frequencies, which leads to a dephasing of the muon spins. We then consider the case of dynamic fields, where the time evolution of the polarization is caused by an exchange of energy between the muon spin and the system under study (relaxation of the muon spins). The term polarization function includes both cases. In the literature the terms depolarization and relaxation functions are also used, depending on the physical phenomena involved in the time evolution of the muon spin polarization.

4.1   Static Internal Fields

We first consider the simplest case in ZF geometry, where all muons of the ensemble experience the same local field $\mathbf{B}_\mu$ forming an angle θ with respect to the initial polarization $\mathbf{P}(0)$, which we take antiparallel to the beam momentum as in a surface beam.

Due to the torque of the field on the muon moment, Eq. 1.24, the polarization vector precesses in a cone around the internal field, see Fig. 1.16. The backward detector in Fig. 4.1 detects a time evolution of the polarization component along the z-axis given by

$$
\begin{aligned}
P_z(t) &= P(0)\big[\cos^2\theta + \sin^2\theta\ \cos(\omega_\mu t)\big] \\
&= P(0)\big[\cos^2\theta + \sin^2\theta\ \cos(\gamma_\mu B_\mu t)\big] \ ,
\end{aligned}
\tag{4.1}
$$

© Springer Nature Switzerland AG 2024
A. Amato, E. Morenzoni, *Introduction to Muon Spin Spectroscopy*,
Lecture Notes in Physics 961, https://doi.org/10.1007/978-3-031-44959-8_4

Fig. 4.1 Muon ensemble
experiencing a local field $\mathbf{B}_\mu$
forming an angle θ with
respect to the initial
polarization $\mathbf{P}(0)$

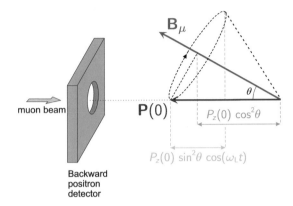

where $B_\mu \equiv |\mathbf{B}_\mu|$. We note that $P_z(t)$ does not depend on the azimuthal angle ϕ, see Fig. 4.2 top panel. In Cartesian coordinates Eq. 4.1 becomes

$$P_z(t) = P(0) \left[\frac{B_{\mu,z}^2}{B_\mu^2} + \frac{B_{\mu,x}^2 + B_{\mu,y}^2}{B_\mu^2} \cos(\gamma_\mu B_\mu t) \right] \ . \tag{4.2}$$

The μSR signal consists of a non-oscillating part proportional to the longitudinal field component (i.e., along $P_z(0)$) and an oscillating part proportional to the transverse field components, see Fig. 4.2 and Exercise 4.1.

For completeness, we give also the time evolution of $P_x(t)$ and $P_y(t)$

$$P_x(t) = P(0) \left[\frac{1}{2} \sin 2\theta \cos \phi [1 - \cos(\gamma_\mu B_\mu t)] - \sin \theta \sin \phi \sin(\gamma_\mu B_\mu t) \right]$$

$$P_y(t) = P(0) \left[\frac{1}{2} \sin 2\theta \sin \phi [1 - \cos(\gamma_\mu B_\mu t)] + \sin \theta \cos \phi \sin(\gamma_\mu B_\mu t) \right] \ . \tag{4.3}$$

In vector form the time evolution of $\mathbf{P}(t)$ can generally be written by decomposing the initial polarization vector $\mathbf{P}(0)$ into components parallel and perpendicular to the direction of the local field $\hat{\mathbf{b}} = \mathbf{B}_\mu / B_\mu$.

The time evolution $\mathbf{P}(t)$ then reads[1]

$$\mathbf{P}(t) = [\mathbf{P}(0) \cdot \hat{\mathbf{b}}]\hat{\mathbf{b}} + [\hat{\mathbf{b}} \times [\mathbf{P}(0) \times \hat{\mathbf{b}}]] \cos(\gamma_\mu B_\mu t) + [\mathbf{P}(0) \times \hat{\mathbf{b}}] \sin(\gamma_\mu B_\mu t)$$

$$= [\mathbf{P}(0) \cdot \hat{\mathbf{b}}]\hat{\mathbf{b}} + [\mathbf{P}(0) - [\mathbf{P}(0) \cdot \hat{\mathbf{b}}]\hat{\mathbf{b}}] \cos(\gamma_\mu B_\mu t) + [\mathbf{P}(0) \times \hat{\mathbf{b}}] \sin(\gamma_\mu B_\mu t) \ . \tag{4.4}$$

---

[1] Since it involves only projections, the equation is valid for any $\mathbf{P}(0)$.

Fig. 4.2 Time evolution of
the muon spin polarization in
a single valued vector field.
For $\theta \neq \pi/2$ a non-oscillating
part of amplitude $\cos^2\theta$ is
present in the μSR signal and
the amplitude of the
oscillation is reduced to
$\sin^2\theta$. The spin precesses on a
cone around the field. We
assume $P_z(0) = 1$

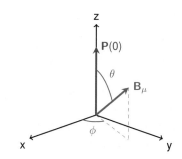

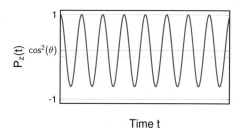

The signal observed in a detector, whose direction is given by $\hat{\mathbf{n}}$ is the projection of
$\mathbf{P}(t)$ onto that direction, i.e., $\mathbf{P}(t) \cdot \hat{\mathbf{n}}$.[2] The component $P_z(t)$ in Eq. 4.1 is obtained
by projecting $\mathbf{P}(t)$ onto the z-axis, which in this case is the direction of the detector,
$\hat{\mathbf{n}} \,\|\, \hat{\mathbf{z}}$.

   In many cases, the muon ensemble will not sense a unique value and direction
of $\mathbf{B}_\mu$ but rather a distribution of fields $f(\mathbf{B}_\mu)$. If the field distribution is known
or guessed, one can calculate the corresponding polarization function by averaging

---

[2] Equation 4.4 can also be written in matrix form (Belousov et al. 1979; Smilga & Belousov 1994)

$$P_i(t) = \sum_{k=1}^{3} M_{ik}(t) P_k(t) \ ,$$

where, expressing the cross product with the Levi-Civita tensor ϵ_{ikl} ($\epsilon_{ikl} = 0$ if two or more indices
are equal, $= 1$ if the set $\{ikl\}$ is obtained from $\{123\}$ by an even permutation, and $= -1$ by an odd)

$$M_{ik} = \hat{b}_i \hat{b}_k + (\delta_{ik} - \hat{b}_i \hat{b}_k) \cos(\gamma_\mu B_\mu t) + \sum_{l=1}^{3} \epsilon_{ikl} \hat{b}_l \sin(\gamma_\mu B_\mu t) \ . \tag{4.5}$$

Eq. 4.1 over the field distribution. With $\int f(\mathbf{B}_\mu)\mathbf{dB}_\mu = 1$ we have

$$P_z(t) = P_z(0) \int f(\mathbf{B}_\mu)\left[\cos^2\theta + \sin^2\theta \cos(\gamma_\mu B_\mu t)\right]\mathbf{dB}_\mu \quad . \tag{4.6}$$

It is useful to introduce the distribution of the modulus of the vector field $|\mathbf{B}_\mu| = B_\mu$. We denote the function $f_m(|\mathbf{B}_\mu|)$ with the subscript m, to emphasize that it represents a different functional dependence. Its relation to $f(\mathbf{B}_\mu)$, written in polar coordinates, is given by

$$f_m(|\mathbf{B}_\mu|) = \int f(B_\mu, \theta, \phi) B_\mu^2 \sin\theta \, d\theta \, d\phi \quad . \tag{4.7}$$

If the field distribution is isotropic, i.e., $f(\mathbf{B}_\mu)$ depends only on B_μ, following relation holds

$$f(B_\mu) = \frac{f_m(B_\mu)}{4\pi B_\mu^2} \quad . \tag{4.8}$$

4.1.1 Single Valued Field

4.1.1.1 Single Crystal

A muon polarization following Eq. 4.1 is the ideal situation, realized in a single crystal exhibiting a monodomain magnetic phase. For a specific θ the polarization function is plotted in Fig. 4.2. The case where an arbitrary number of discrete fields are sensed by the muon can be easily generalized from Eq. 4.1; each field contribution must be weighted according to the occupancy probability of the corresponding muon site.

Two special cases:

- $\theta = \pi/2$ (Fig. 4.3)

$$P_z(t) = P_z(0) \cos(\gamma_\mu B_\mu t) \quad . \tag{4.9}$$

- $\theta = 0$, $\mathbf{B}_\mu \parallel \mathbf{P}(0)$. The muon spin does not precess and the μSR signal remains constant over time

$$P_z(t) = P(0) \quad . \tag{4.10}$$

From Eq. 4.1, and more generally from Eq. 4.4, we expect, that for a single crystal the μSR signal varies with the orientation of the crystal with respect to $\mathbf{P}(0)$ (or the beam direction). Hence, in addition to the field strength, which is provided by the Larmor frequency, one can determine the direction of the local field from the angular

Fig. 4.3 Time evolution of the muon polarization in an ideal magnetic single crystal. For $\theta = \pi/2$ a full amplitude precession is observed in the μSR signal when looking in the z direction defined by the $\mathbf{P}(0)$ direction

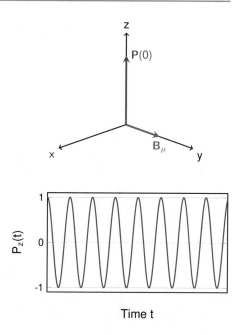

dependence of the amplitude of the oscillating signal. See below for an example of such a determination.

Some Examples The heavy fermion[3] compound UGe$_2$ is a remarkable system exhibiting a coexistence of ferromagnetism and superconductivity under applied pressure (Sakarya et al. 2010).[4] In this system (orthorhombic structure $Cmmm$), there are two muon stopping sites ($2b$ and $4i$ Wyckoff positions), with one fraction of the muon ensemble stopping at one site and the other at a second site. From ZF-μSR measurements performed at ambient pressure, one clearly determines that the internal fields at both sites are directed along the crystallographic a-axis (Fig. 4.4).

Another example can be found in the antiferromagnetic system (spin chain) Li$_2$CuO$_2$ (Staub et al. 2000). In the magnetic state, no precession is detected when the initial muon polarization is along the a-axis (not shown), whereas a precession signal around the zero line is detected when $\mathbf{P}(0) \perp \hat{\mathbf{a}}$, see Fig. 4.5. The μSR spectrum and its Fourier transform reflect three stopping sites.[5]

---

[3] See footnote 25, Chap. 5.

[4] Usually in a conventional BCS superconductor, where the Cooper pairs are formed by $\mid \uparrow \rangle$ and $\mid \downarrow \rangle$ spin states, the magnetic field tends to align the spins and thus destroy the superconducting state.

[5] The μSR spectrum represented by an asymmetry histogram in the time domain can be mapped to the (spin precession) frequency domain by a discrete Fourier transform. Numerically, the discrete Fourier transform can be efficiently implemented by the so-called Fast Fourier Transform (FFT), an algorithm developed by Cooley and Tukey (1965) that reduces the number of computations for

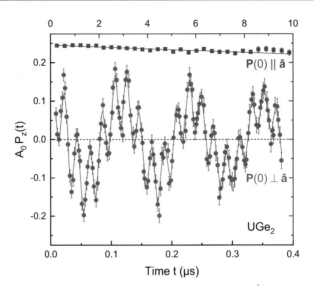

Fig. 4.4 Low temperature ZF-μSR signal in the ferromagnetic state of UGe$_2$ ($T_C \simeq 50$ K). When the initial muon polarization is along the a-axis, no precession is observed (blue curve). This corresponds to $\theta = 0$. When the initial muon polarization is perpendicular to the a-axis, a signal corresponding to the superposition of two oscillations around the zero line is seen. This is the situation for $\theta = \pi/2$. The two frequencies are due to the two muon stopping sites with different local magnetic fields. Modified from Sakarya et al. (2010), © American Physical Society. Reproduced with permission. All rights reserved

A detailed determination of the local fields at several muon sites is discussed in Schenck et al. (2002), where ZF measurements[6] were performed on a single crystal of the tetragonal heavy fermion compound CeRhIn$_5$ ($T_N = 3.8$ K). ZF measurements below T_N find five components in the μSR signal reflecting relatively

---

k points from $2k^2$ to $2k \log_2 k$ (Rao et al. 2010; Walker 1996). Since the precession frequency is directly proportional to the local field, the FFT of a μSR spectrum, ideally, provides directly the field distribution sensed by the muon ensemble. However, there are some limitations to the FFT of μSR spectra that are directly related to the properties of a μSR histogram. First, the limited number of counts at later times (with an exponentially increasing relative error, see footnote 20, Chap. 3) introduces noise into the frequency spectrum. Second, the finite time window of μSR produces the so-called "ringing" in the FFT histogram. These distortions can be reduced by applying apodization, see footnote 43, Chap. 6, but at the expense of frequency resolution. The side effects of the FFT sometimes make fitting in the frequency domain more difficult than in the time domain. The FFT is often used to assess whether expected features are present in the data (e.g., presence, position, and multiplicity of discrete frequencies, spectral weight regions, ...) and for visualization purposes, such as a field distribution expected from the vortex state of a superconductor, see Sect. 6.2.1. See also footnote 33, Chap. 8 for another approach to Fourier analysis.

[6] The muon sites were determined from TF Knight shift measurements in the paramagnetic phase. See Sect. 5.6.3 about the use of Knight shift measurements to determine muon stopping sites.

Fig. 4.5 Top panel: μSR asymmetry measured in the antiferromagnetic state of Li_2CuO_2. When the initial muon polarization is along the a-axis, no precession is observed (not shown). When the initial muon polarization is perpendicular to the a-axis, three spontaneous frequencies are seen, with a signal oscillating around the zero line. The observed signal corresponds to three frequencies reflecting three muon stopping sites. The bottom panel is the Fast Fourier Transform (FFT) of the μSR signal showing the three precession frequencies. Modified from Staub et al. (2000), © Elsevier. Reproduced with permission. All rights reserved

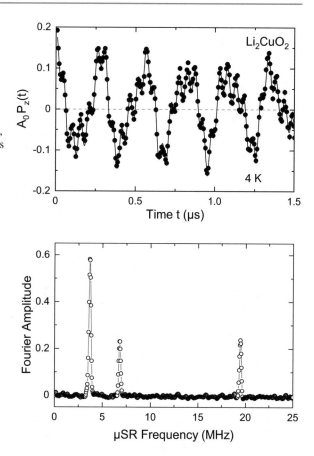

well defined internal fields, resulting from the expected incommensurate helical magnetic structure and the different muon sites (Fig. 4.6).

The crystal was originally mounted with the a axis along the beam momentum $\phi = 0$. The initial polarization $\mathbf{P}(0)$ was rotated by $\sim 47°$ in the vertical direction by a spin rotator, from the original antiparallel direction to the muon beam. For the measurements, the sample was rotated around the b-axis by an angle ϕ, thus changing the direction of $\mathbf{P}(0)$ in the a-c plane and changing the angle(s) between local fields and $\mathbf{P}(0)$. Using the data from the U, D, F, B and R detectors and taking into account the possible muon sites, a detailed analysis permitted to establish that the four different internal fields[7] have distinct components around the c axis and randomly oriented components in the basal plane, i.e., the vectors are confined to the surfaces of cones. Three of these cones with different apertures are associated

---

[7] The fifth component has zero internal field and does not oscillate.

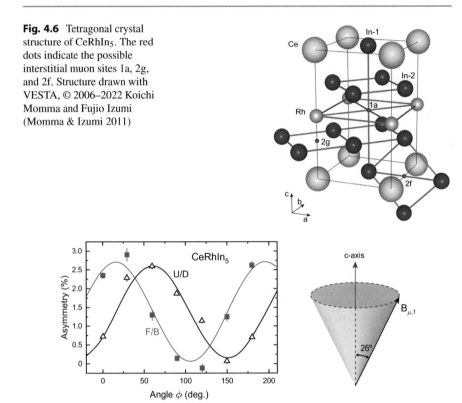

Fig. 4.6 Tetragonal crystal structure of CeRhIn$_5$. The red dots indicate the possible interstitial muon sites 1a, 2g, and 2f. Structure drawn with VESTA, © 2006–2022 Koichi Momma and Fujio Izumi (Momma & Izumi 2011)

Fig. 4.7 Angular dependence of one of the four amplitudes of the oscillating signals in Forward / Backward and Up / Down directions at 2.1 K in CeRhIn$_5$. ϕ is the angle between the a-axis and the beam axis. For the measurements the sample is rotated about the crystallographic b-axis by ϕ. The solid lines represent fits to the expected angular dependence. The right drawing shows the angular direction of the associated local field, which is confined to a cone around the c-axis. Modified from Schenck et al. (2002), © American Physical Society. Reproduced with permission. All rights reserved

with the a site ($1/2$, $1/2$, $1/2$), where $\sim$ 70% of the muons stop.[8] Figure 4.7 shows the angular dependences of the amplitude of one of the oscillating signals and the corresponding internal field vector. The precise determination of the local field vectors allows to establish that the known incommensurate helical magnetic structure of the Ce moments cannot explain the results; additional small ordered magnetic moments on the Rh sublattice are required.

4.1.1.2 Polycrystal

A first example of field distribution can be found in the case of a magnetic sample, such as the one described in Sect. 4.1.1.1, but in polycrystalline form. Such a sample

---

[8] Because of the helical magnetic structure, we expect to find different values of $\mathbf{B}_\mu$ for the same site, so-called magnetically inequivalent site, see Sect. 5.1.2.

Fig. 4.8 Isotropic
distribution of a single valued
local field in a polycrystalline
sample

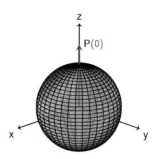

consists of a multitude of randomly oriented crystallites, corresponding to a powder averaging. Therefore, the initial muon polarization vector points randomly with respect to the crystallographic axes of the crystallites. In a polycrystalline sample of a magnetic system the muon ensemble probes a constant value of the local field B_μ, but $\mathbf{B}_\mu$ is isotropically distributed on a sphere with random direction with respect to $\mathbf{P}(0)$, Fig. 4.8.

A similar situation is found in a single crystal sample that has been zero field cooled to a ferromagnetic state, which ideally organizes into multiple isotropically oriented magnetic domains.

In this case the field modulus distribution is a δ-function at B_μ, see Fig. 4.9, and we simply obtain

$$P_z(t) = \frac{1}{3} + \frac{2}{3}\cos(\gamma_\mu B_\mu t) \;, \tag{4.11}$$

plotted in Fig. 4.10. The 1/3 and 2/3 components can be understood qualitatively by considering that for a local field that is random in all directions, 1/3 of the field is parallel or antiparallel to the initial muon spin direction and 2/3 is perpendicular. The parallel field component corresponds to the situation $\theta = 0$ and keeps the polarization aligned, whereas the perpendicular one corresponds to $\theta = \pi/2$ and thus produces a spin precession around the field.[9] For the case of an isotropically distributed field on a plane, see Exercise 4.2.

So far we have assumed that all the muons probe exactly the same value of the field, though possibly with different directions. This is of course the ideal case. In a real sample, a magnetic structure has some disorder producing a distribution of the fields around a mean value $\langle B_\mu \rangle$.

---

[9] Experimentally, the observation of Eq. 4.11 or more generally of a non-oscillating 1/3 term, could in principle also reflect the situation where there is a well defined direction of the field with $\cos^2\theta = 1/3$, see Eq. 4.1. However, the most likely case is the presence of crystallites in the sample. The two cases can be distinguished by rotating the sample, or otherwise see Exercise 4.3.

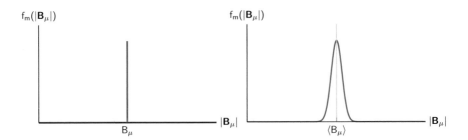

Fig. 4.9 Left: Ideal δ-distribution for the field value in a polycrystalline sample. Right: More realistic distribution (Gaussian) with a width due to disorder, Eq. 4.12

Fig. 4.10 Time evolution of the muon spin polarization in an ideal polycrystalline magnetic system. Note the non-oscillating component with $1/3$ amplitude

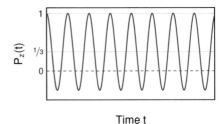

Time t

If we assume that this effect can be described by a Gaussian distribution, with $\sqrt{\langle \Delta B_\mu^2 \rangle} \ll \langle B_\mu \rangle$, we have

$$f_m(B_\mu) = \frac{1}{\sqrt{2\pi \langle \Delta B_\mu^2 \rangle}} \exp\left[-\frac{(B_\mu - \langle B_\mu \rangle)^2}{2\langle \Delta B_\mu^2 \rangle} \right] , \qquad (4.12)$$

where $\langle \Delta B_\mu^2 \rangle$ is the variance (i.e., the square of the standard deviation) of the Gaussian field modulus distribution. $\sqrt{\langle \Delta B_\mu^2 \rangle}$ is sometimes referred to as the field width[10]

$$\langle \Delta B_\mu^2 \rangle = \langle (B_\mu - \langle B_\mu \rangle)^2 \rangle$$
$$= \langle B_\mu^2 \rangle - \langle B_\mu \rangle^2$$
$$\equiv \int (B_\mu - \langle B_\mu \rangle)^2 f_m(B_\mu) dB_\mu . \qquad (4.13)$$

---

[10] The brackets $\langle \rangle$ indicate the weighting over the field distribution as well as the average over the muon ensemble, since ideally the muon ensemble performs the same operation.

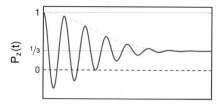

Fig. 4.11 Time evolution of the muon spin polarization in a realistic polycrystalline magnetic system. A Gaussian field distribution causes a Gaussian depolarization (curve envelope). By measuring the depolarization rate σ, one directly determines the variance $\langle \Delta B_\mu^2 \rangle$ of the field distribution

Inserting, with the use of Eq. 4.8, the Gaussian field distribution Eq. 4.12, into the equation for the time dependence of the polarization Eq. 4.6, and performing the integration we obtain

$$P_z(t) = \frac{1}{3} + \frac{2}{3} \cos(\gamma_\mu \langle B_\mu \rangle t) \ \exp\left[-\frac{1}{2} \gamma_\mu^2 \langle \Delta B_\mu^2 \rangle t^2 \right]$$

$$= \frac{1}{3} + \frac{2}{3} \cos(\gamma_\mu \langle B_\mu \rangle t) \ \exp\left[-\frac{\sigma^2 t^2}{2} \right] , \qquad (4.14)$$

where we have introduced the Gaussian depolarization rate $\sigma = \sqrt{\gamma_\mu^2 \langle \Delta B_\mu^2 \rangle}$.

By fitting Eq. 4.14 to the data, one can determine the depolarization rate and therefore directly obtain the width of the field distribution in the investigated compound.

Comparing Fig. 4.10 with Fig. 4.11 we see that in the latter case there is a loss of polarization with time, whereas in Fig. 4.10 the value of the polarization is conserved and only precession is observed. Therefore, one often speaks, instead of polarization function, of depolarization function.

In Eq. 4.14 we have introduced the depolarization rate σ which is a measure of the decay of polarization with time. σ has the dimension of 1/time and is generally expressed in μs^{-1} (inverse microseconds) or, somewhat improperly, in MHz. The depolarization we observe is a direct consequence of the field not being perfectly single valued, but distributed. Some muons sense a field larger (or smaller) than $\langle B_\mu \rangle$ so that their spin will precess faster (or slower), thus producing a loss of polarization (i.e., a loss of spin coherence) of the muon ensemble. This is the so-called static dephasing, where the depolarization is a consequence of the spreading of phases in the cosine term of Eq. 4.6.

Some Examples The vast majority of measurements on magnetic systems are performed on polycrystals. We present here a few examples.

Fig. 4.12 ZF-μSR signal of a polycrystalline sample of MnP in the ferromagnetic state. Note the spontaneous oscillation around $1/3$ of the total amplitude and the very weak depolarization (albeit on a rather short time window) indicating a rather high quality sample. This is also evidenced by the very narrow width of the peak in the Fourier transform. Modified from Khasanov et al. (2017), © IOP Publishing Ltd. Reproduced with permission. All rights reserved

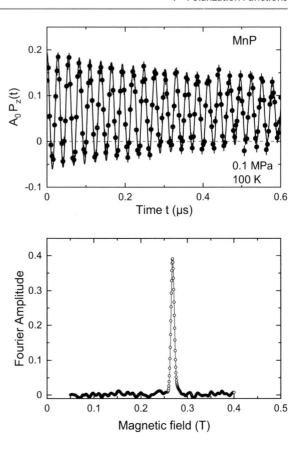

MnP has been extensively studied using polycrystalline samples. At ambient pressure it exhibits a ferromagnetic state at high temperature and a helical state at low temperature. Under pressure the magnetic state changes and even superconductivity appears. Figure 4.12 displays a typical ZF-μSR and its Fourier spectrum in the ferromagnetic state at zero pressure (Khasanov et al. 2017).

Another example is provided by the iron-based high-T_c superconductors, a new class of superconductors discovered in 2008 (Kamihara et al. 2008), see Chap. 6. In these systems, magnetism competes with superconductivity and the mutual interplay can be well studied with μSR, see Sect. 6.4. μSR has been instrumental in determining the magnetic properties since the ordered moment is small $\sim 0.3\,\mu_B$ but still quite large for μSR, see Exercise 3.2. Figure 4.13 shows a measurement of a polycrystalline sample of LaFeAsO, a member of the so-called "1111" family that exhibits magnetism in the absence of doping (Luetkens et al. 2009). ZF spectra above and below the Néel temperature are reported. In the antiferromagnetic state, the buildup of magnetic order produces a rather well-defined local field which is probed by the muons whose spins spontaneously (i.e., without the application of an external field) precess.

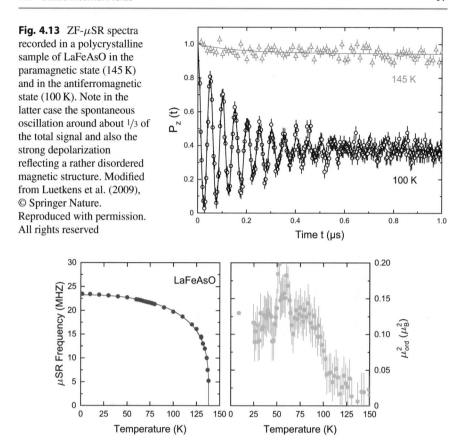

Fig. 4.13 ZF-μSR spectra recorded in a polycrystalline sample of LaFeAsO in the paramagnetic state (145 K) and in the antiferromagnetic state (100 K). Note in the latter case the spontaneous oscillation around about 1/3 of the total signal and also the strong depolarization reflecting a rather disordered magnetic structure. Modified from Luetkens et al. (2009), © Springer Nature. Reproduced with permission. All rights reserved

Fig. 4.14 Temperature dependence of the precession frequency from ZF-μSR (left panel) and of the Bragg peak intensity from neutron diffraction (right panel), showing the appearance of magnetic order upon decreasing the temperature. Note the small error of the μSR points. The neutron data have been obtained by assuming that the entire volume is magnetic, which is indeed the case here. See Sect. 5.2 for a discussion about volume fraction determination by μSR vs. neutron diffraction. Modified from Luetkens et al. (2009) (left panel) and (Cruz et al. 2008) (right panel), © Springer Nature. Reproduced with permission. All rights reserved

The plot of the precession frequency versus temperature is a measure of the local magnetization. Figure 4.14 exemplifies the role of the ZF-μSR technique as an accurate, sensitive, and fast local magnetometer.

4.1.2   Continuous Field Distributions

In this section, we discuss two very common field distributions: the Gaussian and the Lorentzian distribution. This situation arises when the static moments sensed by the muon are randomly oriented.

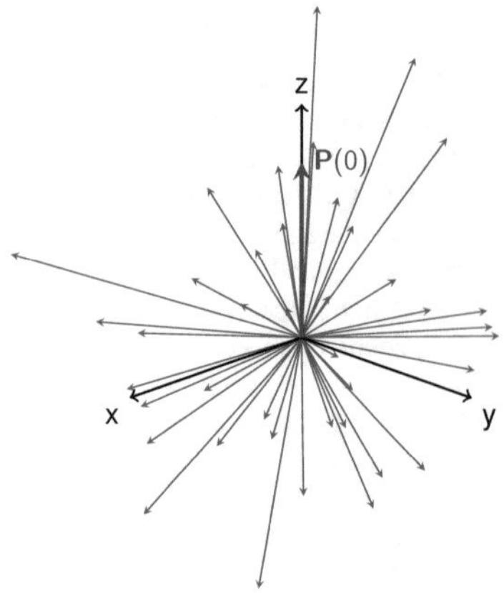

Fig. 4.15 Pictorial view of
the Gaussian distributed fields
sensed by the muon ensemble

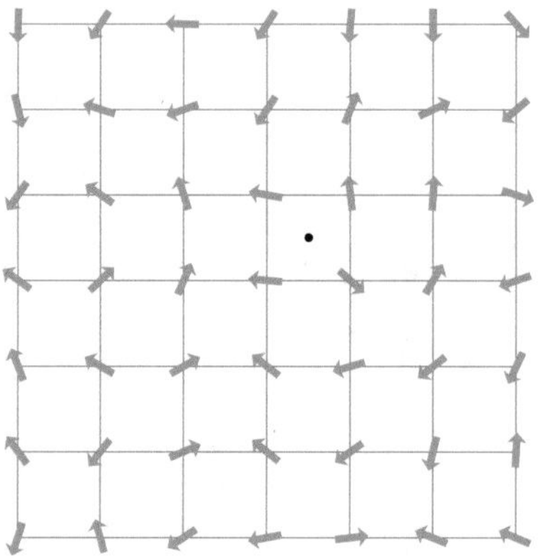

Fig. 4.16 Muon (black dot)
stopping at an interstitial site
of a lattice of randomly
oriented magnetic moments

4.1.2.1 Gaussian Distribution

A Gaussian distribution of the field components along each Cartesian direction with
$\langle B_{\mu,\alpha} \rangle = 0$, $\alpha = x, y, z$, Fig. 4.15, is obtained in the case of a dense arrangement
of randomly oriented static magnetic moments, see Fig. 4.16, and is justified by the
central limit theorem.[11,12] This is the case of the nuclear moments, which are not
only randomly oriented but also static on the μSR time scale.

---

[11] The central limit theorem states that, when independent random variables (in this case, the fields
produced by the individual moments) are added, their properly normalized sum tends to a Gaussian
distribution, regardless of the underlying random distribution, provided it has a finite variance.

[12] Note that this situation is different from the one encountered in Sect. 4.1.1.2, where we had only
a small field variation around a nonzero value with $\sqrt{\langle \Delta B_\mu^2 \rangle}/\langle B_\mu \rangle \ll 1$. For a generalization, see
Sect. 4.1.3.

Fig. 4.17 Gaussian field
distribution along one
Cartesian direction

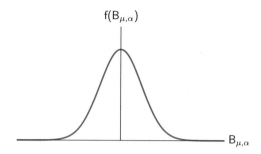

Each Cartesian field component follows a Gaussian distribution (Fig. 4.17)

$$f(B_{\mu,\alpha}) = \frac{1}{\sqrt{2\pi \langle \Delta B_{\mu,\alpha}^2 \rangle}} \exp\left[-\frac{B_{\mu,\alpha}^2}{2\langle \Delta B_{\mu,\alpha}^2 \rangle}\right] \quad , \tag{4.15}$$

with $\langle \Delta B_{\mu,\alpha}^2 \rangle$ the variance of the field distribution along a Cartesian direction, which
we assume to be equal for the three directions, and $\langle B_{\mu,\alpha} \rangle = 0$, so that

$$\langle \Delta B_{\mu,\alpha}^2 \rangle = \langle (B_{\mu,\alpha} - \langle B_{\mu,\alpha} \rangle)^2 \rangle \equiv \langle \Delta B_\mu^2 \rangle = \langle B_\mu^2 \rangle \quad .$$

The three components of $\mathbf{B}_\mu$ are uncorrelated so that for the distribution of the
vector field

$$f(\mathbf{B}_\mu) = f(B_{\mu,x}) f(B_{\mu,y}) f(B_{\mu,z}) \quad . \tag{4.16}$$

Since $f(\mathbf{B}_\mu)$ is isotropic, from Eq. 4.8 we have

$$f(\mathbf{B}_\mu) = \frac{f_m(|\mathbf{B}_\mu|)}{4\pi B_\mu^2} \quad . \tag{4.17}$$

The distribution function of the field modulus, $f_m(|\mathbf{B}_\mu|)$, is given by the Maxwell-
Boltzmann distribution[13] (Fig. 4.18)

$$f_m(B_\mu) = \frac{1}{(2\pi \langle \Delta B_\mu^2 \rangle)^{\frac{3}{2}}} 4\pi B_\mu^2 \exp\left(-\frac{B_\mu^2}{2\langle \Delta B_\mu^2 \rangle}\right) \quad . \tag{4.18}$$

---

[13] The Maxwell-Boltzmann distribution, sometimes simply called the Maxwell distribution,
describes for example the distribution of the molecule's velocity modulus in an ideal gas.

Fig. 4.18
Maxwell-Boltzmann
distribution of the field value
$|\mathbf{B}_\mu|$ for Gaussian distributed
$B_{\mu,x}$, $B_{\mu,y}$ and $B_{\mu,z}$. The
maximum occurs at
$B_\mu^{\max} = \sqrt{2\langle \Delta B_\mu^2 \rangle}$ and the
mean value is
$\langle B_\mu \rangle = \sqrt{8\langle \Delta B_\mu^2 \rangle / \pi}$

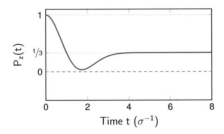

Fig. 4.19 Gaussian Kubo-Toyabe function corresponding to a Gaussian field distribution (or equivalently, a Maxwell-Boltzmann field modulus distribution) and describing the time evolution of the muon spin polarization in a system where dense magnetic moments are randomly oriented. The minimum occurs at $t^{\min} = \sqrt{3}/\sigma$

Again, to obtain the time dependence of the polarization, we insert the modulus distribution, Eq. 4.18, into Eq. 4.6 using Eq. 4.8. By performing the integration we obtain[14] (Fig. 4.19)

$$
\begin{aligned}
P_z^{\text{GKT}}(t) &= \frac{1}{3} + \frac{2}{3}\left(1 - \gamma_\mu^2 \langle \Delta B_\mu^2 \rangle t^2\right) \exp\left[-\frac{1}{2}\gamma_\mu^2 \langle \Delta B_\mu^2 \rangle t^2\right] \\
&= \frac{1}{3} + \frac{2}{3}(1 - \sigma^2 t^2)\exp\left[-\frac{\sigma^2 t^2}{2}\right] ,
\end{aligned}
\tag{4.19}
$$

where we again use the Gaussian depolarization rate

$$
\sigma^2 = \gamma_\mu^2 \langle \Delta B_\mu^2 \rangle .
\tag{4.20}
$$

This well-known polarization function is called Gaussian Kubo-Toyabe (GKT)[15] (Kubo & Toyabe 1967). As previously mentioned, the 1/3 and 2/3 components can be qualitatively understood by considering that a randomly oriented local field corresponds to a 1/3 component parallel or antiparallel to the initial muon spin direction and 2/3 perpendicular to it.

---

[14] Note that in the chosen geometry $P_x(t) = P_y(t) = 0$.

[15] See remark in footnote 95, Chap. 5.

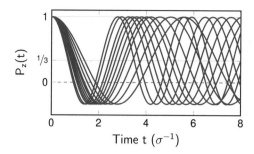

Fig. 4.20 Time evolution of the muon spin polarizations according to Eq. 4.11 for different values of the Maxwell-Boltzmann distributed magnitude of the local field. Adapted from Blundell (1999), © Taylor & Francis. Reproduced with permission. All rights reserved

We can qualitatively understand the shape of the Kubo-Toyabe function by looking at Fig. 4.20. In the figure, the time dependence of the muon spin polarization is plotted for different field values, with each polarization curve following Eq. 4.11. If these field values obey a Maxwell-Boltzmann distribution, initially, all polarization curves will behave similarly, decreasing from one to a minimum value before increasing. However, after a short time, they become out of phase. Thus the average of the curves falls from unity to a minimum and then recovers to an asymptotic value of 1/3.

For very short times ("st") or very low depolarization rates, the Gaussian Kubo-Toyabe function can be approximated by a simple Gaussian, see Exercise 4.4,

$$P_z^{GKT}(t) \simeq \exp(-\sigma_{st}^2 t^2/2) \quad \text{for } t \ll t^{min} = \frac{\sqrt{3}}{\sigma}$$

$$\sigma_{st}^2 = 2\sigma^2 = 2\gamma_\mu^2 \langle \Delta B_\mu^2 \rangle \ . \tag{4.21}$$

This behavior is related to the fact that only the field components perpendicular to the initial muon polarization ($B_{\mu,x}$ and $B_{\mu,y}$) contribute to the muon spin precession and thus to the dephasing.

Some Examples In the paramagnetic phase of a system, the electronic moments fluctuate very rapidly and have no observable effect on the muon spin, see Sect. 4.3. On the other hand, as mentioned before, the nuclear moments are small and can be taken as static during the μSR observation time window[16] so that μSR only senses the field distribution created by the nuclear moments, which is generally of Gaussian shape.

The first example is a measurement in the paramagnetic phase of MnSi, Fig. 4.21. MnSi has a so-called non-centrosymmetric crystallographic structure.[17] Below about 30 K the system exhibits a helical magnetic state. This system has attracted

---

[16] See, however, the remarks in Sect. 4.1.2.1.

[17] A crystal structure is non-centrosymmetric when it lacks a center of symmetry for an inversion operation.

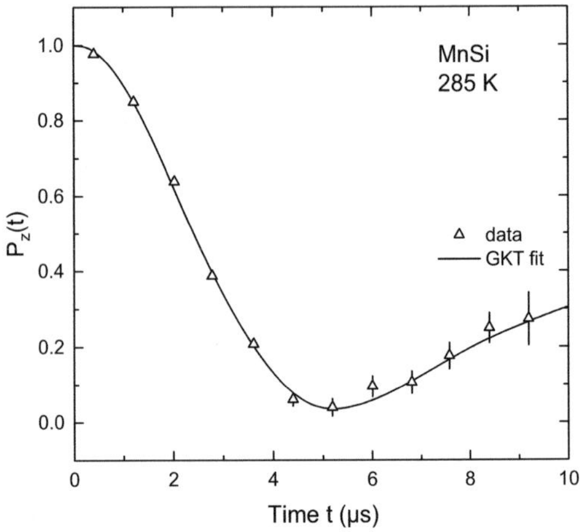

Fig. 4.21 The ZF-μSR signal of MnSi in the paramagnetic phase is well fitted by a Gaussian Kubo-Toyabe function. Modified from Hayano et al. (1979), © American Physical Society. Reproduced with permission. All rights reserved

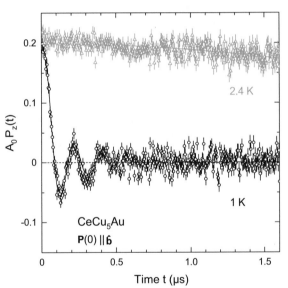

Fig. 4.22 ZF-μSR signal in the compound CeCu$_5$Au. The lower curve is a measurement in the magnetic state and represents a spontaneous precession due to the internal field created by the ordered electronic moments. The upper curve is a measurement in the paramagnetic state where the depolarization is only due to the nuclear moments. This curve is well fitted by a Kubo-Toyabe function with a very small σ. Modified from Amato et al. (1995), © American Physical Society. Reproduced with permission. All rights reserved

increasing interest due to the occurrence of magnetic skyrmions[18] if an external field is applied in the magnetic phase (Mühlbauer et al. 2009). In the paramagnetic phase, a typical Kubo-Toyabe depolarization of the muon ensemble is visible in the ZF-μSR data.

The second example, is a measurement in the nonmagnetic phase of the compound CeCu$_5$Au, which orders magnetically below 2.3 K. This system is a close parent of the heavy fermion system CeCu$_6$, Fig. 4.22. A Gaussian decay of

---

[18] A skyrmion is a topologically stable field configuration with a complex noncoplanar spin structure resembling a magnetic vortex. It has particle-like properties. For an introduction see (Lancaster 2019).

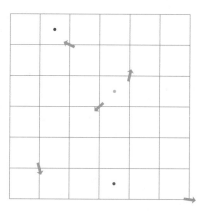

Fig. 4.23 Diluted static magnetic moments in a lattice. Muons stopping at different interstitial sites (dots) sense different fields. A Lorentzian field distribution has a much higher probability for high fields than a Gaussian distribution. For instance the "blue" muon in the upper left corner, close to a single magnetic moment, senses a large field. Such a situation is not found in the case of a dense distribution of moments, where the muon is always surrounded by many moments whose fields tend to compensate

the polarization is observed corresponding to a weak Kubo-Toyabe depolarization of the muon ensemble with no visible minimum in the observation-time window.

4.1.2.2 Lorentzian Distribution

A Lorentzian distribution of fields is obtained in the case of a dilute arrangement of randomly oriented static moments, for example static electronic moments of magnetic impurities (Held & Klein 1975; Walker & Walstedt 1980) (see Fig. 4.23). Such a situation occurs, for instance, in spin glasses (Mydosh 1993) (see Fig. 4.23), such as the so-called canonical spin glasses, which consist of a few percent of magnetic ions embedded in a metallic matrix, e.g., AgMn 1%.

For each Cartesian direction, $\alpha = x, y, z$, the Lorentzian field distribution is given by (see Fig. 4.24)

$$f(B_{\mu,\alpha}) = \frac{\gamma_\mu}{\pi} \frac{a}{a^2 + \gamma_\mu^2 B_{\mu,\alpha}^2} \ , \qquad (4.22)$$

a/γ_μ is the half width at half maximum (HWHM) of the field distribution along one Cartesian direction. As with the Gaussian case, we only consider here the case where the width of the distribution is the same in all three directions.

Fig. 4.24 Lorentzian
distribution of the field along
a Cartesian direction

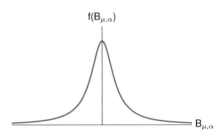

$f(B_{\mu,\alpha})$

$B_{\mu,\alpha}$

Fig. 4.25
Squared-Lorentzian
distribution of the field value
$|\mathbf{B}_\mu|$ for Lorentzian
distributed $B_{\mu,x}$, $B_{\mu,y}$ and
$B_{\mu,z}$. The maximum occurs at
$B_\mu^{\max} = a/\gamma_\mu$

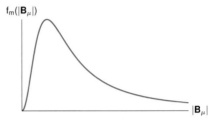

$f_m(|\mathbf{B}_\mu|)$

$|\mathbf{B}_\mu|$

The distribution function for the absolute value of the field (Fig. 4.25) is given by
a squared-Lorentzian distribution (Fiory 1981)[19]

$$f_m(|\mathbf{B}_\mu|) = \frac{\gamma_\mu^3}{\pi^2} \frac{a}{(a^2 + \gamma_\mu^2 B_\mu^2)^2} \, 4\pi B_\mu^2 \ . \tag{4.24}$$

---

[19] Note that the distribution of the modulus cannot be obtained from the product of the individual
Cartesian field distributions as in the Gaussian case. The three field components are not independent (Wan et al. 1999). Conversely, the distribution of each field component Eq. 4.22 can be
obtained from Eq. 4.24. For example, to calculate $f(B_{\mu,x})$ we integrate $f(\mathbf{B}_\mu)$ over $B_{\mu,y}$ and $B_{\mu,z}$

$$f(B_{\mu,x}) = \int_{-\infty}^{\infty} \int_{-\infty}^{\infty} f(\mathbf{B}_\mu) d B_{\mu,y} \, d B_{\mu,z}$$

$$= \int_{-\infty}^{\infty} \int_{-\infty}^{\infty} \frac{f_m(B_\mu)}{4\pi B_\mu^2} d B_{\mu,y} \, d B_{\mu,z}$$

$$= \int_{-\infty}^{\infty} \int_{-\infty}^{\infty} \frac{\gamma_\mu^3}{\pi^2} \frac{a}{\left[a^2 + \gamma_\mu^2 \left(B_{\mu,x}^2 + B_{\mu,y}^2 + B_{\mu,z}^2\right)\right]^2} d B_{\mu,y} \, d B_{\mu,z}$$

$$= \int_{-\infty}^{\infty} \frac{\gamma_\mu^2}{2\pi} \frac{a}{\left[a^2 + \gamma_\mu^2 \left(B_{\mu,x}^2 + B_{\mu,y}^2\right)\right]^{3/2}} d B_{\mu,y}$$

$$= \frac{\gamma_\mu}{\pi} \frac{a}{a^2 + \gamma_\mu^2 B_{\mu,x}^2} \ . \tag{4.23}$$

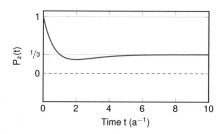

Fig. 4.26 Time evolution of the muon spin polarization in a system with randomly oriented diluted magnetic moments. The squared-Lorentzian field distribution gives the so-called Lorentzian Kubo-Toyabe polarization function. For very short times, the function approaches an exponential function (see text). The minimum occurs at $t_{min} = 2/a$. The Lorentzian KT has a shallower dip than the Gaussian KT because there is a wider range of fields leading to faster damping

Inserting Eq. 4.24 into Eq. 4.6 one obtains

$$P_z^{\mathrm{LKT}}(t) = \frac{1}{3} + \frac{2}{3}(1 - at)\,e^{-at} \,, \tag{4.25}$$

which is called the Lorentzian Kubo-Toyabe function (LKT), Fig. 4.26.

Again, the $1/3$ and $2/3$ components can be qualitatively understood by considering that a randomly oriented local field corresponds to $1/3$ components parallel or antiparallel to the initial muon spin direction and $2/3$ perpendicular.

At early times or for a very small depolarization rate, the Lorentzian Kubo-Toyabe function approaches an exponentially decaying function with $P_z^{\mathrm{LKT}}(t) \simeq \exp(-4at/3)$, Exercise 4.4.

4.1.2.3 Stretched and Gaussian-Lorentzian Kubo-Toyabe Functions
Sometimes a more general Kubo-Toyabe function with an additional stretch parameter β_{KT} is better suited to fit the data

$$P_z^{\beta_{\mathrm{KT}}\mathrm{KT}}(t) = \frac{1}{3} + \frac{2}{3}(1 - (\sigma t)^{\beta_{\mathrm{KT}}}) \exp\left(-\frac{(\sigma t)^{\beta_{\mathrm{KT}}}}{\beta_{\mathrm{KT}}}\right) \,. \tag{4.26}$$

The parameter β_{KT} provides a continuous interpolation between the diluted moment Lorentzian case $\beta_{\mathrm{KT}} = 1$, Eq. 4.25, and the dense moments Gaussian case $\beta_{\mathrm{KT}} = 2$, Eq. 4.19. The function reflects an intermediate (pseudo-Voigtian) field distribution resulting from a superposition of independent Gaussian and Lorentzian components (Crook & Cywinski 1997).

A related function is the Gaussian-Lorentzian Kubo-Toyabe function

$$P_z^{\text{LGKT}}(t) = \frac{1}{3} + \frac{2}{3}(1 - at - \sigma^2 t^2)\exp\left(-at - \frac{\sigma^2 t^2}{2}\right) \quad . \tag{4.27}$$

These functions are expected to provide a reasonable description of the muon spin depolarization in cases beyond the ideal dense or dilute limit, for example in cases where dilute electronic moments are embedded in a dense matrix of nuclear dipoles.

4.1.3 Generalizations of the Kubo-Toyabe Functions

The case where the local magnetic field consists of a sum of two fields, one, $\mathbf{B}_1$, with isotropic Gaussian distribution and the second, $\mathbf{B}_2$, with constant modulus B_0 and random orientation, which can be described by a δ-function, leads to a generalized Gaussian Kubo-Toyabe polarization function (Kornilov & Pomjakushin 1991). Such a case is found in a polycrystalline or ceramic magnetic sample below the transition temperature, where $\mathbf{B}_2$ represents the local spontaneous magnetization and $\mathbf{B}_1$ accounts for the disorder.

The two fields are statistically independent, so that we can write the distribution function of $\mathbf{B}_\mu = \mathbf{B}_1 + \mathbf{B}_2$ as

$$f(\mathbf{B}_\mu) = \int f_1(\mathbf{B}_1) f_2(\mathbf{B}_\mu - \mathbf{B}_1) d\mathbf{B}_1 \quad , \tag{4.28}$$

where f_1 is given by Eq. 4.16 with variance σ^2/γ_μ^2 and $f_2(\mathbf{B}_2) = \delta(|\mathbf{B}_2| - B_0)/(4\pi B_0^2)$. Solving the equation yields

$$f(\mathbf{B}_\mu) = \frac{1}{(2\pi)^{3/2}}\frac{\gamma_\mu}{\sigma B_0 B_\mu}\exp\left[-\frac{\gamma_\mu^2(B_0^2 + B_\mu^2)}{2\sigma^2}\right]\sinh\left[\frac{\gamma_\mu^2 B_0 B_\mu}{\sigma^2}\right] . \tag{4.29}$$

Inserting this expression into Eq. 4.6 and taking advantage of the fact that Eq. 4.29 represents an isotropic distribution, one finds an analytical expression for the polarization corresponding to a generalized Kubo-Toyabe function

$$P_z^{\text{gen GKT}}(t) = \frac{1}{3} + \frac{2}{3}\left(\cos(\gamma_\mu B_0 t) - \frac{\sigma^2 t}{\gamma_\mu B_0}\sin(\gamma_\mu B_0 t)\right)\exp\left[-\frac{\sigma^2 t^2}{2}\right] . \tag{4.30}$$

Considering some limiting cases, we recover functions that we have already encountered before

- for $B_0 \gg \sigma/\gamma_\mu$, the second term in the parentheses disappears, $B_0 = \langle B_\mu \rangle$, and Eq. 4.30 converges to Eq. 4.14.

- for $B_0 \to 0$, Eq. 4.30 converges to the Gaussian Kubo-Toyabe function Eq. 4.19.
- for $\sigma/\gamma_\mu \to 0$, Eq. 4.30 converges to Eq. 4.11.

A similar expression and analogous limits are obtained for a Lorentzian broadening (Larkin et al. 2000)

$$P_z^{\text{gen LKT}}(t) = \frac{1}{3} + \frac{2}{3}\left(\cos(\gamma_\mu B_0 t) - \frac{a}{\gamma_\mu B_0}\sin(\gamma_\mu B_0 t)\right)\exp(-at) \ . \qquad (4.31)$$

4.2 Polarization Functions for Applied External Fields

We now consider how the polarization functions for the zero field configuration that we established in the previous section, are modified when an external field is applied.

4.2.1 Longitudinal Field

We first discuss the case of the internal static Gaussian field distribution, Sect. 4.1.2.1, where we add an external field in LF geometry, i.e., the field is applied along the z-axis $\mathbf{B}_{\text{ext}} \parallel \mathbf{P}(0) \parallel z$-axis (Fig. 4.27).

With respect to the ZF case, Eq. 4.15, the total field distribution is shifted by $\mathbf{B}_{\text{ext}}$ along the z-axis. With $B_{\text{ext}} \equiv |\mathbf{B}_{\text{ext}}|$ we have

$$f(B_{\mu,z}) = \frac{1}{\sqrt{2\pi\langle \Delta B_\mu^2\rangle}}\exp\left[-\frac{(B_{\mu,z} - B_{\text{ext}})^2}{2\langle \Delta B_\mu^2\rangle}\right] \ , \qquad (4.32)$$

while those along the x- and y-axes remain unchanged.

Introducing this distribution into Eq. 4.6 and integrating, we obtain Hayano et al. (1979)

$$P_z^{\text{LF-GKT}} = 1 - \frac{2\sigma^2}{(\gamma_\mu B_{\text{ext}})^2}\left[1 - e^{-\frac{\sigma^2 t^2}{2}}\cos(\gamma_\mu B_{\text{ext}} t)\right]$$

$$+ \frac{2\sigma^4}{(\gamma_\mu B_{\text{ext}})^3}\int_0^t e^{-\frac{\sigma^2 t'^2}{2}}\sin(\gamma_\mu B_{\text{ext}} t')dt' \ . \qquad (4.33)$$

It is easy to check that for $B_{\text{ext}} \to 0$ the ZF Gaussian Kubo-Toyabe function is obtained. Figure 4.28 shows how the ZF function is modified by the application of an external longitudinal field and Fig. 4.29 shows an experimental example in MnSi in the paramagnetic phase, see also Fig. 4.21.

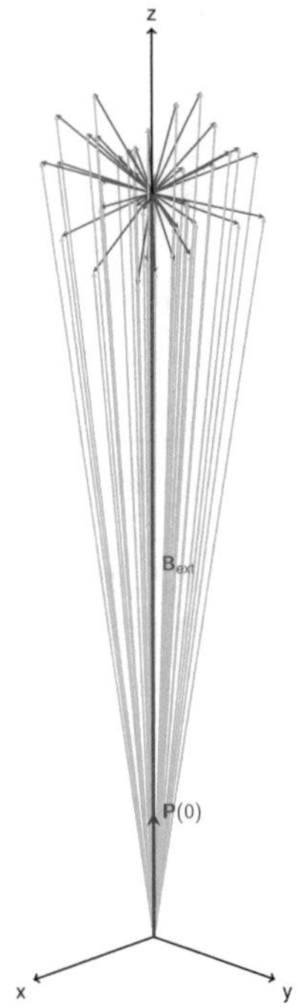

Fig. 4.27 Schematic view of the fields in the LF configuration. The muon ensemble senses the vector sum of the external field $\mathbf{B}_{ext}$ applied along the initial muon polarization $\mathbf{P}(0)$ and of the randomly oriented internal fields. The resulting local fields are shown as green vectors. Depicted is the case of B_{ext} much larger than the internal fields. This leads to the so-called decoupling of the muon spin from the internal fields

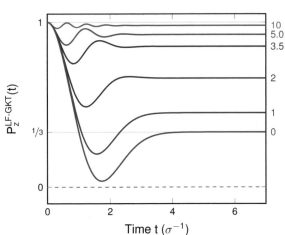

Fig. 4.28 Longitudinal field dependence of the polarization function for isotropic Gaussian distributed internal fields. The time scale is in units of $1/\sigma$. The numbers near the lines give the strength of B_{ext} in units of σ/γ_μ. The zero field curve corresponds to the Gaussian Kubo-Toyabe function. The application of a longitudinal field repolarizes the muon ensemble (decoupling)

Similarly, if a Lorentzian field distribution is present at the muon site, Sect. 4.1.2.2, the addition of an external field along the z-axis modifies the ZF distribution to

$$f(B_{\mu,z}) = \frac{\gamma_\mu}{\pi} \frac{a}{a^2 + \gamma_\mu^2(B_{\mu,z} - B_{ext})^2} \ , \tag{4.34}$$

Fig. 4.29 Example of longitudinal field decoupling in the paramagnetic phase of MnSi. The (static) local field is mainly produced by the Mn nuclear moments. The electronic Mn moments fluctuate very quickly and do not contribute to the muon spin depolarization, $\nu \gg 1$ in Eq. 4.61. Modified from Hayano et al. (1979), © American Physical Society. Reproduced with permission. All rights reserved

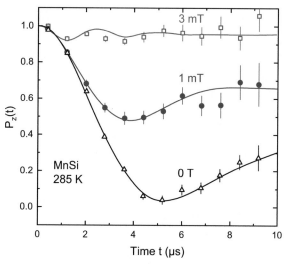

Fig. 4.30 Muon spin polarization for a squared-Lorentzian distribution of the field modulus for different longitudinal external fields, showing the decoupling at large fields. The numbers near the curves give B_{ext} in units of a/γ_μ

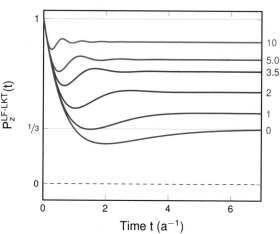

and we obtain the polarization function (Uemura et al. 1985)

$$P_z^{\text{LF-LKT}} = 1 - \frac{a}{\gamma_\mu B_{ext}} j_1(\gamma_\mu B_{ext}\, t)\, e^{-at} - \left(\frac{a}{\gamma_\mu B_{ext}}\right)^2 \left[j_0(\gamma_\mu B_{ext}\, t)\, e^{-at} - 1 \right]$$

$$- \left[1 + \left(\frac{a}{\gamma_\mu B_{ext}}\right)^2 \right] a \int_0^t j_0(\gamma_\mu B_{ext}\, t')\, e^{-at'} dt' \, , \qquad (4.35)$$

Fig. 4.31 Early time behavior of the muon spin polarization for a squared-Lorentzian distribution of the field modulus for different external fields. The numbers next to the curves give B_{ext} in units of a/γ_μ. The dashed line is the function $\exp(-4at/3)$, which is the early time evolution of the ZF P_z^{LKT} function, see Sect. 4.1.2.2, and which represents the initial trend of the polarization also in the field

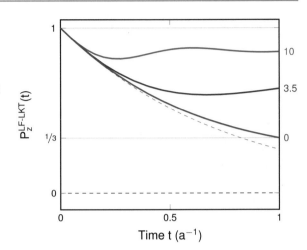

where j_0 and j_1 are spherical Bessel functions. Figures 4.30 and 4.31 show the function for different external fields and the early time behavior, respectively.

LF Decoupling   Figures 4.28 and 4.30 show that by applying a longitudinal field about 10 times stronger than the internal field (expressed either by the HWHM a/γ_μ for Lorentzian or by the standard deviation σ/γ_μ for Gaussian distributions), the muon spin depolarization of the ZF case recovers and becomes essentially time independent. What we observe is a "decoupling" of the muon spin from the static internal fields. As pictured in Fig. 4.27, when the value of the external field is considerably larger than that of the internal field, the total field probed by the muon spin has a very reduced angle with respect to the initial direction of the polarization (i.e., θ small). The non-oscillating "1/3"-component of Eq. 4.1 corresponding to the muons with spin parallel or antiparallel to the local field increases with field, whereas the oscillating "2/3"-component is reduced, while still showing a hint of a precession around the external field.

As we will see in Sect. 4.3 this decoupling effect is not effective when the internal fields are fluctuating. Therefore, LF measurements are used to distinguish between static and dynamic contributions to the depolarization of the muon spin ensemble.

Figure 4.32 shows as an example μSR spectra of an iron-based superconductor taken in the paramagnetic phase at 240 K under different longitudinal fields. The spectra can be well fitted by Lorentz-Kubo-Toyabe polarization functions. The external field decouples the ferromagnetic impurities that produce the underlying Lorentz field distribution.

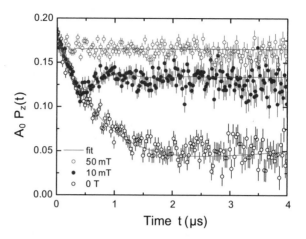

Fig. 4.32 Muon spin depolarization in the paramagnetic phase (240 K) of LaFeAsO$_{1-x}$H$_x$ for $x = 0.05$ as an example of a Lorentzian Kubo-Toyabe depolarization with longitudinal field decoupling. Modified from Lamura et al. (2014), © IOP Publishing. Reproduced with permission

4.2.2   Transverse Field

In this section we treat the case of an external field applied perpendicular to the initial muon spin polarization (TF geometry). The z direction is again defined by the external field and with the Up and Down detectors of Fig. 4.33 we measure the time evolution of the $P_x(t)$ component of the muon spin polarization.

We assume that the applied field is much larger than the internal field width $B_{ext} \gg \sqrt{\langle \Delta B_\mu^2 \rangle}$. Therefore, the total field sensed by the muons is to a good approximation along the z-axis,[20] so we can take $\theta = 90°$ and $B_\mu = B_{\mu,z}$. Equation 4.6 simplifies to

$$P_x(t) = \int f(\mathbf{B}_\mu) \cos(\gamma_\mu B_{\mu,z} t) d\mathbf{B}_\mu \ . \tag{4.36}$$

We first consider a Gaussian distribution of internal fields, with equal field variance $\langle \Delta B_\mu^2 \rangle = \sigma^2 / \gamma_\mu^2$ in the x, y, and z directions

$$f(\mathbf{B}_\mu) = f(B_{\mu,x}) f(B_{\mu,y}) f(B_{\mu,z})$$

$$= \frac{1}{\sqrt{2\pi \langle \Delta B_\mu^2 \rangle}} e^{-\frac{B_{\mu,x}^2}{2\langle \Delta B_\mu^2 \rangle}} \ \frac{1}{\sqrt{2\pi \langle \Delta B_\mu^2 \rangle}} e^{-\frac{B_{\mu,y}^2}{2\langle \Delta B_\mu^2 \rangle}} \ \frac{1}{\sqrt{2\pi \langle \Delta B_\mu^2 \rangle}} e^{-\frac{(B_{\mu,z} - B_{ext})^2}{2\langle \Delta B_\mu^2 \rangle}} \ . \tag{4.37}$$

Since the field sensed by the muons is practically along the z-axis, only the field distribution along the applied field is relevant and we can write

---

[20] This condition is met, for instance, in measurements in the paramagnetic phase or in bulk studies of the vortex state of superconductors, see Sect. 6.2.1.

Fig. 4.33 Schematic of the TF geometry

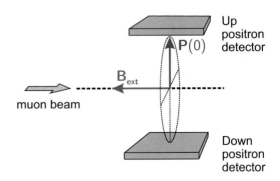

$$P_x^{\text{TF-G}}(t) = \frac{\gamma_\mu}{\sqrt{2\pi}\,\sigma} \int e^{-\frac{\gamma_\mu^2 (B_{\mu,z}-B_{\text{ext}})^2}{2\sigma^2}} \cos(\gamma_\mu B_{\mu,z}\, t) dB_{\mu,z} \tag{4.38}$$

$$\times \underbrace{\frac{\gamma_\mu}{\sqrt{2\pi}\,\sigma} \int e^{-\frac{\gamma_\mu^2 B_{\mu,x}^2}{2\sigma^2}} dB_{\mu,x}}_{1} \; \underbrace{\frac{\gamma_\mu}{\sqrt{2\pi}\,\sigma} \int e^{-\frac{\gamma_\mu^2 B_{\mu,y}^2}{2\sigma^2}} dB_{\mu,y}}_{1} \; . \tag{4.39}$$

Using a variable substitution $B_{\mu,z} - B_{\text{ext}} \to w$, $P_x^{\text{TF-G}}(t)$ can be calculated

$$
\begin{aligned}
P_x^{\text{TF-G}}(t) &= \frac{\gamma_\mu}{\sqrt{2\pi}\,\sigma} \int e^{-\frac{\gamma_\mu^2 (B_{\mu,z}-B_{\text{ext}})^2}{2\sigma^2}} \cos(\gamma_\mu B_{\mu,z}\, t) dB_{\mu,z} \\
&= \frac{\gamma_\mu}{\sqrt{2\pi}\,\sigma} \int e^{-\frac{\gamma_\mu^2 w^2}{2\sigma^2}} \cos[\gamma_\mu (w + B_{\text{ext}})t] dw \\
&= \frac{\gamma_\mu}{\sqrt{2\pi}\,\sigma} \left[\int e^{-\frac{\gamma_\mu^2 w^2}{2\sigma^2}} \cos(\gamma_\mu w\, t) dw \right] \cos(\gamma_\mu B_{\text{ext}}\, t) \\
&\quad - \frac{\gamma_\mu}{\sqrt{2\pi}\,\sigma} \left[\int e^{-\frac{\gamma_\mu^2 w^2}{2\sigma^2}} \sin(\gamma_\mu w\, t) dw \right] \sin(\gamma_\mu B_{\text{ext}}\, t) \\
&= \frac{\gamma_\mu}{\sqrt{2\pi}\,\sigma} \left[\int e^{-\frac{\gamma_\mu^2 w^2}{2\sigma^2}} \cos(\gamma_\mu w\, t) dw \right] \cos(\gamma_\mu B_{\text{ext}}\, t) \; . \tag{4.40}
\end{aligned}
$$

Note that this expression for high transverse fields is quite general and does not depend on the exact form of the field distribution, the last step only requiring that it is symmetric.

The depolarization term of $P_x^{\text{TF-G}}(t)$ (i.e., the part in the square brackets of Eq. 4.40) is the cosine Fourier transform of the field distribution along the direction of the externally applied magnetic field.

The time evolution of the polarization in TF for a Gaussian internal field broadening becomes

$$P_x^{\text{TF-G}}(t) = \cos(\gamma_\mu \, B_{\text{ext}} \, t) \, \exp\left[-\frac{1}{2}\gamma_\mu^2 \langle \Delta B_{\mu,z}^2 \rangle t^2\right]$$

$$= \cos(\gamma_\mu \, B_{\text{ext}} \, t) \, \exp\left[-\frac{\sigma^2 t^2}{2}\right] \, . \tag{4.41}$$

Here, differently from Eq. 4.14, $\theta \simeq 90°$ so that the full initial muon polarization precesses. The analytical expression for the TF setup has been obtained under the assumption $B_{\text{ext}} \gg \sqrt{\langle \Delta B_\mu^2 \rangle}$. Comparisons with numerical solutions of Eq. 4.6 show that for $B_{\text{ext}} = 5\sqrt{\langle \Delta B_\mu^2 \rangle}$ the expression is still a fair approximation (Dalmas de Réotier & Yaouanc 1992).

Similarly to the Gaussian case, for a Lorentzian distribution of fields one obtains

$$P_x^{\text{TF-L}}(t) = \cos(\gamma_\mu \, B_{\text{ext}} \, t) \, \exp(-at) \, , \tag{4.42}$$

where a/γ_μ is the HWHM of the Lorentzian distribution along the z-axis.

An important point to remember, which we will use later, is that the depolarization rate obtained from a TF measurement is a direct measure of the width of the field distribution along the direction of the applied field.[21] Therefore, in the absence of dynamics, a TF experiment allows to determine the distribution of the field component parallel to the external field. In Eqs. 4.41 and 4.42, we have assumed that the average value of the field seen by the muon is the transverse external field $\langle B_\mu \rangle = B_{\text{ext}}$. In reality, depending on the system and the physics involved, different contributions can determine this field and modify it from B_{ext}. An example is the so-called the Knight-shift, which slightly shifts the field sensed by the muon ensemble through the polarization of the conduction electrons (producing a contact hyperfine contribution) and of possible local electronic moments (creating a dipolar field at the muon site), see Sect. 5.6. Another example are measurements of the vortex state of a type-II superconductor, where B_{ext} is replaced by the average field generated by the vortex-lattice, see Sect. 6.2.2. Moreover, in paramagnetic and diamagnetic compounds, the shape of the sample may also play a role through the demagnetization factor, see Appendix D.

These effects are taken into account by replacing in Eq. 4.41 B_{ext} with $\langle B_\mu \rangle$. Hence, a TF measurement yields information about the first and second central moment (or variance) of the underlying microscopic field distribution. This consideration is general for symmetric field distributions and applies by analogy also to

---

[21] This is a general property. If, on top of a field distribution, a well-defined, sufficiently strong field, not parallel to the initial polarization is present, e.g., in a magnetic state in ZF or in TF measurements, only the variance along this field direction contributes to the muon spin depolarization.

the Lorentzian case, Eq. 4.42, where, instead of the undefined variance, information about the HWHM is obtained.

4.2.3 Some Special Polarization Functions

In the previous sections, the polarization function have been obtained classically by calculating the time evolution of the muon spin polarization in an effective well-defined field distribution that may be generated by host spins, such as the nuclear spins.

A dense matrix of nuclear spins creates a Gaussian field distribution and the time evolution in ZF is given by the Kubo-Toyabe expression. This approach describes the time evolution of the muon spin by the Larmor torque equation, Eq. 1.26, but neglects the nuclear spin dynamics. For an exact treatment one should calculate quantum mechanically the time evolution of the entire spin system, consisting of muon and host spins, which are coupled by the magnetic dipolar interaction (see e.g., Sect. 5.7) (Celio & Meier 1983; Meier 1984; Celio 1986).

The muon spin polarization can be determined in the density matrix formalism by solving (see Appendix F and Eq. F.13)

$$\mathbf{P}(t) = \mathrm{Tr}\left(\rho \exp\left(\frac{i}{\hbar}\mathcal{H}t \right) \sigma \exp\left(-\frac{i}{\hbar}\mathcal{H}t \right) \right) \ , \tag{4.43}$$

where σ are the Pauli matrices, Eq. A.17, ρ is the density matrix of the whole spin system and $\mathcal{H}$ is the interaction Hamiltonian.

Because of the size of the Hilbert space, the problem can generally only be handled numerically. If the number of spins involved is N_K the dimension of the Hilbert space for a nuclear spin species I_K is $2 \times (2I_K + 1)^{N_K}$. If N_K is large, the Kubo-Toyabe expression is a good approximation.[22]

When the muon is close to one or two spins and the interaction with the residual spins can be neglected, an analytical solution of Eq. 4.43 can be obtained. In this case the Kubo-Toyabe function is not appropriate because it does not describe the dominant character of the interaction with the few spins.

The solution for a muon spin interacting with a single spin in ZF can be found in Vogel et al. (1986). Experimentally, the predicted oscillations in the polarization due to the μ^+-H nuclear dipole interaction have been observed in hydrogen-bond systems (Nishiyama et al. 2001).

A system formed by a muon spin and two ^{19}F nuclear spins (spin $I_F = 1/2$, abundance 100%), so-called F-μ-F center, with the muon forming a hydrogen bond between two F$^-$ ions, was originally found in alkali fluorides (Brewer et al. 1986).[23]

---

[22] Calculations including the interaction with the six nearest neighbor Cu nuclear spins (spin 2/3) show deviations from the KT behavior. For instance, in ZF, the long time behavior of the polarization slightly decays instead of displaying the typical asymptotic 1/3 tail (Celio 1986).

[23] The localization of the muon close to the F$^-$ ion is a consequence of the large electronegativity of fluorine.

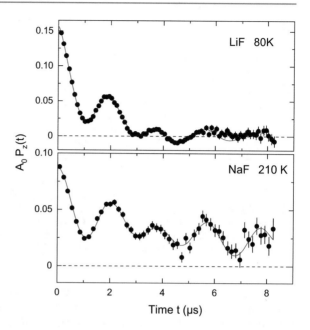

Fig. 4.34 ZF-μSR asymmetry spectra for LiF and NaF crystals. The initial muon polarization is parallel to a crystalline $\langle 100 \rangle$ axis. The solid line is a fit to Eq. 4.45 multiplied by a phenomenological stretched exponential to take into account the relaxation. Modified from Brewer et al. (1986), © American Physical Society. Reproduced with permission

In a simple model, the ZF polarization $P_{F\mu F}(t)$ can be calculated quantum mechanically, assuming a static collinear geometry, with the muon at the center of the line connecting the F nuclei chosen as the quantization axis, and considering only the μ-F dipole interaction.[24]

The Hamiltonian is in this case

$$\mathcal{H} = \hbar^2 \gamma_\mu \gamma_F \frac{\mu_0}{4\pi |\mathbf{r}_i|^3} \sum_{i=1,2} \mathbf{I}_{F,i} \cdot \mathbf{I}_\mu - 3(\mathbf{I}_{F,i} \cdot \hat{\mathbf{r}}_i)(\mathbf{I}_\mu \cdot \hat{\mathbf{r}}_i) \, , \qquad (4.44)$$

where γ_F is the fluorine gyromagnetic ratio and $\mathbf{r}_i$ is the vector connecting the ith fluorine spin $\mathbf{I}_{F,i}$ with the muon. The data taken in crystal samples, Fig. 4.34, can be fitted fairly well by the solution averaged over equivalent directions in the cubic crystal (Brewer et al. 1986)

$$P_{F\mu F}(t) = \frac{1}{6} \left[3 + \cos(\sqrt{3}\,\omega_d t) + (1 - \frac{1}{\sqrt{3}}) \cos(\frac{3-\sqrt{3}}{2}\omega_d t) \right.$$
$$\left. + (1 + \frac{1}{\sqrt{3}}) \cos(\frac{3+\sqrt{3}}{2}\omega_d t) \right] \, , \qquad (4.45)$$

where $\hbar\omega_d = \hbar^2 \mu_0 \gamma_\mu \gamma_F / (4\pi r^3)$ is the dipolar interaction energy, and $r = 0.117(6)$ nm the μ-F distance determined from the experiment.

---

[24] The F-F interaction is smaller by a factor of $\gamma_F / (8\gamma_\mu) = 0.037$.

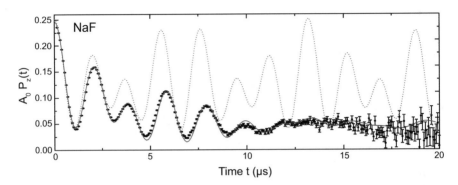

Fig. 4.35 μSR asymmetry data for NaF at 100 K. The dotted line shows the theoretical muon polarization of the F-μ-F system, see Eq. 4.45. The solid red line includes the decoherence effect due to the coupling with the environment. Modified from Wilkinson and Blundell (2020), © American Physical Society. Reproduced with permission

The typical oscillations have been observed in numerous fluorine-containing compounds. In fluorinated molecular magnets, the observation of the oscillations related to the entangled states of the muon and F nuclear spins has been used for an accurate determination of the muon sites in these novel magnetic materials (Lancaster et al. 2007). Electronic structure calculations by the density functional theory (DFT)[25] combined with experimental data have also shown to be able to predict site and shape of the F-μ-F complex and to assess the perturbation of the host lattice caused by the muon (Bernardini et al. 2013; Möller et al. 2013).

The time evolution of the three-spin system F-μ-F described by the Hamiltonian 4.44 is unitary and conserves coherence. The theoretical expression 4.45 reflects the coherent exchange of spin polarization between the muon, which is initially fully polarized, and the fluorine nuclei, which are initially unpolarized. However, the experimental data, Fig. 4.34, show a relaxing signal, or loss of coherence, which has been accounted for phenomenologically by a stretched exponential.

A treatment including the known coupling between the muon spin and more distant F or other nuclear spins is able to reproduce the relaxation of the coherent precession signal, without resorting to phenomenological multiplicative functions (Wilkinson & Blundell 2020), see Fig. 4.35. This result shows that F-μ-F and similar few spin arrangements also represent a model system to study the loss of coherence of a quantum system because of its interaction with the environment, a fundamental issue for the operation of quantum computers (McArdle 2021).

---

[25] See footnote 18, Chap. 5.

4.3   Dynamical Effects

So far we have limited the discussion to cases where the field probed by the muon ensemble is static (at least during the μSR observation time window). This is the usual situation, for instance, in a magnetic phase at low temperature where the magnetic moments are static, or when probing the superconducting state at low temperatures.

In this section we discuss the effect of dynamical fields on the muon spin polarization.[26] We will see that the time evolution of the polarization gives information about the dynamics of the local fields, and that parameters such as fluctuation (or correlation) times and rates can be determined from the depolarization of the μSR signal.

We assume that the dynamical processes are stationary stochastic processes, i.e., that their properties do not vary with the origin of time (i.e., they depend only on time differences), and that they are random. The relevant random variable in our case is generally the local field sensed by the muon $\mathbf{B}_\mu$, which is a consequence of fluctuating spins, local currents or of the muon diffusing in the lattice. Random means that $\mathbf{B}_\mu$ does not depend in a well-defined way on the time t as in a casual process (Wang & Uhlenbeck 1945). An important class of these stochastic processes are the so-called Markovian processes. The Markov property implies that once the state of the dynamical process is known at some time t, the probability law of the future state change of the process depends only on the present state of the system, independently of its history up to that time, i.e., the process is memoryless.

4.3.1   The Strong Collision Approximation

A mathematical description of stochastic processes and their effect on the muon spin polarization can be obtained within the strong collision approximation model (SCA) (Hayano et al. 1979). In this model, the field dynamics is described by a memoryless, Markovian process in which the local field $\mathbf{B}_\mu$ suddenly changes its orientation and magnitude with a fluctuation rate ν. The random fluctuation process proceeds in steps and is sometimes also described in terms of collisions or jumps, where ν is the collision or jump probability per unit time. In this formulation the local field keeps a value for some time, then, after a fluctuation occurring on an average time of $\tau_c = 1/\nu$, the local field suddenly takes a new value with probability defined by

---

[26] Dynamical fields can be due to real fluctuations of the internal field (e.g., caused by fluctuating electronic moments), or to muon diffusion, which can occur at high temperature. In this case, the muon experiences, for example, the field produced by nuclear moments as a time-varying field due to the diffusion from site to site after implantation.

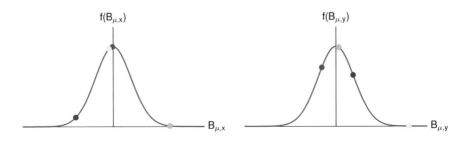

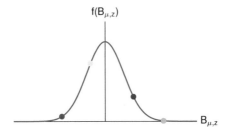

Fig. 4.36 Schematic of the strong collision model. After each fluctuation (represented by the different colored points), the field probed by the muon is randomly chosen from the snapshot field distribution (here assumed to be Gaussian)

an instantaneous ("snapshot") field probability distribution $f(\mathbf{B}_\mu)$.[27] The new value has no correlation with the field before the fluctuation (Fig. 4.36).

The number of fluctuations and the relative probability at time t can be constructed from the following diagram, where each row represents a time step dt

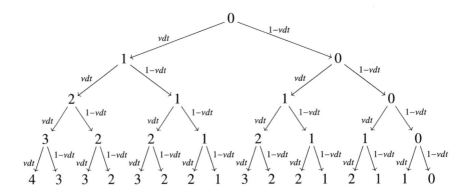

---

[27] The term strong collision expresses the fact that the equilibrium distribution $f(\mathbf{B}_\mu)$ is reached immediately after a field change or collision. An alternative treatment, based on the assumption that the change of $\mathbf{B}_\mu$ is not sudden but gradual, is sometimes named weak collision model (Kubo 1954).

We first consider the fractions of muons that have experienced a well-defined number of fluctuations until the time $t = n\, dt$:

- The fraction that has not experienced any fluctuations is given by

$$p_0(t) = (1 - v\, dt)(1 - v\, dt)(1 - v\, dt)... = (1 - v\, dt)^n$$

$$= (1 - \frac{v\, t}{n})^n \ ,$$

  hence, for $n \to \infty$

$$p_0(t) = e^{-vt} \ , \tag{4.46}$$

  which is the probability that $\mathbf{B}_\mu$ has not changed until the time t.
- The fraction that experienced exactly one fluctuation, is given by the probability for a muon to experience zero fluctuations until a time t_1, times the probability to have one fluctuation at that time, times the probability to have no additional fluctuations until the time t. The time t_1 can be in the interval $[0, t]$ corresponding to different paths of the previous diagram

$$p_1(t) = \int_0^t \underbrace{e^{-vt_1}}_{\text{0 fluct. in } [0, t_1]} \times \overbrace{v\, dt_1}^{\text{fluct. prob. between } t_1 \text{ and } t_1 + dt_1} \times \underbrace{e^{-v(t-t_1)}}_{\text{0 fluct. in } [t_1, t]}$$

$$= e^{-vt} v \int_0^t dt_1$$

$$p_1(t) = e^{-vt} v t \ . \tag{4.47}$$

- For two fluctuations, the fraction is the probability for a muon to experience a fluctuation until a time t_2, which is $p_1(t_2)$, times the probability of having a (second) fluctuation at that time, times the probability of having no additional fluctuation up to the time t, where the time t_2 can vary from 0 to t. Hence

$$p_2(t) = \int_0^t \underbrace{p_1(t_2)}_{\text{1 fluct. in } [0, t_2]} \times \overbrace{v\, dt_2}^{\text{fluct. prob. between } t_2 \text{ and } t_2 + dt_2} \times \underbrace{e^{-v(t-t_2)}}_{\text{0 fluct. in } [t_2, t]}$$

$$= e^{-vt} v^2 \int_0^t t_2\, dt_2$$

$$p_2(t) = e^{-vt} \frac{v^2 t^2}{2} \ . \tag{4.48}$$

- Generalizing for n fluctuations

$$p_n(t) = \int_0^t p_{n-1}(t_n)\, v\, dt_n\, e^{-v\,(t-t_n)}$$

$$p_n(t) = e^{-vt}\frac{v^n\, t^n}{n!}\quad . \tag{4.49}$$

- The sum of all possible fluctuation events is one, as expected

$$p(t) = \sum_{n=0}^{\infty} e^{-vt}\frac{v^n\, t^n}{n!} = e^{-vt}\sum_{n=0}^{\infty}\frac{v^n\, t^n}{n!} = e^{-vt}e^{+vt}$$

$$p(t) = 1\quad . \tag{4.50}$$

4.3.1.1 The Muon Spin Polarization

As an example, we outline the strong collision model to calculate the spin polarization for the case where the underlying snapshot static field distribution is a Gaussian along each Cartesian direction, Eq. 4.15. In this case, the random process is not only Markovian but also Gaussian.[28] Without dynamics ($v = 0$) such a distribution in zero field leads to the Kubo-Toyabe polarization function $P_z^{\mathrm{GKT}}(t)$, Eq. 4.19.

In the dynamical case, the muon spin polarization at time t, $P_{z,\,\mathrm{dyn}}^{\mathrm{GKT}}(t, v)$, is the sum of the subensemble contributions $g_n(t)$, $n = 0, 1, 2, \ldots$, where $g_n(t)$ is the polarization function of those muons that have experienced exactly n fluctuations in the interval $[0, t]$

$$P_{z,\,\mathrm{dyn}}^{\mathrm{GKT}}(t, v) = \sum_{n=0}^{\infty} g_n(t)\quad . \tag{4.51}$$

- $n = 0$ corresponds to the situation without fluctuation. The polarization function at time t is given by the static GKT function, Eq. 4.19, times the probability that a muon will not experience any fluctuation until that time

$$g_0(t) = e^{-vt}\, P_z^{\mathrm{GKT}}(t)\quad . \tag{4.52}$$

- $n = 1$ corresponds to the case, where just one fluctuation up to time t takes place. Let us assume that this occurs at time $0 \le t_1 \le t$. The polarization function will consist of a GKT evolution until time t_1 and a new GKT function starting at time t_1, but with the same σ (note that the snapshot field distribution before and after

---

[28] A Gaussian random process is characterized by the property that all basic probability distributions of the random variables are Gaussian (Wang & Uhlenbeck 1945).

the fluctuation is the same). Since t_1 can be any time between 0 and t, we have to integrate over all possible t_1

$$g_1(t) = \int_0^t \underbrace{P_z^{\mathrm{GKT}}(t_1)e^{-\nu t_1}}_{0 \text{ fluct. in } [0,t_1]} \times \overbrace{\nu\, dt_1}^{\text{fluct. prob. between } t_1 \text{ and } t_1+dt_1} \times \underbrace{P_z^{\mathrm{GKT}}(t-t_1)e^{-\nu(t-t_1)}}_{0 \text{ fluct. in } [t_1, t]}$$

$$g_1(t) = \nu \int_0^t g_0(t_1)\, g_0(t-t_1)\, dt_1 = e^{-\nu t}\nu \int_0^t P_z^{\mathrm{GKT}}(t_1)\, P_z^{\mathrm{GKT}}(t-t_1)\, dt_1 \quad . \tag{4.53}$$

- For $n = 2$, the time evolution of the polarization consists of the just calculated function $g_1(t)$ corresponding to one fluctuation until time t_2, and a new GKT function with the same depolarization rate but starting at time t_2, again to be integrated over t_2 in the interval $[0, t]$

$$g_2(t) = \int_0^t \underbrace{g_1(t_2)}_{1 \text{ fluct. in } [0,t_2]} \times \overbrace{\nu\, dt_2}^{\text{fluct. prob. between } t_2 \text{ and } t_2+dt_2} \times \underbrace{P_z^{\mathrm{GKT}}(t-t_2)e^{-\nu(t-t_2)}}_{0 \text{ fluct. in } [t_2, t]}$$

$$g_2(t) = \nu \int_0^t g_1(t_2)\, g_0(t-t_2)\, dt_2 \quad . \tag{4.54}$$

The functions $g_n(t)$ obey the recursive relation

$$g_n(t) = \nu \int_0^t g_{n-1}(t_n)\, g_0(t-t_n)\, dt_n \quad , \tag{4.55}$$

and we can therefore write[29]

---

[29] We have worked out here an example with the static polarization function $P_z^{\mathrm{GKT}}(t)$ for a Gaussian field distribution. Note that the strong collision model can be used to introduce the dynamics into any type of static polarization function or corresponding snapshot field distribution. In such a case, the GKT function above must be replaced by the corresponding static polarization function $P_z^{\mathrm{stat}}(t)$, and one obtains

$$P_{z,\,\mathrm{dyn}}(t, \nu) = e^{-\nu t}\left[P_z^{\mathrm{stat}}(t) + \nu \int_0^t P_z^{\mathrm{stat}}(t_1)\, P_z^{\mathrm{stat}}(t-t_1)\, dt_1 \right.$$

$$\left. + \nu^2 \int_0^t \int_0^{t_2} P_z^{\mathrm{stat}}(t_1)\, P_z^{\mathrm{stat}}(t_2-t_1)\, P_z^{\mathrm{stat}}(t-t_2)\, dt_1\, dt_2 + \cdots \right] , \tag{4.56}$$

which can be written as an integral equation

$$P_{z,\,\mathrm{dyn}}(t, \nu) = e^{-\nu t} P_z^{\mathrm{stat}}(t) + \nu \int_0^t P_{z,\,\mathrm{dyn}}(t-t', \nu)\, P_z^{\mathrm{stat}}(t')e^{-\nu t'}\, dt' \quad . \tag{4.57}$$

$$P_{z,\,\text{dyn}}^{\text{GKT}}(t, \nu) = \sum_{n=0}^{\infty} g_n(t)$$

$$= e^{-\nu t} P_z^{\text{GKT}}(t) + \nu \int_0^t g_0(t_1)\, g_0(t - t_1)\, dt_1$$

$$+ \nu \int_0^t g_1(t_2)\, g_0(t - t_2)\, dt_2 + \cdots$$

$$+ \nu \int_0^t g_{n-1}(t_n)\, g_0(t - t_n)\, dt_n + \cdots \qquad \text{or explicitely}$$

$$= e^{-\nu t}\Bigg[P_z^{\text{GKT}}(t) + \nu \int_0^t P_z^{\text{GKT}}(t_1)\, P_z^{\text{GKT}}(t - t_1)\, dt_1 +$$

$$+ \nu^2 \int_0^t \int_0^{t_2} P_z^{\text{GKT}}(t_1) P_z^{\text{GKT}}(t_2 - t_1) P_z^{\text{GKT}}(t - t_2)\, dt_1\, dt_2 + \cdots \Bigg].$$

$$(4.58)$$

which can be written as an integral equation. Figure 4.37 shows the polarization functions and how they are constructed from the subensemble functions $g_n(t)$, $n = 0, 1, \cdots$, for $\nu/\sigma = 0.1$, slow fluctuations, $\nu/\sigma = 1$, and $\nu/\sigma = 10$, fast fluctuations. Expressing the time in units of σ^{-1} the relevant parameter determining the behavior of $P_{z,\,\text{dyn}}^{\text{GKT}}(t, \nu)$ is ν/σ.

Analytical solutions can be found using the Laplace transform: the Laplace transform of $P_{z,\text{dyn}}(t, \nu)$ can be expressed in terms of the Laplace transform of the static depolarization function (Hayano et al. 1979)

$$F_z(s) = \int_0^\infty P_{z,\text{dyn}}(t, \nu)\, e^{-st}\, dt = \frac{f_z(s + \nu)}{1 - \nu f_z(s + \nu)}$$

with

$$f_z(s) = \int_0^\infty P_z^{\text{stat}}(t)\, e^{-st}\, dt \ . \qquad\qquad (4.59)$$

Depolarization Function for Some Limiting Cases

- For sufficiently fast fluctuations (i.e., in the case of a Gaussian static distribution, when $\nu \gtrsim 3\sigma$) the depolarization can be approximated by[30] (Dalmas de Réotier

---

[30] This relaxation function is equivalent to the Abragam relaxation function for the transverse field configuration, Eq. 4.70, apart from a factor of two in the exponent. See discussion in Sect. 4.3.1.3.

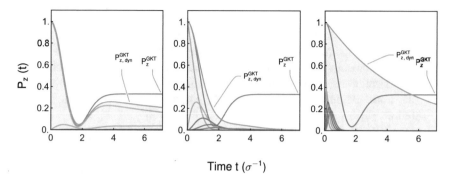

Time t (σ^{-1})

Fig. 4.37 Time evolution of the dynamical polarization function $P_{z,\,\mathrm{dyn}}^{\mathrm{GKT}}(t, \sigma, \nu)$ (dark blue) for $\nu/\sigma = 0.1, 1$ and 10 (from left to right). The time is given in units of $1/\sigma$. For comparison the static GKT function, $P_z^{\mathrm{GKT}}(t, \sigma)$, is shown (blue), as well as the different contributions of the subensemble functions g_0 (orange), g_1 (green), g_2 (red), g_3 (magenta), g_4 (brown). Their sum (for $n \to \infty$) yields $P_{z,\,\mathrm{dyn}}^{\mathrm{GKT}}$, Eq. 4.58. Note that for $\nu/\sigma = 0.1$ basically only the tail differs from the static case, Eq. 4.62, and that $g_0(t)$ mostly contributes. For $\nu/\sigma = 1$ in the displayed time interval $T = t/\sigma = 7$ and the functions $g_n(t)$ with n up to $\sim \nu T = 7$ contribute. For $\nu/\sigma = 10$, since $\nu T = 70$, many $g_n(t)$ are necessary to reproduce the full dynamics

& Yaouanc 1992)

$$P_{z,\,\mathrm{dyn}}^{\mathrm{GKT}}(t, \sigma, \nu) \simeq \exp\left[-2\frac{\sigma^2}{\nu^2}\left[\exp(-\nu t) - 1 + \nu t\right]\right]. \qquad (4.60)$$

In the limit of very fast fluctuations (so-called extreme motional narrowing limit)[31] Eq. 4.60 simplifies to

$$P_{z,\,\mathrm{dyn}}^{\mathrm{GKT}}(t, \sigma, \nu) \simeq \exp\left(-\frac{2\sigma^2}{\nu}t\right). \qquad (4.61)$$

Note that in the fast fluctuation limit the depolarization function is simply exponential and that the depolarization rate depends on the fluctuation rate ν and on the field strength, expressed by $\sigma = \sqrt{\gamma_\mu^2 \langle \Delta B_\mu^2 \rangle}$.

This equation explains the fact that in the paramagnetic regime there is generally no sign of depolarization due to electronic moments in the μSR signal. This behavior can be understood by considering that in solids the fluctuation rate of electronic moments in the paramagnetic state is typically

---

[31] The term narrowing limit comes from NMR, where the resonance lines narrow in the presence of fast atomic motion.

$\sim \nu = 10^{12} - 10^{16}\ \text{s}^{-1}$. To estimate σ we start by taking a typical depolarization rate for nuclear moments $\sigma_N \simeq 0.3\ \mu\text{s}^{-1}$. Taking into account that the magnetic moment of the electron is about 1000 times larger than a typical nuclear moment and assuming a fluctuation rate of $10^{14}\ \text{s}^{-1}$, one obtains a depolarization rate due to fluctuating moments of $2\sigma^2/\nu \simeq 1.8 \times 10^{-3}\ \mu\text{s}^{-1}$. This translates into a negligible loss of polarization of $\sim 0.4\%$ in a muon lifetime.

• In the quasi-static limit ($\nu \ll \sigma$) one obtains

$$P_{z,\,\text{dyn}}^{\text{GKT}}(t, \sigma, \nu) \simeq \frac{1}{3}\exp\left(-\frac{2}{3}\nu t\right) + \frac{2}{3}\left(1 - \sigma^2 t^2\right)\exp\left(-\frac{\sigma^2 t^2}{2}\right)\ , \qquad (4.62)$$

which shows, that for slow fluctuations only the 1/3 tail is affected by the fluctuations. In addition, unlike the case of fast fluctuations, the relaxation rate does not depend on the field strength, which governs σ. This fact can be used to study slow dynamics in the 10 μs wide observation window even in systems where the local field is large.[32] The 2/3 factor in front of ν reflects the fact that only the transverse components of $\mathbf{B}_\mu$ depolarize. Fluctuations along the z-axis (initial direction of the polarization in ZF geometry) do not lead to a depolarization.

4.3.1.2 Dynamical Effects for Gaussian Distributions in a Longitudinal Field

We now consider the depolarization function of a dynamical system with an internal isotropic Gaussian field distribution when a longitudinal field is applied. The corresponding static depolarization function is given by $P_z^{\text{LF-GKT}}$, Eq. 4.33, and has been discussed in Sect. 4.2.1. The task of dynamizing the function $P_z^{\text{LF-GKT}}$ using the strong collision model approximation, Sect. 4.3.1, can only be solved numerically.

Using perturbation theory, an approximate analytical solution for $P_z^{\text{LF-GKT}}$ has been provided by Keren (1994), which can be seen as a generalization of the TF Abragam relaxation function (see Sect. 4.3.1.3) to a longitudinal field. It reads

$$P_{z,\,\text{dyn}}^{\text{LF-GKT}}(t, \sigma, \nu, \omega_\mu) = \exp\left[-\Gamma(t)\,t\right]\ , \qquad (4.63)$$

---

[32] Some caution is in order. As discussed in Sect. 4.2.3, the Kubo-Toyabe expression assumes that the nuclear moments producing the Gaussian field distribution are static and do not couple to the muon moment. Classical and quantum mechanical treatments of the nuclear spin dynamics have shown that deviations from a flat 1/3 tail can occur even in the absence of dynamics, (Dalmas de Réotier et al. 1992; Celio 1986). See also footnote 22.

with

$$\Gamma(t)\,t = 2\sigma^2 \frac{\left\{\left(\omega_\mu^2 + v^2\right) vt + \left(\omega_\mu^2 - v^2\right)\left[1 - e^{-vt}\cos(\omega_\mu t)\right] - 2v\omega_\mu e^{-vt}\sin(\omega_\mu t)\right\}}{\left(\omega_\mu^2 + v^2\right)^2}$$

(4.64)

where $\omega_\mu = \gamma_\mu B_{\text{ext}}$.

In limiting cases it simplifies as follows:

- In zero applied field ($\omega_\mu = 0$) the depolarization function becomes

$$P_{z,\,\text{dyn}}^{\text{LF-GKT}}(t, \sigma, v, \omega_\mu = 0) = P_{z,\,\text{dyn}}^{\text{GKT}}(t, \sigma, v)$$

$$= \exp\left[-\frac{2\sigma^2}{v^2}\left[\exp(-vt) - 1 + vt\right]\right],$$

(4.65)

which reproduces Eq. 4.60.

- In the fast fluctuation (or extreme narrowing) limit, where $v \gg 0.1\mu s^{-1}$, one finds

$$P_{z,\,\text{dyn}}^{\text{LF-GKT}}(t, \sigma, v, \omega_\mu) = \exp\left[-\frac{2\sigma^2 v}{\omega_\mu^2 + v^2}t\right].$$

(4.66)

Again, in the fast fluctuation limit the depolarization has a simple exponential behavior. Moreover, setting the external field to zero one recovers the ZF limit of fast fluctuations Eq. 4.61, previously obtained within the strong collision model.

Figure 4.38 compares curves of analytical solutions with numerical SCA and Monte Carlo solutions (Keren 1994). The Monte Carlo solutions have been obtained by direct integration of Eq. 1.26 with a total field composed of a randomly varying internal field chosen from a Gaussian distribution and with exponential autocorrelation function, see footnote 33, and an applied longitudinal field $B_{\text{ext}} \parallel \hat{z}$. The final Monte Carlo polarization function is obtained by averaging over an ensemble of muons.

Inspection of Fig. 4.38 shows that the three approaches give very similar results. While the analytical result better describes the Monte Carlo simulation in high longitudinal fields, the dynamic Kubo-Toyabe SCA solution is closer to the simulation in low fields. In general, Eq. 4.63 is a good approximation of the strong collision approach when the dynamics is relatively fast.

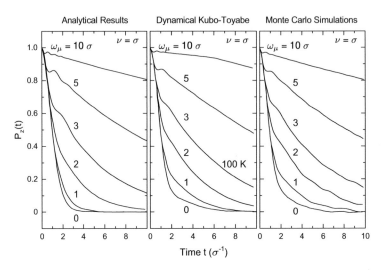

Fig. 4.38 Comparison of the dynamic muon spin polarization in a longitudinal field obtained with three models, for $\nu = \sigma$. Left panel: Analytical result, Eq. 4.63. Middle panel: Dynamical Kubo-Toyabe model with internal Gaussian field distribution, calculated with the SCA. Right panel: Monte Carlo simulations. The longitudinal field is given by ω_μ in units of σ. Modified from Keren (1994), © American Physical Society. Reproduced with permission. All rights reserved

4.3.1.3 Dynamical Effects for Gaussian Distributions in a Transverse Field

An analytical expression (often named Abragam expression) is available in this case. It was originally derived for NMR, where an external field is generally applied to study the broadening or narrowing effects of the magnetic resonance spectral lines as a consequence of the exchange interaction of electronic moments or of motions of the nuclear spins, e.g., in a liquid (Anderson 1954; Abragam 1961). Anderson called this model "random frequency modulation model". It can be applied to treat the effects of field (or Larmor frequency) fluctuations in a TF geometry.

One makes the same high-field assumption as for the static case, so that only the z component of the fluctuating field has to be considered.

The muon spin precesses in a total field $B_{\mu,z}(t) = B_{\text{ext}} + \delta B_{\mu,z}(t)$, where $\delta B_{\mu,z}(t)$ is a small perturbation of the external field changing in a random way. To determine the time evolution of the polarization $P_{x,\,\text{dyn}}^{\text{TF-G}}(t)$, the statistical average of the precession in the transverse field has to be computed

$$P_{x,\,\text{dyn}}^{\text{TF-G}}(t) = \langle \cos \int_0^t \gamma_\mu B_{\mu,z}(t')dt' \rangle \ . \tag{4.67}$$

Obviously, in the absence of fluctuations we find the static result $P_x(t) = \cos(\gamma_\mu B_{\text{ext}} t)$. Assuming that $\delta B_{\mu,z}(t)$ is a Gaussian random function, the integral can be written in terms of the field autocorrelation function

$$\langle \delta B_{\mu,z}(t') \delta B_{\mu,z}(t'+t) \rangle \ , \tag{4.68}$$

which for a Gaussian-Markovian stationary process takes the simple exponential form[33]

$$\langle B_\alpha(t') B_\alpha(t'+t) \rangle = \langle B_{\mu,z}(0) B_{\mu,z}(t) \rangle = \langle B_{\mu,z}(0)^2 \rangle \exp(-t/\tau_c) \ , \tag{4.69}$$

Equation 4.69 can be calculated explicitly taking into account the random Gaussian and stationary character of the fluctuations, see Appendix C for details. One obtains the so-called Abragam relaxation function

$$P_{z,\,\mathrm{dyn}}^{\mathrm{TF\text{-}G}}(t, \sigma, \nu, \gamma_\mu B_{\mathrm{ext}}) = \exp\left[-\frac{\sigma^2}{\nu^2} \left[\exp(-\nu t) - 1 + \nu t \right] \right] \cos(\gamma_\mu B_{\mathrm{ext}}\, t) \ , \tag{4.70}$$

where $\sigma = \sqrt{\gamma_\mu^2 \langle \Delta B_{\mu,z}(0)^2 \rangle}$ and the fluctuation rate of the process is taken as $1/\tau_c = \nu$. In the limit of very slow fluctuations ($\nu \to 0$) the function correctly reduces to the static Gaussian depolarization function, Eq. 4.41. In the fast dynamic limit ($\nu \to \infty$), an exponential decay of the polarization with rate $\lambda = \sigma^2/\nu$ is obtained. Comparing this relaxation rate with the SCA expression for very fast fluctuations in the zero, Eq. 4.61, and in the longitudinal field case, Eq. 4.66, we notice that there is a factor of two missing here.[34] In reality the fast fluctuating TF expression Eq. 4.70 is incomplete, since it does not take into account fluctuations transverse to the polarization vector. It has to be complemented by additional terms, which produce spin flips and hence relaxation, see Sect. 4.4.1.

Figure 4.39 shows the behavior of the envelope of the Abragam relaxation function in TF, Eq. 4.70, for different values of the fluctuation rate ν. In contrast to ZF and LF measurements, one notices that it is difficult to extract information about slow fluctuations ($\nu/\sigma \lesssim 1$) from TF measurements.

---

[33] For Gaussian-Markovian processes, it can be shown that the field-autocorrelation function (i.e., the correlation between field values taken at different times) can be written by the Doob theorem (Doob 1942) as

$$\langle \delta B_\alpha(t') \delta B_\alpha(t'+t) \rangle = \langle \delta B_\alpha(0) \delta B_\alpha(t) \rangle = \langle \delta B_\alpha(0)^2 \rangle \exp(-t/\tau_c) \ ,$$

where $\alpha = x, y, z$. The average is over the ensemble, and τ_c, called here correlation time, is the typical time characterizing how fast a well-defined configurations of fields (or of spins producing these fields) decays, i.e., how fast the correlation disappears.

[34] A factor of two is also missing when comparing Eq. 4.70 with Eqs. 4.60 and 4.65.

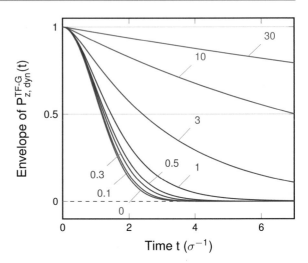

Fig. 4.39 Envelope of the dynamic muon spin polarization for a Gaussian distribution of internal fields in a transverse field, Eq. 4.70, for different fluctuation rates ν expressed as ν/σ and $\sigma = 1\,\mu s^{-1}$

4.3.1.4 Dynamical Effects for Lorentzian Fields

If we apply the SCA procedure to introduce dynamical effects for squared-Lorentzian field distributions, Eqs. 4.22 and 4.24, i.e., we simply insert the ZF function, Eq. 4.25, into Eq. 4.58, we obtain the unphysical result that the initial decay rate of the polarization function is independent of the fluctuation rate ν/a: it is the same for the static $\nu/a = 0$ and the extreme dynamical case $\nu/a \to \infty$ (Uemura 1999); no motional narrowing appears.

This procedure corresponds to the following order of operations

spatial averaging of field "subdistributions" $\to$ static Lorentzian distribution

$\to$ dynamics via SCA.

This is not correct because it corresponds to a mean-field approach, i.e., it averages over inhomogeneities or disorder before considering the effects of dynamics and neglects variations in the spatial cluster configurations with their own dynamics. The Lorentzian field distribution makes a spatial averaging of "subdistributions". However, each of them has its own dynamics. For the physical situation of diluted moments, different muon sites have different ranges of local fields. When the moments fluctuate, the range of fields sensed by the muons depends on their site, i.e., they experience different dynamic subdistributions. The situation is schematically pictured in Fig. 4.40, where we see that the variable range of the fields is much wider for the muon at the blue site than for the one at the red site.[35]

---

[35] In the Gaussian case of dense fluctuating moments the variable range of the field modulation sensed by the muons is identical for all sites and is given by Eq. 4.15. Hence the dynamics is identical for all sites.

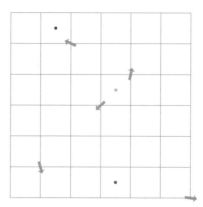

Fig. 4.40 Schematic situation of a diluted moment system picturing the different ranges of local fields at crystallographically equivalent muon sites corresponding to different spatial impurity configurations surrounding the muon. When the moments fluctuate, the local field at the green and blue sites (closer to the moments) varies over a wider range than at the red site

This unphysical result can be avoided by first expressing the Lorentzian field distribution typical of the diluted moment situation as a superposition of Gaussian subdistributions with proper weights. Assuming that each Gaussian subdistribution has a variance $\langle \Delta B_\mu^2 \rangle = \sigma^2/\gamma_\mu^2$, the probability $\rho(\sigma)$ of having a site characterized by σ must satisfy[36]

$$f^{\,\mathrm{L}}(B_{\mu,\alpha}) = \int_0^\infty f^{\,\mathrm{G}}(B_{\mu,\alpha})\,\rho(\sigma)d\sigma \quad , \tag{4.71}$$

where

$$f^{\,\mathrm{L}}(B_{\mu,\alpha}) = \frac{\gamma_\mu}{\pi} \frac{a}{a^2 + \gamma_\mu^2 B_{\mu,\alpha}^2} \quad . \tag{4.72}$$

One can easily verify that the weight function $\rho(\sigma)$, Fig. 4.41, is given by

$$\rho(\sigma) = \sqrt{\frac{2}{\pi}} \frac{a}{\sigma^2} \exp\left(-\frac{a^2}{2\sigma^2}\right) \quad . \tag{4.73}$$

In a second step the dynamic relaxation for Lorentzian fields is obtained by performing the spatial averaging (with $\rho(\sigma)$) of the dynamical polarization function

---

[36] We emphasize here the Gaussian and Lorentzian character of the field distributions with the indices G and L.

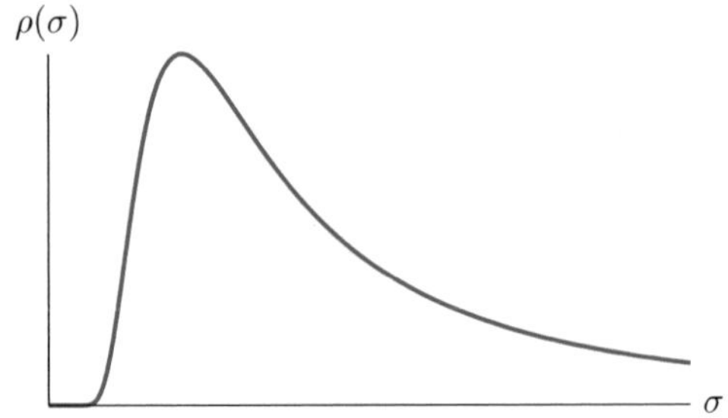

Fig. 4.41 Probability distribution $\rho(\sigma)$. The maximum is at $a/\sqrt{2}$

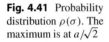

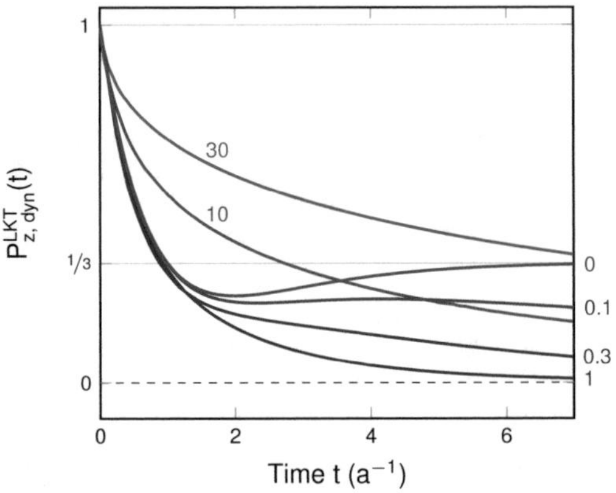

Fig. 4.42 Evolution of the polarization function for dynamical Lorentz fields with fluctuation rates in units of a

for Gaussian fields (Uemura 1981; Uemura et al. 1985) $P_{z,\,\mathrm{dyn}}^{\mathrm{GKT}}$, Eq. 4.58

$$P_{z,\,\mathrm{dyn}}^{\mathrm{LKT}} = \int_0^\infty P_{z,\,\mathrm{dyn}}^{\mathrm{GKT}}\,\rho(\sigma)d\sigma \quad . \tag{4.74}$$

Therefore, the correct order of the dynamical operation is[37]

spatial field "subdistribution" $\rightarrow$ dynamic via SCA

$\rightarrow$ spatial averaging

Figure 4.42 shows the evolution of the polarization function for dynamical Lorentzian fields. The function evolves from the static Lorentzian Kubo-Toyabe curve for $\nu = 0$ to an exponential decay for large ν/a.

---

[37] This case is an example of the general rule that spatial averaging should be the last operation in a disordered system.

Coming back to the fast fluctuation limit, we note that with the procedure just discussed we ensure that each subdistribution has the correct fast fluctuation limit with an exponential relaxation rate $\lambda = 2\sigma^2 \tau_c$. We further note that all subdistributions have the same single fluctuation time τ_c.[38] The variable range of dynamics is only a consequence of the variable σ.

By inserting the exponential fast fluctuation limit of $P_{z,\,dyn}^{GKT}(t, \sigma, \nu)$, Eq. 4.61, in Eq. 4.74, one obtains the fast fluctuation limit ($\nu/a \gtrsim 20$) for Lorentzian fields in ZF

$$P_{z,\,dyn}^{LKT} = \int_0^\infty \exp\left(-\frac{2\sigma^2}{\nu} t\right) \rho(\sigma) d\sigma = \exp\left[-\left(\frac{4a^2}{\nu} t\right)^{1/2}\right] , \qquad (4.75)$$

which is usually referred to as "root" or "square-root" exponential. This function has been used to analyze dilute spin glass compounds (Uemura et al. 1985).

4.3.2 Stretched Exponential Function

The stretched exponential function can be seen as a generalization of the square-root function, Eq. 4.75,

$$P_z^{SE} = \exp\left[-(\lambda t)^\beta\right] . \qquad (4.76)$$

Stretched exponential behavior is observed in many fields of science, including social and economic sciences. In physics it is often used for a phenomenological description of a large variety of relaxation phenomena in disordered systems (Phillips 1996). Since it has an additional parameter β that can depend on physical parameters such as temperature, magnetic field, ... it is sometimes used to fit μSR data that cannot be handled with the functions we have considered so far. However, its physical interpretation is not always unambiguous.

A plausible interpretation can be given within the fast fluctuation limit. Formally we can write a stretched exponential as a distribution of independent exponential relaxation channels with rate λ_i, which is proportional to a fluctuation time $\tau_{c,i} = 1/\nu_i$, Eq. 4.61. For a specific value of β, there is a probability distribution Λ of λ_i values. If λ_i is normalized to the characteristic relaxation rate λ appearing in the stretched exponential function, with $s \equiv \lambda_i/\lambda$, we can write

$$P_z^{SE} = \exp\left[-(\lambda t)^\beta\right] = \int_0^\infty \Lambda(s, \beta) \exp(-s\lambda t) ds . \qquad (4.77)$$

---

[38] The situation where we may have a distribution of correlation times is treated in the next Sect. 4.3.2.

Closed analytical solutions for Λ can be found for many rational values of β. We give here the expressions for $\beta = 1/2$ and $\beta = 1/3$ [39]

$$\Lambda(s, 1/2) = \frac{1}{\sqrt{4\pi s^3}} \exp\left(-\frac{1}{4s}\right) \tag{4.78}$$

$$\Lambda(s, 1/3) = \frac{1}{3\pi s^{\frac{3}{2}}} K_{1/3}\left(\frac{2}{\sqrt{27s}}\right) , \tag{4.79}$$

where $K_{1/3}$ is the modified Bessel function of the second kind.

For $\beta = 1$, $\Lambda(s, 1) = \delta(s - 1)$ corresponding to a single relaxation rate and an exponential muon spin depolarization.

A value of $\beta < 1$ causes the δ-function distribution to broaden, reflecting a widening of the relaxation spectrum. This can be related to inhomogeneous spin dynamics with a distribution of fluctuation times, a situation encountered in spin glasses.[40] Experiments on weakly or moderately concentrated spin glasses show depolarization functions above the freezing temperature, which can be fitted very satisfactorily by a stretched exponential function with β decreasing with temperature and reaching $1/3$ at the freezing temperature, (Campbell et al. 1994; Keren et al. 1996), see Sect. 5.5 for details and experimental examples.

4.4   A Quantum Mechanical Approach to the Muon Spin Relaxation

There are several quantum mechanical approaches to describe the time evolution of the polarization function of the muon ensemble. Most of them have their analogues in NMR theory, where the polarization (or magnetization) of a nuclear spin ensemble is studied. The muon ensemble is a quantum subsystem interacting with its environment (acting as a thermal "bath", which consists of the lattice atoms and conduction electrons) and possibly external fields. Such a subsystem is best described by a density matrix, which allows the description of mixed quantum states, see Appendix F.

A muon can have two possible spin states up $| \uparrow \rangle$ and down $| \downarrow \rangle$ with respect to the quantization axis. Taking it as the z-axis with $\mathbf{B}_{ext} \| \hat{\mathbf{z}}$, the Zeeman effect lowers

---

[39] For more details about properties, systematic behavior of $\Lambda(s, \beta)$, and physical significance of the λ_i and β parameters, see (Johnston 2006).

[40] The stretched exponential encompasses the case of a distribution of fluctuation times, since $\lambda_i \propto \tau_{c,i}$. Note that this situation is different from the case of the square-root exponential, Eq. 4.75, where one assumes a single fluctuation time and it is the variation of the field width at the muon site that is responsible for a range of relaxation rates. An even more general case is when one allows for a distribution of local field widths and fluctuation times, see (Keren et al. 1996).

Fig. 4.43 Splitting of the
muon energy levels $|\uparrow\rangle$ and
$|\downarrow\rangle$ in a magnetic field B_{ext}.
The level with lowest energy
is the one where the spin (and
magnetic moment) of the
muon is parallel to the
magnetic field

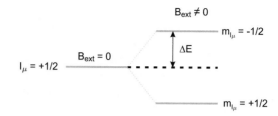

the energy of the $|\uparrow\rangle$ state by

$$\Delta E = -\boldsymbol{\mu}_\mu \cdot \mathbf{B}_{\text{ext}} = -\gamma_\mu \hbar \mathbf{I}_\mu \cdot \mathbf{B}_{\text{ext}} = -\gamma_\mu \hbar \frac{1}{2} B_{\text{ext}} \ , \tag{4.80}$$

whereas the $|\downarrow\rangle$ level has its energy raised by the same amount, resulting in an
energy split, Fig. 4.43

$$2|\Delta E| = \hbar\gamma_\mu B_{\text{ext}} = \hbar\omega_\mu \ . \tag{4.81}$$

Note that the Zeeman splitting of the spin levels in the magnetic field is equal to the
Larmor precession frequency of the muon spin in that field. As we will see later in
this section, this fact implies that the muon spin relaxation rate contains information
about the frequency spectrum of the field fluctuations.

Now consider the muon in an environment of randomly fluctuating fields $\delta\mathbf{B}_\mu(t)$,
with ensemble average $\langle\delta B_{\mu,\alpha}\rangle = 0$, $\alpha = x, y, z$.[41]

Since $|\delta\mathbf{B}_\mu(t)| \ll |\mathbf{B}_{\text{ext}}|$, the fluctuating field can be treated as a perturbation.
The total Hamiltonian of the system is

$$H = -\gamma_\mu \hbar \mathbf{I}_\mu \cdot \left[\mathbf{B}_{\text{ext}} + \delta\mathbf{B}_\mu(t)\right] \ . \tag{4.82}$$

With a beam of 100% polarized muons with polarization parallel to the magnetic
field, all the muons at the time of implantation are in the $|\uparrow\rangle$ state. The interaction
with the fluctuating field will induce transitions between the two Zeeman levels,
leading to a decay of polarization. Although at $t = 0$ all the muons of the ensemble
are in a pure quantum state, the interaction with the environment leads to a mixed
state, which can be described by a density matrix.

4.4.1   Redfield Expressions

The behavior of a spin subsystem weakly interacting with a bath has been considered
for NMR by several authors (Slichter 1978; Abragam 1961). A general procedure

---

[41] Since we are dealing with stationary processes, the ensemble average is equivalent to a time
average.

to express the equation of motion of the density matrix of the spin system has been given by Redfield (1957). This formalism can be applied to describe the time evolution of a density matrix and to quantify the expectation values of the statistically averaged operators in the case of a random field Hamiltonian, representing the fluctuating bath environment. The spin relaxation rates can be expressed as a function of the field autocorrelation function.

The Redfield theory is valid for times t greater than the correlation time τ_c of the field fluctuations, i.e., for fast fluctuations, where the muon spin relaxation is exponential. In the language of NMR the relaxation rates are expressed in terms of the relaxation times T_1 and T_2. For LF (and ZF) $\lambda_L = T_1^{-1}$ and for TF $\lambda_T = T_2^{-1}$. Redfield theory implies that $T_1, T_2 \gg \tau_c$.

The different components of a fluctuating field are assumed to be uncorrelated

$$\langle \delta B_{\mu,i}(t) \delta B_{\mu,j}(t+t') \rangle = 0 \quad \text{for } i \neq j \ ,$$

as in the strong collision approach, Sect. 4.3.1.1.

For the relaxation rates,[42] one finds[43]

- LF (and ZF) geometry $\mathbf{P}(0) \parallel \mathbf{B}_{\text{ext}} \parallel z$-axis, $P_z(t) \sim e^{-\lambda_L t}$

$$\lambda_L = \frac{1}{T_1} = \frac{\gamma_\mu^2}{2} \int_{-\infty}^{+\infty} \Big[\langle \delta B_{\mu,x}(t) \delta B_{\mu,x}(t+t') \rangle$$
$$+ \langle \delta B_{\mu,y}(t) \delta B_{\mu,y}(t+t') \rangle \Big] e^{-i\omega_\mu t'} dt' \qquad (4.83)$$

- TF geometry $\mathbf{P}(0) \parallel x$-axis, $\mathbf{B}_{\text{ext}} \parallel z$-axis, $P_x(t) \sim e^{-\lambda_T t}$

$$\lambda_T = \frac{1}{T_2} = \frac{\gamma_\mu^2}{2} \int_{-\infty}^{+\infty} \Big[\langle \delta B_{\mu,z}(t) \delta B_{\mu,z}(t+t') \rangle +$$
$$+ \frac{1}{2}[\langle \delta B_{\mu,x}(t) \delta B_{\mu,x}(t+t') \rangle + \langle \delta B_{\mu,y}(t) \delta B_{\mu,y}(t+t') \rangle] e^{-i\omega_\mu t'} \Big] dt' \ . $$
$$(4.84)$$

A closer look at Eq. 4.83 reveals that the LF and ZF relaxations depend only on the fluctuations, which are transverse to the direction of observation $\mathbf{P}(0)$, i.e., fluctuations along the x and y directions. Equation 4.83 and the second part of Eq. 4.84 show that the relevant Fourier component of the fluctuations is the one with frequency ω_μ, which is determined by the applied field (here $\omega_\mu = \gamma_\mu B_{\text{ext}}$) and corresponds to the Zeeman splitting of the spin levels, Fig. 4.43.

---

[42] The subscripts L and T underline that the relaxation affects polarization components along or transverse to the external field.

[43] For a derivation of these formulas, we refer to Slichter (1978, Chapter 5).

The expressions manifest the intrinsic[44] resonant character of μSR. If the Fourier spectrum of the transverse fluctuations has a significant weight at the frequency corresponding to the energy difference between the up and down states, the transition between these two states becomes more probable, which translates into a larger muon spin relaxation rate.

Further insight can be gained by assuming an exponential behavior of the field autocorrelation function, as in Sect. 4.3.1.3, $\langle \delta B_{\mu,\alpha}(t')\delta B_{\mu,\alpha}(t'+t)\rangle = \langle \delta B_{\mu,\alpha}(0)\delta B_{\mu,\alpha}(t)\rangle = \langle \delta B_{\mu,\alpha}(0)^2\rangle \exp(-|t|/\tau_c)$.

With this, Eqs. 4.83 and 4.84 become

$$\lambda_L = \frac{1}{T_1} = \gamma_\mu^2 \big[\langle \delta B_{\mu,x}^2 \rangle + \langle \delta B_{\mu,y}^2 \rangle \big] \frac{\tau_c}{1+\omega_\mu^2 \tau_c^2} \tag{4.85}$$

$$\lambda_T = \frac{1}{T_2} = \gamma_\mu^2 \left[\langle \delta B_{\mu,z}^2 \rangle \tau_c + \frac{\langle \delta B_{\mu,x}^2 \rangle + \langle \delta B_{\mu,y}^2 \rangle}{2} \frac{\tau_c}{1+\omega_\mu^2 \tau_c} \right]. \tag{4.86}$$

If we compare the expressions for the TF relaxation rate, Eqs. 4.84 and 4.86, with the TF Abragam formula, Eq. 4.70, we note that in the Redfield expression fluctuations in all Cartesian directions affect the muon spin relaxation, while the Abragam expression accounts only for the effect of the fluctuations parallel to the external field. We have already remarked that Eq. 4.70 is incomplete. The fluctuating z component of the magnetic field causes the Larmor precession rate to be faster or slower, but cannot induce spin-flips. It corresponds to the dephasing effect, that we discussed in the static case, when a spread of fields is present (inhomogeneous broadening).

This static or "quasistatic" character of the z component is also visible in the first (z) term of Eq. 4.86, which becomes dominant for long correlation times τ_c. The Redfield formula also shows that relaxation in the TF case can be caused by static as well as by dynamic effects.[45] For this reason, TF measurements are not as informative as LF measurements for dynamic investigations.

On the other hand, the λ_L of the LF expression is of purely dynamical origin. It is due to an exchange of energy between the muon spin and the crystal lattice environment (the "bath"). This motivates the name spin-lattice relaxation rate often given to the longitudinal relaxation (or to the so-called T_1 processes), although other components such as the conduction electrons can also contribute to this relaxation. To be effective, the fluctuating field must have a component transverse

---

[44] There is no external RF source, as is generally the case in NMR.

[45] Strictly speaking, these considerations stretch out the range of validity of the Redfield expressions, but still give a qualitative understanding.

to the external field and the fluctuation process must supply energy to the muon corresponding to the Zeeman splitting, so that the muon spin can flip.[46]

Assuming isotropic fluctuations and writing $\langle \delta B_{\mu,\alpha}^2 \rangle = \gamma_\mu^2 \sigma^2$ for $\alpha = x, y, z$, the Redfield expression for LF and ZF gives the previously obtained fast fluctuation limit, see Eq. 4.66,

$$\lambda_L = \frac{2\sigma^2 \tau_c}{1 + \omega_\mu^2 \tau_c^2} \ , \tag{4.87}$$

and the TF expression becomes

$$\lambda_T = \sigma^2 \tau_c + \frac{\sigma^2 \tau_c}{1 + \omega_\mu^2 \tau_c^2} \ . \tag{4.88}$$

From Eqs. 4.87 and 4.88 it is clear that in the extreme fast fluctuating limit ($\tau_c \to 0$) the exponential relaxation rates in ZF, LF and TF tend to the same value $2\sigma^2 \tau_c^2$.

4.4.2 Spectral Density

We have already remarked that the integral 4.83 is related to a Fourier transform. In fact, the depolarization rate can be expressed as a function of the Fourier transform $J(\omega)$ of the autocorrelation function. This quantity is often referred to as the spectral density (or weight), since it gives the distribution of frequencies, which are present in the fluctuation process.

With an exponentially decaying correlation, and keeping in mind that $\tau_c = 1/\nu$, we have

$$J(\omega) = \frac{1}{2} \int_{-\infty}^{+\infty} \left[\langle \delta B_{\mu,x}(0) \delta B_{\mu,x}(t') \rangle + \langle \delta B_{\mu,y}(0) \delta B_{\mu,y}(t') \rangle \right] e^{-i\omega t'} dt'$$

$$= \langle B_{\mu,\alpha}^2(0) \rangle \int_{-\infty}^{\infty} e^{-|t'|/\tau_c} e^{-i\omega t'} dt'$$

$$= \frac{2\sigma^2}{\gamma_\mu^2} \frac{\tau_c}{1 + \omega^2 \tau_c^2} = \frac{2\sigma^2}{\gamma_\mu^2} \frac{\nu}{\omega^2 + \nu^2} \ . \tag{4.89}$$

Equation 4.87 can therefore be written as

$$\lambda_L = \gamma_\mu^2 J(\omega_\mu) \ . \tag{4.90}$$

---

[46] We can understand this condition by noting that the longitudinal fluctuations only slightly modulate the Zeeman splitting (since the spin eigenstates are not a function of the field strength), but do not induce transitions between the levels.

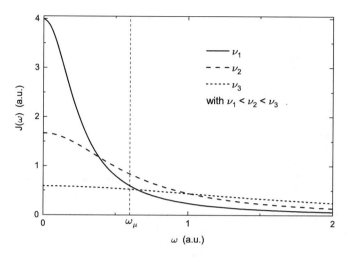

Fig. 4.44 Spectral density functions $J(\omega)$ for different values of the fluctuation rate $\nu = 1/\tau_c$ with $\nu_1 < \nu_2 < \nu_3$. Muon spin relaxation is an intrinsic resonance phenomenon that picks up the component at ω_μ (chosen with the applied LF) from a possibly much wider spectrum. $J(\omega_\mu)$ is maximum for the value of the fluctuation rate ν_2 satisfying $\nu_2 = \omega_\mu$, see text

We see that the depolarization rate probes the spectral density of the autocorrelation function of the fluctuations at the value of the Larmor frequency ω_μ, with $\nu_\mu = \omega_\mu/2\pi = (\gamma_\mu B_{\text{ext}})/2\pi$ typically in the kHz to GHz range.[47, 48]

The spectral density for an exponential autocorrelation function is a Lorentzian. It is shown in Fig. 4.44 for three different fluctuation rates.

We can now quantitatively understand the different behavior in the presence of static or dynamic fields in an LF experiment. With a typical LF of 0.1 T, $\omega_\mu = 85.1 \times 10^6$ s^{-1}, a much smaller value than typical fluctuation rates in solids, which range between $\nu = 10^{13} - 10^9$ s^{-1}. The depolarization rate caused by fluctuations shows little dependence on the applied field, since $\omega_\mu \ll \nu$ in Eq. 4.87. This situation corresponds to the flat behavior of $J(\omega)$ for ν_3 in Fig. 4.44.

This behavior is markedly different from that of static internal fields, where a strong dependence of the depolarization and decoupling is observed at modest LF values, see for example Fig. 4.28. Figure 4.45 depicts qualitatively the static versus the dynamic case.

---

[47] The relationship between autocorrelation function and spectral density is also valid for non-exponential autocorrelation functions. One can show that for a stationary random process the autocorrelation and spectral density functions are the Fourier transform of each other. This is also known as the Wiener-Khintchine theorem (Wiener 1930; Khintchine 1934; Wang & Uhlenbeck 1945).

[48] Note that from λ_L one can obtain information about frequencies much larger than ω_μ. When frequency sum rules apply, as is often the case in spin systems, the low-frequency weight of the spectral density is related to the characteristic frequency of the fluctuations $\omega \gg \omega_\mu$.

Fig. 4.45 Comparison of the effect of a longitudinal field on two depolarization functions. The blue curves are without field, and the red curves with a typical longitudinal field. Top panel: The ZF curve is a Lorentz-Kubo-Toyabe due to a static Lorentzian field distribution, Eq. 4.25. At early times, this function shows an exponential behavior, which can be quenched by applying a relatively low LF (decoupling). Bottom panel: Exponential ZF depolarization due to the presence of fast fluctuating electronic moments, Eq. 4.61. The application of a longitudinal field has only a very small effect on the depolarization due to fast fluctuations, $\nu \gg \omega_\mu$

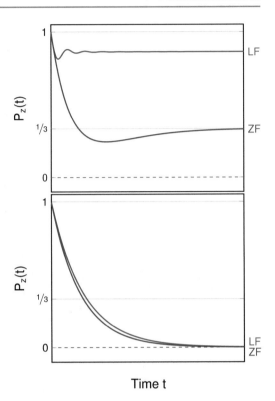

The use of an LF scan to demonstrate the presence of spin fluctuations is shown in Fig. 4.46. The absence of decoupling effects can be observed at very low temperatures, indicating the absence of a static magnetic ground state in a so-called quantum spin liquid (QSL) (Balz et al. 2016).[49] For a material to be a QSL candidate, it must be shown that no magnetic order occurs down to the lowest temperatures. For this μSR is an essential technique because, as we have seen, it is sensitive to very small fractions of μ_B of static magnetism (ordered or disordered) (Mendels et al. 2007; Mendels & Bert 2016). We refer to Sect. 5.3 for more examples of magnetic fluctuation studies based on Eqs. 4.83 and 4.84.

---

[49] A quantum spin liquid (QSL) is a quantum state of matter in which the spins are highly entangled and do not order magnetically down to the lowest temperatures (Broholm et al. 2020). The long-range quantum entanglement is a key feature of QSL, but it is difficult to prove experimentally (as well as theoretically) and can be obscured by disorder or other interactions. Spin liquid behavior may follow from the competition between frustrated interactions that suppress long-range magnetic order. The quest for QSLs involves the identification of new materials.

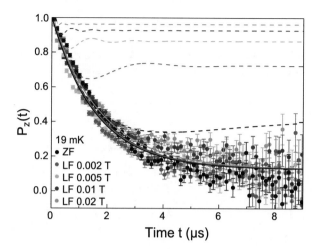

Fig. 4.46 Example of the persistence of spin fluctuations at very low temperatures (19 mK) in $Ca_{10}Cr_7O_{28}$. This system is a so-called quantum spin liquid. The data are taken in increasing longitudinal fields. The ZF data are modeled alternatively by a static LKT function, Eq. 4.25 (dashed black line) and a dynamic exponential function, Eq. 4.61, (solid black line). Using the fitted parameters, the corresponding static (dashed lines) and dynamic (solid lines) depolarization functions are simulated for the applied longitudinal fields using Eqs. 4.35 and 4.66, respectively. The color code of the simulated curves corresponds to the color code of the data. Clearly, the dynamic scenario expected for a spin liquid ground state reproduces the data much better. Modified from Balz et al. (2016), © Springer Nature. Reproduced with permission. All rights reserved

Exercises

4.1 Time evolution of the polarization
Derive the time evolution of the polarization, Eq. 4.1.

4.2 Time evolution of the polarization for planar field isotropy
Derive the time evolution of the polarization for the case of an isotropic local field confined to a plane.

4.3 Discriminate between single valued field and isotropic distribution when measuring a polarization function given by Eq. 4.11
Suppose your μSR experiment gives for the muon spin polarization Eq. 4.11. In principle, such a result can be obtained for a single valued field vector with $\theta = \arccos(1/\sqrt{3}) = 54.736°$ (e.g., in a crystal) or for the same field isotropically distributed (e.g. in crystallites), see footnote 9. How do you distinguish between the two possibilities?

4.4 Short time behavior of Gaussian and Lorentzian Kubo-Toyabe functions
Compare the short-time ZF polarization with the envelope of the TF polarization
for a Gaussian field distribution. Explain the difference qualitatively. Compute the
short-time behavior of the Lorentz-Kubo-Toyabe function.

References

Abragam, A. (1961). *The Principles of Nuclear Magnetism*. Clarendon Press. ISBN: 978-0198520146.
Amato, A., Feyerherm, R., Gygax, F. N., et al. (1995). *Physical Review B, 52*, 54.
Anderson, P. W. (1954). *Journal of the Physical Society of Japan, 9*, 316.
Balz, C., Lake, B., Reuther, J., et al. (2016). *Nature Physics, 12*, 942.
Belousov, Y. M., Gorelkin, V. N., Mikaélyan, A. L., et al. (1979). *Soviet Physics Uspekhi, 22*, 679.
Bernardini, F., Bonfà, P., Massidda, S., et al. (2013). *Physical Review B, 87*, 115148.
Blundell, S. J. (1999). *Contemporary Physics, 40*, 175.
Brewer, J. H., Kreitzman, S. R., Noakes, D. R., et al. (1986). *Physical Review B, 33*, 7813.
Broholm, C., Cava, R. J., Kivelson, S. A., et al. (2020). *Science, 367*, eaay0668.
Campbell, I. A., Amato, A., Gygax, F. N., et al. (1994). *Physical Review Letters, 72*, 1291.
Celio, M. (1986). *Physical Review Letters, 56*, 2720.
Celio, M., & Meier, P. F. (1983). *Physical Review B, 27*, 1908.
Cooley, J. W., & Tukey, J. W. (1965). *Mathematics of Computation, 19*, 297.
Crook, M. R., & Cywinski, R. (1997). *Journal of Physics: Condensed Matter, 9*, 1149.
de la Cruz, C., Huang, Q., Lynn, J. W., et al. (2008). *Nature, 453*, 899.
Dalmas de Réotier, P., & Yaouanc, A. (1992). *Journal of Physics: Condensed Matter, 4*, 4533.
Dalmas de Réotier, P., Yaouanc, A., & Meshkov, S. (1992). *Physics Letters A, 162*, 206.
Doob, J. L. (1942). *Annals of Mathematics, 43*, 351.
Fiory, A. T. (1981). *Hyperfine Interactions, 8*, 777.
Hayano, R. S., Uemura, Y. J., Imazato, J., et al. (1979). *Physical Review B, 20*, 850.
Held, C., & Klein, M. W. (1975). *Physical Review Letters, 35*, 1783.
Johnston, D. C. (2006). *Physical Review B, 74*, 184430.
Kamihara, Y., et al. (2008). *Journal of the American Chemical Society, 130*, 3296.
Keren, A. (1994). *Physical Review B, 50*, 10039.
Keren, A., Mendels, P., Campbell, I. A., et al. (1996). *Physical Review Letters, 77*, 1386.
Khasanov, R., Amato, A., Bonfà, P., et al. (2017). *Journal of Physics: Condensed Matter, 29*, 164003.
Khintchine, A. (1934). *Mathematische Annalen, 109*, 604.
Kornilov, E., & Pomjakushin, V. (1991). *Physics letters A, 153*, 364.
Kubo, R. (1954). *Journal of the Physical Society of Japan, 9*, 935.
Kubo, R., & Toyabe, T. (1967). In R. Blinc (Ed.), *Magnetic resonance and relaxation* (Vol. 810). North-Holland.
Lamura, G., Shiroka, T., Bonfà, P., et al. (2014). *Journal of Physics: Condensed Matter, 26*, 295701.
Lancaster, T. (2019). *Contemporary Physics, 60*, 246–261.
Lancaster, T., Blundell, S. J., Baker, P. J., et al. (2007). *Physical Review Letters, 99*, 267601.
Larkin, M., Fudamoto, Y., Gat, I., et al. (2000). *Physica B: Condensed Matter, 153*, 289–290, 153.
Luetkens, H., Klauss, H.-H., Kraken, M., et al. (2009). *Nature Materials, 8*, 305.
McArdle, S. (2021). *PRX Quantum, 2*, 020349.
Meier, P. F. (1984). *Hyperfine Interactions, 18*, 427.
Mendels, P., & Bert F. (2016). *Comptes Rendus Physique* (Vol. 17, pp. 455–470). Physique de la matière condensée au XXIe siècle: l'héritage de Jacques Friedel.
Mendels, P., Bert, F., de Vries, M. A., et al. (2007). *Physical Review Letters, 98*, 077204.

Möller, J. S., Ceresoli, D., Lancaster, T., et al. (2013). *Physical Review B, 87*, 121108.

Momma, K., & Izumi, F. (2011). *Journal of Applied Crystallography, 44*, 1272–1276.

Mühlbauer, S., Binz, B., Jonietz, F., et al. (2009). *Science, 323*, 915.

Mydosh, J. (1993). *Spin Glasses: An Experimental Introduction*. Taylor & Francis. ISBN: 978-0748400386.

Nishiyama, K., Higemoto, W., Shimomura, K., et al. (2001). *Hyperfine Interactions, 136*, 717.

Phillips, J. C. (1996). *Reports on Progress in Physics, 59*, 1133.

Rao, K., Kim, D., & Hwang, J.-J. (2010). *Fast Fourier Transform: Algorithms and Applications*. Signals and Communication Technology (1st ed.). Springer Netherlands. ISBN: 1-4020-6629-5.

Redfield, A. (1957). *IBM Journal of Research and Development, 1*, 19.

Sakarya, S., Gubbens, P. C. M., Yaouanc, A., et al. (2010). *Physical Review B, 81*, 024429.

Schenck, A., Andreica, D., Gygax, F. N., et al. (2002). *Physical Review B, 66*, 144404.

Slichter, C. P. (1978). *Principles of Magnetic Resonance*. Springer-Verlag. ISBN: 978-3662127841.

Smilga, V. P., & Belousov, Y. M. (1994). *The Muon Method in Science*. Nova Science Publishers. ISBN: 978-1560721611.

Staub, U., Roessli, B., & Amato, A. (2000). *Physica B: Condensed Matter, 299*, 289–290.

Uemura, Y. J. (1981). *Hyperfine Interactions, 8*, 739.

Uemura, Y. J. (1999). In S. Lee, S. Kilcoyne, & R. Cywinski (Eds.), *Muon Science: Muons in Physics, Chemistry, and Materials* (Vol. 85). Institute of Physics Publishing. ISBN: 978-0750306300.

Uemura, Y. J., Yamazaki, T., Harshman, D. R., et al. (1985). *Physical Review B, 31*, 546.

Vogel, S., Celio, M., & Meier, P. F. (1986). *Hyperfine Interactions, 31*, 35.

Walker, J. S. (1996). *Fast Fourier Transforms*. Studies in advanced mathematics (2nd ed.). CRC Press. ISBN: 1-351-44888-9.

Walker, L. R., & Walstedt, R. E. (1980). *Physical Review B, 22*, 3816.

Wan, X., Kossler, W. J., Stronach, C. E., et al. (1999). *Hyperfine Interactions, 122*, 233.

Wang, M. C., & Uhlenbeck, G. E. (1945). *Reviews of Modern Physics, 17*, 323.

Wiener, N. (1930). *Acta Mathematica, 55*, 117.

Wilkinson, J. M., & Blundell, S. J. (2020). *Physical Review Letters, 125*, 087201.

Study of Magnetism

5

In Chap. 4 we presented the evolution of the muon spin polarization in the presence of different magnetic field configurations and distributions, without discussing the physical origin of the fields probed by the muon.

In the present chapter, we first introduce the various contributions to the local field that arise in magnetic materials.[1] The magnetic field is mainly due to the dipolar contributions of the moments of localized electrons and to the possible presence of a spin density of nonlocalized electronic moments at the muon site. We will then give typical examples of investigations of magnetic materials, emphasizing the specific information that can be obtained by μSR.

Roughly summarizing (i) from the amplitude (or amplitudes in the case of a multicomponent signal) of the μSR signal one can differentiate between magnetic and nonmagnetic regions and determine their volume ratio; (ii) from the temperature dependence of the Larmor frequency (or frequencies) one obtains information about order parameters and magnetic phase transitions; (iii) from the value of the obtained frequency one gets information about the size of the magnetic moments and, if the muon site is known, also about the magnetic structure; (iv) finally, the relaxation rate gives information about disorder and magnetic inhomogeneities, signalizes phase transitions and quantifies dynamical parameters such as correlation times and critical exponents. We also present μSR investigations of the magnetic response of paramagnetic and diamagnetic systems. Finally, we discuss the magnetic field produced by nuclear moments. The field strength created by these moments is small, but the signal is well detectable by μSR and cannot always be ignored.

---

[1] There are countless books on magnetism and magnetic materials. A good introduction can be found for example in Blundell (2001) and Coey (2019). The state of the art is well covered in Coey and Parkin (2022).

© Springer Nature Switzerland AG 2024
A. Amato, E. Morenzoni, *Introduction to Muon Spin Spectroscopy*,
Lecture Notes in Physics 961, https://doi.org/10.1007/978-3-031-44959-8_5

5.1   Local Magnetic Field in Magnetic Materials

5.1.1   The Muon-Electron Interaction

To obtain the main contributions to the local field sensed by a muon in a magnetic material, we first consider the interaction between a muon and a single electron. We assume a static muon located at the origin interacting with an electron (charge $-e$, spin $\mathbf{S}$, in units of $\hbar$) at position $\mathbf{r}$. This interaction is most generally described by the Dirac equation (see Appendix G.2) but in the low-energy nonrelativistic limit the Hamiltonian of the muon-electron system can be written as

$$\mathcal{H} = \frac{1}{2m_e}(\mathbf{p} + e\mathbf{A})^2 + g_e\,\mu_B\,\mathbf{S}\cdot(\nabla\times\mathbf{A}) + V(r)\ , \tag{5.1}$$

where the kinetic energy term is obtained by introducing the canonical momentum $\mathbf{p}$, $\mathbf{A}$ is the magnetic vector potential, and $\mathbf{B} = \nabla\times\mathbf{A}$ is the magnetic field created by the muon magnetic moment $\boldsymbol{\mu}_\mu$. The second term is the Zeeman energy of the electron in this field[2] $-\boldsymbol{\mu}_e\cdot\mathbf{B}$ and $V(r) = -e^2/(4\pi\epsilon_0 r)$ the Coulomb potential energy. The vector potential of a magnetic moment can be written as (Jackson 1998)[3]

$$\mathbf{A} = \frac{\mu_0}{4\pi}\frac{\boldsymbol{\mu}_\mu\times\mathbf{r}}{r^3} = \frac{\mu_0}{4\pi}\left(\nabla\times\frac{\boldsymbol{\mu}_\mu}{r}\right)\ . \tag{5.2}$$

---

[2] We recall that the magnetic moment of the electron points opposite to its spin (or angular momentum).

[3] This is obtained by using the 4th Maxwell equation without electric field and the vector identity

$$\nabla\times\mathbf{B} = \nabla\times(\nabla\times\mathbf{A})$$

$$= \nabla(\nabla\cdot\mathbf{A}) - \nabla^2\mathbf{A}$$

$$= \mu_0\mathbf{J}\ .$$

Choosing the Coulomb gauge ($\nabla\cdot\mathbf{A} = 0$) we have

$$\nabla^2\mathbf{A} = -\mu_0\mathbf{J}\ ,$$

which is a Poisson equation, with solution

$$\mathbf{A}(\mathbf{r}) = \frac{\mu_0}{4\pi}\int\frac{\mathbf{J}(\mathbf{r}')}{|\mathbf{r}-\mathbf{r}'|}\,d\mathbf{r}'\ .$$

Equation 5.2 is obtained by integrating over a current loop producing the moment and by taking the dipole term of the multipole expansion of $\mathbf{J}(\mathbf{r}')/|\mathbf{r}-\mathbf{r}'|$ with Legendre polynomials.

By neglecting the quadratic term in $\mathbf{A}$ (which gives a small diamagnetic contribution), one can rewrite the Hamiltonian as

$$\mathcal{H} = \mathcal{H}_0 + \mathcal{H}'$$

$$= \mathcal{H}_0 + \frac{e}{2m_e}(\mathbf{p} \cdot \mathbf{A} + \mathbf{A} \cdot \mathbf{p}) + g_e \mu_B \mathbf{S} \cdot (\nabla \times \mathbf{A}) \ , \tag{5.3}$$

where $\mathcal{H}_0 = \mathbf{p}^2/2m_e + V(r)$ contains the terms that are not of interest for our purposes and do not couple to $\boldsymbol{\mu}_\mu$.

Orbital Contribution Let us consider the first term in $\mathcal{H}'$. With $\hbar\mathbf{L} = \mathbf{r} \times \mathbf{p}$ ($\mathbf{L}$ is the orbital angular momentum of the electron in units of $\hbar$) and the fact that $\boldsymbol{\mu}_\mu$ commutes with $\mathbf{r}$ and $\mathbf{p}$, $\mathbf{p} \cdot \mathbf{A} \propto \mathbf{p} \cdot (\boldsymbol{\mu}_\mu \times \mathbf{p}) = \boldsymbol{\mu}_\mu \cdot (\mathbf{r} \times \mathbf{p}) = \boldsymbol{\mu}_\mu \cdot \mathbf{L}$ and similarly for $\mathbf{A} \cdot \mathbf{p}$, the first term of $\mathcal{H}'$ becomes

$$\frac{e}{2m_e}(\mathbf{p} \cdot \mathbf{A} + \mathbf{A} \cdot \mathbf{p}) = \frac{\mu_0}{4\pi}\frac{e}{2m_e}2\frac{\boldsymbol{\mu}_\mu \cdot \hbar\mathbf{L}}{r^3} = -\boldsymbol{\mu}_\mu \cdot \mathbf{B}_{\text{orb}} \ , \tag{5.4}$$

with

$$\mathbf{B}_{\text{orb}} = -\frac{\mu_0}{4\pi}2\mu_B\frac{\mathbf{L}}{r^3} \ . \tag{5.5}$$

$\mathbf{B}_{\text{orb}}$ is the field created by an electron orbiting around the muon at distance r with angular momentum $\mathbf{L}$.[4]

Dipolar Contribution The second term in $\mathcal{H}'$ depends only on the electron and muon spins. We first consider the magnetic field operator $\mathbf{B} = \nabla \times \mathbf{A}$, where $\mathbf{A}$ is given by Eq. 5.2. Using the well known vector identity $\nabla \times (\nabla \times \mathbf{V}) = \nabla(\nabla \cdot \mathbf{V}) - \nabla^2\mathbf{V}$ one finds

$$\mathbf{B} = \frac{\mu_0}{4\pi}\nabla(\nabla \cdot \frac{\boldsymbol{\mu}_\mu}{r}) - \frac{\mu_0}{4\pi}\nabla^2\frac{\boldsymbol{\mu}_\mu}{r} \ . \tag{5.6}$$

This expression appears to diverge at the muon position, $r = 0$. To deal with this singularity, one splits the ∇^2 term into a $1/3$ and $2/3$ term and apply the identity $\nabla^2(\frac{1}{r}) = \Delta(\frac{1}{r}) = -4\pi\delta(\mathbf{r})$ to the latter. This allows the components of $\mathbf{B}$ to be written as

$$B_i = \frac{\mu_0}{4\pi}\sum_{j=1}^{3}(\nabla_i \nabla_j - \frac{1}{3}\nabla^2\delta_{ij})\frac{\mu_{\mu,j}}{r} + \frac{2}{3}\mu_0 \mu_{\mu,i}\,\delta(\mathbf{r}) \ . \tag{5.7}$$

---

[4] The expression can also be obtained classically from the Biot-Savart law with a current generated by an electron on a circular orbit.

One can show that the first term of 5.7 is not singular at $r = 0$,[5] see Exercise 5.1. It can be computed explicitly to give for $\mathbf{r} \neq 0$

$$\mathbf{B}_{\mu-\mathrm{dip}} = \frac{\mu_0}{4\pi} \frac{3(\boldsymbol{\mu}_\mu \cdot \mathbf{r})\mathbf{r} - r^2 \boldsymbol{\mu}_\mu}{r^5} \; , \tag{5.8}$$

which is the dipolar field created by the muon magnetic moment at $\mathbf{r} \neq 0$. Inserting this expression into 5.3, we notice that the corresponding energy term is symmetric in the magnetic moments of both particles so that we can write

$$\mathcal{H}'_{\mathrm{Dip}} = -\frac{\mu_0}{4\pi} \, \boldsymbol{\mu}_\mu \cdot \frac{3(\boldsymbol{\mu}_e \cdot \mathbf{r})\mathbf{r} - r^2 \boldsymbol{\mu}_e}{r^5} = -\boldsymbol{\mu}_\mu \cdot \mathbf{B}_{e-\mathrm{dip}}$$

$$\mathbf{B}_{e-\mathrm{dip}} = \frac{\mu_0}{4\pi} \frac{3(\boldsymbol{\mu}_e \cdot \mathbf{r})\mathbf{r} - r^2 \boldsymbol{\mu}_e}{r^5} \; . \tag{5.9}$$

$\mathbf{B}_{e-\mathrm{dip}}$ is the dipolar field created by the magnetic moment of an electron at position $\mathbf{r}$ with respect to the muon.

Contact Contribution The singularity at $r = 0$ is contained in the second term of Eq. 5.7, named Fermi contact term

$$\mathcal{H}'_{\mathrm{F.\,cont}} = -\frac{2\,\mu_0}{3} \, \boldsymbol{\mu}_\mu \cdot \boldsymbol{\mu}_e \, \delta(\mathbf{r}) = \frac{2\,\mu_0}{3} g_e \, \mu_\mathrm{B} \, g_\mu \, \mu_\mathrm{B}^\mu \, \mathbf{I}_\mu \cdot \mathbf{S} \, \delta(\mathbf{r}) \; . \tag{5.10}$$

Summarizing all contributions, we obtain for the hyperfine Hamilton operator $\mathcal{H}'$[6]

$$\mathcal{H} = \frac{\mu_0}{4\pi} g_e \, \mu_\mathrm{B} \, g_\mu \, \mu_\mathrm{B}^\mu \, \mathbf{I}_\mu \cdot \left[\frac{\mathbf{L}}{r^3} + \left(\frac{3(\mathbf{S} \cdot \mathbf{r})\mathbf{r}}{r^5} - \frac{\mathbf{S}}{r^3} \right) + \frac{8\pi}{3} \mathbf{S} \, \delta(\mathbf{r}) \right] \; . \tag{5.11}$$

The Hamiltonian of Eq. 5.11 has the form of a Zeeman interaction of $\mathbf{I}_\mu$ with different effective fields. The first term represents the coupling of the muon spin with the angular momentum of the electron. The second term is the dipole-dipole interaction for $r > 0$, i.e., when the electron is not at the muon site. Finally, the last contact term takes into account the case when there is a nonzero electron spin density at the muon site. The interaction described by Eq. 5.11 is called hyperfine interaction. It is the interaction beyond the Coulomb attraction, which is taken into account by the term $V(r)$. In an atom it is the interaction between the magnetic field

---

[5] $(\nabla_\alpha \nabla_\beta - \frac{1}{3}\nabla^2 \delta_{\alpha\beta})/r$ is a traceless tensor of rank two, which transforms under rotation as the spherical harmonic $Y_2^m(\theta, \phi)$.

[6] To combine the terms we use the fact that, up to the very small electron magnetic anomaly, we can set here $g_e = 2$.

of the electrons and the nuclear spin and leads to the very small hyperfine splittings in the atomic spectra.

Consider now the expectation value of the contact contribution, Eq. 5.10. Due to the δ-function it contributes to $\mathcal{H}'$ only if the electron has a nonzero probability to reside at the muon site (hence the name contact interaction), which is the case for an electron in an s-wave state. The wave function of the muon-electron system can be written as the tensor product of an orbital and a spin part. For an electron in the s state

$$\psi_{e-\mu}(\mathbf{r}) = |n = 1, L = 0, m_S, m_{I_\mu}\rangle = |n = 1, L = 0\rangle |m_S, m_{I_\mu}\rangle \ , \qquad (5.12)$$

where n is the principal quantum number of the electron shell, L the orbital quantum number of the electron (since $L = 0$ $m_L = 0$), and m_S and m_{I_μ} the electron and muon spin quantum numbers, respectively. For the expectation value of the contact term we have to evaluate the matrix element

$$\langle \psi_{e-\mu} | \mathcal{H}'_{\text{F. cont}} | \psi_{e-\mu}\rangle$$

$$= \langle n = 1, L = 0, m_S, m_{I_\mu} | \ -\frac{2\mu_0}{3} \boldsymbol{\mu}_\mu \cdot \boldsymbol{\mu}_e \, \delta(\mathbf{r}) \ |n = 1, L = 0, m_S, m_{I_\mu}\rangle \ . \qquad (5.13)$$

Expressing the matrix element in terms of the spins we have

$$\langle \psi_{e-\mu} | \mathcal{H}'_{\text{F. cont}} | \psi_{e-\mu}\rangle = A \langle m_S, m_{I_\mu} \ | \ \mathbf{I}_\mu \cdot \mathbf{S} \ | m_S, m_{I_\mu}\rangle \ , \qquad (5.14)$$

where, with $\langle \mathbf{r} \,|\, n = 1, L = 0\rangle = \psi_{1s}(r)$

$$A = \frac{2\,\mu_0}{3} g_e \,\mu_B \, g_\mu \,\mu_B^\mu \,\underbrace{\langle n = 1, L = 0| \ \delta(\mathbf{r}) \ |n = 1, L = 0\rangle}_{|\psi_{1s}(0)|^2} \ . \qquad (5.15)$$

Because of the δ-function, the integration over the spatial coordinate gives the probability of finding an electron at the muon site, $|\psi_{1s}(0)|^2$. A is the (Fermi contact) hyperfine coupling constant and has here the dimension of energy.

We can write the Fermi contact Hamiltonian as an effective spin-spin Hamiltonian[7, 8]

$$\mathcal{H}'_{\text{F. cont}} = A\, \mathbf{I}_\mu \cdot \mathbf{S} = -\boldsymbol{\mu}_\mu \cdot \mathbf{B}_{\text{F. cont}} \tag{5.16}$$

which has the form of a Zeeman interaction of the muon magnetic moment $\boldsymbol{\mu}_\mu$ with an effective Fermi contact field

$$\mathbf{B}_{\text{F. cont}} = -\frac{2\mu_0}{3}\,|\psi_{1s}(0)|^2\, g_e\mu_B \mathbf{S} = \frac{2\mu_0}{3}\,|\psi_{1s}(0)|^2\, \boldsymbol{\mu}_e \;. \tag{5.17}$$

5.1.2 Hyperfine Contributions in a Solid

Although Eq. 5.11 represents the interaction of a muon spin with a single electron, it contains the main interactions that we have to consider when studying the behavior of the muon spin in a solid where many electrons, localized around the lattice sites or not, may be present in its surrounding.[9]

Orbital Field in a Solid $\mathbf{B}_{\text{orb}}$ is the hypothetical field created by an electron orbiting around the muon with angular momentum $\mathbf{L}$. In a solid such a situation has not been observed yet and one can safely discard this contribution.[10]

Dipolar Field in Solid The electrons in the solid creating a magnetic field sensed by the muon can be roughly divided in localized and nonlocalized electrons. Electrons orbiting around ions at lattice sites can produce localized electronic moments that create a dipolar field at the muon position $\mathbf{r}_\mu$, which is simply the sum over the electronic moments $\boldsymbol{\mu}_j$ at the lattice sites $\mathbf{r}_j$.[11] With $\mathbf{R}_j = \mathbf{r}_j - \mathbf{r}_\mu$ we

---

[7] Note that in this Hamiltonian only the spin operators remain, and that the result of the spatial integration is contained in A and $\mathbf{B}_{\text{F. cont}}$, respectively.

[8] Alternatively, written as a moment-moment interaction, $\mathcal{H}'_{\text{F. cont}} = -A'\boldsymbol{\mu}_\mu \cdot \boldsymbol{\mu}_e$ with $A = -g_e\,\mu_B\,g_\mu\,\mu_B^\mu\,A'$.

[9] Within a mean-field approximation, the problem in the solid, can be dealt with by replacing the electronic operators by their expectation values (i.e., $\mathbf{S} \to \langle \mathbf{S} \rangle$).

[10] In the magnetic materials considered here there are no muon-electron bound states, i.e., no muonium formation. Moreover, "free" muonium in the ground state has no orbital contribution ($L = 0$). See Sect. 7.6 for an example of weakly muon-electron bound states in certain magnetic materials.

[11] These electrons have neither an orbital component with respect to the muon nor a contact component, since the muon is generally at an interstitial or substitutional site. They can, however, polarize conduction electrons, see Sect. 5.1.2.

can write this contribution as

$$\mathbf{B}_{\mathrm{dip}}(\mathbf{r}_\mu) = \frac{\mu_0}{4\pi} \sum_j \left(\frac{3\mathbf{R}_j (\boldsymbol{\mu}_j \cdot \mathbf{R}_j)}{R_j^5} - \frac{\boldsymbol{\mu}_j}{R_j^3} \right) \, , \tag{5.18}$$

or with the dipolar field tensor $\mathcal{D}_{j,\alpha\beta}(\mathbf{r}_\mu)$[12]

$$\mathcal{D}_{j,\alpha\beta}(\mathbf{r}_\mu) = \frac{\mu_0}{4\pi R_j^3} \left(\frac{3 R_{j,\alpha} R_{j,\beta}}{R_j^2} - \delta_{\alpha\beta} \right) \tag{5.19}$$

$$B_{\mathrm{dip},\alpha}(\mathbf{r}_\mu) = \sum_j \sum_{\beta=1}^{3} \mathcal{D}_{j,\alpha\beta}(\mathbf{r}_\mu)\mu_{j,\beta} \, . \tag{5.20}$$

In an atom or lattice ion the electronic magnetic moment is determined by the orbital and spin moments of the electrons in the open shells. In many-electron atoms there are several ways in which the angular momenta associated with the orbital and spin motions combine together to form a total angular moment $\mathbf{J}_j$. The two principal coupling schemes are the Russell-Saunders (or $L - S$) coupling and the jj coupling (for heavy atoms).[13]

The intrinsic magnetism of lanthanide and actinide ions is essentially determined by the filling of the $4f$ and, to a lesser extent, of the $5f$ electronic shell, respectively. For the lanthanides, where the $L - S$ coupling holds well, $\mathbf{J}_j = \sum_i \mathbf{L}_{j,i} + \sum_i \mathbf{S}_{j,i}$ where the sum is over the unpaired electrons in the $4f$ or $5f$ shells of the atom at site j. $\boldsymbol{\mu}_j = -g_J \mu_\mathrm{B} \mathbf{J}_j$ (g_J is the Landé g-factor). For transition elements,[14] the total orbital momentum $\sum_i \mathbf{L}_i$ is usually small and to a good approximation, $\mathbf{J}_j = \sum_i \mathbf{S}_{j,i} = \mathbf{S}_j$.
Inspecting Eq. 5.18 we remark a few points:

- The dipolar field sensed by the muon at its stopping site scales linearly with the value of the electronic moment.
- The value and direction of the dipolar field are strongly dependent on the muon stopping site $\mathbf{r}_\mu$ with respect to the positions $\mathbf{r}_j$ of the electronic moments $\boldsymbol{\mu}_j$, Fig. 5.1.

---

[12] The dipolar field tensor is traceless and symmetric.

[13] In our notation we would write $J_{j,i} J_{j,i}$ coupling, where $\mathbf{J}_{j,i} = \mathbf{L}_{j,i} + \mathbf{S}_{j,i}$.

[14] According to IUPAC, transition elements are elements or ions with incompletely filled d subshells. Practically, also the elements of the zinc group with d^{10} configuration are denoted as transition elements.

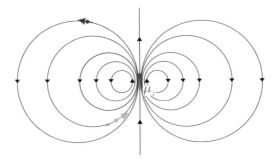

Fig. 5.1 Dipolar field of a single electronic moment. Direction and value (sketched) at two different muon sites are shown

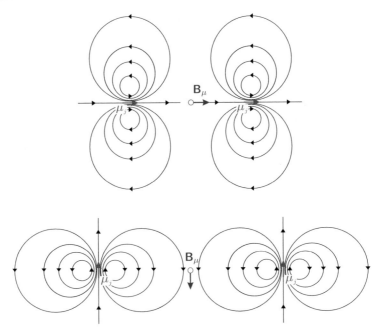

Fig. 5.2 Direction of the dipolar field sensed by a muon located at two different sites with respect to two electronic moments. Note that in the lower figure the local field seen by the muon is antiparallel to the orientation of the electronic moments

- The dipolar field is anisotropic. Therefore, the field at the muon site is not necessarily parallel to the direction of the electronic moments or, for example, in the case of a ferromagnet, to the direction of the magnetization, Fig. 5.2.
- For muons stopping at equivalent crystallographic sites, the direction and value of the field seen by the muon depends on the magnetic structure and is often not unique. This means that crystallographic equivalent sites may be magnetically inequivalent (Fig. 5.3).

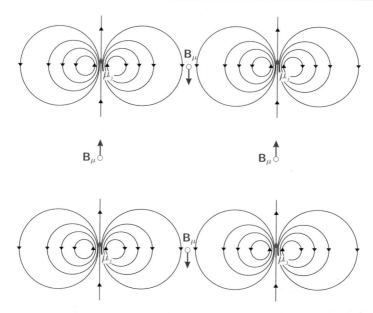

Fig. 5.3 Simple square arrangement of ferromagnetically aligned moments schematically indicating how the dipolar field experienced by an interstitial muon can be parallel and antiparallel to the bulk magnetisation. The stopping sites between two electronic moments are crystallographically equivalent but magnetically inequivalent

Contact Field in Solids A nonzero electronic spin density at the muon site generates a contact field. The density can be induced by an external magnetic field, (e.g. Pauli paramagnetism of the conduction electrons, Sect. 5.6.1) or by spontaneous magnetic order, see examples in Sect. 5.1.4. The first contribution is the Fermi contact term[15]

$$\mathbf{B}_{\text{F. cont}}(\mathbf{r}_\mu) = -\frac{2\mu_0}{3}\,|\psi_s(\mathbf{r}_\mu)|^2\,g_e\mu_{\text{B}}\mathbf{S} = \frac{2\mu_0}{3}\,|\psi_s(\mathbf{r}_\mu)|^2\,\boldsymbol{\mu}_e\;, \qquad (5.21)$$

where ψ_s is the wave function of spin polarized conduction (delocalized) electrons at the muon position $\mathbf{r}_\mu$ (see Exercise 5.2 for a simple derivation of this formula).

The muon may perturb its local environment. For instance, the density of the conduction electrons may be locally enhanced (and their charges screen the muon charge, see Sect. 2.3) and a local lattice expansion may take place, see example in Sect. 5.7.2. To account for this perturbation, the bare contact field is often multiplied

---

[15] The Fermi contact field is sometimes also called, following NMR nomenclature, hyperfine field.

by a charge enhancement factor $\eta(\mathbf{r}_\mu)$ and Eq. 5.21 is modified to[16]

$$\mathbf{B}_{\text{F. cont}}(\mathbf{r}_\mu) = \frac{2\mu_0}{3} \eta(\mathbf{r}_\mu) \, |\psi_s(\mathbf{r}_\mu)|^2 \, \boldsymbol{\mu}_e \ . \tag{5.22}$$

In principle, the field at the muon site arising from polarized conduction electrons can contain other contributions. To understand this one has to go back to Eq. 5.7 and consider the expectation value of the dipolar term $\langle \psi(\mathbf{r})|(\nabla_\alpha \nabla_\beta -\frac{1}{3}\nabla^2 \delta_{\alpha\beta})/r \, |\psi(\mathbf{r})\rangle$ for a general wave function $\psi(\mathbf{r})$. If the electron density distribution is not of spherical or cubic symmetry with respect to the muon site, the polarized conduction electrons can also produce dipolar fields. By contrast, for a spherically symmetric screening cloud only the contact term contributes to the local field since the dipolar tensor transforms as a spherical harmonics of order two and the matrix element is zero.

An additional possible contribution to the contact field is through the Ruderman-Kittel-Kasuya-Yosida (RKKY) interaction, $\mathbf{B}_{\text{cont}, fd}$ (Ruderman & Kittel 1954; Kasuya 1956; Yosida 1957), see Eqs. 5.82 and 5.83.

The RKKY mechanism represents an indirect exchange interaction, mediated by the conduction electrons, between the muon and the local magnetic moments. In this mechanism, the magnetic moments of the conduction electrons are coupled to the magnetic moments of the localized electron through a contact-type exchange interaction. These electrons, in turn, are coupled to the muon spin via the previously discussed Fermi contact interaction. The exact strength of the RKKY interaction depends on the electronic density of states at the Fermi level, the volume enclosed by the Fermi surface, the exchange interaction between the moments of the localized and conduction electrons, and the electronic density at the muon site. The direct (Fermi) and the RKKY-enhanced contact interaction mediated by the conduction electrons are expected to vanish in the insulating state.[17]

For ferromagnetic materials a more general expression for the contact field is often used

$$\mathbf{B}_{\text{cont}}(\mathbf{r}_\mu) = -\frac{2\mu_0}{3} \mu_B \left(n^\uparrow(\mathbf{r}_\mu) - n^\downarrow(\mathbf{r}_\mu) \right) \ . \tag{5.23}$$

The sign convention used in this expression is such that the contact field is positive if parallel to the bulk magnetization in the positive z direction. $n^\downarrow \, (n^\uparrow)$ is the density

---

[16] The expression $\eta(\mathbf{r}_\mu)|\psi_s(\mathbf{r}_\mu)|^2$ is the square of the s-wave like states at the muon site averaged over the electrons at the Fermi surface, including the local electronic structure modification due to the muon charge.

[17] However, a contact field may still arise in the insulating state from the overlap of the muon with the wavefunction of localized magnetic moments, or from the bonding of the muon to an ion covalently bonded to a local magnetic moment. We consider these contact contributions to be included in the $\mathbf{B}_{\text{cont}, fd}$ term.

of electrons with spin antiparallel (parallel)). The contact field at the muon site in magnetic substances can be quite high, on the order of one tenth of a Tesla, see the iron and nickel examples in Sect. 5.1.4.

Recently, state-of-the-art DFT calculations[18] have shown that it is possible to calculate the electronic and magnetic properties of the host material and to quantify the lattice distortion around the muon and its influence on the magnetic and electronic properties of the nearest atoms. Using the electron spin polarization obtained from the DFT simulations, from Eq. 5.23 one can then evaluate the contact field at the muon site (Onuorah et al. 2018), even including the (relatively small) effects of the muon zero point motion (Onuorah et al. 2019).

Summarizing, the main contributions to the local field are the dipole and the contact fields arising from the localized and the conduction electrons, respectively. For s and p electron metals, the dipolar field basically vanishes and solely the contact field is present. On the other hand, in systems with unpaired d and f electrons, which can be considered localized at the ion sites,[19] the dipolar field is dominant.

5.1.3 Demagnetizing and Lorentz Fields

In this section we show how to calculate the local field $\mathbf{B}_\mu$ in a magnetic sample. To obtain the dipolar field contribution one has to sum over all local magnetic moments in the sample, Eq. 5.18, i.e., its value it depends on the sample shape. Although the dipolar field of a single moment decreases as $1/R^3$, since the number of moments increases with R^2, for a ferromagnetic alignment of localized moments the sum is expected to diverge logarithmically.

To address this problem one applies the so-called Lorentz construction, introduced for ferroelectric samples and, by analogy, applied to ferromagnetic ones (Jackson 1998).[20] A homogeneously magnetized sample is ideally divided into two parts. A hollow volume (Lorentz sphere) is ideally cut out around the muon and the dipolar sum is performed only over the lattice sites inside the sphere yielding $\mathbf{B}'_{\mathrm{dip}}$. The radius of the sphere R_{Lor} is taken to be large enough so that the sum is over a sufficiently large number of dipoles and the result does not depend on the size of the sphere. The magnetic moments contained in the rest of the sample are treated as

---

[18] DFT is a computational quantum mechanical modeling method for investigating the electronic structure of many-body systems (principally the ground state). In DFT, the properties of the many-electron system are determined by using functionals of the spatially dependent electron density.

[19] This can be verified by measurements of the magnetic form factor by neutron scattering.

[20] The results shown here for a saturated ferromagnetic sample can also be used for paramagnetic materials magnetized by an external field. In paramagnets with typical fields used in μSR (a few tenths to a few T), the magnetization is usually small. In some cases, however, it is nonnegligible and of relevance for Knight shift measurements, see Sect. 5.6. For diamagnets $\mathbf{B}_{\mathrm{L}}$ and $\mathbf{B}_{\mathrm{dem}}$ are negligibly small.

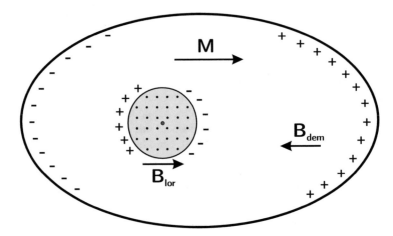

Fig. 5.4 Schematic representation of the contributions to the local field $\mathbf{B}_\mu$ according to the Lorentz construction in a single domain ferromagnetic sample of magnetization $\mathbf{M}$. The muon site is indicated by the red point at the center of the Lorentz sphere. The magnetization is accompanied by a demagnetizing field $\mathbf{B}_{\mathrm{dem}}$, which depends on the shape of the sample. In addition, there is the so-called "Lorentz field" inside the hollow sphere. The fictitious surface magnetic charges responsible for the demagnetizing and the Lorentz field are shown. The black points inside the sphere are the lattice sites of the magnetic moments contributing to the dipolar field. If a contact field is present at the muon site, it has to be added

forming a continuous medium of magnetization $\mathbf{M}$ and contribute to the local field with the demagnetizing field $\mathbf{B}_{\mathrm{dem}}$ and the so-called Lorentz field $\mathbf{B}_{\mathrm{Lor}}$, Fig. 5.4.

As shown in Appendix D, a single-domain ferromagnetic sample with magnetization $\mathbf{M}$ is accompanied by a demagnetizing field, which can be understood as arising from the uncompensated "magnetic charges" at the surface of the sphere[21]

$$\mathbf{B}_{\mathrm{dem}} = -\mu_0 \tilde{N} \, \mathbf{M} \ . \tag{5.24}$$

The demagnetization tensor $\tilde{N}$ depends on the geometrical shape of the sample.[22] In addition, uncompensated "magnetic charges" at the surface of the cavity (originating from carving the Lorentz sphere out of the continuum) create the Lorentz field

$$\mathbf{B}_{\mathrm{Lor}} = \frac{1}{3} \, \mu_0 \mathbf{M} \ . \tag{5.25}$$

---

[21] Alternatively, if the dipole is pictured as a small current loop, the demagnetizing and Lorentz fields can be understood as arising from the currents flowing at the sample surface and at the inner surface of the Lorentz cavity, respectively.

[22] Only for ellipsoidal samples a simple demagnetization tensor can be given and the demagnetizing field is uniform. $\mathbf{B}_{\mathrm{dem}}$ is aligned with $\mathbf{M}$, only if the magnetization is along one of the principal axes of the sample.

The factor $1/3$ is a consequence of the spherical shape of the Lorentz cavity.

The contribution to $\mathbf{B}_\mu$ from the magnetic moment inside the Lorentz sphere is calculated by summing over the corresponding sites

$$\mathbf{B}'_{\mathrm{dip}}(\mathbf{r}_\mu) = \frac{\mu_0}{4\pi} \sum_{j \in \mathrm{Lor}} \left(\frac{3\mathbf{R}_j(\boldsymbol{\mu}_j \cdot \mathbf{R}_j)}{R_j^5} - \frac{\boldsymbol{\mu}_j}{R_j^3} \right) \quad , \quad \mathbf{R}_j = \mathbf{r}_j - \mathbf{r}_\mu \ . \qquad (5.26)$$

Additionally, the contact field must be added.

For the total local field at the muon site in the ferromagnet we obtain

$$\mathbf{B}_\mu = \mathbf{B}'_{\mathrm{dip}} + \mathbf{B}_{\mathrm{cont}} + \mathbf{B}_{\mathrm{Lor}} + \mathbf{B}_{\mathrm{dem}} \ . \qquad (5.27)$$

In a saturated magnet (but also, as mentioned, in a paramagnet), if the sample is a sphere, Lorentz field and demagnetizing field cancel each other and we have

$$\mathbf{B}_\mu = \mathbf{B}'_{\mathrm{dip}} + \mathbf{B}_{\mathrm{cont}} \ . \qquad (5.28)$$

In zero field, a ferro- or ferrimagnet generally displays multiple domains. If there is no remanent magnetization, the sample magnetization, which determines $\mathbf{B}_{\mathrm{dem}}$, is zero ($\mathbf{M} = 0$) and therefore $\mathbf{B}_{\mathrm{dem}} = 0$. In this case the Lorentz sphere is best chosen in a domain of magnetization $\mathbf{M}_{\mathrm{dom}}$ to have a well defined Lorentz field $\mathbf{B}_{\mathrm{Lor}} = \frac{1}{3}\mu_0\mathbf{M}_{\mathrm{dom}}$. The field at the muon site is then given by

$$\mathbf{B}_\mu = \mathbf{B}'_{\mathrm{dip}} + \mathbf{B}_{\mathrm{cont}} + \mathbf{B}_{\mathrm{Lor}} \ . \qquad (5.29)$$

In an antiferromagnet the bulk sample and the domain magnetization vanish $\mathbf{M} = \mathbf{M}_{\mathrm{dom}} = 0$. Therefore, in this case, $\mathbf{B}_{\mathrm{dem}} = \mathbf{B}_{\mathrm{Lor}} = 0$ and the field at the muon site is given by

$$\mathbf{B}_\mu = \mathbf{B}'_{\mathrm{dip}} + \mathbf{B}_{\mathrm{cont}} \ . \qquad (5.30)$$

If the muon site has cubic symmetry with respect to the surrounding magnetic moments $\mathbf{B}'_{\mathrm{dip}} = 0$.[23] In other cases it may average out (see as an example Fe, next section).

The standard μSR measurement for studying magnetic order is a zero field measurement, ideally in a spherical sample. This avoids inconvenient $\mathbf{B}_{\mathrm{dem}}$ contributions.[24] In magnetic materials the dipolar field is generally larger than the

---

[23] This follows from the symmetry properties of the dipolar tensor. The use of the symmetry properties is a very powerful tool to analyze dipolar field contributions, see Seeger (1978).

[24] However, note that, even if $\langle B_{\mathrm{dem}} \rangle = 0$, sample inhomogeneities (e.g., in polycrystalline samples with irregularly shaped grains) may lead to nonhomogeneous fields and increase the width of the local field $\sqrt{\langle \Delta B_\mu^2 \rangle}$.

contact field. An order of magnitude can be estimated by noting that $B_{\text{dip}} \approx \mu_0 \mu_j/(4\pi R_j^3) \approx (\mu_j[\mu_B])/(R_j^3[\text{Å}^3])$ [T] giving, e.g., for a $1\mu_B$ electronic moment of a neighbor atom with $R_j = 2$ Å, typical fields in the tenths of T (or kG) range.

5.1.4   Examples of Local Field Determination

Elemental Magnets The transition elements nickel and iron provide examples where there is no dipolar field contribution to the local field and only the contact contribution is present (Denison et al. 1979). Measurements of $B_{\text{cont}}(T)$ are of interest in the physics of metals, because they provide information about local magnetic properties such as the interstitial magnetization or spin density.

Elemental Nickel Nickel, with a Curie temperature $T_C = 631.0$ K and a magnetic moment $\mu_j = 0.616 \mu_B$, crystallizes in the face-centered cubic structure (FCC). The easy axes of magnetization are along the cube diagonals. At the octahedral and tetrahedral interstitial sites, Fig. 5.5, as well as at the Ni substitutional sites, the dipolar field is zero due to the cubic symmetry. A single muon spin precession frequency is observed in temperature dependent ZF measurements, meaning that there is the same local field at all muon sites. From these data we obtain the contact field as a function of temperature

$$\mathbf{B}_{\text{cont}} = \mathbf{B}_\mu - \mathbf{B}_{\text{Lor}} = \mathbf{B}_\mu - \frac{1}{3}\mu_0 \mathbf{M} \ . \tag{5.31}$$

A measurement of $|\mathbf{B}_\mu|$ as a function of the external field for a single crystal spherical sample (fabricated to high precision by electroerosion) shows that the local field and the domain magnetization have the same direction. From the extrapolated value at $T = 0$ K, $B_\mu(0) = 0.148(1)$ T and the value of the saturation magnetization $\mu_0 M_S(0) = 0.66$ T, one obtains $B_{\text{cont}}(0) = -0.071$ T, see Fig. 5.6. The contact field, which is proportional to the local magnetization, points opposite to the total magnetization. Figure 5.6 indicates that the temperature dependence of the contact field does not simply follow that of the magnetization.

Elemental Iron An additional example is provided by α-Fe. It crystallizes at room temperature in a body-centered cubic structure (BCC), Fig. 5.7. The order is ferromagnetic, with the moments pointing, for instance, in a single domain sample along the c-axis. Interestingly, all the dipolar fields are aligned along the magnetization direction. The dipolar sums at the possible muon sites (6 equivalent octahedral interstitial sites and 12 tetrahedral interstitial sites in the crystallographic unit cell) are not individually zero as in Ni. However, the mean of the site-averaged

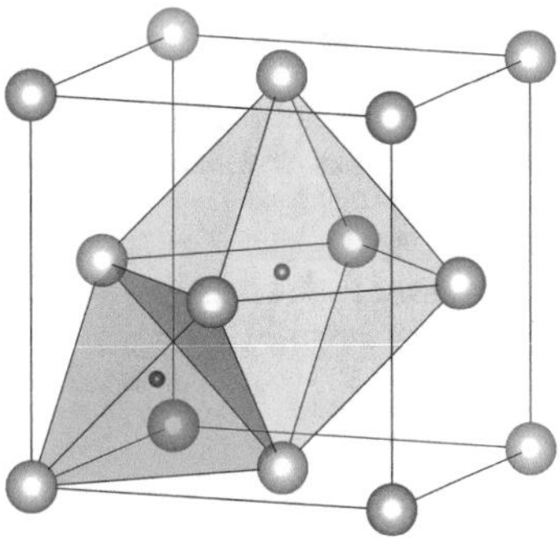

Fig. 5.5 FCC structure of
nickel with an octahedral and
a tetrahedral muon interstitial
site. There are 4 octahedral
and 8 tetrahedral interstitial
sites per unit cell. Structure
drawn with VESTA (Momma
& Izumi 2011)

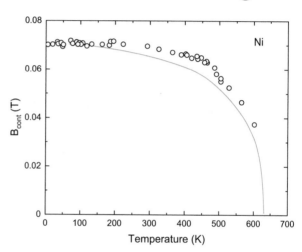

Fig. 5.6 Absolute value of
the contact field in nickel as a
function of temperature,
obtained from ZF
measurements in single
crystal and polycrystalline
samples. Note that the bulk
saturation magnetization
(solid line), normalized at
$T = 0$ K, is below $B_{cont}(T)$.
Data from Denison et al.
(1979), where a detailed
account of μSR studies of
various ferromagnetic
elements can be found

dipolar field values, weighted by equal occupation, is zero. This is attributed to
muon diffusion, which is sufficiently fast at all temperatures to average out $\mathbf{B}_{dip}$.

Therefore, also in Fe the local field at the muon site is simply the sum of $\mathbf{B}_{cont}$
and $\mathbf{B}_{Lor}$ and knowing the Lorentz contribution, the contact field can be extracted,
see Fig. 5.8. As in Ni, the temperature dependence of the contact field is not strictly
proportional to $M(T)$. This is attributed to the fact that the muon as a charge and
spin impurity modifies the intrinsic local electronic and spin density, see Sect. 2.3.
The measured values of $B_{cont}(0)$ in transition elements such as Ni, Fe, and Co,
but also in non-centrosymmetric metallic magnetic compounds with more complex
spiral magnetic structures, such as MnSi and MnGe, can be well reproduced by DFT
calculations (Onuorah et al. 2018, 2019).

It should be stressed that the contact field is generally much smaller than the
dipolar field at the muon site. In this case, $B_\mu(T)$ is expected to be an accurate
measure of the magnetic order parameter, because the influence of the muon on the
dipolar arrangement is much smaller than on the local spin density.

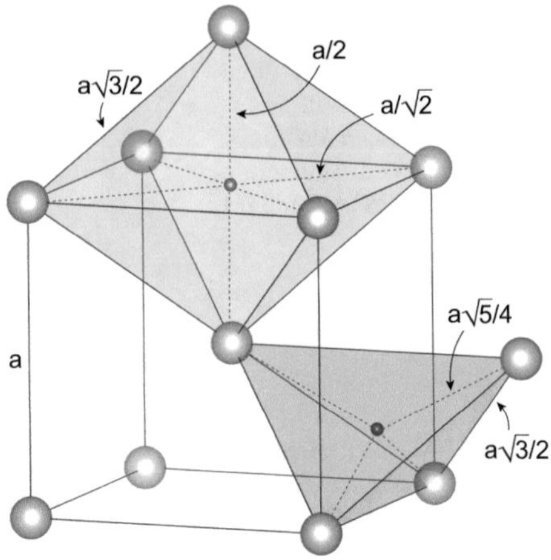

Fig. 5.7 Example of an octahedral (blue) and a tetrahedral (red) interstitial site of BCC iron. Structure drawn with VESTA, © 2006–2022 Koichi Momma and Fujio Izumi (Momma & Izumi 2011)

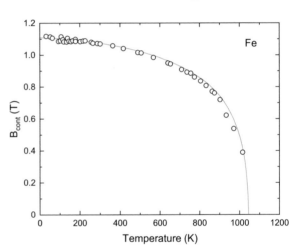

Fig. 5.8 $B_{\mathrm{cont}}(T)$ compared with the normalized magnetization curve $M(T)$ (solid line) for iron. Data taken from Denison et al. (1979)

Nickel and iron are examples of ferromagnetic systems, where, for different reasons, the dipolar contribution vanishes. In an antiferromagnet, with the muon sitting at a high-symmetry site, there is a high probability that not only the dipolar field but also the contact field cancels.

Heavy Fermion Compound UPd$_2$Al$_3$ UPd$_2$Al$_3$, a heavy fermion system[25] with hexagonal structure is an example where both dipolar and contact field cancel. Heavy fermions have attracted particular attention, because of the possibility to realize unconventional superconducting states[26] with a close proximity of super-

---

[25] Heavy fermion systems are compounds containing elements with $4f$ or $5f$ electrons in unfilled electron bands, which, below certain temperatures, have charge carriers with very large effective electron masses (sometimes up to 1000 times larger than the free electron mass).

[26] About unconventional superconductors see Chap. 6 and footnote 58.

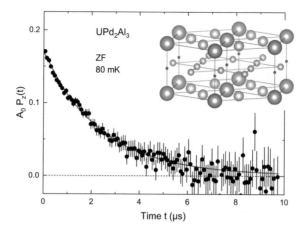

Fig. 5.9 ZF-μSR signal in UPd$_2$Al$_3$ at 80 mK and crystallographic structure. The muon sites are indicated by small red spheres. The large spheres represent the U ions, the Pd ions are located in the same plane. The Al ions are in the same plane as the muon. Modified from Amato et al. (1992), © EDP Sciences. Reproduced with permission. All rights reserved. Structure drawn with VESTA, © 2006–2022 Koichi Momma and Fujio Izumi (Momma & Izumi 2011)

conductivity and magnetism. UPd$_2$Al$_3$ undergoes an antiferromagnetic transition at $T_N = 14$ K followed by a superconducting transition at $T_c = 2$ K (Geibel et al. 1991). Thermodynamic measurements suggest the coexistence of superconductivity and long-range magnetism at low temperatures.

While neutron scattering data indicate an antiferromagnetic structure with propagation vector $\mathbf{k} = (0, 0, 1/2)$ and magnetic moments of U lying in the basal plane, no spontaneous muon spin precession frequency below T_N is observed. Figure 5.9 shows the behavior of the ZF-μSR signal at 80 mK, which is best fitted by an exponential decay with relaxation rate $\lambda = 0.42\,\mu\text{s}^{-1}$. Between 80 mK and 2.5 K this value is temperature independent.

The nonobservation of spontaneous muon spin precession frequencies in such a simple magnetic structure suggests that all muons stop at a highly symmetric site, where the local fields arising from the ordered $5f$ moments cancel.

The only interstitial candidate site of proper symmetry is the b site,[27] see Fig. 5.9. Below T_N part of the depolarization arises from the random local fields due to the nuclear magnetic moment of the ^{27}Al-nuclei, see Sect. 4.1.2.1. For the b site, the calculated value of the nuclear depolarization rate is $0.175\,\mu\text{s}^{-1}$, smaller than the observed rate. The appearance of a weakly relaxing μSR signal as well as the observation of an exponential relaxation indicate that the cancellation of the dipole and contact fields is not perfect. The result suggests that the increase of the depolarization rate above the nuclear value is due to distortions of the magnetic

---

[27] See Sect. 5.6.4 about the site determination in this compound.

sublattice at the muon site, which lead to a distribution of internal fields around the zero value.

5.2   Magnetic Volume Fraction and Magnetic Transitions

With a so-called weak transverse field[28] experiment (wTF, typically 5–10 mT), it is possible to quickly determine the sample fraction of a magnetic phase and the transition temperature without making any assumptions about the magnetic structure.

   The measurement is usually performed as a function of temperature starting from the high temperature paramagnetic phase, see Fig. 5.10. The amplitude of the muon signal precessing at a frequency corresponding to the applied field B_{ext} reflects the volume fraction of the sample that is paramagnetic or not magnetically ordered.[29] As the temperature is lowered and the magnetic phase is entered, muons stopping in the magnetic regions begin to experience a broader distribution of fields, leading to a rapid decay of the muon spin asymmetry at early times. Therefore, the amplitude of the muon signal precessing at the frequency corresponding to B_{ext} starts to decrease at the magnetic transition and reaches a level determined by the nonmagnetic phase (which may also include contributions from the background signal).

   The magnetic volume fraction is given by

$$\frac{V_{magn}(T)}{V_{tot}} = 1 - \frac{A_{0,wTF}(T)}{A_0} \tag{5.32}$$

with

$$A_{wTF}(t, T) = A_{0,wTF}(T)\, G(t) \cos(\gamma_\mu B_{ext}\, t + \varphi) \ . \tag{5.33}$$

A_0 is the experimental asymmetry of the μSR spectrometer and $G(t)$ takes into account the field broadening.

   It is also possible to determine the magnetic volume fraction from ZF measurements. However, one needs to know the polarization function appropriate for the physical situation. In contrast, wTF measurements are extremely fast and straightforward to interpret.

   Besides the possibility of an independent determination of magnetic volume and moment size, an advantage of the μSR technique is that polarized muons can be implanted in any material. This allows the study of magnetic properties also in cases that are not so easily accessible by other means, e.g., when magnetism occurs at

----

[28] This field must be weak with respect to the typical local field probed by the muon in the magnetic phase.

[29] Actually, the precession does not occur exactly at B_{ext}, but at a field value slightly modified by the Knight-shift, see Sect. 5.6. For low fields this difference can be safely neglected.

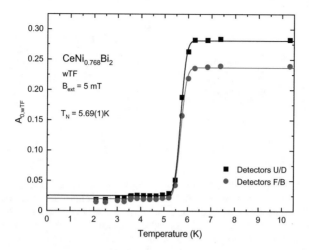

Fig. 5.10 Plot of $A_{0,\text{wTF}}(T)$ for two pairs of positron detectors Up-Down and Left-Right in a wTF of 5 mT in $CeNi_{0.768}Bi_2$. Note the different maximum experimental asymmetry for the two pairs and the background baseline due to muon stopping in nonmagnetic regions around the sample. From the error function fit (solid line) one can determine the magnetic transition and the corresponding volume fraction (Morenzoni et al., unpublished data)

very low temperature (sub-Kelvin range) or is weak. Below we briefly present some examples of magnetic studies making use of these specific properties.

5.2.1   Examples

Quantum Phase Transition in Mott Insulators $RE\text{NiO}_3$ (RE rare earth element) is an archetypal Mott insulator system.[30] When tuned by chemical substitution[31] of the RE element the antiferromagnetic Mott insulating phase can be suppressed resulting in a zero-temperature quantum phase transition (QPT), see footnote 51, to a paramagnetic metallic state. Since novel physics often appears near a Mott QPT, details of this transition, such as whether it is first or second order, are important.

ZF and wTF μSR measurements show (Frandsen et al. 2016) that the QPT in $RE\text{NiO}_3$ is first order: the magnetically ordered volume fraction (determined from wTF measurements, Fig. 5.11) decreases steadily until it reaches zero at the

---

[30] Mott insulators are a class of materials that, according to conventional band theories, should have metallic properties, but are actually insulators. This can occur when, for given concentrations of electrons, the Coulomb repulsion between the electrons and the electron correlations, which are not properly accounted for in conventional band theory, are strong.

[31] This procedure is called sometimes quantum tuning, i.e., a nonthermal parameter such as chemical composition or pressure is varied.

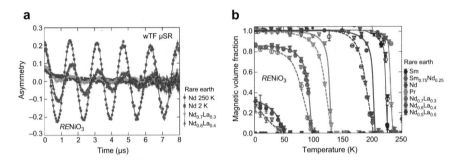

Fig. 5.11 (**a**) wTF asymmetry spectra for three RENiO$_3$ compounds at 2 K near the quantum phase transition (QPT). A spectrum measured at a higher temperature (250 K) is shown for comparison. (**b**) Temperature dependence of the magnetic volume fraction in RENiO$_3$ derived from wTF and ZF measurements. Circles (triangles) represent wTF (ZF) measurements, and filled (open) symbols represent heating (cooling) sequences. Solid and dashed curves are guides to the eye, with solid corresponding to cooling and dashed to warming. The compounds near the QPT (Nd$_{0.7}$La$_{0.3}$NiO$_3$ and Nd$_{0.6}$La$_{0.4}$NiO$_3$, Nd$_{0.5}$La$_{0.5}$NiO$_3$ is paramagnetic at all temperatures) have a significantly reduced ordered volume fraction at low temperature, indicating phase separation between magnetic and paramagnetic regions of the sample. Adapted from Frandsen et al. (2016), © Springer Nature. Reproduced with permission. All rights reserved

QPT, resulting in a wide region of the phase diagram of intrinsic phase separation, while the ordered magnetic moment (obtained from ZF measurements, Fig. 5.12) retains its full value across the phase diagram until it suddenly disappears at the QPT. Similar results are found for the Mott insulator V$_2$O$_3$ when tuned by pressure (Frandsen et al. 2016).

Pressure Dependence of Magnetic Moment and Volume Fraction   The heavy fermion compound URu$_2$Si$_2$ exhibits coexistence of unconventional superconductivity ($T_c \simeq 1.5$ K) with antiferromagnetism. Neutron scattering Bragg data suggested a very weak ordered moment (0.03 μ_B) below 17.5 K. The significant increase of the magnetic Bragg peak intensity with pressure above ~0.5 GPa was interpreted as an increase of the staggered moment up to 0.33 μ_B/U, see left panel of Fig. 5.13 (Bourdarot et al. 2005).

The Bragg peak intensity is actually proportional to $V_{ord}\,\mu_{ord}^2$ but a constant full magnetic volume fraction was assumed for the determination of the ordered moment. μSR (and NMR), which allow to distinguish between size and volume of the ordered moments, lead to a different conclusion. The data show that the transition becomes magnetic under pressure and that solely the magnetic volume fraction is strongly pressure dependent, while the staggered magnetic moment is essentially constant. A magnetic state in the full sample volume is only obtained at about 1 GPa, right panel of Fig. 5.13 (Amato et al. 2004).

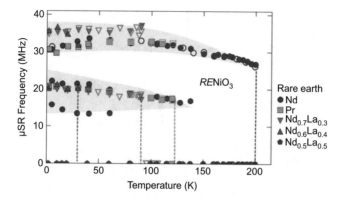

Fig. 5.12 Temperature dependence of the spontaneous frequencies in ZF for various RENiO$_3$ compounds, showing the first-order thermal magnetic phase transitions. Filled (open) circles represent data taken in a cooling (heating) sequence. All magnetically ordered compounds have two or three frequencies lying along two common bands (shaded gray regions), indicating that changing the chemical pressure via RE substitution keeps the ordered moment size similar. The colored dashed lines are guides to the eye indicating the approximate transition temperature for each compound, which tends to zero when plotted as a function of the RE radius (not shown). Adapted from Frandsen et al. (2016), © Springer Nature. Reproduced with permission. All rights reserved

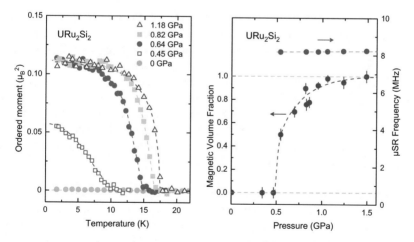

Fig. 5.13 Comparison of neutron scattering (left) and μSR data (right) under pressure in URu$_2$Si$_2$. The neutron scattering data show the temperature dependence of the (100) magnetic Bragg peak intensity for different applied pressures. The data indicate an ordered moment increasing with pressure. The Bragg peak intensity is proportional to $V_{ord}\,\mu_{ord}^2$ but, for the analysis, a constant full magnetic volume fraction was assumed. From the μSR data (right panel) the pressure dependence of the magnetic volume fraction (proportional to the amplitude) and of the ordered moment (proportional to the μSR frequency) are obtained separately. The μSR data show that the pressure does not increase the size of the electronic moment, but does increase the magnetic volume fraction. Modified from Bourdarot et al. (2005) (left) and Amato et al. (2004) (right), © Elsevier (left) and IOP Publishing (right). Reproduced with permission. All rights reserved

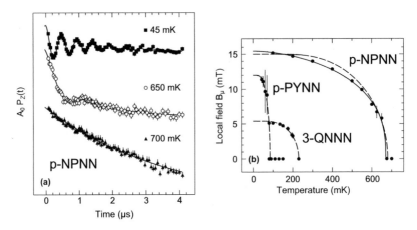

Fig. 5.14 Magnetic phase transition in organic compounds. (**a**) ZF spectra at different temperatures for p-NPNN. At ∼0.67 K there is a clear change between the high temperature paramagnetic
state, with no oscillations, and the low temperature ordered state, in which well-defined oscillations
can be seen. The solid lines are fits to the data. (**b**) Temperature dependence of the local field
(obtained from the spontaneous μSR frequency) in p-NPNN, p-PYNN and 3-QNNN. The solid
line is a fit described in the text. Adapted from Blundell (1997), © Springer Nature. Reproduced
with permission. All rights reserved

Magnetism in Organic Compounds μSR experiments at very low temperatures
(mK) led to the discovery of ordered ferromagnetism in organic materials. The
search for purely organic molecular ferromagnets containing only light elements
(carbon, hydrogen, oxygen and nitrogen) is of great interest, because of the chemical
flexibility of organic compounds. Many organic radicals[32] exist, but only a few are
stable enough to be assembled into crystalline structures. Moreover, even when this
is feasible, it is usually not possible to align the unpaired spins ferromagnetically.
Ferromagnets are rather rare even among the elements and are exclusively found in
the d or f series. It is therefore remarkable that ferromagnetism can be found in
molecular crystals that only have s and p electrons.

μSR data provided the first direct evidence of spontaneous magnetic order in
the so-called β crystal phase of the para-nitrophenyl nitronyl nitroxide radical (p-
NPNN, $C_{13}H_{16}N_3O_4$) with a Curie temperature $T_C \simeq 0.6$ K (Le et al. 1993).
Figure 5.14, right panel, shows the local magnetization of p-NPNN and related
molecular compounds. The ordering was later confirmed by neutron diffraction.
An unpaired electron is associated with the two NO groups present in each
molecule. The residual molecules ensure the overlap of the correct orbitals with
the neighboring molecules, so that 3D ferromagnetic order appears.

---

[32] A radical is an atom, molecule, or ion that has an unpaired valence electron. Generally, these
unpaired electrons make radicals highly chemically reactive, and many radicals spontaneously
dimerize.

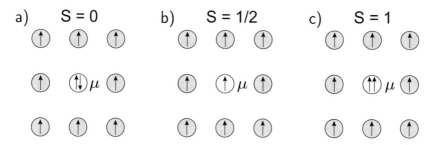

Fig. 5.15 Schematic drawing showing the possible local electronic spin states around the muon in p-NPNN after muon implantation. The muon spin is not shown. (**a**) and (**c**) are formed by muonium addition, (**b**) by a "bare" muon addition. In each case the muon is surrounded by nearest neighbors molecule each with an unpaired electronic spin (gray circles). Modified from Blundell et al. (1995), © EDP Sciences. Reproduced with permission. All rights reserved

Whereas in the ferromagnetic elements discussed previously the crystal structure is simple and the muon sits at highly symmetrical sites, in an organic ferromagnet the muon position is not likely to be so symmetrical because the crystal structure is much more complex and there are many more possible sites. However, even without this knowledge it is possible to obtain detailed information about the magnetic order.

The observation of a clear oscillating signal at low temperature in the ZF spectra of Fig. 5.14a evidences the existence of long-range static magnetism. Fitting the $B_\mu(T)$ data with a function $\propto [1 - (T/T_C)^\alpha]^\beta$ one obtains for p-NPNN $\alpha = 1.7(4)$ and $\beta = 0.36(5)$ (Blundell et al. 1995; Le et al. 1993). This dependence is close to a magnon-like behavior of $M(0) - M(T) \sim T^{3/2}$ for $T \ll T_C$ as well is consistent with a critical magnetization exponent around T_C of about 1/3, the value expected for an isotropic 3D Heisenberg ferromagnet (Blundell 2001).

The muon has been used intensively not only to study the magnetic and superconducting properties of organic materials (for reviews see Lancaster et al. (2013), Nuccio et al. (2014), Pratt (2016) and references therein) but also to study the chemistry of radical.[33]

As an example of the processes that can occur in organic materials, we consider the muon state and the local environment of muons implanted in a molecular radical such as p-NPNN (Fig. 5.15). Following the implantation of a muon in the sample, muonium formed via electron capture may attach to a particular p-NPNN molecule, forming a so-called muoniated radical with muonium bringing a single electronic spin. On its own, each p-NPNN molecule also contains an unpaired spin so that the resulting local electronic spin state around the muon in the muoniated radical can be a singlet, $(S = 0)$ or a triplet $(S = 1)$. Another muon fraction may not thermalize as muonium but as a "free" muon. Attached to a p-NPNN molecule,

---

[33] In this book the muon applications in chemistry are not treated. We refer to books (Roduner 1988; Walker 1983) and review articles (Percival 1979; McKenzie 2013; Rhodes 2002; McKenzie & Roduner 2008) for an overview.

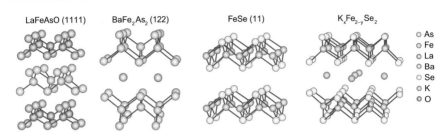

LaFeAsO (1111)   BaFe₂As₂ (122)   FeSe (11)   KₓFe₂₋ᵧSe₂

○ As
○ Fe
○ La
○ Ba
○ Se
○ K
○ O

Fig. 5.16 Crystal structure of some iron-based superconductor families: LaFeAsO (1111), BaFe$_2$As$_2$ (122), FeSe (11) and K$_x$Fe$_{2-y}$Se$_2$. Modified from Si et al. (2016), © Springer Nature.

the resulting electronic spin state would be a doublet ($S = 1/2$).[34] The doublet and triplet states produce a very large hyperfine field at the muon site, leading to a very high unobservable muon spin precession frequency, see Chap. 7. The observation of a muon spin precession signal therefore implies a singlet state. In this diamagnetic state (with the local electronic environment of the muon in an $S = 0$ state) the muon spin precesses in the same way as a free muon, see Sect. 2.3.

Study of Low-Moment Magnetism and Sublattice Magnetization With μSR it is possible to determine the magnetic properties when the ordered moment is small and to distinguish contributions from different sublattice magnetizations. An example is given by the study of the Fe and RE magnetism of the nonsuperconducting REFeAsO, the parent compound of REFeAsO$_{1-x}$F$_x$, an iron-based high-temperature superconductor belonging to the so-called "1111" family. Besides the cuprates, the iron-based superconductors are an important class of high-T_c superconductors. These compounds are based on conducting layers of iron and a pnictogen (typically arsenic or phosphorus) alternating with rare earth oxygen sheets, Fig. 5.16. Similarly to the cuprate family, the properties of iron-based superconductors change dramatically with doping. The parent undoped compounds are metals (unlike the cuprates) but are ordered antiferromagnetically (like the cuprates). The superconducting state emerges upon either hole or electron doping, which suppresses the magnetic order. In the absence of doping the 1111 family shows Fe magnetism with a rather low Fe magnetic moment of about 0.4 μ_B, see Fig. 4.14.[35]

ZF-μSR measurements provide insight about the iron and rare earth magnetism and their interplay (Maeter et al. 2009). At temperatures below T_N^{Fe}, which ranges between 123 and 139 K, a well-defined spontaneous muon spin precession is

---

[34] Such a state could also form when Mu undergoes a substitution reaction with a bound hydrogen, provided the liberated free hydrogen escapes.

[35] This is to be compared with metallic iron with a moment of about 2.2 μ_B.

observed in all compounds indicating a well-defined magnetic field at the muon sites, Fig. 5.17.[36] This indicates that long-range static magnetic order with a commensurate magnetic structure is realized in all investigated compounds and excludes spin glass or incommensurate magnetism, which are characterized by different ZF spectra, see Sect. 5.4. Whereas in the LaFeAsO system only the iron magnetism is detected (La does not have $4f$ electrons), the REFeAsO systems exhibit a second magnetic transition T_N^{RE} at low temperature much below the magnetic ordering of iron T_N^{Fe}.

With the exception of the Ce based system, which shows a higher frequency and a stronger temperature dependence below T_N^{Fe}, all the REFeAsO compounds present a similar temperature dependence of the spontaneous μSR frequency reflecting the magnetization of the Fe sublattice. In LaFeAsO, the observables from the three probes μSR precession frequency, Mössbauer Fe hyperfine field, and square root of the neutron Bragg peak intensity, scale with each other since they all measure the size of the ordered Fe moment. The onset of the rare earth magnetic order for RE=Pr and Ce, at 11(1) and 4.4(3) K, respectively, causes a smooth decrease of the muon precession frequency, as can be seen in Fig. 5.17.

For Sm the transition is first-order, as indicated by the discontinuous Sm magnetic order parameter at T_N^{Sm} = 4.66(5) K. Contrary to the Ce and Pr case, the Sm magnetic order causes the additional appearance of two satellite muon precession frequencies of equal weight. This is an indication that the magnetic order of Sm has a different symmetry than that of Fe and causes a change of the magnetic unit cell.

A qualitatively different behavior is observed for the CeFeAsO compound. The magnetic Bragg intensity as well as the internal magnetic field at the muon site do not scale with the Mössbauer hyperfine field at the Fe site but continuously increase below T_N. ^{57}Fe Mössbauer spectroscopy provides a very accurate measurement of the onsite Fe sublattice magnetization without sizable contribution from the rare earth moments. In contrast, the μSR as well as the neutron data are not only sensitive to the Fe sublattice magnetization, but also contain a significant contribution from the Ce sublattice. A magnetization on the rare earth site induced by the Fe subsystem has the same symmetry as the Fe order and therefore contributes to the same Bragg peaks.[37]

The temperature dependence of the μSR frequency $f_\mu(T)$ in the CeFeAsO system below T_N^{Fe} reflects not only the Fe magnetization but also the magnetization of the Ce moments produced by the field of the magnetized Fe sublattice, i.e., for Ce the Fe field acts as an "external" field inducing a Ce moment. In turn, the induced Ce moments couple to the muon spin (via dipolar and contact fields), so that the local field at the muon site is the sum of the Fe and Ce contributions. We can write

---

[36] In LaFeAsO a second frequency with lower amplitude is observed in addition to the main one. In all systems, the amplitude of precession signal shows that magnetic order develops in the entire sample below the Néel temperature.

[37] In principle, neutron scattering can distinguish between the different contributions from the Fe and the Ce sublattices by taking into account the different magnetic form factors.

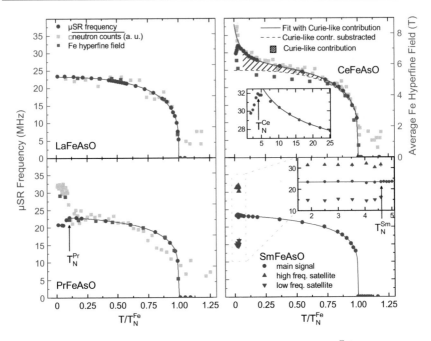

Fig. 5.17 Main μSR frequency as a function of reduced temperature T/T_N^{Fe} from ZF measurements in REFeAsO compounds. Also plotted are the average magnetic hyperfine field at the Fe site from ^{57}Fe Mössbauer spectroscopy and the square root of the magnetic Bragg peak intensity from neutron scattering data. The error bars of the μSR frequency are smaller than the data points. Note: (i) the RE magnetism at low temperatures (with the exception of the La-based system) and (ii) the different temperature dependence of the μSR frequency for the Ce based system. Modified from Maeter et al. (2009), © American Physical Society. Reproduced with permission. All rights reserved

for $T_N^{Ce} < T < T_N^{Fe}$

$$
f_\mu(T) = f_0 \underbrace{\left[1 - \left(\frac{T}{T_N^{Fe}}\right)^\alpha\right]^\gamma}_{\text{Fe-sublattice } f_0(T)} \times \underbrace{\left[1 + \frac{\tilde{C}}{T - \Theta}\right]}_{\text{Curie-like Ce magnetization}} , \qquad (5.34)
$$

where $\tilde{C}$ describes the coupling constant of the Ce moments with the muon spin.

The exceptionally strong coupling of Ce can be understood qualitatively by considering the ground state and the susceptibility of the free RE ions. The ground state of Ce^{3+} is the multiplet $^2F_{5/2}$, that of Sm^{3+} is $^6H_{5/2}$, and Pr^{3+} is in the $^3H_4$ multiplet.[38]

---

[38] The ground state multiplet of the f electron ions is characterized by the total angular momentum J, total orbital momentum L, and total spin S, determined by Hund's rules, $^{2S+1}L_J$. In a compound,

The susceptibility is given by $g_J^2 \mu_B^2 J(J+1)/3T$. Because of the different Landé factors g_J, $(g_{J,\text{Ce}} = 6/7, g_{J,\text{Sm}} = 2/7)$ one expects about a factor of 10 smaller induced Sm magnetic moment compared to Ce for equal magnitudes of Fe-Sm and Fe-Ce coupling constants. For Pr it is more difficult to predict the behavior, since the exact sequence of crystal electric field levels in PrFeAsO is not known.

A fit of Eq. 5.34 to the μSR frequency $f_\mu(T)$ in CeFeAsO is shown in Fig. 5.17 with the two contributions highlighted. This simple model describes the data reasonably well between 10 K and $T_N^{\text{Fe}} = 137$ K. Note that this information has been obtained without precise knowledge of the muon site. If the muon site is known, more details about the Fe and RE magnetic order and their interplay can be obtained. The procedure can be well illustrated for this compound.

The electronic structure of the compound is described by a modified Thomas-Fermi model under inclusion of structural data.[39] A self-consistent distribution of the valence electron density is obtained, from which the electrostatic potential is deduced. The local minima of the potential at interstitial positions are the possible stopping sites of the muons.[40] Two types of possible muon positions are identified, labeled A and B, see Fig. 5.18. The A-type position is located on the line connecting the Ce and As ions along the c direction. The B positions are located at oxygen ions, which are typical muons site in many oxides.[41]

Comparing the calculated the dipole field at the muon sites for an experimentally determined Fe magnetic moment of 0.36 μ_B with the measured $B_\mu(T \to 0)$ for LaFeAsO, one concludes that site A is the main muon site since it gives the correct order of magnitude for the field.[42]

The μSR experiments indicate an antiferromagnetic commensurate order with the iron magnetic moments directed along the a-axis above T_N^{Fe}. This is consistent with previous neutron scattering results. The magnetic field caused by the Fe magnetic order is also directed along the crystallographic c-axis. The calculations additionally show that Fe dipole-Ce dipole interactions are much too weak to account for the Fe-Ce couplings. Instead, the large Fe-Ce effective field is a consequence of a (non-Heisenberg)[43] anisotropic exchange.

---

the magnetic ion is surrounded by other ions producing a static electric field (crystal field), which represents a perturbation with respect to the Coulomb and spin-orbit interactions. Depending on the local symmetry, the degeneracy of the ground state is lifted.

[39] The Thomas-Fermi model can be seen as a predecessor of DFT.

[40] Here the calculations have been done without taking into account the host-lattice relaxation around the muon. This effect can be also included in the calculations, see for instance Onuorah et al. (2018).

[41] The O^{2-} state leads, in analogy to the hydroxide compound OH, to an Oμ compound with the muon close to an oxygen ion.

[42] One has to introduce a renormalization since the calculation neglects the contact field contribution.

[43] By symmetry arguments, a Heisenberg magnetic interaction between the Fe and the RE subsystem is absent in REFeAsO.

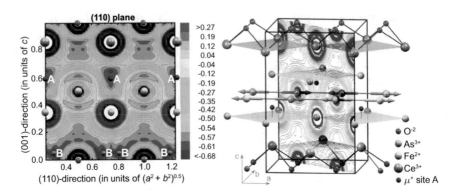

Fig. 5.18 Left: Structure and electrostatic potential energy map of the (110) plane for a muon in the *Cmma* orthorhombic phase of CeFeAsO. Muon positions A and B are indicated. The potential energy is given in atomic units. Right: Unit cell showing the ions, the collinear Fe magnetic order and the A muon position. The origin of the unit cell has been shifted for clarity. Modified from Maeter et al. (2009), © American Physical Society. Reproduced with permission. All rights reserved

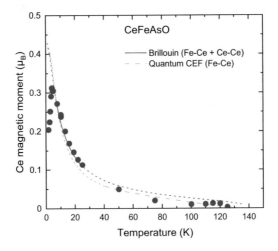

Fig. 5.19 Temperature dependence (for $T > T_N^{Ce}$) of the induced Ce magnetic moment extracted from the μSR frequency data by subtracting the contribution from the Fe subsystem. The data are fitted by a Brillouin and a quantum model approach. Modified from Maeter et al. (2009), © American Physical Society. Reproduced with permission. All rights reserved

The exceptionally large Fe-Ce coupling in CeFeAsO together with the large paramagnetic Ce magnetic moment explain the sizable Ce magnetization observed in the μSR experiment in the Fe antiferromagnetic ordered phase. The temperature dependence of the Ce moment in the effective field, calculated quantum mechanically or using a Brillouin function reproduces the data quite well, Fig. 5.19.

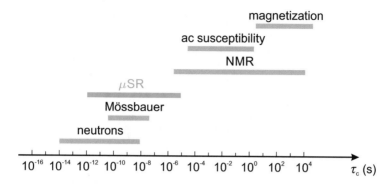

Fig. 5.20 Dynamical range of a μSR experiment compared to other methods $\tau_c = 1/\nu$ is the characteristic fluctuation/correlation time of the field sensed by the probe or of the spin correlations

5.3   Magnetic Fluctuations

μSR is a well suited method to study magnetic fluctuations. The observation window of the technique is typically 10–20 μs. However, μSR is sensitive to a large dynamical range of magnetic fluctuation phenomena with fluctuation/correlation time τ_c between 10^{-4}–10^{-5} and 10^{-11}–10^{-12} s. Figure 5.20 compares the μSR dynamical range with other techniques and shows the complementarity with nuclear probe techniques such as neutron scattering and NMR. We leave it to an exercise, Exercise 5.3, to justify the range limits.

In general, the fluctuation rate changes with temperature. For instance, when approaching a magnetic transition from a paramagnetic phase upon cooling, a slowing down of the fluctuation rate[44] or an increasing of the correlation time are expected, with times that usually become observable in the μSR time window shown in Fig. 5.20. Generally, there are two procedures to determine the fluctuation rate ν, which work well if ν is not too large:

- Measurements as a function of the applied field at a fixed temperature in the paramagnetic phase. The field dependence of the depolarization (ν is assumed to be field independent), can be fitted using Eq. 4.87, which corresponds to a Lorentzian spectral density $J(\omega)$, Sect. 4.4.2. Figures 5.21 and 5.22 show two examples of such measurements.
- Measurements as a function of temperature at a fixed applied field. As already pointed out and also visible at the dashed line of Fig. 4.44, for a given applied field B_{ext} corresponding to a Larmor frequency (or Zeeman splitting) $\omega_\mu = \gamma_\mu B_{ext}$ the muon spin relaxation is small if the spectrum of the fluctuation is too narrow (low ν or slow fluctuations) or too broad (high ν or fast fluctuations).

---

[44] This corresponds to a narrowing of the spectral density $J(\omega)$, Fig. 4.44.

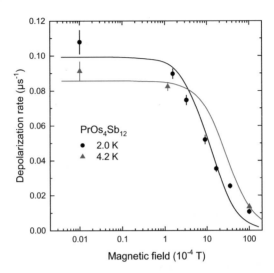

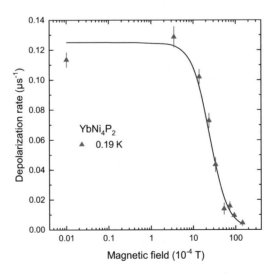

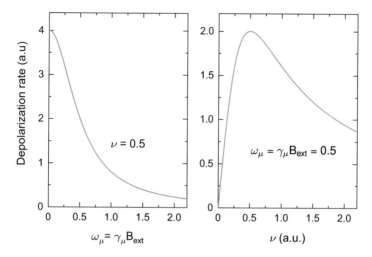

Fig. 5.23 Right panel: Dependence of the depolarization rate on the fluctuation rate at constant LF, Eq. 4.87. This situation occurs when performing measurements as a function of the temperature. Note the relaxation rate maximum at $\nu = \omega_\mu$. In the curve the temperature (hidden parameter) decreases from right to left. Left panel: Dependence of the depolarization rate on the applied LF field with fixed fluctuation rate, Eq. 4.87. This situation is obtained by performing measurements at a given temperature. In this case the spectral density function is mapped

A maximum is reached when the fluctuation rate matches the Larmor frequency $\omega_\mu = \nu = 1/\tau_c$.

The dependence of the depolarization rate on ν at fixed LF according to Eq. 4.87 is plotted in the right panel of Fig. 5.23. First, in the slow fluctuation regime, the depolarization rate is proportional to the fluctuation rate, then it goes though the maximum defined by the condition $\nu = \omega_\mu$, and finally, in the fast fluctuation regime, becomes inversely proportional to the fluctuation rate.[45] The maximum of the depolarization rate is often referred in the literature, borrowing from the NMR language, as T_1 minimum.

Although based on Eq. 4.87 and therefore strictly valid only in the fast fluctuation regime, the behavior of the right panel of Fig. 5.23 qualitatively explains the peaked behavior observed, for example, at a magnetic phase transition, when the depolarization rate is plotted as a function of temperature, see for example Fig. 5.24. At high temperatures, in the paramagnetic phase, the magnetic fluctuation rates (consequence of electronic spin fluctuations) are high and therefore the muon spin depolarization low. On reducing the temperature and approaching the phase transition, the fluctuations slow down; first an increase in the depolarization is observed up to the maximum and then, as the fluctuations die out, the region of linear decrease is reached. On further cooling, the magnetically ordered regime is entered

---

[45] The decrease of the depolarization with increasing fluctuation rate can be also understood by noticing that the muon spin cannot respond when the fluctuations become much faster than ω_μ.

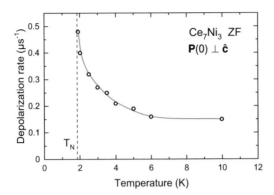

Fig. 5.24 Temperature dependence of the exponential ZF relaxation rate in the heavy fermion system Ce_7Ni_3 above the antiferromagnetic transition at T_N = 1.85 K. The temperature dependence for $T > T_N$ follows a critical power law typical for a second order transition. Modified from Kratzer et al. (2002), © Elsevier. Reproduced with permission. All rights reserved

where the depolarization rate is mainly determined by the field inhomogeneity of the ordered phase (inhomogeneous broadening). A peaklike behavior of the muon spin relaxation rate at a phase transition is also expected in ZF experiments since the slowing down of the magnetic fluctuations with decreasing temperature leads to an increase of the spectral weight at $\omega_\mu = 0$.[46]

5.3.1   Examples

Here are some examples that characterize typical μSR investigations of the dynamical properties of magnetic materials.

Anisotropic Fluctuations   The transverse character of the fluctuations inducing spin flips in LF and ZF, Eq. 4.83, can be used to detect and study anisotropic fluctuations. An example is found in erbium, which has an antiferromagnetic transition at $T_N \approx$ 85 K. Figure 5.25 shows the depolarization rate measured in ZF for two different crystal orientations with respect to the initial muon spin polarization (Hartmann et al. 1991). When the sample is oriented so that the muon spin is parallel to the c-axis, the depolarisation rate is almost constant over the measured temperature range. In contrast, when the sample is mounted with the c-axis perpendicular to the muon spin, the depolarisation rate increases very rapidly as the Néel temperature is approached. This indicates the slowing down of the magnetic field fluctuation components parallel to the c-axis, while the (weaker) components perpendicular to it show no freezing process. This behavior reflects the antiferromagnetic structure where the moments are aligned along the c-axis but completely disordered in the perpendicular plane.

---

[46] A Zeeman splitting of the spin levels is also present in ZF due to fluctuating field components.

Fig. 5.25 ZF muon spin depolarization rate in an erbium single crystal for two different orientations showing the anisotropy of the fluctuations. Initial muon polarization perpendicular to the c-axis (squares) and muon polarization parallel to the c-axis (circles). Modified from Hartmann et al. (1991), © Springer Nature. Reproduced with permission. All rights reserved

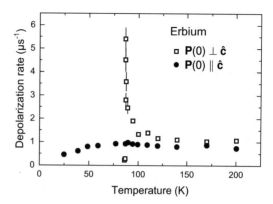

Magnetic Fluctuations and Critical Phenomena   In several uranium-based heavy fermion compounds superconductivity and ferromagnetism coexist. The relevant role is played by the U $5f$ electrons, which show localized as well as itinerant character (necessary to carry superconductivity).

Generally, in these compounds, the Curie temperature is significantly higher than the superconducting critical temperature and the upper critical field B_{c2} at low temperatures greatly exceeds the Pauli paramagnetic limiting field. These features indicate that the superconducting phase is associated with spin triplet Cooper pairing[47] and likely mediated by low-energetic magnetic fluctuations in the ferromagnetic phase.

The heavy fermion superconductor UTe_2 ($T_c \simeq 1.6$ (Ran et al. 2019)) is an example of how the measurement of very short fluctuation times on the order of nanoseconds or shorter can be used to detect dynamical effects and critical phenomena near phase transitions in magnetic systems (Sundar et al. 2019). This information is obtained by ZF and LF measurements. In Chap. 6 we will see how the coexistence of superconductivity and magnetism can be investigated by a combination of TF- and ZF-μSR experiments.

ZF spectra down to 0.025 K do not show oscillations indicative of magnetic order. The ZF (and LF) signals, Fig. 5.26, can be fitted by the sum of two exponentially relaxing components $A(t) = A_1 e^{-\lambda_1 t} + A_2 e^{-\lambda_2 t}$.[48] A_1 and A_2 are nearly equal, reflecting the two muon stopping sites of this and similar compounds. As a function

---

[47] In conventional and in many unconventional superconductors, Sect. 6.1.4, the electrons (or holes) constituting the Cooper pairs have spins pointing in opposite directions (spin singlet state). Spin singlet pairs can be broken by a magnetic field, which, via Zeeman splitting of the pair, aligns the two spins of the Cooper pair in the same direction (Pauli paramagnetic effect). The associated critical field is named Pauli limiting field. In contrast, superconductors, where the two spins are parallel in the triplet state, can be much more resilient to magnetic fields and coexist with ferromagnetism, see Chap. 6.

[48] There is also a nonnegligible background contribution, as it can be guessed from the spectra.

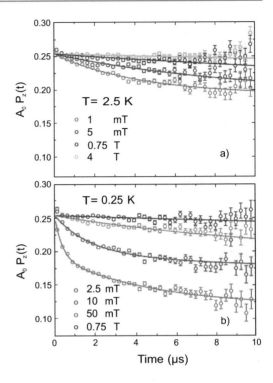

of temperature λ_1 and λ_2 increase monotonically with decreasing temperature. This
indicates that the local magnetic field sensed at each muon site is dominated by a
slowing down of magnetic fluctuations coexisting with superconductivity.

The ZF and LF spectra and the field dependence of the fitted relaxation rates
are reasonably well described by Eq. 4.66, or equivalently by the Redfield equation,
Eq. 5.35, indicating the dynamical nature of the magnetism. For λ_1, see Fig. 5.27,

$$\lambda_1(B_{\text{ext}}) = \frac{2\gamma_\mu^2 \, \Delta B_\mu^2 \, \tau_{\text{c}}}{1 + (\gamma_\mu \, B_{\text{ext}} \, \tau_{\text{c}})^2} \quad . \tag{5.35}$$

In the Redfield expression $\Delta B_\mu^2 = \langle B_\mu^2 \rangle$ is the width of the transverse components
of the time-varying local magnetic field at the muon site, and τ_{c} is the characteristic
fluctuation time. The fit for 2.5 K yields $\tau_{\text{c}} = 8(3) \times 10^{-10}$ s, and $B_\mu = \sqrt{\Delta B_\mu^2} =$
7.6(2.2) mT. For 0.25 K the fit yields $\tau_{\text{c}} = 9(2) \times 10^{-8}$ s and $B_\mu = 2.3(4)$ mT.

Although no evidence for magnetic order is found down to 0.025 K, the behavior
of the gradual slowing down of magnetic fluctuations with decreasing temperature
below 5 K suggests that the weak ferromagnetic fluctuations approach a magnetic
instability at $T = 0$ K.

Fig. 5.27 Field dependence
of the relaxation rates λ_1 and
λ_2 from the fits of the LF
asymmetry spectra at (**a**)
2.5 K, and (**b**) 0.25 K. The
solid red curves are fits of
$\lambda_1(B_{ext})$ to Eq. 5.35. The
different magnitude of the
relaxation rates reflects
different hyperfine couplings
of the U 5f electrons to the
muon at the two stopping
sites. Adapted from Sundar
et al. (2019), © American
Physical Society. Reproduced
with permission. All rights
reserved

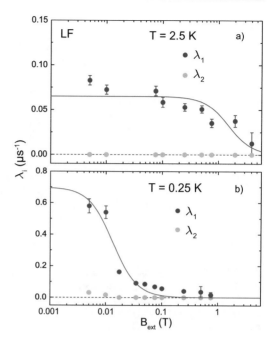

This is shown in Fig. 5.28 which plots λ_1/T as a function of temperature. Down to $T = 0.4$ K one observes $\lambda_1/T \propto T^{-(4/3)}$ as predicted by the phenomenological self-consistent renormalization (SCR) theory for itinerant three-dimensional metals near a ferromagnetic critical point (Moriya 1991; Ishigaki & Moriya 1996).[49] The change in the critical exponent below ~ 0.3 K suggests a breakdown of the SCR theory close to the presumed ferromagnetic critical point. As the inset of Fig. 5.28 shows, T/λ_1[50] goes to zero at $T \to 0$, indicating that the ground state of UTe$_2$ is close to a ferromagnetic quantum critical point.[51]

Time Scales of Magnetic Order The complementary use of μSR, neutron scattering and other techniques with their different timescale sensitivities, see Fig. 5.20,

---

[49] The theory relating the spin relaxation of a nuclear probe to critical phenomena in magnetic materials was originally developed for NMR. NMR experiments require the application of an external magnetic field, which tends to suppress the critical behavior. In this respect, ZF-μSR is advantageous.

[50] For magnetic fluctuations, the muon spin relaxation can be expressed in terms of susceptibility functions instead of correlation functions, Eq. 4.83. λ_1/T is then proportional to the imaginary part of the dynamical local spin susceptibility. Thus, $T/\lambda_1 \to 0$ means that the susceptibility diverges, indicating a quantum phase transition, see Yaouanc and Dalmas de Réotier (2011) for details.

[51] A quantum critical point (QCP) is a point in the phase diagram of a material where a continuous phase transition (quantum phase transition) takes place at $T = 0$ K. Critical exponents describe the behavior of physical quantities near continuous phase transitions.

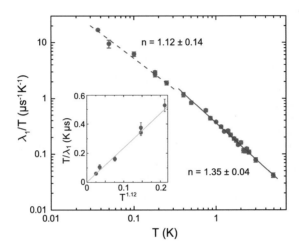

Fig. 5.28 ZF measurements of the temperature dependence of λ_1/T. The solid blue line is a fit for $T \geq 0.4$ K with the power-law expression $\lambda_1/T \propto T^{-n}$, which yields the exponent $n = 1.35(4)$. The dashed line is a similar fit between $0.037 \leq T \leq 0.3$ K, yielding $n = 1.12(14)$. The inset shows that the intercept of T/λ_1 versus $T^{1.12}$ at $T = 0$ K is consistent with a zero value. Modified from Sundar et al. (2019), © American Physical Society. Reproduced with permission. All rights reserved

allows a better understanding of the dynamics of magnetic order. An example is given by the hexagonal heavy fermion compound UPt$_3$.[52]

Many signatures of unconventional superconductivity have been reported for UPt$_3$. For instance, careful measurements on high-quality crystals reveal the presence of a double superconducting transition (at about 0.5 K) in zero external field (for a review see Joynt and Taillefer (2002)). Since unconventional superconductivity is often linked to magnetism and magnetic fluctuations, it is important to understand the magnetic properties. Peculiar weak magnetic properties have been observed, such as the occurrence of the so-called small-moment antiferromagnetic order (SMAF). Below $T_N^{SMAF} \simeq 6$ K, neutron scattering observes antiferromagnetic order with an unusually small magnetic moment ($\mu_s \sim 0.02$ μ_B/U) (Aeppli et al. 1988). This putative "phase" is also inferred from magnetic X-ray scattering data. In contrast, macroscopic thermodynamic and transport studies, NMR as well

---

[52] The low-temperature normal state properties of UPt$_3$ are those of a (strongly renormalized) Fermi liquid system with a quasiparticle mass on the order of 200 times the free electron value. In a Fermi liquid, the dynamics and thermodynamics of the system at low temperatures can be described by substituting the noninteracting particles of a Fermi gas with nearly independent fermionic quasiparticles, each of which has the same spin, charge, and momentum as the original particles. The single-particle energies are renormalized, which involves, for example, a change in the effective mass and allows excitations to be described using quasiparticles.

Fig. 5.29 ZF relaxation rate measured on two UPt$_3$ crystalline samples with different orientation of the muon spin polarization relative to the crystal axes. The results show that the relaxation rate is independent of the temperature in both samples. Modified from Dalmas de Réotier et al. (1995), © Elsevier. Reproduced with permission. All rights reserved

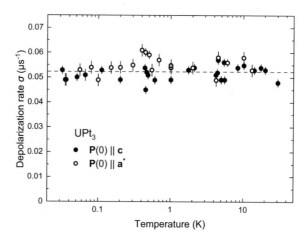

as ZF measurements on high-quality samples fail to detect the SMAF order, see Fig. 5.29.[53]

Besides the SMAF order, conventional antiferromagnetism with rather large moments, the so-called large-moment antiferromagnetism (LMAF) can be induced upon substituting small amounts of Pd or Au for Pt, or by replacing U with small amounts of Th. For both pseudobinary systems $U_{1-x}Th_xPt_3$ and $U(Pt_{1-x}Pd_x)_3$ this state can be induced in the concentration range of about $0.01 \leq x \leq 0.1$, with a Néel temperature maximum $T_{N,max}^{LMAF} \simeq 6$ K at about 5 at.% Th or Pt. Although the LMAF state is characterized by magnetic moments at least one order of magnitude larger than those observed in the SMAF state, the magnetic structure is identical for both "phases". Moreover, the transition temperature of the SMAF "phase" and the highest transition temperature of the LMAF phase appear to be strikingly similar, pointing to a close relationship between the two states, see Fig. 5.30.

The LMAF phase is characterized by the appearance of a two-component function in the μSR ZF signal: one showing an oscillatory behavior, Eq. 4.11, and the other one showing a Lorentzian Kubo-Toyabe behavior, Eq. 4.25. The two components are indicative of two magnetically inequivalent muon stopping sites, one with a finite local field B_μ, the second characterized by an isotropic Lorentzian distribution of local fields with field components averaging to zero producing a Lorentzian Kubo-Toyabe decay of the polarization.

As shown by neutron diffraction studies of $U(Pt_{1-x}Pd_x)_3$ the SMAF state is quite robust upon alloying and a similar situation is expected for the Th-substituted analog. Measurements in these related compounds allow to understand the absence of any anomaly in the muon spin depolarization rate when cooling the sample into the SMAF "phase", while neutron scattering data show Bragg peaks.

---

[53] Note that these data on high-quality crystals differ from formerly published data measured on first generation samples. Subtle effects in heavy fermion systems are known to be extremely dependent on the sample quality.

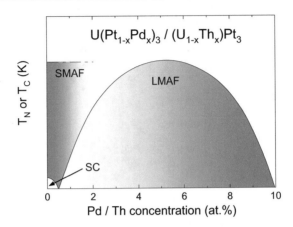

Fig. 5.30 Schematic magnetic and superconducting phase diagram for $U_{1-x}Th_xPt_3$ and $U(Pt_{1-x}Pd_x)_3$ alloys. Whereas the SMAF state, which occurs below $T_N^{SMAF} \simeq 6$ K, is observed only by neutron diffraction and magnetic X-ray scattering, the LMAF and the superconducting (SC) states are detected by several techniques including μSR. Modified from Amato et al. (2004), © Institute of Physics. Reproduced with permission. All rights reserved

In principle, an explanation for the lack of magnetic signature in the μSR signal of UPt$_3$ could invoke the cancellation of the internal fields at the muon stopping site as in the UPd$_2$Al$_3$ example, presented earlier. This can be discarded here since a cancellation does not hold for the Pt site and is incompatible with the NMR results. The μSR and the NMR studies lead to the conclusion that this state is not truly static but possesses moments fluctuating at rates typically greater than 10 MHz, i.e., $2\gamma_\mu^2 \Delta B_\mu^2/\nu \ll 1$. This is too fast to produce muon spin depolarization whereas, such a state appears static and ordered for the nearly instantaneous time scales of neutron and X-ray scattering.

μSR results indicating the dynamic nature of the SMAF "phase" are obtained not only on pure UPt$_3$ samples, but also on samples with low ($x \leq 0.005$) Pd concentration.

In the $U(Pt_{0.99}Pd_{0.01})_3$ system (Keizer et al. 1999) μSR detects only the LMAF phase, while the neutron scattering data are sensitive to both magnetic states, see Fig. 5.31. The comparison of the μSR with the neutron scattering data shows that T_N^{SMAF} should be considered a crossover temperature signaling a slowing down of magnetic fluctuations rather than being a true phase transition temperature.

The μSR sensitivity to small moments and the different time window compared to neutron diffraction studies demonstrate that in these systems static magnetism does not coexist with unconventional heavy fermion superconductivity.

Fig. 5.31 (a) Temperature variation of the magnetic intensity measured at the magnetic Bragg peak $\mathbf{k} = (0.5, 0, 1)$ for an annealed $U(Pt_{0.99}Pd_{0.01})_3$ sample. (b) Spontaneous μSR frequency obtained on the same sample. Note the absence of any signal above $T_N^{LMAF} \simeq 1.9$ K. The crossover from SMAF to LMAF near 1.9 K appears as a pronounced increase in the Bragg intensity. Modified from Amato et al. (2004), © Institute of Physics. Reproduced with permission. All rights reserved

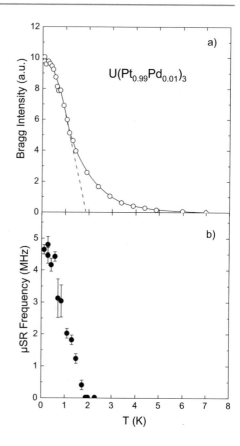

5.4   Incommensurate Magnetic Structures

In general, the periodicity of a magnetic structure can be expressed as a rational fraction of the periodicity of the underlying crystal lattice (commensurate magnetic structure) or as an irrational number (incommensurate magnetic structure). In the incommensurate case the magnetic moments are ordered, but with a periodicity described by the propagation vector that is not commensurate with that of the crystal structure,[54] see Fig. 5.32.

In the commensurate case, for a given interstitial site, a muon experiences a discrete number of local fields, the number of magnetically inequivalent muon sites (which can result from inequivalent or even equivalent crystallographic sites, see

---

[54] The propagation vector $\mathbf{k}$ specifies the translational properties of the moments throughout the lattice. For a commensurate structure $\mathbf{k} = \frac{u}{w}\mathbf{K}$, where $\mathbf{K}$ is the reciprocal lattice vector and $u < w$ are integers. A single propagation vector is often sufficient to describe a simple magnetic structure. Complex structures require several $\mathbf{k}'s$.

Fig. 5.32 (a)
One-dimensional
commensurate
antiferromagnetic structure
with lattice periodicity a and
magnetic structure periodicity
$2a$, i.e., $k = \frac{1}{2}$. (**b**)
Incommensurate magnetic
structure. Here k is not a
rational number

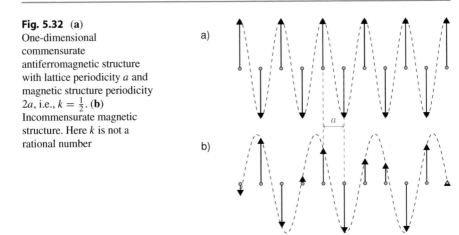

Fig. 5.3) is finite, and the field distribution $f_m(B_\mu)$ obtained from the μSR signal shows well-defined peaks.

In the incommensurate case, the field distribution $f_m(B_\mu)$ is continuous, extending between a minimum and a maximum cutoff field, and thus completely different from the commensurate case.

To derive the time evolution of the muon polarization $P_z(t)$ for a muon ensemble seeing a modulation of the internal field that is incommensurate with the crystallographic structure, we consider the case of a single-**k** amplitude-modulated magnetic structure.[55] The magnetic structure may be described as

$$\boldsymbol{\mu}_j = \tilde{\boldsymbol{\mu}} \cos (\mathbf{k} \cdot \mathbf{r}_j + \varphi) \ , \tag{5.36}$$

where $\boldsymbol{\mu}_j$ is the moment of the atom at position $\mathbf{r}_j$ and $\tilde{\boldsymbol{\mu}}$ is the magnetic moment that is modulated with wave vector **k**. With the muons located at a crystallographic site, each muon will sense a different field.

A Simple Model We assume the local field B_μ to be proportional to the modulated magnetic moment and directed along the y-axis and that **k** and $\mathbf{r}_\mu$ (position of the muon site) are parallel and along the x-axis. The spatial distribution of the field $B_{\mu,y}$ is also cosine-like with

$$B_{\mu,y} = B_{\max} \cos(k x_\mu) \equiv B_{\max} \cos(u) \ . \tag{5.37}$$

---

[55] Incommensurate magnetic structures also occur as helical, conical and other complicated structures. Especially in rare earth compounds very complicated magnetic structures have been found.

For an incommensurate k-structure, the argument kx assumes all possible values between 0 and 2π with equal probability and we can replace it by a continuous phase u with uniform probability.

We are interested in the probability distribution of the absolute field values $f_m(B_\mu)$, since the observed oscillation in this case reflects only the absolute field value. The probability for the muon to probe a value in an interval dB_μ is equal to the number of phase values in the interval du. Taking into account that for a given value of B_μ there are four possible choices of u and that each phase is equally distributed with normalized probability $g(u) = 1/2\pi$, we have

$$f_m(B_\mu)dB_\mu = 4g(u)du = \frac{2}{\pi}\frac{du}{dB_\mu}dB_\mu \ ,$$

we obtain

$$f_m(B_\mu) = \frac{2}{\pi}\frac{\left|d\left(\arccos\left(\frac{B_\mu}{B_{max}}\right)\right)\right|}{dB_\mu} = \frac{2}{\pi\sqrt{B_{max}^2 - B_\mu^2}} \ , \tag{5.38}$$

where $0 \leq B_\mu \leq B_{max}$, see Fig. 5.33. Following our assumption of $\mathbf{B}_\mu$ along y and assuming that the initial muon spin direction is along the z-axis, we can use the probability distribution Eq. 5.38 to calculate the time evolution of the polarization

$$P_z(t) = \int_0^{B_{max}} f_m(B_\mu) \cos(\gamma_\mu B_\mu t)dB_\mu \ , \tag{5.39}$$

and we find

$$P_z(t) = J_0(\gamma_\mu B_{max} t) \ , \tag{5.40}$$

Fig. 5.33 Probability distribution of the field modulus for a simple incommensurate magnetic structure

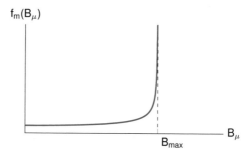

Fig. 5.34 Time evolution of
the μSR signal in the
presence of an
incommensurate magnetic
lattice, see text for details

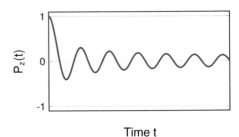

where J_0 is the zero order Bessel function of the first kind, see Fig. 5.34[56] and
Fig. 5.35 for an example.

When the time t is large relative to $1/(\gamma_\mu B_{\max})$, we can use the following
asymptotic expansion

$$J_0(x \to \infty) = \sqrt{\frac{2}{\pi\,x}}\,\cos\left(x - \frac{\pi}{4}\right) \ , \qquad (5.42)$$

and obtain the large time expansion

$$J_0(\gamma_\mu B_{\max} t) \simeq \sqrt{\frac{2}{\pi\,\gamma_\mu B_{\max} t}}\,\cos\left(\gamma_\mu B_{\max} t - \frac{\pi}{4}\right) \ . \qquad (5.43)$$

This expansion already matches the actual polarization function for $\gamma_\mu B_{\max} t \gtrsim$
1. Hence, by fitting the large time interval with a simple cosine function, a negative
phase shift is obtained. Such as result may point to an incommensurate magnetic
structure.

This example shows that the phase parameter (whose significance is often
neglected in the μSR analysis) may contain useful information. Moreover, note that
the period of the oscillation of the μSR signal provides the maximum local field.

Recalling the Bessel function expansion

$$J_0(x) = \sum_{k=0}^{\infty} \frac{(-1)^k}{k!^2}\left(\frac{x}{2}\right)^{2k} \ , \qquad (5.44)$$

---

[56] If the polar angle of $\mathbf{B}_\mu$ is not $90°$, Eq. 5.40 modifies to

$$P_z(t) = \cos^2\theta + \sin^2\theta\, J_0(\gamma_\mu B_{\max} t) \ . \qquad (5.41)$$

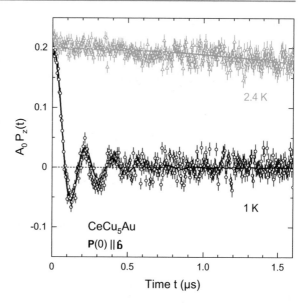

Fig. 5.35 μSR asymmetry in ZF following a Bessel function for CeCu$_5$Au, which has a magnetic state below T_N = 2.2 K. The data were taken on a single crystal, so there is no "1/3" term. Above T_N, the signal reflects the Gaussian field distribution created by the ^{63}Cu and ^{65}Cu nuclear moments. Below T_N the observation of a Bessel function indicates the presence of an incommensurate magnetic structure. Modified from Amato et al. (1995), © American Physical Society. Reproduced with permission.

we can write the short-time expansion as

$$J_0(\gamma_\mu B_{\max} t) \simeq 1 - \frac{(\gamma_\mu B_{\max})^2}{4} t^2 \ . \tag{5.45}$$

For short times, the damping is strong and controlled by the cutoff field $B_{\max}$ of the field distribution and not by the variance of the field distribution as in the usual cases.

A More General Model   Equation 5.37 is a rather drastic simplification, where one assumes that all the fields $\mathbf{B}_\mu$ probed by the implanted muons point in the same well-defined direction. This situation can indeed occur in special cases where the muon sits at a highly symmetric site. However, it does not hold for nonsymmetric sites and for helical incommensurate structures. In both cases, the direction of the field at the muon site is not constant.

A more general expression can be obtained reverting to Eq. 5.36 and calculating the corresponding dipolar field at the muon position r_μ (Andreica 2001; Schenck et al. 2001)

$$\mathbf{B}_{\text{dip}}(\mathbf{r}_\mu) = \frac{\mu_0}{4\pi} \sum_j \left(\frac{3\mathbf{R}_j(\boldsymbol{\mu}_j \cdot \mathbf{R}_j)}{R_j^5} - \frac{\boldsymbol{\mu}_j}{R_j^3} \right) \ , \tag{5.46}$$

where $\mathbf{R}_j = \mathbf{r}_j - \mathbf{r}_\mu$. Inserting $\boldsymbol{\mu}_j = \tilde{\boldsymbol{\mu}} \cos(\mathbf{k} \cdot \mathbf{r}_j + \varphi)$ in the last expression we have

$$
\begin{aligned}
\mathbf{B}_{\text{dip}}(\mathbf{r}_\mu) &= \frac{\mu_0}{4\pi} \sum_j \cos(\mathbf{k} \cdot \mathbf{r}_j + \varphi) \left(\frac{3\mathbf{R}_j(\tilde{\boldsymbol{\mu}} \cdot \mathbf{R}_j)}{R_j^5} - \frac{\tilde{\boldsymbol{\mu}}}{R_j^3} \right) \\
&= \cos(\mathbf{k} \cdot \mathbf{r}_\mu + \varphi) \frac{\mu_0}{4\pi} \sum_j \cos(\mathbf{k} \cdot \mathbf{R}_j) \left(\frac{3\mathbf{R}_j(\tilde{\boldsymbol{\mu}} \cdot \mathbf{R}_j)}{R_j^5} - \frac{\tilde{\boldsymbol{\mu}}}{R_j^3} \right) \\
&\quad - \sin(\mathbf{k} \cdot \mathbf{r}_\mu + \varphi) \frac{\mu_0}{4\pi} \sum_j \sin(\mathbf{k} \cdot \mathbf{R}_j) \left(\frac{3\mathbf{R}_j(\tilde{\boldsymbol{\mu}} \cdot \mathbf{R}_j)}{R_j^5} - \frac{\tilde{\boldsymbol{\mu}}}{R_j^3} \right) \\
&= \mathbf{C}_{\text{s}} \cos(\mathbf{k} \cdot \mathbf{r}_\mu + \varphi) - \mathbf{S}_{\text{s}} \sin(\mathbf{k} \cdot \mathbf{r}_\mu + \varphi) \ .
\end{aligned}
$$

$$ \text{(5.47)} $$

This equation holds for commensurate and incommensurate magnetic structures. The lattice sums $\mathbf{C}_{\text{s}}$ and $\mathbf{S}_{\text{s}}$ are the same for all magnetically equivalent sites and the modulation depends only on $\mathbf{r}_\mu$. Equation 5.47 describes an ellipse in the $\mathbf{C}_{\text{s}} - \mathbf{S}_{\text{s}}$ plane with the $\mathbf{B}_{\text{dip}}(\mathbf{r}_\mu)$ vector connecting the origin with a point on the ellipse.

For an incommensurate $\mathbf{k}$ and a sufficiently large number of muon sites, every possible point on the ellipse is realized, each of them only once. For a commensurate $\mathbf{k}$ only a few distinct values of $\mathbf{B}_{\text{dip}}(\mathbf{r}_\mu)$ occur. This calculation has to be performed for each magnetically inequivalent muon site. Considering $\mathbf{C}_{\text{s}}$ and $\mathbf{S}_{\text{s}}$ perpendicular to each other (and to $\hat{\mathbf{z}}$), and denoting the minimum and maximum length of the two vectors as B_{min} and B_{max} the distribution of the field modulus reads[57]

$$
f_m^{\text{inc}}(B_\mu) = \frac{2}{\pi} \frac{B_\mu}{\sqrt{B_\mu^2 - B_{\text{min}}^2} \sqrt{B_{\text{max}}^2 - B_\mu^2}} \ , \qquad \text{(5.48)}
$$

which is characterized by two peaks corresponding to the minimum and maximum cutoff field values.[58] The polarization function associated with this field distribution cannot be calculated analytically. One often approximates Eq. 5.48 with a shifted distribution of the type 5.38

$$
f_m^{\text{inc, approx}}(B_\mu) \simeq \frac{1}{\pi} \frac{1}{\sqrt{\Delta B^2 - (B_\mu - B_{\text{av}})^2}} \ , \qquad \text{(5.49)}
$$

---

[57] Note that, with a redefinition of B_{min} and B_{max}, this distribution is also valid in the general case where $\mathbf{C}_{\text{s}}$ and $\mathbf{S}_{\text{s}}$ are not perpendicular to each other (Yaouanc & Dalmas de Réotier 2011).

[58] The field distribution 5.38 is a special case when one of the lattice sums $\mathbf{C}_{\text{s}}$ or $\mathbf{S}_{\text{s}}$ is zero.

Fig. 5.36 Upper panel:
Exact (Eq. 5.48, blue line)
and approximated (Eq. 5.49,
red line) field distribution at
the muon site for an
incommensurate structure.
Lower panel: Corresponding
muon polarization functions,
see text for details. The field
values correspond to the
helical structure of MnSi , see
Fig. 5.37. Modified from
Amato et al. (2014),
© American Physical Society.
Reproduced with permission.
All rights reserved

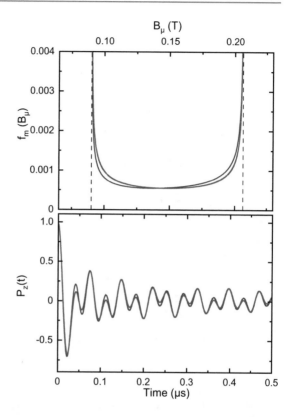

where $\Delta B = (B_{max} - B_{min})/2$ and $B_{av} = (B_{max} + B_{min})/2$. The upper panel of Fig. 5.36 shows the difference between the exact and the approximated field distribution, which is symmetric with respect to the singularities, i.e., some weight is transferred from the upper to the lower cutoff field. The lower panel of Fig. 5.36 shows the oscillating muon polarization function for both field distributions for the case where the local field is perpendicular to the initial muon spin. For the exact field distribution a numerical calculation was performed, while for the approximate field distribution the muon polarization function can be obtained analytically and is given by

$$P_{z,\text{approx}}(t) = J_0(\gamma_\mu \Delta B \, t) \, \cos(\gamma_\mu B_{av} \, t) \, , \qquad (5.50)$$

where J_0 is a Bessel function of the first kind. We remark that the approximate function $P_{z,\text{approx}}(t)$ captures the essential features of the polarization function calculated with the exact field distribution and constitutes a good approximation.

An example of a field distribution given by Eq. 5.48 is found in MnSi (Amato et al. 2014). The spins in MnSi form a left-handed incommensurate helix with a propagation vector $k \simeq 0.36 \, \text{nm}^{-1}$ in the [111] direction (Ishikawa et al. 1976). The static Mn moments ($\sim 0.4 \mu_B$ for $T \to 0$) point in a plane perpendicular to the

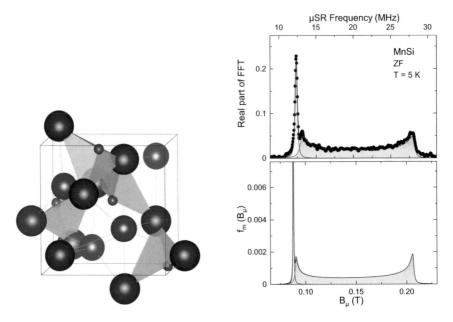

Fig. 5.37 μSR studies of MnSi. Left: Sketch of the crystallographic structure. The Mn ions are in magenta and the Si ions in blue. Note that six Mn ions that do not belong to the primary unit cell, are also plotted. The muon sites at the $4a$ position are in red. There are four crystallographically equivalent sites in the unit cell at: I (x, x, x), II $(1/2 - x, \bar{x}, 1/2 + x)$, III $(1/2 + x, 1/2 - x, \bar{x})$, and IV $(\bar{x}, 1/2 + x, 1/2 - x)$, where $x = 0.532$. Upper right panel: Fast Fourier Transform of the ZF-μSR spectrum at 5 K. Three of the four crystallographic muon sites (II, III, IV) produce a field distribution given by Eq. 5.48 (shadowed orange). The muon site I at $(0.532, 0.532, 0.532)$, gives a narrow field distribution (shadowed blue). The blue and red curve components are the transforms of a time-domain fit with a single value precession cosine and with Eq. 5.50, respectively. Bottom right panel: Computed field distributions for the I site (blue) and for the II, III and IV sites (orange). For better visualization, the field distributions have been folded with a Lorentzian function with FWHM of 0.5 and 2.0 mT, respectively. Structure drawn with VESTA, © 2006–2022 Koichi Momma and Fujio Izumi (Momma & Izumi 2011). Modified from Amato et al. (2014), © American Physical Society. Reproduced with permission. All rights reserved

propagation vector. The period, which is incommensurate to the lattice constant, is about 18 nm. Due to the helical magnetic structure, three of the four crystallographic equivalent muon sites produce a field distribution as given by Eq. 5.48, see Fig. 5.37.

Finally, note that while a single-k amplitude-modulated incommensurate magnetic structure gives rise to Bessel-like oscillations, the converse is not always true, i.e., the observation of such oscillations does not unambiguously guarantee that the magnetic structure is incommensurate.

5.5   Dynamics of Spin Glasses

We have seen in Sect. 4.1.2.2 that in the case of a dilute arrangement of randomly oriented static magnetic moments, a Lorentzian distribution of fields along each Cartesian direction is obtained. Such a situation is realized in so-called spin glasses, alloys with dilute magnetic impurities, where the interaction between their spins is randomly ferromagnetic or antiferromagnetic. They display a complex dynamic behavior with a wide spectrum of fluctuations, see Sect. 4.3.2

Well-studied familiar spin glass systems are the alloys consisting of a few percent of a magnetic transition element (such as Fe or Mn) in a nonmagnetic metallic host (Cu, Ag, Au) (canonical spin glasses (Cannella & Mydosh 1972)). In these metals the interaction between localized moments is mediated by the conduction electrons. This type of indirect exchange (RKKY interaction) (Ruderman & Kittel 1954; Kasuya 1956; Yosida 1957) has been introduced in Sect. 5.1.2. It couples moments over relatively large distances and is the dominant exchange interaction in metals where there is little or no direct overlap between adjacent magnetic ions. The RKKY exchange coefficient oscillates from positive to negative as the separation of the ions changes and has the damped oscillatory behavior shown in Fig. 5.38. Therefore, depending on the separation between one pair of ions, their magnetic coupling can be randomly ferromagnetic or antiferromagnetic and can be frustrated (i.e., not satisfied) with respect to another pair.

The frustration makes it difficult for these systems to find an optimal configuration, so that they exhibit nontrivial thermodynamic and dynamic properties that are different and richer than those observed in their nondisordered counterparts. For temperatures below T_g (glass or freezing temperature), spin glasses show an amorphous magnetic state, or quenched disorder.

Spin glass systems have been extensively studied as paradigmatic examples of frozen disorder and as prototypes of complex systems not only in physics, but also in disciplines such as biology and economics (Mydosh 1993; Fischer & Hertz 1993). The term "glass" relates to the analogy between the magnetic disorder in a spin glass and the spatial disorder in a conventional glass, which possesses numerous metastable low-energetic states.[59]

These spin glasses exhibit Curie-Weiss susceptibilities at high temperature. At low temperatures they form an unusual antiferromagnetic-like state, which is considered a third distinct type of low-temperature magnetic arrangement in solids, in addition to the ferromagnetic and antiferromagnetic states. There is no sign of a sharp feature in the specific heat; only sometimes a broad bump. However, low-field ac susceptibility shows a sharp cusp as a function of temperature, evidence of a

---

[59] When a liquid such as a silica compound is cooled rapidly, it is possible for it to become blocked in a metastable state (not the ground state) at T_g. This blocking results from an abrupt lack of thermal vibrations. Below T_g, the viscosity becomes so large (increasing by up to 10 orders of magnitude) that the liquid appears to stop flowing. The situation is similar for spin glasses, whose energy landscape also consists of many low-energy metastable states corresponding to local free energy minima.

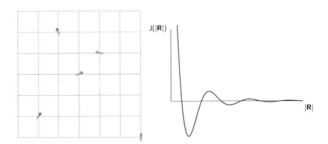

Fig. 5.38 Random distribution of magnetic moments in a metallic matrix and RKKY exchange interaction plotted as a function of distance between two magnetic moments

well-defined transition temperature T_g. There is now a consensus that the spin glass transition is a "true" thermodynamic phase transition.

The apparent freezing of the spin dynamics below T_g leads to random but static[60] order of the spin orientation. Above T_g the relaxation of spin glasses is highly anomalous compared to that of standard paramagnets, with a single correlation or fluctuation time τ_c, see Eq. 4.69.

μSR can probe slowly relaxing spin systems with a wide spectrum of fluctuation rates. In the temperature range above the freezing temperature, the initially polarized muons are gradually depolarized by the fluctuating dipolar fields produced by the neighbor local moments.

Experiments on relatively highly concentrated spin glasses (5–10% atomic concentration of magnetic impurities) show muon depolarization functions above T_g, which can be fitted very satisfactorily by stretched exponentials (Campbell et al. 1994)[61]

$$P_z^{SE} = \exp\left[-(\lambda t)^\beta\right] \ . \tag{5.51}$$

The depolarization rate λ as well as the exponent β are temperature dependent. Figures 5.39 and 5.40 show the temperature dependence of the two parameters for AgMn spin glasses at two different Mn concentrations (Campbell et al. 1994).

For both samples it is clear that $\beta(T)$ varies strongly, decreasing from a value close to 1 at high temperatures to a limiting value close to $1/3$ on approaching T_g. The $\beta \approx 1$ plateau extends up to temperatures about 4 times T_g. The depolarization rate λ tends to diverge as the temperature is lowered toward T_g, indicating a phase

---

[60] Static on the time scale of μSR. Standard magnetization measurements show that the magnetic relaxation extends to essentially infinite long times, similar to the extremely slow glass flow sometimes visible in windows of very old churches.

[61] Note that measurements of AgMn (0.5 at.%) suggest that a stretched exponential can be used in a wider range of concentrations (Keren et al. 1996).

Fig. 5.39 The stretched exponent β as a function of temperature for ZF data on *Ag*Mn 5 at.% and *Ag*Mn 7 at.%. The dashed line indicates the $\beta = 1/3$ value. Modified from Campbell et al. (1994), © American Physical Society. Reproduced with permission. All rights reserved

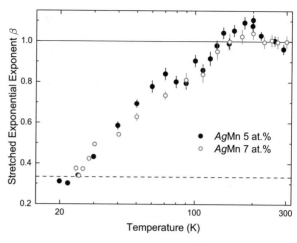

Fig. 5.40 The temperature dependence of the muon depolarization rate λ for ZF data on *Ag*Mn 5 at.% and *Ag*Mn 7 at%. Modified from Campbell et al. (1994), © American Physical Society. Reproduced with permission. All rights reserved

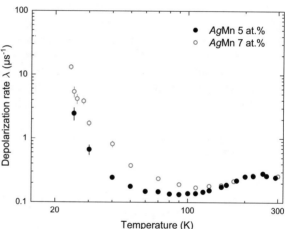

transition. This behavior seems to be very general and a number of other spin glasses or glassy systems have been found to follow the same pattern.

In Sect. 4.3.2 we have shown that we can write a stretched exponential as a distribution of independent exponential relaxations λ_i given by a function $\Lambda(\frac{\lambda_i}{\lambda}, \beta)$, Eq. 4.77.

At high temperatures, where $\beta \approx 1$, we have to a good approximation a simple paramagnetic regime. In this case, $\Lambda(\frac{\lambda_i}{\lambda}, \beta = 1) = \delta(\frac{\lambda_i}{\lambda} - 1)$ and the simple exponential muon spin relaxation with rate λ reflects an exponential autocorrelation behavior of the impurity spin with a single correlation time τ_c, see Sect. 4.4.1

$$P_z(t) = \exp(-\lambda t) \longleftrightarrow q(t) = \frac{\langle S_i(t) S_i(0) \rangle}{\langle S_i(0)^2 \rangle} = \exp(-t/\tau_c) \ . \tag{5.52}$$

The decrease of β for temperatures below about 4 times T_g reflects a broadening of the relaxation spectrum with a distribution of λ_i, which sets in well above the glass transition temperature and mirrors the complex dynamics of a spin glass system. Assuming the fast fluctuation relation $\lambda_i \propto \tau_{c,i}$, the distribution of λ_i can be related to a distribution of correlation times of the impurity spins. The corresponding global autocorrelation function is then given by

$$q(t) \propto \int_0^\infty \Lambda(\frac{\lambda_i}{\lambda}, \beta) \, \exp\left(-\frac{t}{\alpha\lambda_i}\right) d\lambda_i \;, \tag{5.53}$$

where $\tau_{c,i} = \alpha\lambda_i$. For small values of β (i.e., nonexponential $P_z(t)$) numerical calculations give a strongly nonexponential relaxation for $q(t)$. Furthermore, at any temperature, $q(t)$ has a form similar to the polarization function, i.e., it is close to a stretched exponential with the same $\beta(T)$ (Campbell et al. 1994).

In summary, the μSR measurements of spin glasses show that the distribution of spin correlation times has essentially a unique value τ_c for temperatures above about 4 times T_g. It then widens as the temperature is lowered; when T_g is reached the spin correlation times are broadly distributed. The form of this distribution controls the shape of $P(t)$, and appears to be general in all spin glasses.

Generally, determining the exact form of the distribution of correlation times is a very difficult problem, but we can keep in mind here that the observation of a stretched exponential for the muon spin depolarization function can be taken as an indication of a wide spectrum of relaxation channels, or more specifically of correlation times τ_c.

5.6   Magnetic Response in the Paramagnetic or Diamagnetic State: The Knight-Shift

We consider here the case where the sample is in a paramagnetic or a diamagnetic state. When we apply a magnetic field a magnetization proportional to B_{ext} is induced.[62] If we perform a TF experiment we observe a muon spin precession corresponding to a field B_μ. The field or frequency shift relative to the applied field B_{ext}[63] is called Knight-shift[64]

$$K_{exp} = \frac{|\mathbf{B}_\mu| - |\mathbf{B}_{ext}|}{|\mathbf{B}_{ext}|} = \frac{\omega_\mu - \omega_L}{\omega_L} \;. \tag{5.54}$$

---

[62] The proportionality is valid if the applied field strength is not too large (see Brillouin function).

[63] $\omega_L = \gamma_\mu B_{ext}$ is the muon spin angular frequency in "free space", i.e., without the condensed matter effects induced by the applied field.

[64] Originally, the term "Knight-shift" denoted in NMR the shift of the resonance frequency in metals relative to insulators, i.e., essentially only the frequency shift due to the paramagnetism of the conduction electrons in metals (Knight 1949). In μSR is generally used for the total shift in paramagnetic and diamagnetic materials.

This ratio contains important information about the microscopic magnetic response of the investigated compound since the shift originates from the hyperfine fields produced by the field-induced polarization of the conduction electrons (Fermi contact field contribution) and the localized electronic moments. The local moments contribute to the Knight-shift via two mechanisms: the dipolar interaction between the local moments and the μ^+, and the indirect RKKY interaction, in which the exchange interaction between local moment and conduction electrons produces an additional spin polarization of the conduction electrons and thus a hyperfine contact field at the muon site, see Sect. 5.1.2. Since the muon behaves like a light proton isotope, muon Knight-shifts can be used to test electronic structure calculations of charged impurities in a metal.

As outlined in Sect. 5.1.3, $\mathbf{B}_\mu$ is given by Eq. 5.27[65]

$$\mathbf{B}_\mu = \mathbf{B}_{\text{ext}} + \mathbf{B}'_{\text{dip}} + \mathbf{B}_{\text{cont}} + \mathbf{B}_{\text{Lor}} + \mathbf{B}_{\text{dem}} \ . \tag{5.55}$$

To evaluate K_{exp} we write

$$|\mathbf{B}_\mu| = \sqrt{(\mathbf{B}_{\text{ext}} + \mathbf{B}'_{\text{dip}} + \mathbf{B}_{\text{cont}} + \mathbf{B}_{\text{Lor}} + \mathbf{B}_{\text{dem}})^2} =$$

$$= B_{\text{ext}} \sqrt{1 + \frac{2}{B_{\text{ext}}^2}(\mathbf{B}'_{\text{dip}} + \mathbf{B}_{\text{cont}} + \mathbf{B}_{\text{Lor}} + \mathbf{B}_{\text{dem}}) \cdot \mathbf{B}_{\text{ext}} + \frac{1}{B_{\text{ext}}^2}(\mathbf{B}'_{\text{dip}} + \mathbf{B}_{\text{cont}} + \mathbf{B}_{\text{Lor}} + \mathbf{B}_{\text{dem}})^2} \ . \tag{5.56}$$

The induced field $\mathbf{B}_\mu - \mathbf{B}_{\text{ext}}$ is not necessarily parallel to the external field, see Fig. 5.41. However, because B_{ext} is much larger than all other terms (the susceptibility is small), one can neglect the quadratic term and simplify to

$$|\mathbf{B}_\mu| \cong B_{\text{ext}} \sqrt{1 + \frac{2}{B_{\text{ext}}^2}(\mathbf{B}'_{\text{dip}} + \mathbf{B}_{\text{cont}} + \mathbf{B}_{\text{Lor}} + \mathbf{B}_{\text{dem}}) \cdot \mathbf{B}_{\text{ext}}}$$

$$= B_{\text{ext}} + \left(\frac{1}{B_{\text{ext}}}(\mathbf{B}'_{\text{dip}} + \mathbf{B}_{\text{cont}} + \mathbf{B}_{\text{Lor}} + \mathbf{B}_{\text{dem}}) \cdot \mathbf{B}_{\text{ext}}\right) \ , \tag{5.57}$$

i.e., to first order only the projections onto the applied field $\mathbf{B}_{\text{ext}}$ of $\mathbf{B}'_{\text{dip}}$, $\mathbf{B}_{\text{cont}}$, $\mathbf{B}_{\text{Lor}}$ and $\mathbf{B}_{\text{dem}}$ contribute to the Knight-shift

$$K_{\text{exp}} = \frac{\mathbf{B}_{\text{ext}} \cdot (\mathbf{B}_\mu - \mathbf{B}_{\text{ext}})}{B_{\text{ext}}^2} \ . \tag{5.58}$$

---

[65] Since the sample has a magnetization induced by the external field, a (small) Lorentz and a diamagnetizing field contribution are present.

Fig. 5.41 The muon frequency shift, see Eq. 5.58, is related to the length of the green vector. $K_{\exp}$ corresponds to the vector difference between the field sensed by the muon and the external field, projected to the direction of the external field and normalized to its value

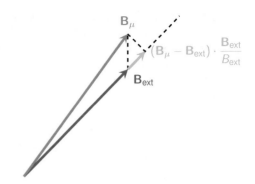

This approximate expression is often used because it simplifies the discussion of Eq. 5.54 and of Knight-shift properties such as the angular dependence, see Sect. 5.6.4. See Exercise 5.4 for an estimate of the error introduced by this approximation.

$K_{\exp}$ is usually in the range of a few tens of parts per million for a diamagnet and at most a few percent for a paramagnet.

Since the Lorentz and the demagnetizing fields do not contain microscopic information, the experimentally measured shift must be corrected for the contribution of these two terms. One defines the so-called muon Knight-shift as

$$
\begin{aligned}
K_\mu &= \frac{\mathbf{B}_{\text{ext}} \cdot \left[(\mathbf{B}_\mu - \mathbf{B}_{\text{Lor}} - \mathbf{B}_{\text{dem}}) - \mathbf{B}_{\text{ext}} \right]}{B_{\text{ext}}^2} \\
&= K_{\exp} - \frac{\mathbf{B}_{\text{ext}} \cdot (\mathbf{B}_{\text{Lor}} + \mathbf{B}_{\text{dem}})}{B_{\text{ext}}^2} \quad .
\end{aligned}
\tag{5.59}
$$

In μSR as well as in NMR the measurement of the Knight-shift also provides a means to determine the local (i.e., at the probe site) susceptibility of a system.

We assume the sample to be an ellipsoid, with one of its principal axes oriented along the direction of the external field, which by definition we take to be the z-axis. Taking into account that the magnetization of the Lorentz sphere M_{dom} is the same as the total sample magnetization M (see Sect. 5.1.3) and that the susceptibility is small, we can rewrite Eq. 5.59 as

$$
K_\mu = K_{\exp} - \left(\frac{1}{3} - N_{zz} \right) \chi_{0,z} \quad ,
\tag{5.60}
$$

where $\chi_{0,z}$ is the sample bulk susceptibility along the z direction.[66] If the sample is a sphere, the demagnetizing factor is $N_{zz} = 1/3$ and the demagnetizing field cancels the Lorentz field. If this is not the case, for a precise determination of the Knight-shift a careful estimation of the demagnetizing factor has to be done, since B_{dem} and B_{Lor} cannot be neglected with respect to B'_{dip} and B_{cont}.

5.6.1   Paramagnetism of the Conduction Electrons: Fermi Contact Term Knight-Shift

The magnetic properties of simple metallic systems are essentially determined by the conduction electrons. The magnetic response and susceptibility are governed by the Pauli paramagnetism (related to the spin of the conduction electrons) and Landau diamagnetism (related to the orbital moment of the conduction electrons).

In systems without localized electronic moments, the shift is determined by the Fermi contact interaction between the μ^+ spin and the spin of the conduction electrons and $\mathbf{B}_{cont} = \mathbf{B}_{F.\,cont}$.[67] With a spherical sample

$$\mathbf{B}_\mu = \mathbf{B}_{ext} + \mathbf{B}_{F.\,cont} \quad . \tag{5.61}$$

The contribution to K_μ is in this case

$$K_{F.\,cont} = \frac{\mathbf{B}_{ext} \cdot \mathbf{B}_{F.\,cont}}{B_{ext}^2} \quad . \tag{5.62}$$

As we will show below, K_μ is proportional to the density of conduction electrons at the muon site (which gives the strength of the contact interaction, see Sect. 5.1.2) and to the Pauli susceptibility (which reflects how much the conduction electrons are polarized by an external magnetic field).

---

[66] Note that the intrinsic susceptibility χ_0 is related to the measured sample susceptibility $\chi_{meas} = M/H_{ext} = \mu_0 M/B_{ext}$ by $\chi_0 = M/(H_{ext} - N_{zz}M) = \chi_{meas}/(1 - N\chi_{meas})$. This is because in a sample with $N \neq 0$ the demagnetizing field opposes the external field, so that the effective magnetizing field inside the sample is $H_{ext} - NM$. In diamagnetic and paramagnetic materials with typical susceptibilities in the range $10^{-5} - 10^{-6}$ and $10^{-3}-10^{-5}$, respectively, the difference is small.

[67] As mentioned in Sect. 5.1.2, in the presence of localized moments, e.g., in rare earths, additional dipolar and RKKY induced contact terms have to be considered.

5.6.1.1 Pauli Susceptibility

To introduce the relevant expressions we first derive the Pauli susceptibility in the simple free electron gas model, where the density of electronic states is

$$D(E) = \frac{3}{2} \frac{n_e}{E_F^{3/2}} \sqrt{E} \ , \tag{5.63}$$

n_e is the electron density and E_F is the Fermi energy.

Without an external field, spin up and spin down states are equally populated. If a field B_{ext} is applied, the states with magnetic moments parallel to B_{ext} (electron spin antiparallel to the applied field $n^\downarrow$) will be lowered by $\mu_B B_{ext}$ and those with antiparallel moment (electron spin parallel to applied field $n^\uparrow$) will be raised by the same amount. Since both subbands are filled up to E_F, there is an overweight of electrons with magnetic moment parallel to B_{ext}, while $n^\downarrow + n^\uparrow = n_e$. As a consequence, the metal develops a weak spin polarization.

The electron density for the two states is

$$n^\downarrow = \frac{1}{2} \int D(E + \mu_B B_{ext}) f(E) dE$$

$$n^\uparrow = \frac{1}{2} \int D(E - \mu_B B_{ext}) f(E) dE \ , \tag{5.64}$$

where $f(E)$ is the Fermi-Dirac distribution

$$f(E) = \frac{1}{e^{(E-\mu)/k_B T} + 1} \ , \tag{5.65}$$

and μ the chemical potential.[68]

For small B_{ext} the density of states can be expanded

$$D(E \pm \mu_B B_{ext}) = D(E) \pm \mu_B B_{ext} D'(E) \tag{5.66}$$

and the magnetization becomes[69]

$$M = \mu_B (n^\downarrow - n^\uparrow)$$

$$\simeq \mu_B^2 \, B_{ext} \int\limits_0^\infty \frac{dD}{dE} f(E) dE$$

---

[68] The chemical potential can be expressed as the change in the free energy of the system when an electron is added while keeping temperature and volume constant: $\mu = \left(\frac{\partial F}{\partial N_e}\right)_{T,V}$. $\mu = E_F \left[1 - \frac{\pi^2}{12} \left(\frac{k_B T}{E_F}\right)^2 + O\left(\frac{k_B T}{E_F}\right)^4 \right]$ (Ashcroft & Mermin 1976). At $T = 0$, $\mu = E_F$. For typical electronic densities, $k_B T / E_F \simeq 0.01$.

[69] Note that since we choose $\mathbf{B}_{ext} \parallel z$-axis, this magnetization as well as the contact field are also parallel to the z-axis.

$$= \mu_B^2 \, B_{ext} \left[\underbrace{D(E) f(E) \Big|_0^\infty}_{D(0)=0, \, f(\infty)=0} - \int_0^\infty \underbrace{\frac{df}{dE}}_{\simeq -\delta(E-E_F)} D(E) dE \right]$$

$$= \mu_B^2 \, B_{ext} \, D(E_F) \ . \tag{5.67}$$

Hence

$$M \simeq \frac{3}{2} \frac{n\mu_B^2}{E_F} B_{ext} \tag{5.68}$$

and the Pauli susceptibility is

$$\chi_P = \frac{M}{H_{ext}} = \frac{\mu_0 M}{B_{ext}} = \mu_0 \frac{3}{2} \frac{n\mu_B^2}{E_F} \ . \tag{5.69}$$

The magnetization of the conduction electrons can also be expressed in terms of their average spin as

$$M = -n_e g_e \mu_B \, \langle S_z \rangle = \chi_P \frac{B_{ext}}{\mu_0}$$

$$\Rightarrow \langle S_z \rangle = -\frac{B_{ext}}{\mu_0 \, n_e \, g_e \, \mu_B} \chi_P \ . \tag{5.70}$$

On the other hand, $\langle S_z \rangle$ can be written in terms of the contact field $B_{F.\,cont}$ at the muon site $\mathbf{r}_\mu$ produced by an electron density $\eta(\mathbf{r}_\mu)|\psi_s(\mathbf{r}_\mu)|^2$ with average spin $\langle S_z \rangle$, see Eq. 5.22

$$B_{F.\,cont} = -\frac{2}{3} \mu_0 \, g_e \, \mu_B \, \langle S_z \rangle \, \eta(\mathbf{r}_\mu)|\psi_s(\mathbf{r}_\mu)|^2 \ . \tag{5.71}$$

With Eq. 5.70, we obtain

$$B_{F.\,cont} = -\frac{2}{3} \mu_0 \, g_e \, \mu_B \, \eta(\mathbf{r}_\mu)|\psi_s(\mathbf{r}_\mu)|^2 \left(-\frac{B_{ext}}{\mu_0 \, n_e \, g_e \, \mu_B} \chi_P \right)$$

$$= \frac{2}{3} \chi_P \frac{\eta(\mathbf{r}_\mu)|\psi_s(\mathbf{r}_\mu)|^2 B_{ext}}{n_e} \ . \tag{5.72}$$

Finally the contact muon Knight-shift due to the conduction electrons can be written as[70]

$$K_{F.\,cont} = \frac{B_{F.\,cont}}{B_{ext}} = \frac{2}{3} \chi_{cond} \frac{\eta(\mathbf{r}_\mu)|\psi_s(\mathbf{r}_\mu)|^2}{n_e} \ . \tag{5.73}$$

---

[70] Note that in the simple free electron gas model $|\psi_s(\mathbf{r}_\mu)|^2 = n_e$, so that Eq. 5.73 simplifies to $K_{F.\,cont} = 2 \, \chi_P \, \eta(\mathbf{r}_\mu)/3$.

$K_{F.cont}$ is therefore proportional to the Pauli susceptibility[71] and to the (enhanced) electron density at the muon site, see Eq. 5.22. Note that, in principle, other (diamagnetic) contributions from the conduction electrons to this Knight-shift should have been considered. One stems from the Landau diamagnetism of the conduction electrons at the Fermi surface and is due to the modification of the electron motion in the magnetic field (orbital magnetism). It is typically only to a few parts per million and is ignored.[72] Another rather small ($\sim$15–20 ppm) but not always negligible effect, also known as chemical shift, originates from the diamagnetism of the screening cloud around the positive muon, which acts as a charged impurity (Zaremba & Zobin 1980).

The electron spin density at the muon site (and the contact field) are proportional to the contact hyperfine coupling constant A, see Eq. 5.15 and Sect. 5.1.2. Therefore $K_{F.cont}$ can also be expressed in terms of A

$$K_{F.\,cont} = \frac{A}{\mu_0\,\hbar^2 \gamma_\mu\,\gamma_e\,n_e}\,\chi_{cond}\ .\qquad (5.74)$$

In NMR the Knight-shift expresses the spin density sensed by the nuclear spin probe at the lattice site and an expression similar to 5.73 holds with $\eta(\mathbf{r}_\mu)\,|\psi_s(\mathbf{r}_\mu)|^2$ replaced by the square of the s-wave contribution at the nucleus.

Therefore, from this simple treatment we expect the muon contact Knight-shift to be proportional not only to the conduction electron susceptibility of the host material but also to the NMR Knight-shift. Figure 5.42 shows a plot of K_{cont} versus the NMR Knight-shift, K_{host}. Note that the proportionality holds relatively well only for simple metals such as Cu and the alkalis Na, K, Rb, and Cs, which are closer to a free electron gas picture. Moreover, $K_{host} \gg K_{F.cont}$, which reflects the fact that the electron density of s-wave electrons is much larger at a lattice nucleus than at an interstitial muon site.

5.6.2   Knight-Shift in Materials with Local Moments

Besides to the previously mentioned contribution to the Fermi contact term, for compounds with local electronic moments, such as the d and f electron compounds, the dipole contribution to K_μ is important.

---

[71] We use the approximately temperature independent spin susceptibility of the conduction electrons χ_{cond} instead of χ_P to express the fact that generally the experimental value is used, which has the advantage of including additional effects such as correlation and exchange effects, and may include other terms. For simple metals $\chi_{cond} \approx \chi_P$.

[72] See also the remark about the orbital hyperfine field in Sect. 5.1.2.

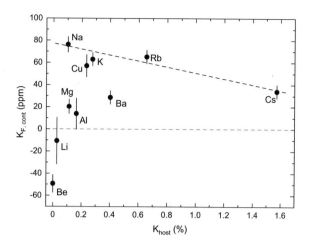

5.6.2.1 The Dipolar Field Contribution

In analogy to the contact term we can write

$$K'_{\text{dip}} = \frac{\mathbf{B}_{\text{ext}} \cdot \mathbf{B}'_{\text{dip}}}{B_{\text{ext}}^2} \quad , \tag{5.75}$$

where K'_{dip} accounts for the shift due to the field produced by the dipoles inside the Lorentz sphere.

The magnetic response of the f or d local moments is described by their susceptibility tensor, $\boldsymbol{\chi}_{fd}$, which, at least at low temperatures is much larger than the Pauli susceptibility. Using the tensor $\boldsymbol{\chi}_{fd}$, we can write for the electronic moments $\boldsymbol{\mu}_j$ induced by the external field

$$\boldsymbol{\mu}_j = \frac{v_0}{\mu_0}\boldsymbol{\chi}_{fd}\mathbf{B}_{\text{ext}} \quad , \tag{5.76}$$

where v_0 is the volume per magnetic ion.[73] With the usual assumption of the external field in the z direction of the laboratory frame, we have for the Cartesian components of the moment $\beta = x, y, z$

$$\mu_{j,\,\beta} = \frac{v_0}{\mu_0}\,\chi_{fd,\,\beta z}\,B_{\text{ext}} \quad . \tag{5.77}$$

In analogy to Eq. 5.20 we write the components of the dipolar field in terms of the dipolar tensor of the j th moment, which is at lattice position $\mathbf{r}_j$ and at position

---

[73] Remember that the magnetization is the volume density of the magnetic moments.

$\mathbf{R}_j = \mathbf{r}_j - \mathbf{r}_\mu$ from the muon site, Eq. 5.19

$$B_{\text{dip}', \alpha} = \sum_{j \in \text{Lor}} \sum_{\beta=1}^{3} D_{j, \alpha\beta} \, \mu_{j, \beta} \ , \tag{5.78}$$

where the sum is over all dipole moments in the Lorentz sphere.

In an applied field all the local moments have the same vector value given by Eq. 5.76 so that the moment vectors $\boldsymbol{\mu}_j = \boldsymbol{\mu}$ can be removed from the sum in Eq. 5.78, which becomes

$$\mathbf{B}'_{\text{dip}} = \left(\sum_{j \in \text{Lor}} \mathcal{D}_j \right) \boldsymbol{\mu} = \left(\sum_{j \in \text{Lor}} \mathcal{D}_j \right) \frac{v_0}{\mu_0} \chi_{fd} \mathbf{B}_{\text{ext}}$$

$$= \tilde{\mathcal{D}}_{\mathbf{r}_\mu} \chi_{fd} \, \mathbf{B}_{\text{ext}} \ , \tag{5.79}$$

where $\mathcal{D}_j$ is the tensor with components $D_{j,\alpha\beta}$ and we have defined the total dipolar tensor $\tilde{\mathcal{D}}_{\mathbf{r}_\mu}$[74]

$$\tilde{\mathcal{D}}_{\mathbf{r}_\mu} = \frac{v_0}{\mu_0} \sum_{j \in \text{Lor}} \mathcal{D}_j \ , \tag{5.80}$$

which depends on the muon stopping site relative to the position of the magnetic moments in the Lorentz sphere.[75] This tensor can be easily calculated for a known crystallographic structure by assuming (or knowing) the muon site.[76]

With 5.79 Eq. 5.75 becomes

$$K'_{\text{dip}} = \frac{\mathbf{B}_{\text{ext}} \cdot \tilde{\mathcal{D}}_{\mathbf{r}_\mu} \chi_{fd} \, \mathbf{B}_{\text{ext}}}{B_{\text{ext}}^2} \ . \tag{5.81}$$

5.6.2.2 The RKKY-Enhanced Contact Field Contribution

In Sect. 5.6.1 we have considered the direct effect of an external field on the spin polarization of the conduction electrons. The presence of local moments provides, via the RKKY interaction, an additional contact mechanism to polarize the conduction electrons, see Sect. 5.1.2. In its simplest form, this interaction produces

---

[74] The volume v_0 has been introduced to have a unitless tensor and the constant μ_0 has been factorized to simply express the Knight-shift terms with $\mathbf{B}_{\text{ext}}$.

[75] Note that $\tilde{\mathcal{D}}_{\mathbf{r}_\mu}$ is traceless but in general $\tilde{\mathcal{D}}_{\mathbf{r}_\mu} \chi_{fd}$ not.

[76] Note that to use Eq. 5.79, all vectors and tensors must be expressed in the same reference frame, usually the crystal or laboratory frame. If this is not the case, the Euler angles must be used to transform the quantities.

an enhanced contact field, which in analogy to Eq. 5.79 can be written as

$$\mathbf{B}_{\text{cont},fd} = \mathcal{A}_{\mathbf{r}_\mu} \, \chi_{fd} \, \mathbf{B}_{\text{ext}} = \frac{\mu_0}{v_0} \, \mathcal{A}_{\mathbf{r}_\mu} \, \mu \; , \tag{5.82}$$

where $\mathcal{A}_{\mathbf{r}_\mu}$ is the enhanced contact tensor. Since the contact coupling is normally independent of the direction of $\mathbf{B}_{\text{ext}}$, we can write $\mathcal{A}_{\mathbf{r}_\mu} = A_{\text{cont},fd,\mathbf{r}_\mu} \cdot \mathcal{E}$, where $\mathcal{E}$ is the unit tensor $\mathcal{E}_{\alpha\beta} = \delta_{\alpha\beta}$ and $A_{\text{cont},fd,\mathbf{r}_\mu}$ is given by

$$A_{\text{cont},fd,\mathbf{r}_\mu} \propto \mathcal{J} \sum_{\text{nn}} \frac{2k_F R_j \cos(2k_F R_j) - \sin(2k_F R_j)}{(2k_F R_j)^4} \; . \tag{5.83}$$

Here, the sum is over the nearest neighbor ions, k_F is the Fermi wave number and $\mathcal{J}$ is the exchange energy between the localized moment and the conduction electrons.

The corresponding contribution to the Knight-shift is then

$$K_{\text{cont},fd} = \frac{\mathbf{B}_{\text{ext}} \cdot \mathbf{B}_{\text{cont},fd}}{B_{\text{ext}}^2} \; . \tag{5.84}$$

5.6.2.3 The Total Knight-Shift

Combining Eqs. 5.73, 5.75 and 5.84, we can write the total Knight-shift for systems with local moments as

$$
\begin{aligned}
K_\mu &= K_{\text{F. cont}} + K'_{\text{dip}} + K_{\text{cont},fd} \\[1em]
&= \frac{\mathbf{B}_{\text{ext}} \cdot \mathbf{B}_{\text{F. cont}}}{B_{\text{ext}}^2} + \frac{\mathbf{B}_{\text{ext}} \cdot \mathbf{B}'_{\text{dip}}}{B_{\text{ext}}^2} + \frac{\mathbf{B}_{\text{ext}} \cdot \mathbf{B}_{\text{cont},fd}}{B_{\text{ext}}^2} \\[1em]
&= \frac{2}{3} \, \chi_{\text{cond}} \, \frac{\eta(\mathbf{r}_\mu)|\psi(\mathbf{r}_\mu)|^2}{n} + \frac{\mathbf{B}_{\text{ext}} \cdot \tilde{\mathcal{D}}_{\mathbf{r}_\mu} \, \chi_{fd} \, \mathbf{B}_{\text{ext}}}{B_{\text{ext}}^2} + \frac{\mathbf{B}_{\text{ext}} \cdot \mathcal{A}_{\mathbf{r}_\mu} \, \chi_{fd} \, \mathbf{B}_{\text{ext}}}{B_{\text{ext}}^2} \; .
\end{aligned}
\tag{5.85}
$$

We emphasize a few points:

- For a d or f metal K_μ is a weighted sum of the (approximately temperature independent) conduction electron and of the (strongly temperature dependent) d or f susceptibilities. At low temperatures, the susceptibility due to the local moments (which varies with $1/T$) is much stronger than the former and the first term on the right-hand side of Eq. 5.85 can be neglected. In the ideal case, the temperature dependence is entirely contained in the susceptibility tensor χ_{fd}.
- K'_{dip} and $K_{\text{cont},fd}$ both depend on the same susceptibility tensor and follow a similar temperature dependence.

- Since the conduction electron susceptibility χ_{cond} and the enhanced contact tensor $\mathcal{A}_{\mathbf{r}_\mu}$ are isotropic, the entire angular dependence of the Knight-shift is contained in the dipolar tensor $\tilde{\mathcal{D}}_{\mathbf{r}_\mu}$ and the susceptibility tensor χ_{fd}.[77]
- For a cubic lattice, in the principal axis coordinate system $\chi_{fd,x} = \chi_{fd,y} = \chi_{fd,z}$. Recalling that the dipolar tensor is traceless ($\sum_{i=x,y,z} \tilde{\mathcal{D}}_{\mathbf{r}_\mu}^{ii} = 0$), we note that the contribution of the dipolar field vanishes and only the contact field at the interstitial muon site contributes to the Knight-shift averaged over the three Cartesian directions $K_\mu = \frac{1}{3} \sum_i K_\mu^i$.[78]
- In the above equations, the Knight-shift is expressed in terms of the bulk quantity χ_{fd}. Experimentally, the muon as a local probe measures the local susceptibility, which may differ from the bulk susceptibility for various reasons. The difference can give further insight into the magnetic response of the investigated compound, see example in Sect. 5.6.5. In some systems, deviations from the theoretical $K - \chi$ relationship may be related to muon-induced modifications of the local susceptibility, see the example of a Pr compound below.

The smallness of $K_{F,cont}$ relative to the other two terms reflects the fact that $\chi_{cond} \approx \chi_P \ll \chi_{fd}$. In this case, since $\chi_{fd} \simeq \chi_{bulk}$, by measuring the Knight-shift and the bulk susceptibility independently and plotting the shift versus the susceptibility, with temperature as an implicit parameter,[79] one obtains a straight line with zero intercept, see Fig. 5.43 for such an example. If there are temperature independent contributions to Knight-shift and susceptibility, the K_μ versus χ_{bulk} relation remains linear, but is generally with a nonzero intercept. The $K - \chi$ relation may also be nonlinear, see Sect. 5.6.5.

5.6.3   Determination of the Muon Stopping Site

The dipolar contribution to the Knight-shift depends on the crystal structure of the compound under investigation and on the muon site via the dipolar coupling tensor. Using the relationship defined by Eq. 5.85, we can experimentally determine the muon site in systems possessing localized moments. The task is to determine the diagonal components of the total dipolar tensor $\tilde{\mathcal{D}}_{\mathbf{r}_\mu}$ and to compare them with the calculated values. This requires single crystals, which are not always available.

One procedure is to measure the Knight-shift as a function of the temperature along the principal directions of the crystal, (e.g., a, b, and c), where the suscepti-

---

[77] The angular dependence of the susceptibility is usually determined by bulk magnetization measurements.

[78] The dipolar contribution, however, contributes to the muon line broadening.

[79] Such a plot is called Clogston-Jaccarino plot (Clogston & Jaccarino 1961).

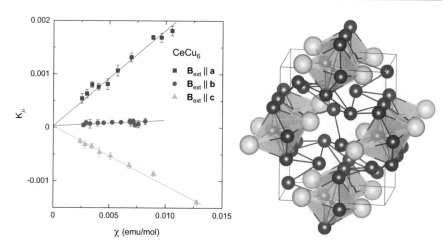

Fig. 5.43 Left: Muon Knight-shift versus the magnetic susceptibility (Clogston-Jaccarino plot) for the principal directions in $CeCu_6$ (orthorhombic structure) with the temperature as an implicit parameter (the temperature decreases going from left to right). From the slopes the main components of the dipolar tensor can be obtained, compared to calculations and the muon site found. Note the nonzero intercept due to a Fermi contact term contribution. Right: Crystal structure of $CeCu_6$. The muon sits between two Ce ions (Ce: yellow spheres, muon sites: small red spheres). Left graph modified from Amato et al. (1997), © Springer Nature. Reproduced with permission. All rights reserved. Structure drawn with VESTA, © 2006–2022 Koichi Momma and Fujio Izumi (Momma & Izumi 2011)

bility tensor is diagonal. From Eq. 5.81 one obtains

$$K_\mu^a(T) = (\tilde{\mathcal{D}}_{\mathbf{r}_\mu}^{aa} + A_{\mathrm{cont},fd,\mathbf{r}_\mu})\,\chi_a(T)$$

$$K_\mu^b(T) = (\tilde{\mathcal{D}}_{\mathbf{r}_\mu}^{bb} + A_{\mathrm{cont},fd,\mathbf{r}_\mu})\,\chi_b(T)$$

$$K_\mu^c(T) = (\tilde{\mathcal{D}}_{\mathbf{r}_\mu}^{cc} + A_{\mathrm{cont},fd,\mathbf{r}_\mu})\,\chi_c(T)\ . \tag{5.86}$$

If the susceptibilities along the principal direction of the crystal are known, then by measuring the Knight-shift as a function of the temperature along these directions and by taking into account that the dipolar tensor is traceless, the diagonal components of $\tilde{\mathcal{D}}_{\mathbf{r}_\mu}$ can be obtained (we have 3 measurements and 1 condition) and compared with calculated values for potential sites such as high-symmetry interstitial sites.

The left-hand sides of Figs. 5.43 and 5.44 show the muon Knight-shift as a function of the macroscopic magnetic susceptibility for the principal directions of orthorombic $CeCu_6$ and tetragonal $CeRu_2Si_2$. The corresponding structures are shown in the respective right panels. The observed scaling indicates that the local susceptibility probed by the muon is comparable to the macroscopic susceptibility (Amato et al. 1997).

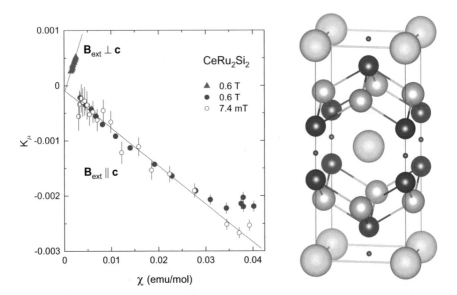

Fig. 5.44 Left: Muon Knight-shift versus magnetic susceptibility for the principal directions in $CeRu_2Si_2$ with the temperature as an implicit parameter. Only two measurements are needed to determine the site because the system is tetragonal. Right: Crystal structure of $CeRu_2Si_2$. The muon sits between the 4 Ce ions in the basal plane (Ce: yellow spheres, muon: small red spheres). Left graph modified from Amato et al. (1997), © Springer Nature. Reproduced with permission. All rights reserved. Structure drawn with VESTA, © 2006–2022 Koichi Momma and Fujio Izumi (Momma & Izumi 2011)

From the data one can extract the dipolar and contact hyperfine coupling constants: $\tilde{\mathcal{D}}_{\mathbf{r}_\mu}^{aa} = 0.89$ kG/μ_B, $\tilde{\mathcal{D}}_{\mathbf{r}_\mu}^{bb} = -0.10$ kG/μ_B, $\tilde{\mathcal{D}}_{\mathbf{r}_\mu}^{cc} = -0.79$ kG/μ_B and $A_{\mathrm{cont},fd,\mathbf{r}_\mu}$ $= 0.17$ kG/μ_B. Among the possible interstitial sites, only the site $(0, 0, 1/2)$ (b-site with multiplicity 4) has compatible theoretical values for the dipolar contribution: $\tilde{\mathcal{D}}_{\mathbf{r}_\mu}^{aa} = 0.97$ kG/μ_B, $\tilde{\mathcal{D}}_{\mathbf{r}_\mu}^{bb} = -0.11$ kG/μ_B, $\tilde{\mathcal{D}}_{\mathbf{r}_\mu}^{cc} = -0.86$ kG/μ_B.

One may wonder about the kG/μ_B units of the hyperfine constants, as it is often the case in the literature. Since Knight-shift and susceptibility are dimensionless, it would be natural to express the hyperfine constants in Eq. 5.85 or 5.86 also as dimensionless quantities as well. The question of the different definitions of the hyperfine constants and of the choice of units is addressed in Appendix E. We leave it as an exercise (Exercise 5.5) to estimate the dipolar and RKKY contact hyperfine constants from the Fig. 5.43.

5.6.4 Angular Dependence of the Knight-Shift

Another route to determine the dipolar tensor is to perform measurements of the angular dependence at a given temperature. For this, after each measurement, the sample is rotated with respect to the applied field. It is useful to rotate around a principal axis of the crystal and take the crystal frame as the reference frame.[80] Then the external field can be expressed as $\mathbf{B}_{ext} = B_{ext}(\sin\theta\cos\phi, \sin\theta\sin\phi, \cos\theta)$, where θ and ϕ are the polar and azimuth angles of the field in the crystal frame. In this frame, explicit evaluation of Eq. 5.79 shows that the angular dependence of the vector $\mathbf{B}'_{dip}$ projected along the direction of the external field $B'_{dip,\|} = (\mathbf{B}_{ext}\cdot\mathbf{B}'_{dip})/B_{ext}$ has the form (Feyerherm et al. 1994)

$$
\begin{aligned}
B'_{dip,\|} = &\tfrac{1}{3}(\tilde{\mathcal{D}}_{\mathbf{r}_\mu}^{xx}\,\chi_x + \tilde{\mathcal{D}}_{\mathbf{r}_\mu}^{yy}\,\chi_y + \tilde{\mathcal{D}}_{\mathbf{r}_\mu}^{zz}\,\chi_z)\,B_{ext}\\
&+ \tfrac{2}{3}[\tilde{\mathcal{D}}_{\mathbf{r}_\mu}^{zz}\,\chi_z - \tfrac{1}{2}(\tilde{\mathcal{D}}_{\mathbf{r}_\mu}^{xx}\,\chi_x + \tilde{\mathcal{D}}_{\mathbf{r}_\mu}^{yy}\,\chi_y)]\,P_2^0(\cos\theta)\,B_{ext}\\
&- \tfrac{1}{3}\tilde{\mathcal{D}}_{\mathbf{r}_\mu}^{xz}(\chi_x + \chi_z)\,P_2^1(\cos\theta)\cos\phi\,B_{ext}\\
&- \tfrac{1}{3}\tilde{\mathcal{D}}_{\mathbf{r}_\mu}^{yz}(\chi_y + \chi_z)\,P_2^1(\cos\theta)\sin\phi\,B_{ext}\\
&+ \tfrac{1}{6}(\tilde{\mathcal{D}}_{\mathbf{r}_\mu}^{xx}\,\chi_x - \tilde{\mathcal{D}}_{\mathbf{r}_\mu}^{yy}\,\chi_y)\,P_2^2(\cos\theta)\cos 2\phi\,B_{ext}\\
&+ \tfrac{1}{6}\tilde{\mathcal{D}}_{\mathbf{r}_\mu}^{xy}(\chi_x + \chi_y)\,P_2^2(\cos\theta)\sin 2\phi\,B_{ext}\;.
\end{aligned}
\tag{5.87}
$$

We assume here that χ_{fd} is diagonal in the chosen coordinate frame with components χ_x, χ_y, χ_z; the terms P_l^m are the associated Legendre polynomials. The P_l^m that we use[81] are

$$
P_2^0(\cos\theta) = \tfrac{1}{2}(3\cos^2\theta - 1)
$$

$$
P_2^1(\cos\theta) = -3\cos\theta\sin\theta
$$

$$
P_2^2(\cos\theta) = 3\sin^2\theta\;.
\tag{5.88}
$$

Since $K'_{dip} = B'_{dip,\|}/B_{ext}$, Eq. 5.87 expresses the dependence of the dipolar Knight-shift on the orientation of the external field with respect to the Cartesian crystal axes.[82]

Taking into account the symmetry of the system, this expression simplifies. For example, looking again at the compound UPd_2Al_3, Fig 5.9, and assuming,

---

[80] In other cases, the axis of rotation is defined by the applied field. See footnote 76.

[81] Note that in Schenck (1999) and Yaouanc and Dalmas de Réotier (2011) the P_2^1 terms in Eq. 5.87 have positive sign. The difference is due to a different choice of the associated Legendre polynomial. Here we adopt the common definition $P_2^1(\cos\theta) = -3\cos\theta\sin\theta$.

[82] Equation 5.87 is also valid for the contact term $B_{cont,fd}$. One has simply to replace $\tilde{\mathcal{D}}_{\mathbf{r}_\mu}$ with $\mathcal{A}_{cont,fd,\mathbf{r}_\mu}$.

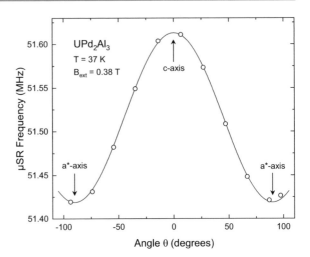

Fig. 5.45 Angular dependence of the muon precession frequency measured for a single crystal sample of the hexagonal compound UPd$_2$Al$_3$ in a transverse field of 0.38 T in the paramagnetic phase at 37 K. The field is rotated in the a^*c- (or yz-) plane and θ denotes the angle between the field and the c-axis. Modified from Amato et al. (1997), © Springer Nature. Reproduced with permission. All rights reserved

as discussed previously in this chapter that the muon occupies solely the b site $(0, 0, 1/2)$ of the hexagonal structure with space group $P6/mmm$, we have a diagonal dipolar tensor with elements satisfying[83]

$$\tilde{\mathcal{D}}_{\mathbf{r}_\mu}^{xx} = \tilde{\mathcal{D}}_{\mathbf{r}_\mu}^{yy} = -\frac{1}{2}\tilde{\mathcal{D}}_{\mathbf{r}_\mu}^{zz} \ , \tag{5.89}$$

since the b site is axially symmetric with respect to z. Also, due to the hexagonal symmetry, we can write for the susceptibility tensor $\chi_x = \chi_y = \chi_\perp$ and $\chi_z = \chi_\parallel$.

If we perform TF measurements and rotate the field around the x-axis (i.e., the field is in the yz-plane and $\phi = \pi/2$) and Eq. 5.87 becomes

$$B'_{\text{dip},\parallel} = \frac{1}{3}B_{\text{ext}}\tilde{\mathcal{D}}_{\mathbf{r}_\mu}^{zz}\left[\left(\chi_\parallel - \chi_\perp\right) + \left(2\chi_\parallel + \chi_\perp\right)P_2^0(\cos\theta)\right] =$$

$$= B_{\text{ext}}\tilde{\mathcal{D}}_{\mathbf{r}_\mu}^{zz}\left[\cos^2\theta\left(\chi_\parallel + \frac{1}{2}\chi_\perp\right) - \frac{1}{2}\chi_\perp\right] \ . \tag{5.90}$$

Dividing this expression by B_{ext} yields K'_{dip}, which contains the entire angular dependence of the Knight-shift. Figure 5.45 shows the raw frequency data, which, when corrected for the demagnetizing and Lorentz fields, give the angular dependence of K'_{dip}.

By fitting Eq. 5.90 to the data, one obtains $\tilde{\mathcal{D}}_{\mathbf{r}_\mu}^{zz}$, which can be compared to a computed value under assumption of a muon site. Excellent agreement with the muon at site b is found. The occurrence of only a single signal in the spectra is a clear

---

[83] Here we assume that the sample is oriented so that $a = x$, $a^* = y$, and $c = z$.

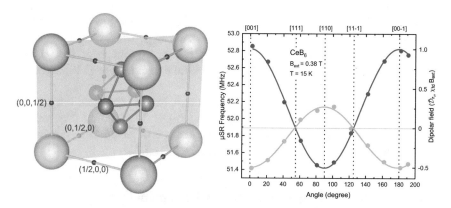

Fig. 5.46 Left panel: Cubic structure of CeB_6. The cerium atoms, carrying the electronic moments, are at $(0,0,0)$ and equivalent positions. The muon site (small red spheres) is located between two cerium atoms. Right panel: Angular dependence of the observed muon precession frequencies measured for a single crystal sample of CeB_6 in a transverse field of 0.38 T at 15 K. The field is rotated around the $[1\bar{1}0]$ axis, i.e., the field lies in the gray plane on the left panel figure. θ denotes the angle between the field and the [001] axis. The two asymmetry signals have an amplitude ratio of 2:1 corresponding on one hand to the $(1/2,0,0)$ and $(0,1/2,0)$ sites (green symbols) and on the other hand to the $(0,0,1/2)$ site (red symbols). At the magic angle $\theta_m \approx 54.7°$ the dipolar field is zero and only one frequency is observed. Right panel modified from Amato et al. (1997), © Springer Nature. Reproduced with permission. All rights reserved. Structure drawn with VESTA, © 2006–2022 Koichi Momma and Fujio Izumi (Momma & Izumi 2011)

indication that a single crystallographic site with an axial symmetry is occupied by the implanted muon.

The system CeB_6, which crystallizes in the simple cubic structure $Pm3m$ provides another example. The muon stops at the d site $(1/2,0,0)$, which has a multiplicity of three, the other two sites being $(0, 1/2, 0)$ and $(0,0,1/2)$, Fig. 5.46. These sites may become magnetically inequivalent when an external field is applied and Knight-shift measurements are made, as they have the following related dipolar tensors, see Exercise 5.6.

$$\tilde{\mathcal{D}}_{(\frac{1}{2},0,0)} = \begin{pmatrix} \tilde{\mathcal{D}} & 0 & 0 \\ 0 & -\frac{1}{2}\tilde{\mathcal{D}} & 0 \\ 0 & 0 & -\frac{1}{2}\tilde{\mathcal{D}} \end{pmatrix}$$

$$\tilde{\mathcal{D}}_{(0,\frac{1}{2},0)} = \begin{pmatrix} -\frac{1}{2}\tilde{\mathcal{D}} & 0 & 0 \\ 0 & \tilde{\mathcal{D}} & 0 \\ 0 & 0 & -\frac{1}{2}\tilde{\mathcal{D}} \end{pmatrix}$$

$$\tilde{\mathcal{D}}_{(0,0,\frac{1}{2})} = \begin{pmatrix} -\frac{1}{2}\tilde{\mathcal{D}} & 0 & 0 \\ 0 & -\frac{1}{2}\tilde{\mathcal{D}} & 0 \\ 0 & 0 & \tilde{\mathcal{D}} \end{pmatrix} . \tag{5.91}$$

Taking into account that we have a cubic system (i.e., $\chi_x = \chi_y = \chi_z$), from Eq. 5.87 we see that if we apply the field along the [001]-axis (z direction) the sites (1/2,0,0) and (0,1/2,0) will have the same $\mathbf{B}'_{dip}$, which is half the value and of opposite direction compared to that felt by the muon at the (0,0,1/2) site. This is true when rotating the external field around the [1$\bar{1}$0] axis, see Fig. 5.46.

When the field is directed along the [111] axis (or an equivalent one), which corresponds to the magic angle $\theta_m = \arccos(1/\sqrt{3}) \approx 54.7°$, all the sites are also magnetically equivalent and only one μSR frequency is observed. Note that for this angle the dipolar contribution is actually zero, Exercise 5.7.

5.6.5   Nonlinear Knight-Shift Versus Susceptibility

The relationship may be nonlinear for a number of reasons. As first observed in NMR (MacLaughlin 1981), it can originate from the presence of multiple species of magnetic electrons with different coupling constants and susceptibilities in the investigated compounds,[84] from temperature dependent coupling constants,[85] or from the onset of new susceptibility components at low temperature.[86]

In selected $4f$ compounds, particularly those containing non-Kramers[87] ions with a singlet ground state, such as some Pr-intermetallics, the muon-Pr interaction locally changes the Pr^{3+} susceptibility. The magnetic properties of singlet $4f$ states are sensitive to small modifications of the crystal electric field. The muon charge locally modifies the Pr^{3+} crystal electric field splitting and alters the magnetic response of at least two of the muon's nearest Pr ions, so that the local susceptibility monitored by the muon is different from the bulk susceptibility (Feyerherm et al. 1994; Tashma et al. 1997). In this case, the muon Knight-shift no longer reflects the bulk susceptibility and deviations from the K versus χ linearity are observed.

---

[84] If there are n different paramagnetic subsystems with separate contributions χ_n to the total bulk susceptibility $\chi_{bulk} = \sum_n \chi_n$, which couple to the muon spin via separate coupling constants $\tilde{\mathcal{D}}_{\mathbf{r}_\mu,n}$ and $A_{cont,fd,n}$, $K_\mu(T) = \sum_n K_{\mu,n}(T) = \sum_n (\mathcal{D}_{\mathbf{r}_\mu,n} + A_{cont,fd,n})\chi_n$. In this case, χ_{bulk} and $K_\mu(T)$ can exhibit a very different temperature dependence, and the Knight-shift versus susceptibility dependence is no longer linear.

[85] For example the thermal depopulation of low-lying f electron crystalline-electric-field states modifies the coupling (Ohishi et al. 2009).

[86] Muon (and NMR) Knight-shift anomalies such as the deviation from a linear K versus χ_{bulk} behavior, have been shown to be a useful tool for studying the onset of the heavy fermion state and the evolution with temperature of the electronic correlations in heavy fermion systems (Ohishi et al. 2009).

[87] Ions with integer J are named non-Kramers ions, those with half integer J Kramers ions. The free Pr^{3+} ion ($4f^2$) ground state is $^3H_4$.

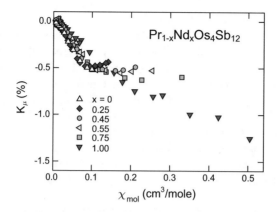

Fig. 5.47 Clogston-Jaccarino plots of muon Knight-shift K_μ vs. magnetic susceptibility χ_{mol} for $Pr_{1-x}Nd_xOs_4Sn_{12}$, where $x = 0, 0.25, 0.45, 0.55, 0.75$, and 1.00. For clarity, the error bars are not shown. In the Pr-substituted compounds $Pr_{1-x}Nd_xOs_4Sb_{12}$, $x \leq 0.75$, the Clogston-Jaccarino plots exhibit an anomalous saturation of K_μ vs. χ for large susceptibilities (low temperatures). This indicates a reduction of the coupling strength between spin of the muon and $4f$ paramagnetism for temperatures below ≈ 15 K. In the unsubstituted $NdOs_4Sb_{12}$ this effect is not observed and K_μ vs. χ is to a fair approximation linear. Modified from Ho et al. (2014), © American Physical Society. Reproduced with permission. All rights reserved

The nonlinearity at low temperatures of K vs. χ in the family of the lanthanide intermetallics $Pr_{1-x}Nd_xOs_4Sn_{12}$ ($0 \leq x \leq 1$) evidences that the coupling strength between the muon spin and the surrounding $4f$ magnetism is anomalously suppressed in the Pr-containing compounds but not in the Pr-free end compound $NdOs_4Sb_{12}$ (Ho et al. 2014), see Fig. 5.47.

Next we show an example of a compound where deviations from the K vs. χ linearity give insight into low-energy magnetic excitations.

The intermediate valence compound SmB_6 is predicted to have topological insulator properties at low temperatures.[88] It behaves like a poor metal at high

---

[88] Topological insulators are materials that have bulk insulating states and symmetry-protected conducting surface states (Hasan & Kane 2010). The surface states fall within the bulk energy gap and allow metallic conduction at the surface. These states are possible due to the combination of spin-orbit interactions and time-reversal symmetry. A conducting surface is not unique to topological insulators, as ordinary band insulators can also support conducting surface states. What is specific about topological insulators is that their surface states are symmetry-protected. For instance, as long as the time-reversal symmetry is preserved (i.e., there is no magnetism), the topological order is not changed by smooth perturbations; the conducting states at the surface are topologically protected. No form of nonmagnetic disorder can destroy them. The surface (or edge) states of a topological insulator lead to a conducting state with properties different from any other known 1D or 2D electronic system. Carriers in these surface states have their spin locked at a right angle to their momentum (spin-momentum locking) by strong spin-orbit coupling. As a consequence, scattering processes are strongly suppressed and conduction on the surface is highly metallic. These special properties could be useful for applications ranging from spintronics to quantum computation.

temperatures and develops an insulating gap at lower temperatures. The gap originates from the hybridization of the localized Sm $4f$ electrons with the itinerant Sm $5d$ electrons (Kondo insulator). The Fermi level is located in the hybridization gap. Rapid fluctuations between nonmagnetic (Sm^{2+} $4f^6$) and magnetic (Sm^{3+} $4f^5 5d^1$) SmB_6 configurations give rise to intermediate valence properties. While there is experimental evidence for SmB_6 being a 3D topological insulator, there are several unexplained properties of the insulating bulk. Low-temperature thermodynamic and transport anomalies have been attributed to various types of magnetic excitations that could negatively affect the topological surface states. Muon Knight-shift, TF, and ZF measurements provide evidence for various forms of fluctuating magnetism at low temperature (Akintola et al. 2018; Biswas et al. 2014; Gheidi et al. 2019).

We discuss here only the Knight-shift measurements as an example of how the nonlinear K vs. χ dependence reflects the evolution of the electronic state and the onset of magnetic fluctuations with temperature. The applied magnetic field polarizes the localized Sm $4f$ magnetic moments and the conduction electrons. The local field $\mathbf{B}_\mu$ sensed by the muon is the vector sum of the dipolar field of the $4f$ moments $\mathbf{B}_{\text{dip}}$ and of the contact field, which has a direct contribution, $\mathbf{B}_{\text{F. cont}}$, and an indirect one, $\mathbf{B}_{\text{cont}, fd}$, via the RKKY interaction with the Sm $4f$ moments polarizing the conduction electrons at the muon site or via a related mechanism in the insulating state, see footnote 17.

Figure 5.48 shows Fourier transforms of TF-μSR time spectra recorded on a single-crystal SmB_6 sample. Three peaks (broadened by the instrumental resolution and the specific Fourier transform procedure) are visible. The central peak is a background contribution arising from muons stopping in an Ag backplate. The left and right peaks have an amplitude ratio of 2:1, corresponding to the μ^+ stopping sites, which are the same as in CeB_6, i.e., at the midpoint of the horizontal or vertical edges of the cubic Sm ion sublattice and therefore perpendicular ($\perp$) or parallel ($\parallel$) to the applied field, see Fig. 5.46 and Sect. 5.6.3.

The two $\perp$ and the single $\parallel$ μ^+ stopping sites are crystallographically equivalent but become magnetically inequivalent with the field applied parallel to the c-axis. From Eq. 5.91 it follows that the dipolar field generated by the polarization of the Sm $4f$ moments along the c-axis direction is antiparallel to $\mathbf{B}_{\text{ext}}$ at the $\perp$-sites $(1/2, 0, 0)$ and $(0, 1/2, 0)$, and parallel to the field and twice as large at the $\parallel$ site $(0, 0, 1/2)$.

Following Eqs. 5.86 and 5.91 the Knight-shifts (corrected for demagnetization and Lorentz fields) at the $\perp$- and $\parallel$-sites are given by

$$K_{\mu,\perp}(T) = (-\frac{1}{2}\tilde{\mathcal{D}} + A_{\text{cont}, fd, \perp})\, \chi_{4f}(T) + K_{\text{F. cont}, \perp}$$

$$K_{\mu,\parallel}(T) = (\tilde{\mathcal{D}} + A_{\text{cont}, fd, \parallel})\, \chi_{4f}(T) + K_{\text{F. cont}, \parallel} \ , \qquad (5.92)$$

where $\chi_{4f}(T) = \chi_{\text{bulk}} - \chi_{\text{P}}$ and we allow for anisotropic contact terms.

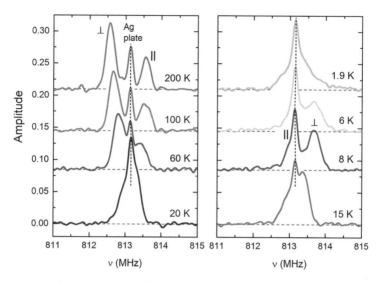

Fig. 5.48 Evolution of the Fourier transforms of TF spectra with temperature. A high field of 6 T is applied parallel to the c-axis. The three peaks are explained in the text. Modified from Akintola et al. (2018), under CC-BY-4.0 license. © The Authors

Figure 5.49 shows the Clogston-Jaccarino plot with the Knight-shifts plotted against χ_{mol} (χ_{bulk} in our notation). Above 110 K, the plot shows a linear relationship between Knight-shift and susceptibility as expected from Eq. 5.92. However, the slope and intercept indicate that the contact terms are anisotropic.[89]

Below 110 K, the Clogston-Jaccarino plot strongly deviates from linearity. This reflects the opening of the hybridization gap and the gradual reduction of both the Pauli susceptibility ($\propto K_{F, cont}$) and the electronic spin density at the muon sites ($\propto A_{cont, fd}$). Near 30 K, both $K_{\mu,\perp}$ and $K_{\mu,\parallel}$ simultaneously approach zero value. At this temperature the Kondo gap is completely formed. Consequently, the dipolar as well as the contact couplings are expected to vanish due to the Kondo screening of the Sm $4f$ moments and the absence of a screening cloud of conduction electrons around the muon. On further lowering the temperature, below 20–25 K, nonzero Knight-shifts reappear. In this temperature range ZF measurements show an increasing muon spin relaxation rate, due to magnetic fluctuations, Fig. 5.50 (Biswas et al. 2014; Gheidi et al. 2019). These magnetic fields are dynamic in nature and may still preserve the time-reversal symmetry of SmB$_6$. However, they appear to be detrimental to the surface conductivity above a few K (Gheidi et al. 2019; Akintola et al. 2018).

---

[89] This anisotropy has been observed in other $4f$ compounds and reflects the influence of the electric quadrupole moment of the nonspherical Sm $4f^5$ shell on the polarization of the conduction electrons at the muon site (Schenck et al. 2003).

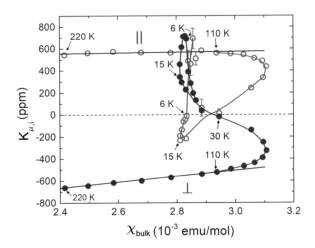

Fig. 5.49 Muon Knight-shift at the two magnetically inequivalent muon sites versus the bulk magnetic susceptibility for $B_{ext} = 6$ T. The temperature is an implicit parameter. The straight black lines are fits to Eq. 5.92. Modified from Akintola et al. (2018), under CC-BY-4.0 license. © The Authors

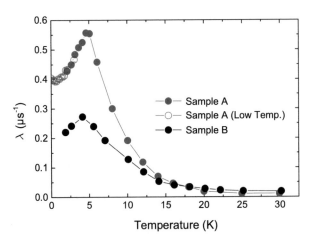

Fig. 5.50 Temperature dependence of the ZF muon spin relaxation due to electronic moments in SmB$_6$ from two differently grown samples. Both samples show a qualitatively similar temperature dependence, but λ is smaller in sample B. The origin of these different microscopic properties may reside in two completely different growth processes of the single crystals and in different degrees of impurities. See also Gheidi et al. (2019) for additional details. Modified from Biswas et al. (2014), © American Physical Society. Reproduced with permission. All rights reserved

5.7 Depolarization Created by Nuclear Moments

5.7.1 Classical Calculation

We have already mentioned that the small dipolar fields created by the nuclear moments around the muon produce a broadening of the μSR signal. In the following, we outline a classical approach to quantify the corresponding field width in the TF and ZF configurations. A quantum mechanical calculation produces the same result (Hayano et al. 1979).

5.7.1.1 The TF Case

As usual the applied field is along the z-axis. The field created by the nuclear magnetic moment is only of dipolar origin. So a nuclear moment $\boldsymbol{\mu}_{N,j}$ at position $\mathbf{r}_j$ creates a field at the muon site (taken here as origin of the coordinate system, $\mathbf{r}_\mu = 0$) given by Eq. 5.18

$$\mathbf{B}_{N,j}(0) = \frac{\mu_0}{4\pi} \frac{3\mathbf{r}_j(\boldsymbol{\mu}_{N,j} \cdot \mathbf{r}_j)}{r_j^5} - \frac{\boldsymbol{\mu}_{N,j}}{r_j^3} \quad . \tag{5.93}$$

The total field is given by the lattice sum of the dipoles

$$\mathbf{B}_N(0) = \frac{\mu_0}{4\pi} \sum_j \left(\frac{3\mathbf{r}_j(\boldsymbol{\mu}_{N,j} \cdot \mathbf{r}_j)}{r_j^5} - \frac{\boldsymbol{\mu}_{N,j}}{r_j^3} \right) \quad . \tag{5.94}$$

As in Sect. 4.2.2, we consider the case where the applied field is much larger than the internal field, $B_{\text{ext}} \gg \frac{\gamma_\mu}{\gamma_N} B_N$.[90] In this case one has only to consider the field distribution along the applied field, i.e., along the z direction.

Expressing the vectors $\mathbf{r}_j$ and $\boldsymbol{\mu}_{N,j}$ in spherical coordinates, see Fig. 5.51, the z component of $\mathbf{B}_{N,j}(0)$ (with $|\boldsymbol{\mu}_{N,j}| = \mu_N = \hbar\gamma_N I_N$ for all nuclei) is given by

$$
\begin{aligned}
B_{N,j,z}(0) &= \frac{\mu_0}{4\pi} \frac{\mu_N}{r_j^3} \left[(3\cos^2\theta_j - 1)\cos\theta_j' + 3\sin\theta_j\cos\theta_j\cos\phi_j'\sin\theta_j' \right] \\
&= \frac{\hbar\mu_0}{4\pi} \frac{\gamma_N}{r_j^3} \left[I_{N,j,z}(3\cos^2\theta_j - 1) + 3I_{N,j,x}\sin\theta_j\cos\theta_j \right] \quad .
\end{aligned} \tag{5.95}
$$

The Larmor precession of the nuclear moments around $\mathbf{B}_{\text{ext}}$ averages out the second term (so-called "nonsecular term")[91] on the right hand-side of Eq. 5.95. The

---

[90] Typical field strengths from nuclear moments are of the order 10^{-4} T or less.

[91] The distinction between secular and nonsecular terms is better understood quantum mechanically. Quantum mechanically the magnetic moment is expressed in terms of spin operators, as in the second line of Eq. 5.95. Neglecting the nonsecular terms and keeping only the secular ones

Fig. 5.51 Parameters of the dipolar interaction between a nuclear moment and the muon. $\mathbf{r}_j$ is the vector between the muon located at the origin and a nucleus with magnetic moment $\boldsymbol{\mu}_{N,j}$. Without loss of generality, the azimuthal angle of $\mathbf{r}_j$ can be chosen to be zero. In a large externally applied field along the z direction, the nuclear moment precesses around the external field describing the trajectory around the blue circle

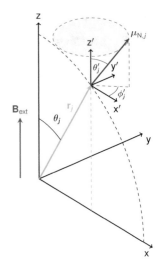

variance of the field distribution along the direction of the external field, is then[92]

$$\sigma_{\mathrm{TF,N}}^2 = \gamma_\mu^2 \, \langle \Delta B_{\mathrm{N},z}^2 \rangle$$

$$= \gamma_\mu^2 \left[\left\langle \left(\sum_j B_{\mathrm{N},j,z} \right)^2 \right\rangle - \left\langle \left(\sum_j B_{\mathrm{N},j,z} \right) \right\rangle^2 \right] . \tag{5.96}$$

The sums are over the nuclear magnetic moments and the average is over all possible orientations of $\boldsymbol{\mu}_{N,j}$, i.e., over all possible θ'_j and ϕ'_j angles.

The second term of Eq. 5.96 on the right-hand side is zero, because the angular averaging corresponds to terms of the form

$$\frac{1}{4\pi} \int\limits_0^{2\pi} \int\limits_0^{\pi} \cos\theta'_j \, \sin\theta'_j \; d\theta'_j d\phi'_j \; .$$

On the other hand, the first term contains only terms of the form

$$\frac{1}{4\pi} \int\limits_0^{2\pi} \int\limits_0^{\pi} \cos^2\theta'_j \, \sin\theta'_j \; d\theta'_j d\phi'_j \; ,$$

---

means that only the matrix elements of $I_{\mathrm{N},j,z}$ are considered. See Yaouanc and Dalmas de Réotier (2011) for a detailed discussion.

[92] Here we use the fact that $\mathrm{Var}(X+a) = \mathrm{Var}(X)$, that is $\mathrm{Var}(B_{\mathrm{N},j,z}(0) + B_{\mathrm{ext}}) = \mathrm{Var}(B_{\mathrm{N},j,z}(0))$.

which is equal to $1/3$. Therefore, we obtain[93]

$$\sigma_{\text{TF,N}}^2 = \frac{1}{3}\left(\frac{\mu_0}{4\pi}\right)^2 \gamma_\mu^2 \mu_N^2 \sum_j \frac{(3\cos^2\theta_j - 1)^2}{r_j^6} . \tag{5.97}$$

The correct quantum mechanical expression is obtained by considering that quantum mechanically the expectation value of the spin operator squares $\langle \mathbf{I}_N^2 \rangle = I_N(I_N + 1)$

$$\sigma_{\text{TF,N}}^2 = \frac{1}{3}\left(\frac{\mu_0}{4\pi}\right)^2 \hbar^2 \gamma_\mu^2 \gamma_N^2 I_N(I_N + 1) \sum_j \frac{(3\cos^2\theta_j - 1)^2}{r_j^6} . \tag{5.98}$$

This variance depends on the muon-nuclear moment distance as r^{-6}. Therefore, the lattice sum converges very rapidly, and the sum is dominated by the nearest neighbor and next nearest neighbor nuclei. Equation 5.98 is the solution for single crystals with a single species of nuclear moment.

In the chosen approximation, $B_{\text{ext}} \gg \frac{\gamma_\mu}{\gamma_N} B_N$, the value of $\sigma_{\text{TF,N}}^2$ is independent of the magnitude of the external field and only depends on the position of the muon relative to the lattice sites of the nuclear moments (expressed by r_j and θ_j) (or in other words, on the orientation of the crystal axes with respect to the field direction). Therefore the muon site can be determined not only by measuring the angular dependence of the Knight-shift, but also by investigating the angular dependence of the nuclear depolarization rate.

Figure 5.52 shows an example of such a determination for the system UPd_2Al_3, where the muon site was previously determined by a Knight-shift measurement, see Sect. 5.6.3.

With a polycrystalline sample, we have to perform the spatial averaging of Eq. 5.98 with

$$\frac{1}{4\pi} \int_0^{2\pi} \int_0^\pi (3\cos^2\theta_j' - 1)^2 \sin\theta_j' \, d\theta_j' d\phi_j' = \frac{4}{5} ,$$

yielding

$$\sigma_{\text{TF,N,poly}}^2 = \frac{4}{15}\left(\frac{\mu_0}{4\pi}\right)^2 \hbar^2 \gamma_\mu^2 \gamma_N^2 I_N(I_N + 1) \sum_j \frac{1}{r_j^6} . \tag{5.99}$$

---

[93] Note that the covariance between the different $B_{N,j}$ is zero. Thus the variance of the total field distribution is equal to the sum of the variances of the dipolar field of each nuclear moment surrounding the muon $\langle (\sum_j B_{N,j,z})^2 \rangle = \sum_j \langle B_{N,j,z}^2 \rangle$.

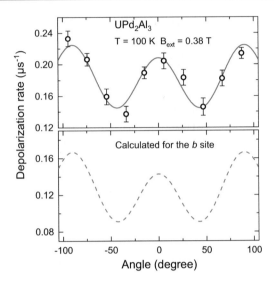

Fig. 5.52 Angular dependence of the depolarization rate measured for a single crystal sample of the hexagonal compound UPd_2Al_3 in a transverse field of 0.38 T at 37 K. This quantity corresponds to the depolarization of the μSR signal for which the angular dependence of the frequency is shown in Fig. 5.45. The field is rotated in the a^*c- (or yz-) plane and θ denotes the angle between the field and the c-axis. Upper panel: The solid line is a fit of the function $\sigma_{TF,N}(\theta) = \sigma_0 + c_0 P_2^0(\cos\theta) + c_1 P_4^0(\cos\theta)$. $P_4^0(\cos\theta)$ is the associated Legendre polynomial $P_4^0(\cos\theta) = \frac{1}{8}(35\cos^4\theta - 30\cos^2\theta + 3)$. Lower panel: Calculated depolarization rate for the b-site, using Eq. 5.98. The fitted parameters c_0 and c_1 are very close to the calculated ones. The slight deviation between the measured and calculated values of σ_0 can be ascribed to a slight misalignment of the different monocrystallites used for the experiment or to lattice imperfections. Modified from Amato et al. (1997), © Springer Nature. Reproduced with permission. All rights reserved

5.7.1.2 The ZF Case

In ZF, the nonsecular terms cannot be neglected. All the three field components of the dipolar field of the nuclear moments, and not just the projection along the z direction, as in the TF case, determine the field width sensed by the muons. The x and y components of the dipolar fields are[94]

$$B_{N,j,x}(0) = \frac{\mu_0}{4\pi} \frac{\mu_{N,j}}{r_j^3} \left[(3\sin^2\theta_j - 1)\sin\theta_j' \cos\phi' + 3\sin\theta_j \cos\theta_j \cos\theta_j' \right]$$

$$B_{N,j,y}(0) = \frac{\mu_0}{4\pi} \frac{\mu_{N,j}}{r_j^3} \left[-\sin\theta_j' \sin\phi' \right] \ . \tag{5.100}$$

---

[94] The z component of the field including the nonsecular terms, is given by Eq. 5.95. The direction of the initial muon polarization is along the z-axis.

The variance of the field components perpendicular to the initial muon polarization is given by

$$\sigma_{ZF,N,x}^2 + \sigma_{ZF,N,y}^2 = \gamma_\mu^2 \left[\langle \Delta B_{N,j,x}(0)^2 \rangle + \langle \Delta B_{N,j,y}(0)^2 \rangle \right]$$

$$= \gamma_\mu^2 \left\{ \left[\langle \left(\sum_j B_{N,j,x} \right)^2 \rangle - \langle \left(\sum_j B_{N,j,x} \right) \rangle^2 \right] \right.$$

$$\left. + \left[\langle \left(\sum_j B_{N,j,y} \right)^2 \rangle - \langle \left(\sum_j B_{N,j,y} \right) \rangle^2 \right] \right\} . \qquad (5.101)$$

As for the TF case, we have to average over all possible orientations of the nuclear magnetic moments. The average value of each field component is zero, so that the second term in the two square brackets does not contribute. The remaining terms give

$$\sigma_{ZF,N,x}^2 + \sigma_{ZF,N,y}^2 = \frac{1}{3} \left(\frac{\mu_0}{4\pi} \right)^2 \hbar^2 \gamma_\mu^2 \gamma_N^2 I_N (I_N + 1) \sum_j \frac{(5 - 3\cos^2\theta_j)}{r_j^6} . \qquad (5.102)$$

Similarly, the variance of the z component becomes

$$\sigma_{ZF,N,z}^2 = \frac{1}{3} \left(\frac{\mu_0}{4\pi} \right)^2 \hbar^2 \gamma_\mu^2 \gamma_N^2 I_N (I_N + 1) \sum_j \frac{(3\cos^2\theta_j + 1)}{r_j^6} . \qquad (5.103)$$

Remembering that for the polycrystalline average $< \cos^2\theta > = \frac{1}{3}$[95,96] we have

$$\sigma_{ZF,N,poly}^2 = \frac{1}{3} \left[\sigma_{ZF,N,x}^2 + \sigma_{ZF,N,y}^2 + \sigma_{ZF,N,z}^2 \right]$$

$$= \frac{2}{3} \left(\frac{\mu_0}{4\pi} \right)^2 \hbar^2 \gamma_\mu^2 \gamma_N^2 I_N (I_N + 1) \sum_j \frac{1}{r_j^6} . \qquad (5.104)$$

In the polycrystalline case the variance is isotropic and the depolarization function is expressed by the Kubo-Toyabe function, Eq. 4.19, with a variance given by Eq. 5.104.

---

[95] Note that the Kubo-Toyabe function Eq. 4.19 with an isotropic variance is valid for an isotropic crystalline environment, e.g., for a site with cubic symmetry. In other cases, the field variances in the three Cartesian directions are not a priori equal. A generalization for a site with uniaxial symmetry (e.g., hexagonal, rhombohedral, tetragonal crystals) is discussed in Solt (1995).

[96] If the muon site has cubic symmetry, summation over each coordination sphere gives an equivalent result. Since $\sum_j (\cos\theta_j)^2 / r_j^6 = \frac{1}{3} \sum_j 1/r_j^6$ we have $\sigma_{ZF,N,x}^2 = \sigma_{ZF,N,y}^2 = \sigma_{ZF,N,z}^2 =$ Eq. 5.104.

Fig. 5.53 Observed ZF
depolarization $P_z(t)$ and
high-field transverse
depolarization envelope $P_x(t)$
in ZrH$_2$ at room temperature.
Modified from Hayano et al.
(1979), © American Physical
Society. Reproduced with
permission. All rights
reserved

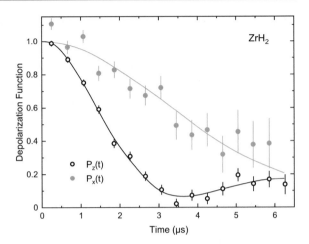

Comparing the polycrystalline expressions of the ZF with the TF cases we see that[97]

$$\sigma^2_{\text{ZF,N,poly}} = \frac{5}{2}\sigma^2_{\text{TF,N,poly}} \ . \tag{5.105}$$

This prediction has been confirmed experimentally in several systems. A good example is ZrH$_2$ (Hayano et al. 1979). This compound is ideal because there is no "disturbing" electric quadrupole moment, see Sect. 5.7.2, the dipolar field created by the protons is large, and the muon diffuses very slowly. By comparing the depolarization rate in ZF obtained by fitting a Kubo-Toyabe function[98] with the depolarization rate in high TF ($B_{\text{ext}} = 0.5$ T), Eq. 4.41, one obtains $\sigma^2_{\text{ZF,N,poly}}/\sigma^2_{\text{TF,N,poly}} = 2.4\pm0.1$, in good agreement with the prediction of Eq. 5.105.

5.7.2 Influence of the Quadrupolar Interaction on the Nuclear Dipolar Width

The previous discussion is about the so-called Van Vleck limit (Van Vleck 1948), where one implies that the axis defined by the applied field (in the TF case) or the initial direction of the muon polarization (in the ZF case) provides a natural quantization axis for the nuclear spin $\mathbf{I}_N$. This is no longer true if the nuclear spin

---

[97] We note that this factor $5/2$, which is due to neglecting the nonsecular contribution in TF, together with the previously mentioned factor of 2 in the squared depolarization rate at short time, σ^2_{st} see page 101, makes the short time decay of the TF polarization a factor of $\sqrt{5}$ smaller compared to the ZF polarization, see Fig. 5.53.

[98] It is actually a dynamical Kubo-Toyabe function with very slow dynamics, Eq. 4.62, to take into account the slow diffusion rate of the muon.

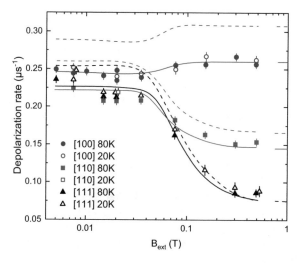

Fig. 5.54 Measured and fitted Gaussian depolarization rates of the μ^+ spin precession signals in single crystal Cu versus external field along different crystal orientations. The dashed curves are calculated from the Hartmann model (Hartmann 1977) for an octahedral site assignment of the muon, assuming no lattice distortion. The solid curves include a radial shift of the nearest neighbor distance by 5%. The broken and solid lines on the right indicate the depolarization rates in the absence of a quadrupole interaction (Van Vleck values for the high-TF limit). Modified from Camani et al. (1977), © American Physical Society. Reproduced with permission. All rights reserved

is subject to other interactions that are not negligible compared to the Zeeman interaction, for instance in the presence of a strong electric field gradient produced by the muon charge acting on the electric quadrupole moment of the nucleus (i.e., for $I_N > 1/2$). Then, for example in the TF case, the nonsecular part cannot be neglected, or in other words, $\langle I_x \rangle$ and $\langle I_y \rangle \neq 0$.

In the ZF case, the quantization axis will be defined by the electric field gradient extending radially from the muon to a given nuclear moment and the static component of $\mathbf{I}_N$ along this axis has to be considered. Contrary to the predictions of Eqs. 5.104 and 5.99, the inclusion of the electric field gradient leads to a dipolar width which depends on the applied magnetic field (Hartmann 1977).[99]

The effect of the electric field gradient on a quadrupole moment has been nicely demonstrated in single crystals of Cu (both stable isotopes ^{63}Cu and ^{65}Cu have spin $3/2$) by measuring the depolarization of the implanted muons as a function of the applied magnetic field and the crystal orientation at low temperatures, where the muon is immobile (Camani et al. 1977).

The data can be well fitted with a Gaussian damped precession function. As shown in Fig. 5.54 the Gaussian relaxation rate σ displays a pronounced field

---

[99] More details of the calculation can be found in Schenck (1985), Yaouanc and Dalmas de Réotier (2011).

dependence, which depends on the crystal orientation. The dependence is weak at low applied fields and approaches very roughly the Van Vleck values at high applied fields, assuming for the muon an octahedral interstitial position.

A closer inspection of the high field limits shows that the experimental values for the [110] and [100] directions are slightly lower than the theoretical Van Vleck values. The results can be explained by a model taking into account the effect of the quadrupolar interaction on σ (Hartmann 1977) and assuming that the distances between muon and nearest-neighbor Cu nuclei slightly increase (local lattice expansion due the positive muon charge), see Fig. 5.54. The dashed curves correspond to the case of no displacement of the nearest neighbor Cu nuclei to the muon, while the solid curves assume an outward displacement of 5% of the nearest neighbors.[100] The solid curves reproduce the data very well.

Summarizing, this experiment shows that (i) the electric field gradient generated by the muon interacts with the Cu quadrupole moment, (ii) the Coulomb force of the muon's positive charge produces a local lattice expansion, and (iii) the muon site in fcc Cu is an octahedral position.

Exercises

5.1. Non-singularity of the dipolar field term, Eq. 5.7
Show that the first term of Eq. 5.7 is nonsingular at $r = 0$.

5.2. Simple derivation of the contact field
Find the expression for the contact field, (e.g. Eq. 5.21), by considering that the muon is inside a sphere magnetized by the magnetic moment of an electron.

5.3. Dynamical range, μSR fluctuation time window
Using what you know about the about the μSR technique justify the typical dynamical range of μSR with fluctuation times between $\sim 10^{-4}$ and $\sim 10^{-12}$ s.

5.4. Error of the first order Knight-shift expression
Estimate the error introduced by the first order Knight-shift expression, Eq. 5.58.

5.5. Hyperfine constants from temperature dependent Knight-shift measurements
Estimate the dipolar hyperfine constants from Fig. 5.43 in the three directions and compare them with the values given in the text.

5.6. Total dipolar tensor of a cubic crystal
Justify the dipolar tensors of Eq. 5.91.

---

[100] Note that the local lattice expansion does not affect σ for $\mathbf{B}_{ext} \parallel [111]$.

5.7. Magic angle for zero dipolar contribution

Show that in the example of Fig. 5.46, when the field is directed along a $\langle 111 \rangle$ axis, the total dipolar contribution is zero.

References

Aeppli, G., Bucher, E., Broholm, C., et al. (1988). *Physical Review Letters, 60*, 615.
Akintola, K., Pal, A., Dunsiger, S. R., et al. (2018). *npj Quantum Materials, 3*, 36.
Amato, A., Dalmas de Réotier, P., Andreica, D., et al. (2014). *Physical Review B, 89*, 184425.
Amato, A., Feyerherm, R., Gygax, F. N., et al. (1992). *Europhysics Letters, 19*, 127.
Amato, A., Feyerherm, R., Gygax, F. N., et al. (1995). *Physical Review B, 52*, 54.
Amato, A., Feyerherm, R., Gygax, F. N., et al. (1997). *Hyperfine Interactions, 104*, 115.
Amato, A., Graf, M. J., de Visser, A., et al. (2004). *Journal of Physics: Condensed Matter, 16*, S4403.
Andreica, D. (2001). *Magnetic Phase Diagram in Some Kondo-Lattice Compounds*, PhD thesis, ETH Zurich.
Aoki, Y., Tsuchiya, A., Kanayama, T., et al. (2003). *Physical Review Letters, 91*, 067003.
Ashcroft, N., & Mermin, N. (1976). *Solid State Physics*. Saunders College. ISBN: 978-0030839931.
Biswas, P. K., Salman, Z., Neupert, T., et al. (2014). *Physical Review B, 89*, 161107.
Blundell, S. J. (1997). *Applied Magnetic Resonance, 13*, 155.
Blundell, S. J., (2001). *Magnetism in Condensed Matter*. Oxford University Press. ISBN: 0-19-158664-1.
Blundell, S. J., Pattenden, P. A., Pratt, F. L., et al. (1995). *Europhysics Letters, 31*, 573.
Bourdarot, F., Bombardi, A., Burlet, P., et al. (2005). *Physica B: Condensed Matter, 359–361*, 986.
Camani, M., Gygax, F. N., Rüegg, W., et al. (1977). *Physical Review Letters, 39*, 836.
Camani, M., Gygax, F. N., Rüegg, W., et al. (1979). *Physical Review Letters, 42*, 679.
Campbell, I. A., Amato, A., Gygax, F. N., et al. (1994). *Physical Review Letters, 72*, 1291.
Cannella, V., & Mydosh, J. A. (1972). *Physical Review B, 6*, 4220.
Clogston, A. M., & Jaccarino, V. (1961). *Physical Review, 121*, 1357.
Coey, J. M. D. (2019). *Magnetism and Magnetic Materials*. Cambridge University Press. ISBN: 1-108-71751-9.
Coey, J. M. D., & Parkin, S. S. P. (Eds.). (2022). *Handbook of Magnetism and Magnetic Materials*. Springer. ISBN: 3-030-63210-5.
Dalmas de Réotier, P., Huxley, A., Yaouanc, A., et al. (1995). *Physics Letters A, 205*, 239.
Denison, A., Graf, H., Kündig, W., et al. (1979). *Helvetica Physica Acta, 52*, 460.
Feyerherm, R., Amato, A., Gygax, F. N., et al. (1994). *Hyperfine Interactions, 85*, 329.
Fischer, K. H., & Hertz, J. A. (1993). *Spin Glasses*. Cambridge University Press. ISBN: 978-0521447775.
Frandsen, B., Liu, L., Cheung, S. C., et al. (2016). *Nature Communications, 7*, 12519.
Geibel, C., Schank, C. Thies, S., et al. (1991). *Zeitschrift für Physik B Condensed Matter, 84*, 1.
Gheidi, S., Akintola, K., Akella, K. S., et al. (2019). *Physical Review Letters, 123*, 197203.
Hartmann, O. (1977). *Physical Review Letters, 39*, 832.
Hartmann, O., et al. (1991). *Hyperfine Interactions, 64*, 381.
Hasan, M. Z., & Kane, C. L. (2010). *Reviews of Modern Physics, 82*, 3045.
Hayano, R. S., Uemura, Y. J., Imazato, J., et al. (1979). *Physical Review B, 20*, 850.
Ho, P.-C., MacLaughlin, D. E., Shu, L., et al. (2014). *Physical Review B, 89*, 235111.
Ishigaki, A., & Moriya, T. (1996). *Journal of the Physical Society of Japan, 65*, 3402.
Ishikawa, Y., Tajima, K., Bloch, D., et al. (1976). *Solid State Communications, 19*, 525.
Jackson, J. D. (1998). *Classical Electrodynamics*. Wiley. ISBN: 978-0471309321.
Joynt, R., & Taillefer, L. (2002). *Reviews of Modern Physics, 74*, 235.

Kasuya, T. (1956). *Progress of Theoretical Physics, 16*, 45.
Keizer, R. J., de Visser, A., Menovsky, A. A., et al. (1999). *Journal of Physics: Condensed Matter, 11*, 8591.
Keren, A., Mendels, P., Campbell, I. A., et al. (1996). *Physical Review Letters, 77*, 1386.
Knight, W. D. (1949). *Physical Review, 76*, 1259.
Kratzer, A., Noakes, D., Kalvius, G., et al. (2002). *Physica B: Condensed Matter, 312–313*, 469.
Lancaster, T., Blundell, S. J., & Pratt F. L. (2013). *Physica Scripta, 88*, 068506.
Le, L., Keren, A., Luke, G., et al. (1993). *Chemical Physics Letters, 206*, 405.
MacLaughlin, D. E. (1981). In L. M. Falicov, W. Hanke, & M. B. Maple (Eds.), *Valence Fluctuations in Solids* (Vol. 321). North-Holland. ISBN: 978-0444862044.
Maeter, H., Luetkens, H., Pashkevich, Y. G., et al. (2009). *Physical Review B, 80*, 094524.
McKenzie, I. (2013). *Annual Reports on the Progress of Chemistry Section C: Physical Chemistry, 109*, 65.
McKenzie, I., & Roduner, E. (2008). *Handbook of Heterogeneous Catalysis*, Chap. 3.2.3.4 (Vol. 1073). Wiley. ISBN: 978-3527610044.
Momma, K., & Izumi, F. (2011). *Journal of Applied Crystallography, 44*, 1272–1276.
Moriya, T. (1991). *Journal of Magnetism and Magnetic Materials, 100*, 261.
Mydosh, J. (1993). *Spin Glasses: An Experimental Introduction*. Taylor & Francis. ISBN: 978-0748400386.
Nuccio, L., Schulz, L., & Drew, A. (2014). *Journal of Physics D: Applied Physics, 47*, 473001.
Ohishi, K., Heffner, R. H., Ito, T. U., et al. (2009). *Physical Review B, 80*, 125104.
Onuorah, I. J., Bonfà, P., & De Renzi, R. (2018). *Physical Review B, 97*, 174414.
Onuorah, I. J., Bonfà, P., De Renzi, R., et al. (2019). *Physical Review Materials, 3*, 073804.
Percival, P. W. (1979). *Radiochimica Acta, 26*, 1.
Pratt, F. L. (2016). *Journal of the Physical Society of Japan, 85*, 091008.
Ran, S., Eckberg, C., Ding, Q.-P., et al. (2019). *Science, 365*, 684.
Rhodes, C. J. (2002). *Journal of the Chemical Society, Perkin Transactions, 2*, 1379.
Roduner, E. (1988). *The Positive Muon as a Probe in Free Radical Chemistry*. Lecture notes in chemistry. Springer. ISBN: 978-3540500216.
Ruderman, M. A., & Kittel, C. (1954). *Physical Review, 96*, 99.
Schenck, A. (1985). *Muon Spin Rotation Spectroscopy*. Adam Hilger Ltd. ISBN: 0-85274-551-6.
Schenck, A. (1999). In S. Lee, S. H. Kilcoyne, & R. Cywinski (Eds.), *Muon Science: Muons in Physics, Chemistry, and Materials* (Vol. 39). Institute of Physics Publishing. ISBN: 978-0750306300.
Schenck, A., Andreica, D., Gygax, F. N., et al. (2001). *Physical Review B, 65*, 024444.
Schenck, A., Gygax, F. N., Andreica, D., et al. (2003). *Journal of Physics: Condensed Matter, 15*, 8599.
Seeger, A. (1978). In G. Alefeld & J. Völkl (Eds.), *Hydrogen in Metals* (Vol. 28, p. 349). Springer topics in current physics. Springer. ISBN: 978-3540358923.
Si, Q., Yu, R., & Abrahams, E. (2016). *Nature Reviews Materials, 1*, 16017.
Solt, G. (1995). *Hyperfine Interactions, 96*, 167.
Spehling, J., Günther, M., Krellner, C., et al. (2012). *Physical Review B, 85*, 140406.
Sundar, S., Gheidi, S., Akintola, K., et al. (2019). *Physical Review B, 100*, 140502.
Tashma, T., Amato, A., Grayevsky, A., et al. (1997). *Physical Review B, 56*, 9397.
Van Vleck, J. H. (1948). *Physical Review, 74*, 1168.
Walker, D. C. (1983). *Muon and muonium chemistry*. Cambridge University Press.
Yaouanc, A., & Dalmas de Réotier, P. (2011). *Muon Spin Rotation, Relaxation and Resonance*. Oxford University Press. ISBN: 978-0199596478.
Yosida, K. (1957). *Physical Review, 106*, 893.
Zaremba, E., & Zobin, D. (1980). *Physical Review B, 22*, 5490.

Study of Superconductivity

6

Magnetism and superconductivity are fundamental and closely interrelated phenomena of condensed matter. Magnetism can destroy superconductivity; on the other hand, superconductivity can expel or exclude a magnetic field from a conductor. A local magnetic probe technique such as μSR, which is sensitive to static and dynamic magnetic fields, can provide detailed information about the superconducting phase(s) of different materials. In particular, μSR can quantify the key parameters characterizing a superconducting material such as the critical temperature T_c, the magnetic penetration depth $\lambda(T)$, the coherence length ξ and the superconducting energy gap $\Delta(T)$. Some of these quantities can be anisotropic and can also be tuned by external parameters such as pressure (applied by means of a device or by chemical doping) and magnetic field.

Depending on their behavior under application of a magnetic field, superconductors are classified as type-I or type-II. We will give examples of both cases, although the focus is on the more topical type-II superconductors,[1] where the interplay between magnetism and superconductivity, both natural fields of application of μSR, is often important. Here we can make use of the local character of μSR to address questions such as the coexistence and competition of the two phenomena.

In this chapter, we first summarize the main concepts of superconductivity[2] relevant in our context, we outline what kind of information can be obtained from μSR measurements in bulk samples, the methodology used for this purpose, and discuss some typical examples of investigations. Additional and complementary

---

[1] Type-II superconductors are also of paramount importance for applications.

[2] There are many books presenting superconductivity from a theoretical, experimental, or phenomenological point of view. We mention first the classical books of Tinkham (1996), Schrieffer (1999), de Gennes (1999), or Parks (editor) (1969). More recent are Annett (2004), Kleiner and Buckel (2016), Mangin and Kahn (2017), Ketterson and Song (1999). Poole et al. (2014) and the two volumes edited by Bennemann and Ketterson (2008) have comprehensive material and literature lists.

© Springer Nature Switzerland AG 2024

A. Amato, E. Morenzoni, *Introduction to Muon Spin Spectroscopy*,
Lecture Notes in Physics 961, https://doi.org/10.1007/978-3-031-44959-8_6

Fig. 6.1 First measurement of the resistance of a mercury sample cooled by liquid helium to temperatures around 4.2 K. Reproduced from Kamerlingh Onnes (1911a). Plotted is the resistance in Ohms versus temperature in K

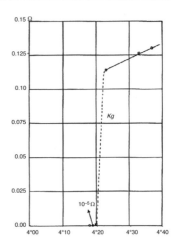

information on superconductors can be obtained by experiments in thin films with low-energy muons, and we refer to Chap. 8 for further studies.

6.1 Concepts of Superconductivity

The first superconductor was discovered on April 8, 1911 by Heike Kamerlingh Onnes in Leiden, The Netherlands (Kamerlingh Onnes 1911b). After producing liquid helium, he studied the resistance of solid mercury at very low temperatures and observed that, upon cooling, at a critical temperature $T_c \simeq 4.2$ K, the electrical resistance vanishes, Fig. 6.1.

Perfect conductivity is the first hallmark of superconductivity. Perhaps even more significant is the observation of perfect diamagnetism: up to a critical value, a magnetic field is expelled from the bulk of a superconductor. This is the Meissner-Ochsenfeld effect[3] discovered in 1933 (Meissner & Ochsenfeld 1933), Fig. 6.2.

It is important to recognize the difference between a perfect conductor and a superconductor. The observation of a zero resistivity and perfect diamagnetism, when a field is applied to an already cooled sample, is also a property of a perfect conductor. But only a superconductor shows perfect diamagnetism when the sample is cooled in field. The Meissner-Ochsenfeld effect cannot be explained merely by infinite conductivity. In a superconductor the diamagnetic state is reached independently of the previous history of the sample, see Fig. 6.3, i.e., the process is reversible and superconductivity is a new state of matter, characterized by a macroscopic wave function.

---

[3] The field penetrates a thin layer at the surface of the superconductor over a length scale determined by the magnetic penetration depth, see Sect. 6.1.1.1.

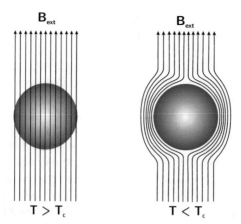

Fig. 6.2 Schematic view of the Meissner-Ochsenfeld effect. A magnetic field is applied above the critical temperature T_c. Upon cooling below T_c, the field is excluded from the bulk of the sample (perfect diamagnetism)

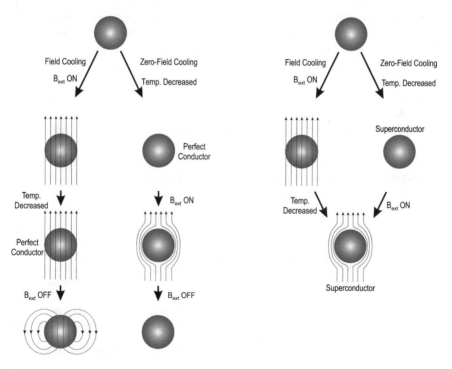

Fig. 6.3 Comparison between the behavior of an ideal perfect conductor (left panel) and a superconductor (right panel), when cooling below T_c after or before applying an external magnetic field

After the discovery, the progress in the theoretical understanding of superconductivity was rather slow. Fritz and Heinz London developed a phenomenological theory describing the Meissner effect (London & London 1935), Sect. 6.1.1.1. Ginzburg and Landau (GL) developed a macroscopic theory (Ginzburg & Landau 1950) based on a phenomenological pseudo-wave function as order parameter of the superconducting state and thermodynamic arguments of second-order phase transitions.[4]

The first theory explaining the microscopic origin of superconductivity was developed in 1957 by Bardeen, Cooper and Schrieffer, BCS theory (Bardeen et al. 1957).[5] The theory predicts that below a critical temperature T_c, in the presence of a no matter how weak electron-electron attractive potential, electrons near the Fermi surface form so-called Cooper pairs. These pairs are bound states of two electrons of opposite spins and momenta, which condense into a coherent macroscopic superconducting ground state. In conventional superconductors the electron-electron attraction is mediated by the phonons. The pairing modifies the density of states at the Fermi energy and an energy gap is formed, see Sect. 6.1.4. The BCS theory provides a microscopic foundation for the London and Ginzburg-Landau theories of superconductivity.[6] It quantitatively predicts many of the properties of elemental and so-called conventional superconductors.[7] In these superconductors the attraction is generally mediated by phonons, the pairing symmetry is spherically symmetric (s-wave), and the spins of the two electrons are in a spin singlet state . Note that the BCS theory requires only an attractive potential independent of its origin. With modifications and extensions it is also used to discuss properties of unconventional superconductors.[8]

Superconductors are obvious candidates for use in efficient energy devices, for applications requiring the transport of high current densities, or for applications requiring very large magnetic fields. Besides current density and critical field, an important parameter for their technological applications is T_c. The search for materials that are superconducting at higher temperatures, possibly even room

---

[4] The London equations follow as a natural consequence of the Ginzburg-Landau theory.

[5] BCS is a so-called weak-coupling theory. This means that the electron-phonon interaction is weak or T_c/Θ_D and $\Delta(0)/k_B\Theta_D \ll 1$, where Θ_D is the Debye temperature and $k_B\Theta_D$ the maximum phonon energy. J. Bardeen, L. N. Cooper and J. R. Schrieffer were awarded the Nobel Prize in Physics in 1972 for their theory of superconductivity.

[6] Gor'kov by using quantum field theory methods and Green functions generalized the BCS theory to a spatially varying order parameter, thus allowing the study of inhomogeneous superconductors. His formalism made possible the description of superconductors beyond the weak-coupling limit of BCS. He also showed that the GL theory can be derived from the microscopic BCS theory (Gor'kov 1959).

[7] Superconductors are called conventional when they can be well explained by BCS theories or extensions like Eliashberg theory, which, based on Gor'kov's formalism, extends BCS theory by including the time dependent electron-phonon interaction. Eliashberg theory can explain the properties of strong-coupled superconductors such as lead (Eliashberg 1960; Carbotte 1990; Marsiglio 2020).

[8] See footnote 58 about the definition of unconventional superconductor.

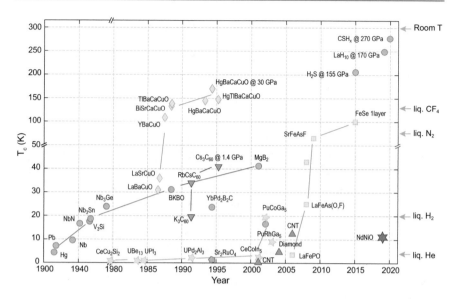

Fig. 6.4 Evolution of the critical temperature with time. On the right side the liquefaction temperature of several gases that could be used as cooling medium, is shown. Adapted from Pia Jensen Ray, Master's Thesis, Niels Bohr Institute, University of Copenhagen, Denmark

temperature, has been ongoing since their discovery and remains an attractive and topical task in condensed matter and materials science. Progress has been slow in this respect, see Fig. 6.4. Notably, two stepwise increases of the critical temperature occurred around 1986 and 2006, following the discovery of cuprates by Bednorz and Müller, working at IBM Zurich, Switzerland (Bednorz & Müller 1986)[9] and of iron-based superconductors discovered by the group of H. Hosono in Japan (Kamihara et al. 2006, 2008), respectively. Already at the early stages of these discoveries, μSR has provided important local information of these new classes of superconducting materials.

The maximum critical temperature was further increased in 2015 when it was discovered that hydrogen sulfide H_3S (considered a conventional superconductor) has a T_c of 203 K when compressed to pressures of 150 GPa (Drozdov et al. 2015). Subsequently, several hydrogen-rich superconductors have been synthesized and near room temperature superconductivity has been detected in LaH_{10} (250 K at 170 GPa) (Drozdov et al. 2019). Due to the very high pressure, experiments on hydride superconductors are technically very challenging and (at least presently, see footnote 2, Chap. 11) not amenable to μSR measurements.

---

[9] J. G. Bednorz and K. A. Müller were awarded the Nobel Prize in Physics in 1987 for this discovery.

6.1.1   The Two Characteristic Length Scales of Superconductors

Two parameters play a central role when describing the superconducting state of a
material:

- The magnetic penetration depth λ is the length scale of the field penetration at the
  surface of a superconductor in the Meissner state. It is related to the density of
  superconducting electrons in the material and provides information at the atomic
  level. The length also plays a fundamental role in superconducting technology.
- The coherence length. There are actually two definitions of the coherence length,
  which are related but not identical. In Ginzburg-Landau theory $\xi_{GL}(T)$ defines
  the length scale of the variation of the superconducting electron density.[10] This
  quantity diverges at $T = T_c$ as $(T_c - T)^{-1/2}$. The other definition was introduced
  by Pippard (1953). It is related to the Fermi velocity for the material, v_F, and the
  energy gap at $T = 0$, $\Delta(0)$, and associated with the condensation energy to the
  superconducting state. It is quantified in the BCS theory as $\xi_{BCS} = \hbar v_F / \pi \Delta(0)$,
  which is temperature independent. Empirically, it represents the spatial extent of
  the Cooper pairs.

  For $T \ll T_c$ in so-called "clean" superconductors (i.e., when the electron
  mean free path $\ell \gg \xi_{BCS}$) $\xi_{GL} \approx \xi_{BCS}$, whereas in "dirty" superconductors
  (i.e., when $\ell \ll \xi_{BCS}$) $\xi_{GL} \approx \sqrt{\xi_{BCS}\ell}$. In both cases, ξ_{GL} diverges near T_c as
  $(T_c - T)^{-1/2}$.

6.1.1.1 The Magnetic Penetration Depth

The London Equations   The equations developed by the London brothers describe
the magnetic field and the current flowing in a superconductor (London & London
1935). They introduced the notion of penetration depth. Although the equations do
not provide an explanation of the origin of superconductivity, they are able to explain
the Meissner-Ochsenfeld effect (Meissner & Ochsenfeld 1933). With modifications,
they are also used to describe the vortex state of a superconductor, see Sect. 6.2.2.2.

The equations can be derived by following considerations. In a sample with no
resistance the electrons feel a force

$$\mathbf{F} = -e\,\mathbf{E} = m_e^* \frac{\partial \langle \mathbf{v} \rangle}{\partial t} \quad . \tag{6.1}$$

where m_e^* is the effective mass of the electron.

---

[10] Consider, for instance, a superconductor interfacing a normal conductor. Well inside the
superconductor the superconducting electron density n_{sc} has a constant positive value, while well
inside the normal conductor n_{sc} is zero. The change to zero cannot be abrupt, but occurs over a
length scale defined by $\xi_{GL}(T)$.

Recalling that the current density is $\mathbf{j}_s = -n_{sc}e\langle\mathbf{v}\rangle$, where n_{sc} is the density of the superconducting carriers and $\mathbf{j}_s$ the associated (super)current, one obtains the first London equation[11]

$$\Lambda \frac{\partial \mathbf{j}_s}{\partial t} = \mathbf{E} \quad \text{with} \quad \Lambda = \frac{m_e^*}{n_{sc}\, e^2} \quad . \tag{6.2}$$

Taking the curl of this equation and using the 3rd Maxwell equation

$$\nabla \times \mathbf{E} = -\frac{\partial \mathbf{B}}{\partial t} \quad , \tag{6.3}$$

one obtains

$$\nabla \times \frac{\partial \mathbf{j}_s}{\partial t} = -\frac{1}{\Lambda}\frac{\partial \mathbf{B}}{\partial t} \quad , \tag{6.4}$$

or

$$\frac{\partial}{\partial t}\left(\nabla \times \mathbf{j}_s + \frac{1}{\Lambda}\mathbf{B}\right) = 0 \quad . \tag{6.5}$$

The quantity in the parentheses must be a constant, since the time derivative vanishes. Up to this point the derivation is fully compatible with classical electromagnetism[12] applied to the frictionless acceleration of electrons in an ideal conductor. The essential new assumption of the London brothers is that the bracket is not an arbitrary constant, but is actually equal to zero. Thus one obtains

$$\nabla \times \mathbf{j}_s = -\frac{1}{\Lambda}\mathbf{B} = -\frac{1}{\mu_0\, \lambda_L^2}\mathbf{B} \quad . \tag{6.6}$$

This is known as London's second equation; λ_L is called the London penetration depth

$$\lambda_L = \sqrt{\frac{\Lambda}{\mu_0}} = \sqrt{\frac{m_e^*}{\mu_0\, e^2\, n_{sc}}} \quad . \tag{6.7}$$

---

[11] This equation reflects the ideal conductivity in a superconductor. From $\mathbf{E} = \Lambda \frac{\partial \mathbf{j}_s}{\partial t} = 0$ it follows that $\mathbf{j}_s$ is constant, i.e., even without an electric field a nonzero stationary current can flow. By contrast, in a normal conductor $\mathbf{j} = \sigma_e\, \mathbf{E}$, σ_e electrical conductivity; one needs an electric field to generate a current.

[12] A quantum mechanical and deeper derivation of the London equations considers that the canonical momentum must be zero in the ground state of the system: $\mathbf{p} = m_e^*\mathbf{v} - e\mathbf{A} = 0$. From this it follows $\mathbf{j}_s = -\mathbf{A}/\Lambda$.

It is directly related to the density of superconducting electrons $\lambda_L(T)^{-2} \propto n_{sc}(T)$ and also determines its temperature dependence.

The London model does not quantify the density of superconducting electrons. In the ideal case (and as an upper limit) it is equal to the density of conduction electrons $n_{sc} = n_e$.

Note that the equation[13]

$$\mathbf{j}_s = -\frac{\mathbf{A}}{\Lambda} = -\frac{\mathbf{A}}{\mu_0 \lambda_L^2} = -\frac{n_{sc} e^2}{m_e^*}\mathbf{A} \tag{6.8}$$

contains both London equations, see also footnote 12.

The London equations do not depend on the coherence length, which is implicitly assumed to be zero. They are valid in the so-called extreme type-II superconductors, Sect. 6.1.2, where the magnetic penetration depth is much larger than the coherence length.[14]

Field and Current Profiles in the Meissner State Starting from the London equations we can calculate the field and current profiles at the surface of a superconductor in the Meissner state. We start with the 4th Maxwell equation

$$\nabla \times \mathbf{B} = \mu_0 \mathbf{j}_s \quad (\text{assuming } \frac{\partial \mathbf{E}}{\partial t} = 0) \ . \tag{6.9}$$

Taking the curl and using the London Eq. 6.6 we obtain

$$\lambda_L^2 \nabla \times (\nabla \times \mathbf{B}) + \mathbf{B} = 0 \ . \tag{6.10}$$

With the well known vector identity $\nabla \times (\nabla \times \mathbf{F}) = \nabla(\nabla \cdot \mathbf{F}) - \nabla^2 \mathbf{F}$ and $\nabla \cdot \mathbf{B} = 0$ we finally have

$$\mathbf{B} - \lambda_L^2 \nabla^2 \mathbf{B} = 0 \ . \tag{6.11}$$

Similarly, if we take the curl of Eq. 6.6 and use the 4th Maxwell equation, we obtain

$$\mathbf{j}_s - \lambda_L^2 \nabla^2 \mathbf{j}_s = 0 \ . \tag{6.12}$$

Let us consider the response of a long superconducting rod of diameter $R \gg \lambda_L$, when we apply a magnetic field B_{ext} along the rod[15] in the z direction.

---

[13] This equation is not gauge invariant and is only correct for the gauge $\nabla \cdot \mathbf{A} = 0$, called London gauge.

[14] As shown by Eq. 6.17, only when $\lambda_L \gg \xi_{BCS}$ and $\ell \gg \xi_{BCS}$, the magnetic penetration depth is given by λ_L.

[15] In this geometry the demagnetization factor N is zero, see Appendix D.

Fig. 6.5 Decay of the field
inside a superconductor over
a characteristic length λ_L.
The screening is due to
supercurrents $j_s(x)$ flowing
around the superconductor
perpendicular to B_z

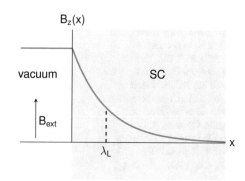

Fig. 6.6 When a current is
applied to a superconductor,
the transport occurs also on a
thin layer near the surface

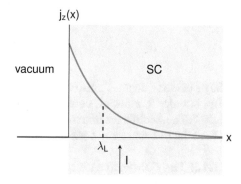

The solution of Eq. 6.11 is

$$B_z(x) = B_{ext}\, e^{-x/\lambda_L} \ , \tag{6.13}$$

where x is the inward distance from the surface of the rod and we have taken
the boundary condition $B_z(x = 0) = B_{ext}$. The superconductor reacts to the
application of an external field by spontaneously generating surface supercurrents
$\mathbf{j}_s$, see Eq. 6.6, which flow around the rod in such a way that the external field
is canceled inside the sample and is only allowed to enter in a thin surface layer,
defined by the London penetration depth, Fig. 6.5.

Similarly, if we apply a current I to the superconductor in the z direction, we
see from Eq. 6.12 that the current only flows in the thin layer at the surface of the
superconductor, Fig. 6.6, and we have

$$j_{s,z}(x) = \frac{I}{2\pi\, R\, \lambda_L}\, e^{-x/\lambda_L} \ . \tag{6.14}$$

Fig. 6.7 Cross section of a superconducting cable made of the high-T_c cuprate material Bi-2212. The superconducting wires are the small black dots, which are embedded in a copper matrix. See Minervini (2014) for details. Courtesy J. Jiang, Applied Superconductivity Center, National High Magnetic Field Laboratory, USA

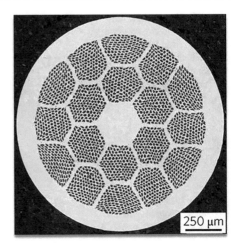

The penetration depth is a also central parameter in applications. A superconducting wire, in order to transport current effectively, does not need a large cross section, Fig. 6.7. This is different from normal conductors, which require large cross sections to limit the resistance and corresponding losses.

6.1.1.2 The Coherence Length

We do not derive the Ginzburg-Landau and the BCS expressions here, but only summarize the main points, which are necessary to understand μSR investigations of superconductors. We refer to books on superconductivity (Tinkham 1996; Poole et al. 2014; Mangin & Kahn 2017; Annett 2004; Parks 1969) for more details. The Ginzburg-Landau theory was developed from a phenomenological point of view. Following the theory of Landau for second-order phase transitions, Ginzburg and Landau proposed that the free energy density f_s of a superconductor near the superconducting transition can be developed in powers of a complex position dependent function $\psi(\mathbf{r})$, which describes the superconducting carriers and represents the superconducting order parameter. ψ is zero above the critical temperature and increases continuously as the temperature is reduced below the phase transition. The squared modulus of ψ is directly related to the local density of the superconducting carriers[16]

$$n_{sc}(\mathbf{r}) = |\psi(\mathbf{r})|^2 \ . \tag{6.15}$$

The position dependence of $\psi(\mathbf{r})$ allows the GL theory to treat spatial inhomogeneities, for instance when superconducting regions are adjacent to normal state

---

[16] No interpretation of the parameter ψ was given in the original paper.

Fig. 6.8 Evolution of the superconducting order parameter near the interface of a semi-infinite superconductor with vacuum. ψ decays to zero with ξ_{GL}. In this geometry the order parameter can be taken to be real

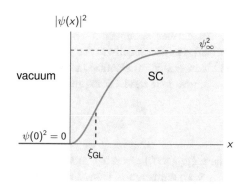

regions.[17] Such a situation is encountered at the vacuum interface of the superconductor, at the interface between a non-superconductor and a superconductor, in the intermediate state of type-I superconductors, Sect. 6.1.3, or in a vortex of type-II superconductors, Sect. 6.2, where the vortex core in normal state is surrounded by supercurrents. In the GL theory the coherence length ξ_{GL} determines the variation of the order parameter in regions of spatial inhomogeneities, see Fig. 6.8.

Besides the coherence length, the GL theory also introduces the magnetic penetration depth. With Eq. 6.15, the GL theory reduces to the London theory in the homogeneous region, where ψ is constant in space $\psi \equiv \psi_\infty$.

London and Ginzburg-Landau are so-called local theories, because they relate the supercurrent density and the vector potential at the same point in space, Eq. 6.8. To explain experimental results in impure tin, Pippard introduced, in analogy to the anomalous skin effect, the concepts of coherence length and of nonlocal response of a superconductor (Pippard 1953). Practically, he generalized the London equations with a nonlocal kernel.[18]

In Pippard theory, if the extent of the superconducting pairs is not negligible with respect to the magnetic penetration depth, the behavior of a paired electron at a position does not depend solely on the magnetic field at that position, as assumed in the London equation, but also depends on the change in the magnetic field over a coherence region of size ξ. In the new equation, Eq. 8.10, the supercurrent is related to an average of the vector potential over ξ around the point of interest. Pippard also took into account the effect of scattering at impurities, so that the parameter ξ also depends on the degree of purity, expressed by the electron mean free path ℓ in the normal metal.

---

[17] Since it is based on expanding the free energy in $|\psi|^2$ and $|\nabla\psi|^2$, the GL theory is strictly speaking only valid in the vicinity of T_c where the order parameter is small. However, various applications show that its region of applicability extends to lower temperatures.

[18] More details on local and nonlocal theories and their effect on the electromagnetic response of a superconductor can be found in Chap. 8 where examples of magnetic field profiling in thin films using LE-μSR are presented. Nonlocal effects modify the magnetic field profile at the surface of a superconductor in the Meissner phase, which is no longer exponential.

The functional dependence of the nonlocal kernel and the expression for ξ and λ, first derived semiphenomenologically by Pippard, are very close to those later derived microscopically in the BCS theory, which includes both nonlocal (Pippard) and local (London) electrodynamics.

We give here the BCS expression for the coherence length

$$\frac{1}{\xi} = \frac{J(0, T)}{\xi_{BCS}} + \frac{1}{\ell} \, , \tag{6.16}$$

where $\xi_{BCS} \equiv \hbar v_F / \pi \Delta(0)$ and $J(0, T)$ is a function slowly increasing monotonically with temperature from 1 at $T = 0$ to 1.33 at $T = T_c$. In the local case, BCS introduces a modification of the London penetration depth, so that the expression for the magnetic penetration depth λ becomes

$$\lambda(T) = \lambda_L(T) \left(1 + \frac{\xi_{BCS}}{J(0, T)\,\ell} \right)^{\frac{1}{2}} \, . \tag{6.17}$$

Since neither London nor GL theories quantify the magnetic penetration depth, this expression can be used with either theories, whenever nonlocality is not important.

Equation 6.17 can be well applied in many cases. For instance, in high-temperature superconductors (where $\xi \ll \lambda$), in dirty superconductors (where $\ell \ll \xi_{BCS}$), but also in pure conventional superconductors of type-I near T_c, where λ diverges but not ξ_{BCS} so that the condition $\xi_{BCS} \ll \lambda(T)$ is fulfilled.

A determination of the effective magnetic penetration depth λ quantifies the superfluid density n_{sc}[19] as a phenomenological parameter via the relation[20]

$$\lambda(T) = \sqrt{\frac{m^*}{\mu_0\, e^2\, n_{sc}(T)}} \, . \tag{6.18}$$

The magnetic penetration depth depends on temperature via the temperature dependence of the density of superconducting electrons. An approximate expression is given by the two-fluid model of superconductivity introduced by Gorter and Casimir (1934a,b). The model assumes that the total density of conduction electrons n_e is divided into a fluid of normal electrons with density $n_n(T)$ coexisting (but not interacting) with a fluid of superconducting electrons with density $n_{sc}(T)$. At all temperatures $n_{sc}(T) + n_n(T) = n_e$. At $T = 0$ all conduction electrons have condensed to superconducting electrons $n_{sc}(0) = n_e$. Based on measurements

---

[19] Actually μSR provides a measure of $\rho_{sc} \equiv n_{sc}/m^*$. This ratio itself, and not just n_{sc}, is also sometimes referred to as superfluid density.

[20] In dirty superconductors $\lambda^2 \approx \lambda_L^2 \xi_{BCS}/\ell \gg \lambda_L^2$. The increase of the magnetic penetration depth with respect to the London value can be interpreted as a reduction of n_{sc}/m^*.

Fig. 6.9 Temperature dependence of the magnetic penetration depth according to the two-fluid model. At low temperatures, most of the conduction electrons have condensed to the superconducting state, effectively screening the applied field

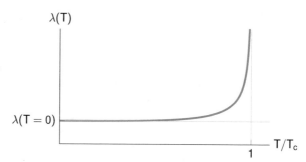

of thermodynamic quantities Gorter and Casimir proposed for the temperature dependence

$$n_{\text{sc}}(T) = n_e \left(1 - \left(\frac{T}{T_c} \right)^4 \right) . \tag{6.19}$$

From Eq. 6.18 one obtains

$$\lambda(T) = \frac{\lambda(T = 0)}{\sqrt{1 - \left(\frac{T}{T_c} \right)^4}} , \tag{6.20}$$

plotted in Fig. 6.9.

The penetration depth increases as the temperature approaches the critical temperature. This reflects the fact that less and less superconducting current is screening the applied field. At T_c and above it becomes effectively infinite corresponding to a uniform field in the material.

The BCS theory provides the temperature dependence of λ_L and $J(0, T)$ and therefore of λ. Although qualitatively similar in all cases, the BCS expression does not scale with T/T_c because the ratio $\lambda_L/\xi_{\text{BCS}}$ varies for different superconductors. However, for a clean (i.e., large ℓ), local (i.e., small ξ_{BCS}) s-wave superconductor the results for $\lambda(T)$ can be well fitted by the expression

$$\lambda(T) = \frac{\lambda(T = 0)}{\sqrt{1 - \left(\frac{T}{T_c} \right)^2}} . \tag{6.21}$$

For an unconventional d-wave superconductor, see Sect. 6.1.4, a good fit is obtained by replacing in the above equation the power of two in T/T_c with $4/3$.[21]

6.1.2   Type-I and Type-II Superconductors

In the previous sections we have seen that at the superconductor/vacuum interface, the magnetic field decays to zero over the length scale λ, Fig. 6.6, and that the superconducting order parameter $|\psi|^2$ is suppressed over the length scale ξ_{GL}, Fig. 6.8. The value of the ratio of these two quantities $\kappa \equiv \lambda/\xi_{GL}$, Ginzburg-Landau parameter,[22] determines whether a superconductor is of type-I or type-II.

Condensation Energy and Energy Balance   Let us consider the condensation energy, which is the energy per unit volume gained by the system going from the normal to the superconducting state, when electrons condense to Cooper pairs. An expression for this energy can be obtained by considering that the superconducting state is destroyed by an applied field $B_{ext} \geq B_c$ where B_c is the thermodynamic critical field of the superconductor. To destroy superconductivity and restore the normal state we have to supply an energy per unit volume equal to $B_c^2/2\mu_0$. Hence we can write

$$f_{condens} = f_n - f_s = \frac{B_c^2}{2\mu_0} \; . \tag{6.22}$$

where f_n and f_s are the Helmholtz free energies per unit volume in the respective phases in zero field.

Now consider two superconductors, one with $\xi_{GL} \gg \lambda$ and the other with $\lambda \gg \xi_{GL}$. We assume that both superconductors see a field close to B_c. Considering the energy at the superconductor/vacuum interface, we have two possibilities:

1. for $\xi_{GL} > \lambda$, left side of Fig. 6.10, we have at the interface a region of thickness $\sim (\xi_{GL} - \lambda)$ where the magnetic field is excluded, which from the point of view of the system requires a positive energy. On the other hand, in this region the order parameter has not reached the bulk value $|\psi_\infty|$, i.e., only a fraction of the electrons have condensed and only a fraction of the possible condensation energy gain (negative energy) is available to compensate the magnetic field energy. Overall, the interface (or surface) energy is positive.

---

[21] The properties of the magnetic penetration depth in unconventional superconductors are discussed in detail in (Prozorov & Giannetta 2006).

[22] κ varies slowly with temperature. At T_c, for a pure superconductor $\kappa = 0.96 \, \lambda_L(0)/\xi_{BCS}$ and for a dirty superconductor $\kappa = 0.715 \, \lambda_L(0)/\ell$ (Tinkham 1996).

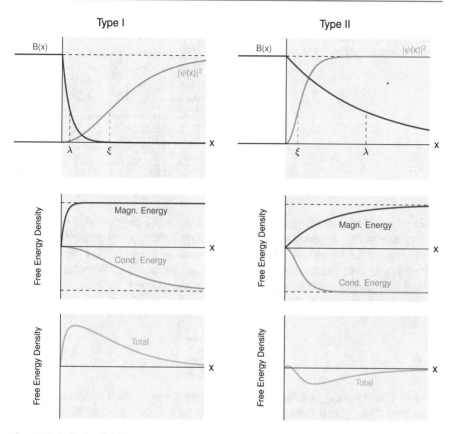

Fig. 6.10 Left: local field, order parameter, and free energy density contributions for $\xi_{GL} > \lambda$ (type-I superconductor). Right: for $\lambda > \xi_{GL}$ (type-II superconductor). A type-II superconductor gains energy by creating interfaces

2. for $\lambda > \xi_{GL}$, right side of Fig. 6.10, we have the opposite case. In the region $\sim (\lambda - \xi_{GL})$ the gain in negative condensation energy is almost fully available, while the energy cost to expel the magnetic field is reduced. This results in a negative total surface energy.

The situation described above distinguishes between two types of superconductors that exhibit different magnetic behavior, see Fig. 6.11. For the superconductors where $\xi_{GL} > \lambda$ (type-I superconductors), the positive interface energy leads the system to have a total interface area as small as possible. In superconductors where $\lambda > \xi_{GL}$ (type-II superconductors), energy is gained by creating interfaces. Hence, we do not have a stable equilibrium of the macroscopic volumes of the two phases (as in a type-I superconductor) and the interfaces proliferate. The partitioning into normal and superconducting volumes can continue until the magnetic flux of each microscopic superconducting region reaches the quantized flux value of $\Phi_0 \equiv h/2e$. This behavior was first predicted by Abrikosov (1957).

Table 6.1 Parameters of some typical type-I and type-II superconductors

Material	T_c (K)	ξ (nm)	λ (nm)	B_{c2} (T)
Type-I				
Al	1.2	1600	16	0.01
Sn	3.7	230	34	0.03
Pb	7.2	83	37	0.08
Type-II				
Nb	9.3	38	39	0.4
Nb$_3$Sn	18	3	80	25
MgB$_2$	37	5	185	14
YBCO	92	1.5	150	150

In a type-II superconductor above a critical value B_{c1} the so-called mixed state occurs (also known as vortex state, Abrikosov state, or Shubnikov phase[23]) in which, as the applied field is raised, an increasing amount of magnetic flux penetrates the material in form of flux tubes each carrying a flux quantum Φ_0. Only when a second critical field B_{c2} is reached, superconductivity is destroyed. By energy arguments one can show that the second critical field and ξ_{GL} are simply related (Tinkham 1996)

$$\xi_{GL} = \sqrt{\frac{\Phi_0}{2\pi\, B_{c2}}} \ . \tag{6.23}$$

Note that between B_{c1} and B_{c2} the system retains zero electrical resistivity. Type-II superconductors play a fundamental technological role because the value of B_{c2} is much higher than that of B_c for type-I superconductors.

A more stringent criterion for classifying the type of superconductors is given by the Ginzburg-Landau theory, which shows that the crossover from positive to negative surface energy occurs for $\kappa = 1/\sqrt{2}$. This value separates type-I from type-II superconductors

$$\text{type-I}: \quad \kappa < \frac{1}{\sqrt{2}}$$

$$\text{type-II}: \quad \kappa > \frac{1}{\sqrt{2}} \ .$$

Typical parameters of type-I and type-II superconductors are given in Table 6.1. The different magnetic behavior of the two superconductor types is shown in Fig. 6.11.

Almost all the elements that are superconductors are of type-I. Only vanadium, technetium, and niobium (close to the type-I border) are type-II. Most alloys and compounds, which generally have a short mean free path compared to pure

---

[23] Abrikosov compared his theoretical predictions about magnetization curves with experimental data on Pb-Tl alloys obtained by Shubnikov in 1937 (Shubnikov et al. 1937).

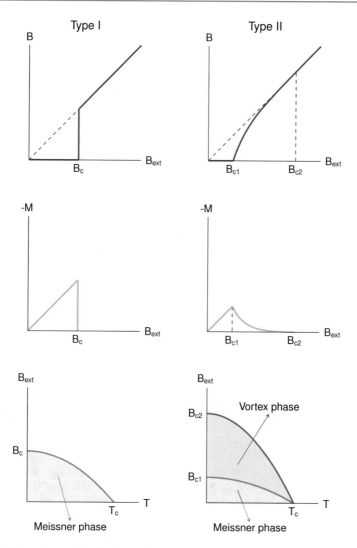

Fig. 6.11 Comparison of the ideal magnetic behavior in type-I (left) and type-II (right) superconductors. In the Meissner state of both kinds of superconductors, the field is expelled from the superconductor except in a thin layer determined by the magnetic penetration depth. In a type-II superconductor, favored by a negative interface energy (see Fig. 6.10), for an applied field above B_{c1} the field penetrates the sample in the form of vortices so that the internal average field $\langle B \rangle$ is no more equal to zero. Above B_{c2} the superconducting state of a type-II superconductor is destroyed. The thermodynamic critical field B_c of a type-II superconductor corresponds to the field for which, in the M vs. B_{ext} plot, the area between B_{c1} and B_c is equal to that between B_c and B_{c2}

metals, are type-II with some exceptions, see, e.g., Sect. 6.1.3. High temperature superconductors and other classes of superconductors such as organic, heavy fermions, chevrel phases and ruthenates are of type-II.

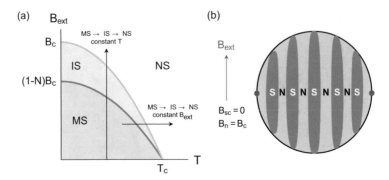

Fig. 6.12 (**a**) Applied field versus temperature phase diagram for a type-I superconducting sphere ($N = 1/3$). "MS", "IS", and "NS" denote the regions of the phase diagram corresponding to the Meissner state (blue region), the intermediate state (pink region), and the normal state (white region), respectively. The vertical and horizontal arrows indicate MS → IS → NS paths at constant temperature and constant B_{ext}, respectively. (**b**) Schematic of the separation of the sphere in normal state (N) and superconducting (S) domains, after (Shapiro & Shapiro 2019). The field in the superconducting domains is equal to zero, $B_{sc} = 0$. The field in the normal state domains is equal to the thermodynamic critical field: $B_n = B_c$. Modified from Karl et al. (2019), © American Physical Society. Reproduced with permission. All rights reserved

6.1.3   The Intermediate State

The magnetic response of a type-I superconductor depends on the demagnetizing factor of the sample. Applying a field to a sample with nonzero demagnetizing factor $0 < N < 1$ a nonuniform field at the surface is created,[24] that can exceed the critical field. For instance, at the equator of a sphere ($N = 1/3$), at the points with polar angles $\theta = \pm 90°$ (measured from the direction of $\mathbf{B}_{ext}$) $B = 3\,B_{ext}/2$, so that the critical field B_c is already reached once $B_{ext} = 2\,B_c/3$ and the superconductivity is locally destroyed. In this case, some regions (but not the totality) of the sphere must go to the normal state and a partial penetration of the field occurs with normal and superconducting domains coexisting, a state called intermediate state.

Generalizing to ellipsoidal samples, the intermediate state is obtained whenever the applied field lies in the range

$$(1 - N)B_c < B_{ext} < B_c \ . \tag{6.24}$$

For bulk samples (i.e., with linear dimensions much larger than the coherence length ξ_{GL}) and for fields not too close to the phase boundaries, the local field inside the normal state domains is practically equal to the thermodynamic critical field $B_n = B_c$ and $B_{sc} = 0$ (Tinkham 1996). As an example, Fig. 6.12a shows

---

[24] The nonuniform surface field is a consequence of the diamagnetism of the superconducting sample, which distorts the applied field.

Fig. 6.13 Ideal magnetic behavior in a type-I superconducting sphere ($N = 1/3$), showing the intermediate state

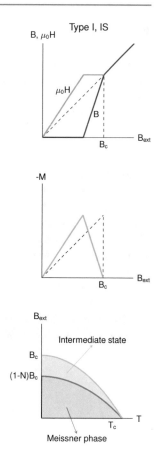

the B_{ext}—T phase diagram of a type-I superconducting spherical sample, and Fig. 6.12b schematically displays the separation of a clean sample into normal state (N) and superconducting (S) domains, obtained from simulations based on the time dependent Ginzburg–Landau equations (Shapiro & Shapiro 2019). In this case, the intermediate state contains a set of tubular domains separated by normal regions.[25] Figure 6.13 shows the magnetic behavior of the sphere as a function of the applied field.

---

[25] Several factors determine the topology of the intermediate state. A tubular topology is destroyed by structural defects and transformed into a laminar pattern. In more disordered samples with significant bulk pinning, a dendritic (tree-like) topology and normal state percolation appear (Prozorov 2007; Shapiro & Shapiro 2019).

6.1.4 Energy Gap and Symmetry of the Pairing State

An important microscopic parameter of a superconductor is the energy gap $\Delta(T)$. The existence of an energy gap between the ground state and the spectrum of the elementary excitations of the superconductor (Bogoliubov quasiparticles) is one of the central predictions of the microscopic BCS theory. The energy gap manifests itself in the low-temperature behavior of $\lambda(T)$, which reflects changes in the superfluid density $n_{sc}(T)$ responsible for the screening of electromagnetic fields.

Breaking a Cooper pair (and hence creating two quasiparticles) requires a minimum energy of $2\Delta(T)$. Figure 6.14 shows qualitatively the evolution with temperature of the density of states, their population, and of the gap at the Fermi surface, and illustrates the interplay between n_{sc} and $\Delta(T)$ in a superconductor. At $T = 0$ all electrons are paired in Cooper pairs and the gap reaches its maximum value $\Delta(0)$. As the temperature increases, more and more pairs can be broken. This reduces the density of superconducting electrons and at the same time the energy gap, which is intimately linked to the coherent many-electron state formed by the Cooper pairs.

The wave function of the two paired electrons can be written as the product of a space (orbital) and a spin part

$$\Psi(\mathbf{r_1}, S_1, \mathbf{r_2}, S_2) = \phi(\mathbf{r_1}, \mathbf{r_2}) \, \chi(S_1, S_2) \ . \qquad (6.25)$$

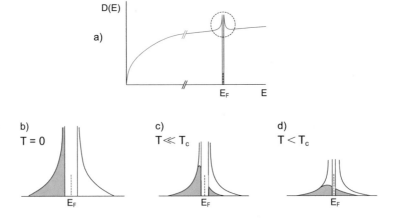

Fig. 6.14 (**a**) Density of states $D(E)$ in an s-wave superconductor. The formation of the gap leads to an increase of D(E) close to the Fermi energy E_F, due to the piling up of states that, in the normal state, are in the gap interval. Bottom figure: Evolution of the state population (gray area) with temperature. (**b**) At $T = 0$ all electrons are condensed in pairs and all the states below $E_F - \Delta(0)$ are occupied. (**c**) A temperature increase leads to pair breaking. This reduces the occupancy of the states below $E_F - \Delta(T)$, creates quasiparticles excitations above $E_F + \Delta(T)$, and reduces the gap value. (**d**) This behavior is enhanced with increasing temperature until at T_c the gap disappears, $\Delta(0) = 0$ and the density of states of a normal conductor is established

For two identical fermions $\Psi(\mathbf{r_1}, S_1, \mathbf{r_2}, S_2)$ must be antisymmetric under the exchange of the two particles. If the spin state is a singlet then $S_{tot} = 0$ and the spin part

$$\chi = \frac{1}{\sqrt{2}}(|\downarrow\uparrow\rangle - |\uparrow\downarrow\rangle) \qquad (6.26)$$

is antisymmetric. Therefore the orbital part, whose parity is given by $(-1)^l$, must be even, i.e., $l = 0$ (s-wave), $l = 2$ (d-wave), ..., while an odd parity of the orbital part is associated to a spin triplet pairing $S_{tot} = 1$.[26] From this perspective a conventional superconductor is sometimes defined as a condensate of spin singlet $l = 0$ Cooper pairs, i.e., with the most symmetric pairing state (s-wave). All other states with $l > 0$ are unconventional.

Most of the known superconductors have singlet spin state. The $l = 0$ case is realized in the conventional BCS superconductors, where the energy gap is the same for all points of the Fermi surface. The $l = 2$ case (d-wave pairing) is found in high-T_c cuprate superconductors. In this case the gap is no longer constant on the Fermi surface but depends on $\mathbf{k}$ and it is said to have $d_{x^2-y^2}$ symmetry.

For singlet spin pairing the gap function can be written as a product of a temperature dependent part $\Delta(T)$ and a function $-1 \leq g(\mathbf{k}) \leq 1$, describing the $\mathbf{k}$- (or angular) dependence of the gap on the Fermi surface

$$\Delta(T, \mathbf{k}) = \Delta(T) g(\mathbf{k}) \; . \qquad (6.27)$$

For s-wave superconductors $g(\mathbf{k}) = 1$ for all $\mathbf{k}$-points. For d-wave pairing, in principle any $g(\mathbf{k})$ satisfying the pairing symmetry is admissible.

The relation between superfluid density and gap function is easily expressed for spherical or cylindrical Fermi surfaces.[27]

For a spherical Fermi surface, a reasonable approximation for a conventional s-wave superconductor, the BCS expression for the normalized superfluid density $\rho_{sc}(T)$ for a clean superconductor reads (Tinkham 1996)

$$\rho_{sc}(T) = \left(\frac{n_{sc}(T)}{n_{sc}(0)}\right) = \left(\frac{\lambda(0)}{\lambda(T)}\right)^2 = 1 + 2 \int_{\Delta(T)}^{\infty} \left(\frac{\partial f}{\partial E}\right) \frac{E}{\sqrt{E^2 - \Delta(T)^2}} dE \; ,$$

$$(6.28)$$

where $E = \sqrt{\varepsilon^2 + \Delta(T)^2}$ is the quasiparticle energy of state $\mathbf{k}$, $\varepsilon = \frac{\hbar^2 k^2}{2m^*} - E_F$ is the corresponding kinetic energy of a free electron measured from the Fermi energy

---

[26] The symmetry of the superconducting state is simply related to the relative orbital angular momentum of the electrons in the Cooper pairs.

[27] More general expressions, relating the superfluid density to Fermi surfaces with arbitrary dispersion relation $\varepsilon(\mathbf{k})$ and energy gap $\Delta(\mathbf{k})$ are given in (Chandrasekhar & Einzel 1993).

and $f(E) = \left(1 + \exp\left(\frac{E}{k_BT}\right)\right)^{-1}$ is the Fermi-Dirac distribution function. Taking the derivative of the distribution and substituting E with ε we obtain[28]

$$\left(\frac{n_{\mathrm{sc}}(T)}{n_{\mathrm{sc}}(0)}\right) = \left(\frac{\lambda(0)}{\lambda(T)}\right)^2 = 1 - \frac{1}{2k_BT}\int_0^\infty \cosh^{-2}\left(\frac{\sqrt{\varepsilon^2 + \Delta(T)^2}}{2k_BT}\right) d\varepsilon \ .$$

(6.29)

For a dirty s-wave superconductor, a closed expression is available (Tinkham 1996)

$$\left(\frac{n_{\mathrm{sc}}(T)}{n_{\mathrm{sc}}(0)}\right) = \left(\frac{\lambda(0)}{\lambda(T)}\right)^2 = \frac{\Delta(T)}{\Delta(0)} \tanh \frac{\Delta(T)}{2k_BT} \ .$$

(6.30)

$\Delta(T)$ is the BCS gap function, which is obtained by selfconsistently solving the BCS gap equation (Tinkham 1996). It is tabulated in Mühlschlegel (1959). A useful parametrization, which has been found to well represent the temperature dependence of the gap of s-wave superconductors at any coupling strength (i.e., also for superconductors with $\alpha \equiv \Delta(0)/k_BT_c \neq \alpha_{\mathrm{BCS}} = 1.764$), and which is also used for unconventional superconductors, is Carrington and Manzano (2003)

$$\Delta(T) = \Delta(0) \tanh\left(1.82[1.018(T_c/T - 1)]^{0.51}\right) \ .$$

(6.31)

The applicability of Eq. 6.31 (or equivalent expression) is the basic assumption of the so-called α-model (Padamsee et al. 1973; Johnston 2013), a semiempirical extension of the BCS theory often employed to analyze μSR data. The model assumes that the normalized gap $\Delta(T)/\Delta(0)$ (or equivalently the temperature dependence of the gap) of a superconductor is the same as in the BCS theory, regardless of the coupling strength. Only the gap ratio $\alpha = \Delta(0)/k_BT_c$ is an adjustable parameter of the model and we can write

$$\Delta(T) = \frac{\alpha}{\alpha_{\mathrm{BCS}}} \Delta_{\mathrm{BCS}}(T) \ .$$

(6.32)

This assumption finds support in experimental data and calculations. Measurements of the temperature dependence of the superconducting energy gap of lead ($\alpha_{\mathrm{Pb}} = 4.5$) show that the gap function normalized to the zero temperature gap deviates only slightly from the BCS curve (Gasparovic et al. 1966). A similar result is found from theoretical calculations of strongly coupled superconductors with Eliashberg theory (Marsiglio 2020). Moreover, the solutions of the self-consistent gap equation for d-wave and other nonisotropic gaps show that the normalized temperature dependent part of the gap versus reduced temperature is well reproduced by the expression 6.31 and hence by the BCS gap. See Exercise 6.1 for more details and remarks in footnote 31.

---

[28] As it can be easily inferred from Eq. 6.33 the expression is also valid in the isotropic 2D case.

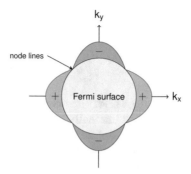

Fig. 6.15 Gap of a superconductor with d-wave symmetry $(d_{x^2-y^2})$ on a cylindrical Fermi surface. The gap has nodes when $|k_x| = |k_y|$ and changes sign between the nodes. k_x is along the a-axis of the crystal and k_y along b. The d-wave gap has a lower symmetry than the Fermi surface, reflecting the unconventional character of the superconductor

For the high-T_c cuprates and many other layered superconductors a cylindrical Fermi surface (2D) is widely used as approximation with the Fermi cylinder along the k_z direction. With a, b (planar axes) and c (axis of the uniaxial symmetry) indicating the principal axis of a crystal and allowing for in-plane anisotropy, Eq. 6.29 becomes [29]

$$\left(\frac{n_{sc}(T)}{n_{sc}(0)}\right)_{\substack{a\\b}} = \left(\frac{\lambda(0)}{\lambda(T)}\right)_{\substack{a\\b}}^2$$

$$= 1 - \frac{1}{2\pi k_B T} \begin{pmatrix} \cos^2(\phi) \\ \sin^2(\phi) \end{pmatrix} \int_0^{2\pi} \int_0^\infty \cosh^{-2}\left(\frac{\sqrt{\varepsilon^2 + \Delta(T,\phi)^2}}{2k_B T}\right) d\varepsilon \, d\phi.$$

$$(6.33)$$

For a d-wave superconductor the gap function generally expressed as $\Delta(T,\mathbf{k}) = \Delta(T)\cos(2\phi)$, where $\phi = \arctan(k_y/k_x)$ and for $\Delta(T)$ the parametrization 6.31 can be used (Fig. 6.15).

The low temperature behavior of the superfluid density (or equivalently of the magnetic penetration depth) gives information about the pairing symmetry. Particularly, it can reveal whether the gap has nodes or not (nodeless).

For the isotropic s-wave pairing where the superconducting gap depends only on temperature, Eq. 6.29 simplifies for $T \ll T_c$ to

$$n_{sc}(T) \simeq n_{sc}(0)\left(1 - \sqrt{\frac{2\pi\Delta(0)}{k_B T}}\, e^{-\frac{\Delta(0)}{k_B T}}\right), \qquad (6.34)$$

---

[29] The a and b subindices indicate the flowing direction of the supercurrents.

and for the penetration depth we obtain

$$\lambda(T) \simeq \lambda(0) \left(1 + \sqrt{\frac{\pi \Delta(0)}{2k_B T}} \, e^{-\frac{\Delta(0)}{k_B T}} \right) . \tag{6.35}$$

The superfluid density decreases very slowly from the $T = 0$ value. The exponential behavior is typical of a thermally activated gapped system, where the gap defines the thermal energy scale. For an example see Fig. 6.33.

In a superconductor with a gap function with nodes (nodal superconductor), at low temperatures, $\lambda(T)$ varies much more rapidly than in the case of s-wave pairing. A d-wave gap such as $\Delta(T) \cos(2\phi)$ disappears along directions of the Fermi surface where $\phi(\bmod \pi/2) = \pi/4$ (line nodes). At these nodes (and generally near the zeros of the gap function) pair breakup and quasiparticle formation can easily occur even at very low temperatures, effectively reducing the superfluid density (and increasing λ). Figure 6.39 shows a comparison between a superconductor without gap nodes and one with nodes.

For a superconductor with $d_{x^2-y^2}$ symmetry, a linear dependence on temperature (Hirschfeld & Goldenfeld 1993) is obtained for $T \ll T_c$, see Fig. 6.40

$$n_{sc}(T) \simeq n_{sc}(0) \left(1 - \frac{2 \ln 2}{\Delta(0)} T \right) , \tag{6.36}$$

respectively,

$$\lambda(T) \simeq \lambda(0) \left(1 + \frac{\ln 2}{\Delta(0)} T \right) . \tag{6.37}$$

6.1.4.1 Multiple Superconducting Gaps

In a superconductor where the electrons occupy different bands each band can have its own gap, but, because of interband coupling, there is only one critical temperature. The presence of a multiple gap structure[30] influences the temperature dependence of various quantities.

Simple cases of multi-gap superconductors can be treated by assuming independent band contributions to the superfluid density, within the α-model, see page 244.[31] The parameters α and $\Delta(0)$ are different for the two bands, but T_c is the

---

[30] This means that the density of states of the superconducting electrons has several peaks. At sufficiently low temperatures, the low-energy excitations are dominated by the smallest gap in the system.

[31] Note that the α-model is not self-consistent (Johnston 2013). Moreover, the assumption of independent band contributions (i.e., neglect of interband coupling) is not consistent with the imposition of same T_c and same temperature dependence of the two gaps. A two-band model taking into account weak inter- and intraband coupling (named γ-model) with isotropic s-wave gaps on the two Fermi surfaces, that self-consistently evaluates the temperature dependence of the two gaps and the associated London penetration depths has been proposed (Kogan et al. 2009). If

same. The superfluid density is calculated for each sheet according to Eq. 6.29 and then added together with a weighting factor, which takes into account the relative Fermi surface contribution.

For example, for two s-wave bands we have

$$\rho_{sc}(T) = \left(\frac{\lambda(0)}{\lambda(T)}\right)^2 = w\,\rho_{sc,1}(T) + (1-w)\,\rho_{sc,2}(T) \;, \tag{6.38}$$

where

$$\rho_{sc,i}(T) = 1 + 2 \int_{\Delta_i(T)}^{\infty} \left(\frac{\partial f}{\partial E}\right) \frac{E}{\sqrt{E^2 - \Delta_i(T)^2}}\, dE \;. \tag{6.39}$$

For an example see page 270.

6.2 Vortex State of a Type-II Superconductor

The field flux that penetrates in a type-II superconductor when the applied field satisfies $B_{c1} < B_{ext} < B_{c2}$, forms flux tubes (vortices) which arrange themselves usually in a triangular pattern forming the so-called "Flux-Line Lattice" (FLL), see Fig. 6.16.[32]

The vortex core, with a radius of the order of ξ_{GL}, is in the normal state. Outside the core, the field and the supercurrent decrease on a length scale determined by the magnetic penetration depth λ, see Fig. 6.17. Each vortex carries a magnetic flux equal to the magnetic flux quantum Φ_0[33]

$$\Phi_0 = \frac{h}{2e} = 2.067833848 \times 10^{-15}\ \mathrm{T\,m^2} \;. \tag{6.40}$$

---

the interband coupling is not too large α- and γ-models give similar results for the gap parameters (Zhang et al. 2015). Due to its simplicity, the α-model is widely used to analyze single and multiple gaps superconductors, in spite of its limitations. When used to analyze specific heat data, it provided convincing evidence for the two-gap superconductivity in the topical MgB_2 compound (Bouquet et al. 2001). Another specific heat comparison of MgB_2 parameters obtained from a two-band α model analysis with results from Eliashberg theory showed that the α-model is sufficiently accurate and can be used to extract gap values (Dolgov et al. 2005).

[32] Subtle effects can favor configurations other than triangular. The energy difference between triangular and quadratic lattices is small.

[33] $1\ \mathrm{T\,m^2} = 1\ \mathrm{Wb}$.

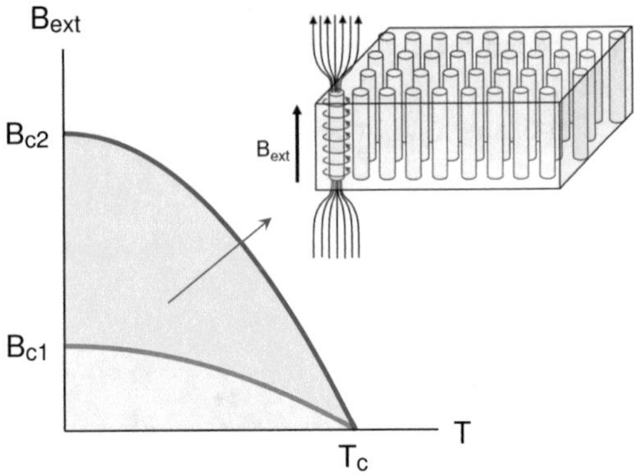

Fig. 6.16 Schematic view of
the phase diagram of a type-II
superconductor with flux-line
lattice in the vortex state

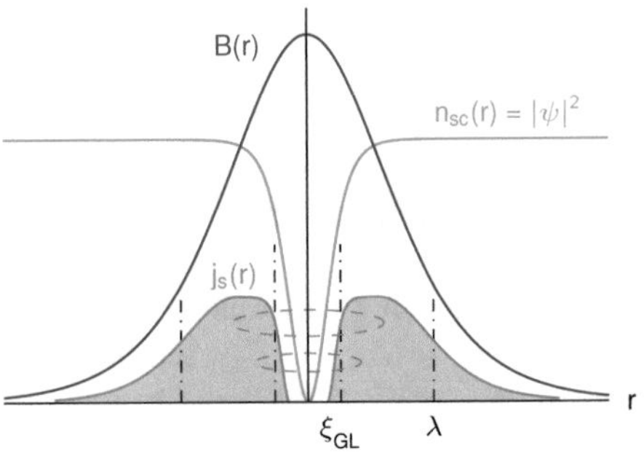

Fig. 6.17 Local magnetic
field and density of
superconducting electrons
near a single vortex.
Superconducting currents $\mathbf{j}_s$
flow around the normal core,
screening the applied field

6.2.1   Principle of a μSR Experiment in the Vortex State

The TF-μSR technique is an ideal and powerful method to study the distribution of the local magnetic fields associated with the vortex lattice in type-II-
superconductors (Brandt & Seeger 1986).

As shown in Sect. 4.2.2, the μSR signal is the cosine Fourier transform of the
distribution of the local field component along the direction of the applied field (z-
axis), Eq. 4.36.[34]

$$P_x(t) = \int_0^\infty f(B_{\mu,z}) \cos\left(\gamma_\mu B_{\mu,z}\, t\right) dB_{\mu,z} \ . \qquad (6.41)$$

---

[34] The condition for the validity of Eq. 4.36, namely that the field width is much smaller than the
average field is generally fulfilled in vortex state measurements.

Fig. 6.18 Spatial
distribution $B_z(\mathbf{r})$ for a
regular triangular vortex
lattice ($\mathbf{B}_{\text{ext}} \parallel \hat{z}$). In an
experiment, muons randomly
sample the field

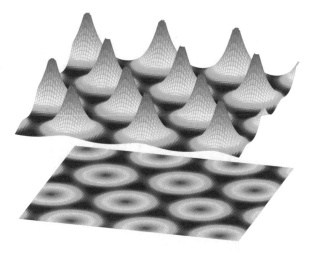

Using the Fourier theorem, we can invert the equation and determine $f(B_{\mu,z})$

$$f(B_{\mu,z}) = \frac{2\gamma_\mu}{\pi} \int\limits_0^\infty P_x(t) \cos\left(\gamma_\mu B_{\mu,z}\, t\right) dt \quad. \tag{6.42}$$

From the field distribution[35] various static and dynamic properties of the vortex state
can be determined; for an overview see (Sonier et al. 2000; Sonier 2007; Dalmas de
Réotier & Yaouanc 1997).

The length scale of the field inhomogeneities of a flux-line lattice is mainly
determined by the magnetic penetration depth and to a lesser extent by the coherence
length. Therefore, from the μSR signal the characteristic length of a superconductor
$\lambda(T)$ and $\xi_{\text{GL}}(T)$ can be determined.

For the measurements a fairly regular flux-line lattice must be established, see
Sect. 6.5.1. Therefore, the magnetic field is first applied above T_c, and then the
temperature gradually lowered below T_c (field-cooling). It is important to realize
that the position of the vortices is not related to the underlying crystallographic
structure. The spacing between them is determined by the value of the applied field
(see Eq. 6.49 below) and is much larger than the crystal lattice parameter. Therefore,
the muons, which stop at well-defined crystallographic sites, randomly sample the
field created by the vortex lattice, see Fig. 6.18.

Figure 6.19 shows ideally how the spatial distribution of the local field, the
asymmetry, and the field distribution evolve when applying B_{ext} (larger than B_{c1}
and not too close to B_{c2}) to a type-II superconductor and then cooling in field.

---

[35] Sometimes also called μSR lineshape.

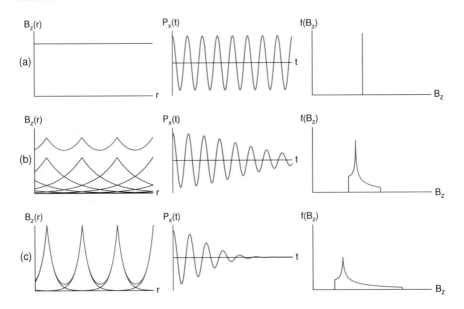

Fig. 6.19 Schematic situation of a μSR vortex state measurement. Left: spatial distribution of the field sensed by the muon in a superconductor. Middle: corresponding asymmetry signal. Right: Fourier transform yielding the field distribution. First row (**a**): in the normal state, essentially the external field is probed (we neglect here small Knight-shift contributions). The μSR signal shows undamped precession, with a frequency corresponding to the applied field, and the field distribution is very narrow. Middle row (**b**): situation just below T_c. The FLL forms and the local field is the vector sum of the fields of isolated vortex (which in the London model is given by Eq. 6.45). The field is maximum at the vortex core position. The magnetic penetration depth is large (since the superfluid density is small) and the range of field variation small. The μSR asymmetry shows a slowly depolarizing precession signal. The precession frequency is determined by the average field, which is very close to the external field (about the condition for this to occur, see footnote 39). Bottom row (**c**): $T \ll T_c$; the penetration depth is short (high superfluid density), and the field variation much larger than in (**b**). This results in a fast depolarization of the μSR signal. Modified from Blundell (1999), © Taylor & Francis. Reproduced with permission. All rights reserved

6.2.2   Local Field in the Vortex State

6.2.2.1 Field Generated by an Isolated Vortex

The spatial dependence of the magnetic field generated by an isolated vortex $\mathbf{B}_v(\mathbf{r})$ can be obtained from the second London equation modified to take into account the presence of a quantized vortex with a two-dimensional δ-function at the core position

$$\mu_0\lambda^2\, \nabla \times \mathbf{j}_s(\mathbf{r}) + \mathbf{B}_v(\mathbf{r}) = \hat{\mathbf{z}}\, \Phi_0\, \delta(\mathbf{r})\,, \tag{6.43}$$

where $\hat{\mathbf{z}} \perp \mathbf{r}$ is the unit vector along the flux line. Using the 4th Maxwell equation, we obtain, in analogy to Eq. 6.11

$$\mathbf{B_v}(\mathbf{r}) - \lambda^2 \nabla^2 \mathbf{B_v}(\mathbf{r}) = \hat{\mathbf{z}}\, \Phi_0\, \delta(\mathbf{r}) \ . \tag{6.44}$$

$\mathbf{B_v}(\mathbf{r})$ has only the component along the z-axis and depends only on $r = |\mathbf{r}|$. Equation 6.44 has an exact solution

$$B_{v,z}(r) = \frac{\Phi_0}{2\pi\lambda^2} K_0\left(\frac{r}{\lambda}\right) \ , \tag{6.45}$$

where K_0 is a modified Bessel function of the second kind (0th order). Note that Eq. 6.44 assumes that the coherence length (and therefore the vortex core radius) is zero (London approximation).[36]

The solution can be approximated near the core and far away from it as

$$B_{v,z}(r) \approx \frac{\Phi_0}{2\pi\lambda^2} \ln\left(\frac{\lambda}{r}\right) + 0.12 \qquad\qquad \xi \ll r \ll \lambda$$

$$B_{v,z}(r) \approx \frac{\Phi_0}{2\pi\lambda^2} \sqrt{\frac{\pi}{2} \frac{\lambda}{r}} \exp\left(-\frac{r}{\lambda}\right) \qquad\qquad r \gg \lambda \ . \tag{6.46}$$

6.2.2.2 Field Distribution from the London Model

The London model, which is valid up to $B_{\mathrm{ext}} \sim B_{c2}/4$, is a good approximation when we can neglect the size of the vortex core; this is the case for the so-called extreme type-II superconductors, where $\kappa \gg 1$. To calculate the local field in the vortex state the vortices are assumed to be separated and noninteracting. Then the field is given by the linear superposition of the singular vortex fields (Pincus et al. 1964).

We consider a triangular (or hexagonal) vortex lattice specified by the vectors **a** and **b**. The spacing between the vortices can be determined from geometrical considerations and the flux quantization of a vortex. The area S of the unit cell containing one vortex (blue area in Fig. 6.20) is

$$S = d^2 \frac{\sqrt{3}}{2} \ ,$$

where d is the distance between the vortices. Since each vortex carries a flux Φ_0 we can write[37]

$$\Phi_0 = S \langle B_z(\mathbf{r}) \rangle \ ,$$

---

[36] The solution diverges for $r \to 0$, but this singularity is not reached because at $r \approx \xi_{\mathrm{GL}}$ the order parameter starts to decrease to zero, see Fig. 6.17.

[37] We simply denote the calculated local field in the FLL state with **B** (without index μ) to emphasize its theoretical character, and use the symbol $\mathbf{B}_\mu$ when referring to values that are probed experimentally.

Fig. 6.20 Triangular vortex
lattice with unit cell

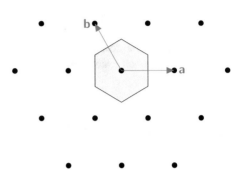

where $\langle B_z(\mathbf{r})\rangle$ is the average field in the unit cell (or equivalently in the whole
superconductor, since B_z is periodic).[38,39] Therefore,

$$d = \sqrt{\frac{2\Phi_0}{\langle B_z\rangle\sqrt{3}}} \ . \tag{6.49}$$

Generalizing Eq. 6.44 for a lattice of vortices at positions $\mathbf{r}_l$, the spatial field
distribution $\mathbf{B}(\mathbf{r})$ must satisfy the modified London equation

$$\mathbf{B}(\mathbf{r}) - \lambda^2\,\nabla^2\mathbf{B}(\mathbf{r}) = \Phi_0\sum_l \delta(\mathbf{r}-\mathbf{r}_l)\,\hat{\mathbf{z}} \ . \tag{6.50}$$

---

[38] Note that spatial averages can be expressed in terms of the field probability distribution $f(B_z)$.
For instance $\langle B_z(\mathbf{r})\rangle = \frac{1}{S}\int_S B_z(\mathbf{r})d\mathbf{r} = \int B_z f(B_z)dB_z$.

[39] In the vortex state there is a bulk (diamagnetic) magnetization and the average field depends
on the sample shape. Therefore, the sample averaged quantities H, magnetization M, magnetic
field $\langle B_{\mu,z}\rangle$ (which is the field given by the average precession frequency measured in the TF
experiment) and demagnetization factor N (with $0 \le N \le 1$) are related to each other by

$$H = \frac{\langle B_{\mu,z}\rangle}{\mu_0} - M = H_{\text{ext}} - NM \ . \tag{6.47}$$

The μ^+ Knight-shift (e.g., in high-T_c materials) is generally negligible, so the observed muon spin
precession is given by

$$\langle B_{\mu,z}\rangle - \mu_0 H_{\text{ext}} = (1-N)\,\mu_0 M \quad (M<0) \ . \tag{6.48}$$

For a (large) flat sample and an applied field perpendicular to the surface $N \simeq 1$ and the average
field in the vortex state is equal to the applied field $\langle B_{\mu,z}\rangle = B_{\text{ext}} = \mu_0 H_{\text{ext}}$. In the commonly
used sample geometry, $\langle B_{\mu,z}\rangle$ is close to B_{ext}. However, the temperature dependence of $\langle B_{\mu,z}\rangle$ still
gives information about the superconducting state, see Fig. 6.43d.

Fig. 6.21 Reciprocal space
of the triangular vortex lattice

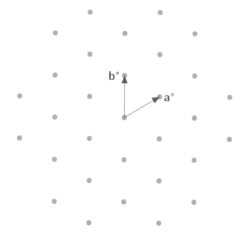

As in Eq. 6.44, $\mathbf{B(r)}$ has only the component along $\hat{\mathbf{z}}$ and $\mathbf{r} \perp \hat{\mathbf{z}}$. In an ideal vortex state the vectors $\mathbf{r}_l$ form a periodic two-dimensional lattice, so that we can work in the reciprocal (Fourier) $\mathbf{k}$-space, Fig. 6.21.

The reciprocal vectors are

$$\mathbf{a}^* = 2\pi \frac{\mathbf{b} \times \mathbf{c}}{\mathbf{a} \cdot (\mathbf{b} \times \mathbf{c})} \qquad \mathbf{b}^* = 2\pi \frac{\mathbf{c} \times \mathbf{a}}{\mathbf{a} \cdot (\mathbf{b} \times \mathbf{c})} \qquad |\mathbf{a}^*| = |\mathbf{b}^*| = \frac{4\pi}{\sqrt{3}d} \quad .$$

A point of the reciprocal space is expressed by

$$\mathbf{k} = m\,\mathbf{a}^* + n\,\mathbf{b}^* \quad ,$$

and we can decompose

$$B_z(\mathbf{r}) = \sum_{\mathbf{k}} b_{\mathbf{k}} \exp(i\,\mathbf{k} \cdot \mathbf{r}) \tag{6.51}$$

with Fourier components

$$b_{\mathbf{k}} = \frac{1}{S} \int_S B_z(\mathbf{r}) \exp(-i\,\mathbf{k} \cdot \mathbf{r}) d\mathbf{r} \quad , \tag{6.52}$$

where the integral is over the unit cell. The modified London equation, with $N_v = 1/S$ vortex density, becomes[40]

$$\sum_{\mathbf{k}} (b_{\mathbf{k}} + \lambda^2 k^2 b_{\mathbf{k}}) \exp(i\,\mathbf{k} \cdot \mathbf{r}) = N_v \Phi_0 \sum_{\mathbf{k}} \exp(i\,\mathbf{k} \cdot \mathbf{r}) \ . \tag{6.53}$$

One finds

$$b_{\mathbf{k}} = \frac{\langle B_z \rangle}{1 + \lambda^2 k^2} \tag{6.54}$$

and

$$B_z(\mathbf{r}) = \sum_{\mathbf{k}} \frac{\langle B_z \rangle}{1 + \lambda^2 k^2} \exp(i\,\mathbf{k} \cdot \mathbf{r}) \ . \tag{6.55}$$

With $b_{\mathbf{k}=0} = \langle B_z \rangle$, the variance of the field distribution $\langle \Delta B_z^2 \rangle = \langle B_z^2 \rangle - \langle B_z \rangle^2$ is given by[41]

$$\langle \Delta B_z^2 \rangle = \sum_{\mathbf{k} \neq 0} |b_{\mathbf{k}}|^2 \ . \tag{6.56}$$

For the ideal triangular lattice

$$k^2 = \frac{16\pi^2}{3d^2} (m^2 - m\,n + n^2) \ , \tag{6.57}$$

with $k^2 \lambda^2 \gg 1$, see Exercise 6.2, we can write

$$\langle \Delta B_z^2 \rangle = \frac{3\Phi_0^2}{64\pi^4 \lambda^4} \sum_{\mathbf{k} \neq 0} \frac{1}{(m^2 - m\,n + n^2)^2} \ , \tag{6.58}$$

and finally

$$\langle \Delta B_z^2 \rangle = 0.00371 \frac{\Phi_0^2}{\lambda^4} \ . \tag{6.59}$$

---

[40] For the right-hand side of the equation we use the fact that $\sum_l \delta(\mathbf{r} - \mathbf{r}_l) \equiv g(\mathbf{r})$ has the periodicity of the space lattice and can therefore be expanded in plane waves of the form $g(\mathbf{r}) = \sum_{\mathbf{k}} g_{\mathbf{k}} \exp(+i\,\mathbf{k} \cdot \mathbf{r})$. The Fourier coefficients $g_{\mathbf{k}}$ are given by $g_{\mathbf{k}} = \frac{1}{S} \int_S g(\mathbf{r}) \exp(-i\,\mathbf{k} \cdot \mathbf{r}) d\mathbf{r} = \frac{1}{S} \int_S \exp(-i\,\mathbf{k} \cdot \mathbf{r}) \sum_l \delta(\mathbf{r} - \mathbf{r}_l) d\mathbf{r} = \frac{1}{S} \int_S \exp(-i\,\mathbf{k} \cdot \mathbf{r}) \delta(\mathbf{r}) d\mathbf{r} = \frac{1}{S}$, because we integrate over the unit cell centered at $\mathbf{r}_l = \mathbf{0}$.

[41] Here we use the identity $\frac{1}{S} \int_S \exp[i(\mathbf{k} - \mathbf{k}') \cdot \mathbf{r}] d\mathbf{r} = \delta_{\mathbf{k},\mathbf{k}'}$.

The variance of the field distribution in the vortex state is directly related to the magnetic penetration depth λ, so a simple TF-measurement in the vortex state, see Sect. 4.2.2, allows the determination of this important characteristic length of a superconductor. Note that Eq. 6.59 predicts a field width independent of the external field.[42]

Equation 6.59 is also valid for an anisotropic superconductor with uniaxial asymmetry, if the field is applied parallel to the symmetry axis. For instance, in a layered superconductor with ab-planes, if the field is applied perpendicular to the planes (i.e parallel to the c-axis), $\lambda = \lambda_{ab}$ with λ_{ab} determined by the supercurrents flowing in the ab-planes, see Eq. 6.66.

For a square FLL the coefficient in Eq. 6.59 changes to 0.00386 (Sidorenko et al. 1991). Note that if, in addition to the field inhomogeneities due to the FLL, other sources of field broadening are present (e.g., vortex lattice disorder in single crystals or demagnetization effects due to irregular grains in polycrystalline samples) and remain unaccounted for, the magnetic penetration depth determined from Eq. 6.59 represents a lower estimate.

Several properties of an ideal flux-line lattice for the whole range of applied fields $B_{c1} \leq B_{ext} \leq B_{c2}$ and a wide range of Ginzburg-Landau parameters $1/\sqrt{2} \leq \kappa \leq 1000$ have been calculated from numerical solutions of the Ginzburg-Landau equations and compared with different analytical approximations (Brandt 2003). The widely used expression 6.59 is really satisfactory only for $\kappa \gtrsim 70$ and over a narrow range of not too large applied fields. Note that the κ condition is not always satisfied in high-T_c superconductors.

From the known spatial distribution Eq. 6.59 we can calculate the corresponding field distribution $f(B_z)$, which is the probability of finding a local magnetic field B_z at a position $\mathbf{r}$ in the xy-plane and is given by

$$f(B_z) = \langle \delta(B_z - B_z(\mathbf{r})) \rangle = \frac{1}{S} \int_S \delta(B_z - B_z(\mathbf{r}))\, d\mathbf{r} \quad . \qquad (6.60)$$

The resulting $f(B_z)$ is shown in Fig. 6.22.

Some distinctive positions and fields of a regular vortex are marked with colored dots in Fig. 6.22. The field distribution is asymmetric and ranges from a minimum B_{min} to a maximum field B_{max}. The minimum field is found at the center of the equilateral triangle formed by three nearest neighbor vortex cores of the FLL (green dots). The maximum field is found at the vortex core (red dots). However, the probability of the muon probing this field is very small, since the area of the core is very small (in fact, in the London model, which assumes a pointlike core, $B_{max} = \infty$ and $f(B_{max}) = 0$). Below B_{max} there is a long tail, since high field values are only found in the small region around the vortex core. Finally, there is a field labeled B_{sad} (blue dot), which is the most probable field, reached at the midpoint (saddle)

---

[42] This follows from the assumption $k^2\lambda^2 \gg 1$, which allows to discard the 1 in the denominator of Eq. 6.55.

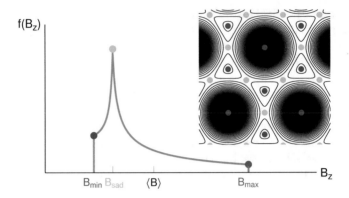

Fig. 6.22 Magnetic field distribution $f(B_z)$ obtained from the London model for an ideal triangular FLL. The inset shows the contour lines of constant field of the corresponding spatial distribution, see Fig. 6.18. The colored dots indicate, in the inset, specific stopping sites of the muons and in the main plot the corresponding probabilities to measure such a field. Note that the maximum field (at the center of the vortex, red dot) is infinite in the London model. Here the profile has been truncated near the vortex center to represent a more realistic situation

between three adjacent vortex cores. The position of these characteristic field values with respect to the average field also scale with $1/\lambda^2$ and can in principle also be used to determine λ (Sidorenko et al. 1991).

In the London model the field distribution at these three points exhibits a so-called van Hove singularity. In real distributions, obtained by Fourier transforming the asymmetry spectra, these singularities are smeared out due to the finite frequency (and hence field) resolution of the μSR spectrum, to the discrete binning of the histograms, to the vortex disorder, to the possible field broadening from nuclear moments and to the apodization of the data in the Fourier transform procedure.[43]

Figure 6.23 shows as an example the Fourier transforms $f(B_{\mu,z})$ of the μSR time spectra at different temperatures of the superconductor LaPt$_4$Ge$_{12}$, a member of the family of Ge-based skutterudites MPt$_4$Ge$_{12}$ (M = Sr, Ba, La, Pr) with T_c ranging from 5 to 8.3 K (Zhang et al. 2015). Above $T_c = 8.3$ K, a narrow and sharp peak at B_{ext} is visible, as expected for a weak nuclear moment depolarization. Below T_c, a significantly broader asymmetric distribution appears with all the typical features of the field distribution of a well ordered flux-line lattice.

---

[43] Apodization is the operation of multiplying the asymmetry spectra by a weighting function that generally smoothly decreases from one at $t = 0$ to zero at the end of the spectrum. The purpose of the apodization is to suppress the late part of the spectrum, which may be dominated by statistical noise or distorted, e.g., because of unknown background contributions. Apodization leads to an increase of the width of the spectrum. For example, assuming that the original spectrum is Gaussian damped with variance σ^2 and that we apodize with another Gaussian with variance σ_{apod}^2, the variance will increase to $\sigma^2 + \sigma_{\text{apod}}^2$. The effect of the apodization is small if we choose $\sigma_{\text{apod}} < \sigma$.

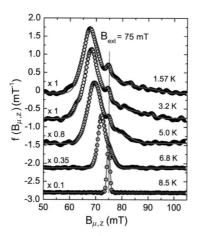

Fig. 6.23 Fourier transforms of the μSR time spectra of LaPt$_4$Ge$_{12}$ at different temperatures, with an applied field $B_{\text{ext}} = 75$ mT. For clarity, the data at each temperature are scaled and shifted vertically. For $T > T_c = 8.3$ K the field distribution is very sharp and corresponds to the applied field, weakly broadened by the magnetic field of the nuclear moments. Below T_c the peak at 75 mT is still visible. It is due to background signal originating from muons missing the sample and stopping nearby. Besides this, a broader asymmetric field distribution is present, with a minimum field greater than zero and a saddle field diamagnetically shifted with respect to the applied field. On lowering the temperature, the distribution becomes broader and the saddle point shifts, reflecting the decrease of the penetration depth. Modified from Zhang et al. (2015), © American Physical Society. Reproduced with permission. All rights reserved

With the London model, one can determine the vortex spacing d provided the symmetry of the FLL is known. In principle the symmetry can be obtained from the exact shape of $f(B_{\mu,z})$. However, the differences between triangular and square symmetry are rather subtle in a real spectrum, where the van Hove singularities are smeared out. On the other hand, the values of $\langle \Delta B^2_{\mu,z} \rangle$ for a triangular lattice $\langle \Delta B^2_{\mu,z} \rangle = 0.00371\,\Phi_0^2/\lambda^4$ and square lattice $\langle \Delta B^2_{\mu,z} \rangle = 0.00386\,\Phi_0^2/\lambda^4$ are so close that a good estimate of λ can be obtained from Eq. 6.59 even without knowing of the FLL symmetry.

6.2.3   Coherence Length and Applied Magnetic Field Dependence

The London model discussed in the previous section assumes $\xi_{\text{GL}} = 0$. This implies neglecting the size of the vortex core and is a good approximation for extreme type-II superconductors, which have very high values of κ. Since the coherence length is related to the critical field B_{c2} by[44]

$$B_{c2} = \frac{\Phi_0}{2\pi\,\xi_{\text{GL}}^2}\ ,\qquad\qquad(6.61)$$

---

[44] This can be derived from Ginzburg-Landau theory (Tinkham 1996).

$\xi_{GL} = 0$ means $B_{c2} \to \infty$ so that for any applied field $B_{ext} \ll B_{c2}$ and corresponds to using Eq. 6.58, where there is no applied field dependence. To treat the case of finite coherence length, one needs to go beyond the original London model. Several approaches are used. We list here those that are most commonly used to analyze μSR spectra in the vortex state (Maisuradze et al. 2009). Since we are dealing with a periodic structure, the models are generally amenable to a modification of the Fourier coefficients b_k in Eq. 6.51 (Yaouanc et al. 1997).

- In the so-called London model with Gaussian cutoff (LG)[45] the original coefficients are multiplied with a Gaussian which reduces the weight of the Fourier components for $k \gtrsim 2\pi/\xi_{GL}$ and yields a finite value of the magnetic field at the vortex core (Brandt 1988a,b, 1972)[46]

$$\mathbf{b_k} = \frac{\langle B_z \rangle \, e^{-k^2 \xi_{GL}^2 / 2}}{1 + k^2 \lambda^2} \, \hat{\mathbf{z}} \; . \tag{6.62}$$

Other possible cutoff functions taking into account the finite size of the vortex core and their range of validity are discussed in Yaouanc et al. (1997).

- The so-called "Modified London Model" (ML) introduces an additional term $(1 - b)$ (where $b \equiv \langle B_z \rangle / B_{c2}$), which reflects the field dependence of the Ginzburg-Landau superconducting order parameter. This model must be used for fields $B_{ext} \gtrsim 0.05 B_{c2}$, because the reduced vortex distance with increasing field makes the area of the vortex core to become significant relative to the unit cell of the vortex lattice (Hao et al. 1991; Riseman et al. 1995)

$$\mathbf{b_k} = \frac{\langle B_z \rangle \, e^{-k^2 \xi_{GL}^2 / [2(1-b)]}}{1 + k^2 \lambda^2 / (1 - b)} \, \hat{\mathbf{z}} \; . \tag{6.63}$$

- Other models are based on solutions of the Ginzburg-Landau equations for a vortex lattice. Although strictly valid only for temperatures close to T_c,[47] the GL theory being based on a spatially varying order parameter, provides a natural description of the magnetic field in the vortex core where the order parameter is suppressed, see Fig. 6.17.

Approximate analytical solutions of the GL equations provide the expressions for the Fourier coefficients of Eq. 6.51 (AGL model). Depending on the field strength with respect to B_{c2}, κ, and crystal symmetry, different expressions for the Fourier coefficients and for the cutoff can be found. For more details we refer

---

[45] To label the models we use the acronyms of Maisuradze et al. (2009).

[46] For $\langle B_z \rangle \lesssim 0.25 B_{c2}$ and $\kappa \gtrsim 2$ this procedure well reproduces the numerical solutions of the GL equations (Brandt 1988b, 1972).

[47] It is generally assumed that its range of applicability can be extended to lower temperatures and arbitrary fields.

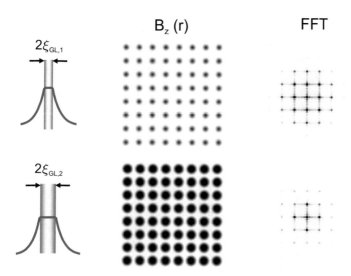

Fig. 6.24 Effect on the Fourier coefficients of $B_z(\mathbf{r})$ for different coherence lengths but same magnetic penetration depth λ and applied field. The Fourier coefficients (right column) have been obtained by Fast Fourier Transform of $B_z(\mathbf{r})$. A squared symmetry is taken for simplification. The upper row exhibits the situation with a short ξ_{GL} (or vortex radius), while the lower row is the situation with longer ξ_{GL}. In the reciprocal lattice, increasing the coherence length leads to an intensity decrease of the $b_\mathbf{k}$ components for large $\mathbf{k}$

here to the literature (Clem 1975; Hao et al. 1991; Yaouanc et al. 1997; Dalmas de Réotier & Yaouanc 2011).[48]

- An efficient iterative method for numerically solving the Ginzburg-Landau equations has been developed by Brandt (1988a) (NGL model). It allows to compute the field profile $B_z(\mathbf{r})$ of a vortex lattice of arbitrary symmetry and is valid for any value of the magnetic field and the GL parameter κ (Brandt 2003, 1997).

Figures 6.24 and 6.25 qualitatively exemplify the effect on the Fourier coefficients $b_\mathbf{k}$ of $f(B_z)$, when the size of the vortex core or the dependence on the external field are included in the vortex state model, respectively.

An example of magnetic field distribution in the vortex state obtained by the numerical solution of the Ginzburg-Landau equations is shown in Fig. 6.26 (Maisuradze et al. 2009). The central region between minimum and mean field $\langle B \rangle$ is the most important for the analysis of the μSR spectra, because the high-field tail is usually below the noise level of the spectra and is generally not observed,

---

[48] We remind here that expressions for the Fourier coefficients, e.g., Eq. 6.57, are only valid for isotropic superconductors or superconductors with axial symmetry with the field applied along the symmetry axis.

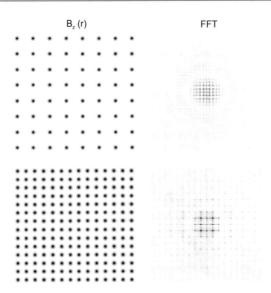

Fig. 6.25 Effect on the Fourier coefficients of $B_z(\mathbf{r})$ for different magnitudes of the external field but same magnetic penetration depth and coherence length. The Fourier coefficients (right column) have been obtained by Fast Fourier transform of $B_z(\mathbf{r})$. The figure compares the case of a low (top row) with a high field (bottom row). Increasing the strength of the applied field leads to smaller vortex spacing in real space, Eq. 6.49, and to an increase of the lattice constant in the reciprocal space, Eq. 6.57. The intensity of the Fourier coefficients $b_\mathbf{k}$ decays faster for the stronger field

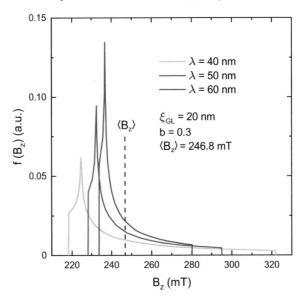

Fig. 6.26 Local magnetic field distribution $f(B_z)$ for an ideal triangular FLL obtained by a numerical solution of the GL expressions. The curves are for different values of the magnetic penetration depth λ but fixed coherence length and an applied field $b \equiv \langle B_z \rangle / B_{c2} = 0.3$, $\langle B_z \rangle \approx B_{\text{ext}}$. Note that the shape of $f(B_z)$ and the values of the characteristic fields B_{min}, B_{sad} and B_{max} depend strongly on the magnetic penetration depth. Modified from Maisuradze et al. (2009), © IOP Publishing. Reproduced with permission. All rights reserved

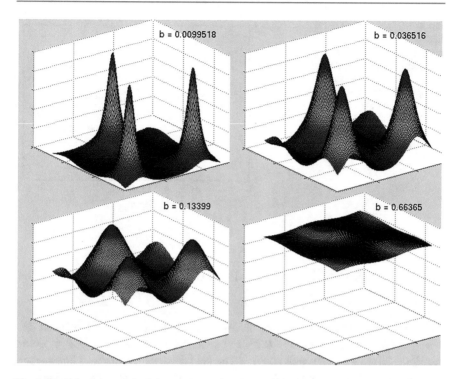

Fig. 6.27 Calculation of the field profile in the vortex state of a type-II superconductor obtained by numerical solution of the GL equations. As in Fig. 6.26 the parameter b represents the strength of the applied field with respect to B_{c2}. The calculations are for $\lambda = 50$ nm and $\xi_{GL} = 20$ nm. The intervortex distance is inversely proportional to the applied field strength, see Eq. 6.49. Note that the distance is normalized in the figures. Courtesy A. Maisuradze, private communication

especially at low fields and for $\kappa \gg 1$, see Fig. 6.23 or left side of Fig. 6.31 as examples.

Figure 6.27 shows how the applied magnetic field modifies the field profile $B_z(\mathbf{r})$ in the vortex state. Clearly, when the condition $B_{ext} \ll B_{c2}$ no longer holds, an increase of the applied field suppresses the field inhomogeneity.

The field dependence of $\langle \Delta B_z^2 \rangle$ obtained from numerical calculations of the GL equations is shown in Fig. 6.28.

From the calculations based on the numerical solution of the Ginzburg-Landau model (NGL) a useful approximate analytical expression, valid in a wide range, has been obtained for the field dependence of $\langle \Delta B_z^2 \rangle$[49]

---

[49] A previous similar approximation, valid for $0 < b < 1$

$$\langle \Delta B_z^2 \rangle = 7.571 \cdot 10^{-4} \left(1 - \frac{\langle B \rangle}{B_{c2}} \right)^2 \left[1 + 3.9 \left(1 - \left(\frac{\langle B \rangle}{B_{c2}} \right)^2 \right) \right] \frac{\Phi_0^2}{\lambda^4} , \qquad (6.64)$$

is sometimes used (Brandt 1988a).

Fig. 6.28 Dependence of the field variance on the magnitude of the external field for an ideal triangular vortex lattice and different κ. The green dashed line represents the field independent value predicted by the original London model, Eq. 6.59

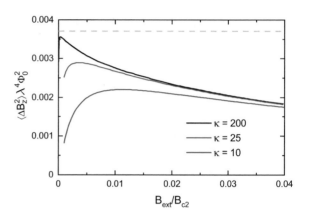

$$\langle \Delta B_z^2 \rangle = 7.596 \cdot 10^{-4} \left(1 - \frac{\langle B \rangle}{B_{c2}} \right)^2 \left[1 + 1.21 \left(1 - \sqrt{\frac{\langle B \rangle}{B_{c2}}} \right)^3 \right]^2 \frac{\Phi_0^2}{\lambda^4} . \qquad (6.65)$$

This expression represents an error of less than 5%, for $\kappa \geq 5$ and a reduced field $b = \langle B \rangle / B_{c2}$ ranging from $b \approx 0.25/\kappa^{1.3}$ (where the maximum of $\langle \Delta B_z^2 \rangle$ occurs, see Fig. 6.28) to $b = 1$ (Brandt 2003).[50] The field dependent terms take into account that the intervortex distance decreases with increasing applied field and that the size of the vortex core, where the superconducting order parameter is suppressed, is finite. Both effects narrow the field distribution. A comparison of the prediction of Eq. 6.65 with values from the numerical solution of the GL equations is shown in Fig. 6.29.

If the external magnetic field is small compared to the upper critical1 field, that is when $B_{ext}/B_{c2} \to 0$, both expressions (Eqs. 6.65 and 6.64) converge to Eq. 6.59.

Equations 6.65 and 6.64 show that a measurement of the field width as a function of temperature and field allows to determine the two characteristic lengths of the superconductor $\lambda(T)$ and $\xi_{GL}(T)$ (or B_{c2}, see Eq. 6.61). An example is given at page 269.

---

[50] For a square FLL, the prefactor changes to 8.370×10^{-4}. This means that the uncertainty associated with not knowing the FLL symmetry results in a 2.5% error in λ.

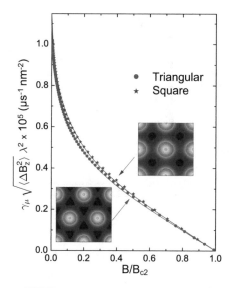

Fig. 6.29 Dependence of $\sqrt{\langle \Delta B_z^2 \rangle}\, \lambda^2$ on the reduced field $b \equiv \langle B_z \rangle / B_{c2}$ for a triangular (circles) and a square (stars) vortex lattice. The values are from the numerical solution of the GL equations for $\kappa = 100$ (Brandt 2003). The solid lines are the predictions of Eq. 6.65 for a triangular and a square FLL. Note the small difference between the two lattices. The insets show contour plots of the field values. Modified from Khasanov et al. (2008), © American Physical Society. Reproduced with permission. All rights reserved

6.2.4   Anisotropy of the Magnetic Penetration Depth

Many unconventional superconductors, such as the high-T_c cuprates and iron-based superconductors, have layered crystal structures that lead to anisotropic parameters, including magnetic penetration depth, coherence length, or critical fields.

In the London equation the anisotropy is taken care of by replacing the scalar $\Lambda = m_e^* / n_{sc} e^2$ in Eq. 6.8 with a tensor $\tilde{\mathbf{\Lambda}} = \tilde{\mathbf{m}}^* / n_{sc} e^2$, where $\tilde{\mathbf{m}}^*$ is the effective mass tensor (Kogan 1981; Thiemann et al. 1989). For the uniaxial anisotropy case the diagonalized $\tilde{\mathbf{\Lambda}}$ tensor has a term Λ_{ab} associated with supercurrents flowing in the ab-plane, and Λ_c associated with supercurrents flowing along the c-axis. Similarly to Eq. 6.7, the corresponding magnetic penetration depths are related to the elements of $\tilde{\mathbf{\Lambda}}$ by $\lambda_{ab} = \sqrt{\Lambda_{ab}/\mu_0}$ and $\lambda_c = \sqrt{\Lambda_c/\mu_0}$.

Following a procedure similar to the one outlined in Sect. 6.2.2.2, the Fourier components for the equilibrium rigid FLL can be calculated in intermediate fields ($B_{c1} \lesssim B_{ext} \ll B_{c2}$) for any orientation of the vortex axis (which is taken to coincide with the direction of the applied field $\mathbf{B}_{ext}$); from the Fourier terms the width of the field distribution can be calculated. If the field is rotated by an angle θ away from the axis perpendicular to the ab-planes, one obtains for the angular dependence of

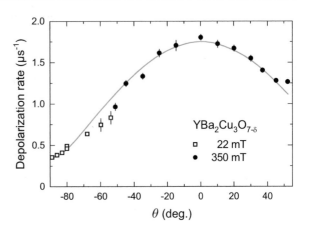

Fig. 6.30 Angular dependence of the μSR depolarization rate ($\propto \lambda^{-2}$) in the vortex state of single crystal $YBa_2Cu_3O_{7-\delta}$. θ is the angle between the applied field and the c-axis. The line is a fit using Eq. 6.66. Modified from Forgan et al. (1991), © Springer Nature. Reproduced with permission.

the field variance (Thiemann et al. 1989; Barford & Gunn 1988)

$$\langle \Delta B_z^2(\theta) \rangle = \langle \Delta B_z^2(0) \rangle \left(\cos^2 \theta + \frac{1}{\gamma^2} \sin^2 \theta \right) \ , \tag{6.66}$$

where $\langle \Delta B_z^2(0°) \rangle$ is the field variance for $\mathbf{B}_{\text{ext}} \parallel \hat{\mathbf{c}}$ and is given by Eq. 6.59 (with $\lambda = \lambda_{ab}$) and $\gamma = \lambda_c/\lambda_{ab} = \sqrt{(m_c^*/m_{ab}^*)}$ is the anisotropy parameter. For $\mathbf{B}_{\text{ext}} \perp \hat{\mathbf{c}}$, we obtain $\langle \Delta B_z^2(90°) \rangle = 0.00371\, \Phi_0^2/(\lambda_{ab}\lambda_c)^2$. In this geometry, the supercurrents flow in the ab-planes and along the c-axis, so that the effective magnetic penetration depth is the geometric mean of λ_{ab} and λ_c.[51,52]

Figure 6.30 exhibits an angle dependent measurement in orthorhombic $YBa_2Cu_3O_{7-\delta}$ (YBCO), which can be well approximated by the London model with uniaxial asymmetry.[53] Fitting Eq. 6.66 to the data yields a value of the anisotropy parameter $\gamma \simeq 5$ which reflects the layered crystal structure of the cuprate with CuO_2 planes (ab-planes) where supercurrents (and normal currents) flow much more easily than perpendicular to them.

---

[51] Note that other vortex structures with more pronounced 2D-character ("pancake" vortices, see footnote 52) have been found. For instance, in the extremely anisotropic $Bi_2Sr_2CaCu_2O_8$ ($\gamma \gtrsim 150$), significant deviations from the predictions of the anisotropic London model have been observed (Cubitt et al. 1993a). See Sect. 6.5 for the phase diagram of this compound.

[52] In highly anisotropic layered materials, in the usual approximation where the weak interlayer (Josephson) coupling is neglected but the electromagnetic interaction between vortices is kept, the vortex lines decompose into so-called "pancake" vortices, when the field is applied perpendicular to the superconducting layers (c-axis). The pancakes are confined to move on a superconducting layer.

[53] Without special treatment (detwinning), single crystals of YBCO grow with twins in the ab-plane, which have boundaries where a- and b-oriented crystallites meet, therefore averaging the two planar lattice constants and penetration depths so that $\lambda_{ab} = \sqrt{\lambda_a \lambda_b}$. For bulk measurements on untwinned YBCO crystal see Ager et al. (2000), for low-energy muon measurements see Sect. 8.5.1.1 and Kiefl et al. (2010).

A measurement of the magnetic penetration depth in a polycrystalline sample of an anisotropic system with uniaxial symmetry corresponds to an angular averaging of Eq. 6.66. The measurement provides an effective value λ_{eff}, which, for an anisotropy ratio $\gamma \gtrsim 5$, is solely determined by the shorter penetration depth λ_{ab} via the simple relation

$$\lambda_{\text{eff}} = 3^{1/4}\lambda_{ab} = 1.316\,\lambda_{ab} \ . \tag{6.67}$$

This indicates that in a polycrystalline sample for a uniaxial superconductor with large anisotropy, the screening of the applied field is mainly determined by the currents flowing in the ab-planes. It also means that even if the value of γ is not known, one can determine λ_{ab}, provided that the system is sufficiently anisotropic (Barford & Gunn 1988).[54] This property was used intensively, especially in the early stages of high-T_c research, and helped to establish the original Uemura plot (Uemura et al. 1989), see the following section and Fig. 6.36.

6.3 Analysis of the μSR Spectra

In this section we describe how to analyze the TF-μSR spectra to extract values of the characteristic lengths and to determine the values and symmetries of single or multiple superconducting gaps and give some typical examples.

To extract λ and ξ_{GL} one of the previously described models based on London or Ginzburg-Landau theory is generally chosen. The model is used to provide either simply the field variance or the full field distribution $f(B_z)$. In principle, an assumption about the vortex lattice symmetry has to be made. However, as already remarked, see Fig. 6.29 and footnote 50, the symmetry has a small influence on the overall error and the more common triangular symmetry is generally assumed. Once the temperature dependence of the magnetic penetration depth is determined, one can extract from it information (value and symmetry) about the gap(s), see Sects. 6.1.4 and 6.1.4.1.

6.3.1 Models of Data Analysis

A simple and widely applied empirical method is to obtain the field width from the data by assuming that the field distribution of the vortex lattice $f(B_z)$ can be represented by a Gaussian or a sum of Gaussian functions (Weber et al. 1993) and relating the field width to λ and possibly ξ_{GL} via Eqs. 6.59 and 6.65.

---

[54] Note that Barford and Gunn (1988) originally gave a factor 1.23 instead of the correct 1.316, see also Fesenko et al. (1991).

6.3.1.1 Single Gaussian Analysis

Let us first consider the simplest case of a single Gaussian. In this case $\langle \Delta B^2 \rangle$ is obtained from the μSR spectrum fitted with a precession signal times a Gaussian function (which is the depolarization function in case of Gaussian field distribution, see Sect. 4.2.2). This approach was widely used in the beginning of the high-T_c superconductor research, where many samples could only be synthetized in polycrystalline form.

In low quality samples, the periodic vortex lattice is easily perturbed by random pinning of the vortices and fluctuations in temperature. Moreover, in polycrystalline samples the signal is the integral over all possible orientations of the crystal grains, which averages out the asymmetric tail of the ideal field distribution, see Sect. 6.2.4. All these effects lead to an almost symmetric $f(B_z)$, which can be satisfactorily approximated by a Gaussian function with a field variance $\langle \Delta B_z^2 \rangle$. As shown in Eq. 4.41, the μSR signal of the muon ensemble sensing this field distribution is characterized by a Gaussian depolarization rate σ_{sc}, which is a direct measure of the Gaussian field variance, i.e.,

$$\sigma_{sc}^2 = \gamma_\mu^2 \langle \Delta B_z^2 \rangle \ . \tag{6.68}$$

To obtain σ_{sc} one has generally first to subtract from the measured σ the temperature independent weakly broadening contribution of the nuclear moments σ_N, which is obtained by a measurement above T_c

$$\sigma_{sc} = \sqrt{\sigma^2 - \sigma_N^2} \ . \tag{6.69}$$

In high-κ superconductors (such as several high-T_c superconductors) and low average magnetic field $\langle B \rangle$ (i.e., $B_{ext} \ll B_{c2}$, London limit) one can directly use Eq. 6.59 to obtain the penetration depth.

Assuming that Eqs. 6.59 and 6.68 are valid, there is a simple numerical relationship between the muon depolarization rate σ_{sc} and the magnetic penetration depth λ, namely

$$\lambda = \frac{327.5}{\sqrt{\sigma_{sc}}} \ , \tag{6.70}$$

where σ_{sc} is expressed in μs^{-1} and λ in nm.

For good-quality single crystals where the characteristic asymmetric field distribution of a well formed ideal FLL is expected, a single Gaussian fit is not accurate enough.[55]

---

[55] A Gaussian fit of an asymmetric field distribution underlying Eq. 6.59 is sensitive only to the central part of the non-Gaussian distribution.

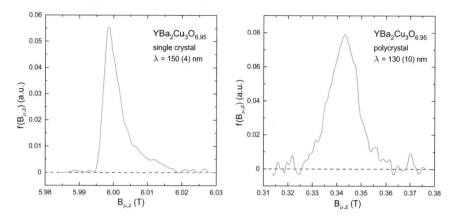

Fig. 6.31 Comparison of $f(B_{\mu,z})$ for a single crystal of $YBa_2Cu_3O_{6.95}$, with the typical asymmetric field distribution, and for a polycrystalline sample exhibiting a symmetric Gaussian-like field distribution. Modified from Sonier et al. (1999) (left) and Pümpin et al. (1990) (right), © American Physical Society. Reproduced with permission. All rights reserved

Figure 6.31 compares the results of the measured field distribution in YBCO for a polycrystalline and a single crystal sample, with the field applied along the c direction.

Examples of Single Gaussian Analysis

Determination of $\lambda(T)$, $\Delta(0)$, and of the Symmetry
The compound Mo_3Al_2C is noncentrosymmetric, i.e., characterized by the absence of inversion symmetry in its crystal structure. Superconductivity in Mo_3Al_2C was discovered in the '60s. Because of its lack of inversion symmetry and because transport, thermodynamic and NMR measurements hinted to a superconducting state possibly deviating from the BCS behavior (Bauer et al. 2010), this compound has regained interest and its superconducting properties revisited by TF-μSR measurements (Bauer et al. 2014). The μSR spectra can be well fitted with

$$A_x(t) = A_{\text{sample}} \exp\left(-\frac{\sigma^2 t^2}{2}\right) \cos\left(\gamma_\mu \langle B_{\mu,z} \rangle t + \varphi\right) + A_{\text{bg}} \cos\left(\gamma_\mu \langle B_{bg} \rangle t + \varphi\right).$$

(6.71)

The first term is the signal from the sample and the second takes into account background contributions. As just explained, see Eq. 6.69, the observed Gaussian muon spin depolarization rate contains the contribution due to the FLL field broadening and the temperature independent weak broadening contribution arising from the nuclear moments. Typical spectra at a transverse field $B_{\text{ext}} = 15$ mT below and above T_c are shown in the left panels of Fig. 6.32. The data are very well fitted

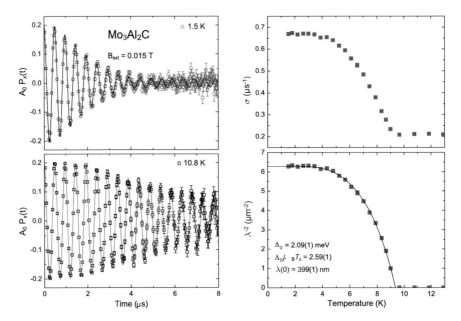

Fig. 6.32 Measurements of the magnetic penetration depth by TF-μSR in the compound Mo$_3$Al$_2$C. The two figures on the left-hand side highlight the difference in depolarization above and below T_c. Above T_c only a very weak depolarization due to the nuclear moments is observed. The increased depolarization below T_c arises from the formation of the FLL. On the right-hand side the temperature dependence of the Gaussian depolarization rate is shown, as well as the temperature dependence of λ^{-2} obtained employing Eq. 6.65. The solid line is a least-squares fit according to the s-wave BCS model. Modified from Bauer et al. (2014), © American Physical Society. Reproduced with permission. All rights reserved

by Eq. 6.71 (solid lines). The temperature dependence of σ is shown in the upper right panel. A large depolarization rate is observed in the superconducting state. The temperature independent contribution of the nuclear moment ($\sigma_N = 0.2~\mu s^{-1}$) is clearly visible above $T_c = 9.1$ K.

Mo$_3$Al$_2$C is a high-κ superconductor with $\kappa \approx 76$. With $B_{c2}(0) = 15.7$ T and $\langle B \rangle / B_{c2} \ll 1$ in principle the conditions for using Eq. 6.59 are given. However, since the temperature dependence of B_{c2} is known from a previous measurement (Bauer et al. 2010), Eq. 6.65 was used to extract $\lambda(T)$. The difference between the two expressions, if any, should only appear around T_c, where B_{c2} tends to zero. The temperature dependence of $\lambda^{-2} \propto n_{sc}$ can be well fitted by the BCS theory of s-wave superconductors, Eq. 6.28. Fitting the $\lambda(T)^{-2}$ curve parameters such as T_c, $\lambda(0)$, superconducting gap and pair symmetry are extracted, see Sect. 6.1.4.

Another example is the study of the superconductor Mo$_3$Sb$_7$ (Khasanov et al. 2008), where μSR measurements were performed to clarify the symmetry of the superconducting order parameter. Figure 6.33 displays the temperature dependence of λ^{-2}. The experiment was analyzed according to the procedure previously

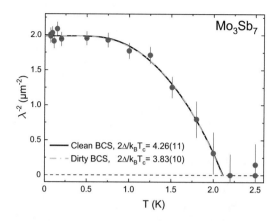

Fig. 6.33 Temperature dependence of λ^{-2} of Mo_3Sb_7 in $B_{ext} = 0.02$ T. The black solid and the dashed-dotted green lines are fits with the clean and dirty weak-coupling BCS models. The plateau below 0.7 K is typical of a gapped system and shows that a sizeable amount of thermal energy is needed to break Cooper pairs and reduce the superfluid density. The value of the gap is found to be $\Delta(0) = 0.39(1)$ meV. Modified from Khasanov et al. (2008), © American Physical Society. Reproduced with permission. All rights reserved

described, i.e.: i) the spectra fitted with a single Gaussian function (Eq. 6.71); ii) the magnetic penetration depth obtained from Eq. 6.65; iii) the superfluid density fitted with the clean, Eq. 6.29, and the dirty BCS expression, Eq. 6.30, for comparison with the temperature dependence of the gap given by Eq. 6.31.

Figure 6.33 shows that at low temperatures $\lambda(T)^{-2}$ saturates and becomes practically constant below $T \simeq 0.3\, T_c$. This is typical of an s−wave superconductor, where the temperature dependence is very weak at low T because the isotropic energy gap exponentially suppresses the quasiparticle excitations as $T \rightarrow 0$, see Eq. 6.35. The data fit well with both the clean and dirty superconductor models.

Determination of the Coherence Length
The coherence length can be obtained by investigating the dependence of the measured field variance on the applied field and using Eqs. 6.65 or 6.64 to determine λ and ξ_{GL} (related to B_{c2} through Eq. 6.61).

We present the iron pnictide superconductor $RbFe_2As_2$ as an example of such a study (Shermadini et al. 2010). Figure 6.34 shows the temperature dependence of σ_{sc} of a polycrystalline sample obtained from TF-μSR measurements in four different external fields. A very good fit is obtained by assuming a Gaussian field distribution. The overall decrease in σ_{sc} observed upon increasing the applied field is the direct consequence of the decrease in width of the internal field distribution when increasing the field toward B_{c2}.

In the experiment, to ensure a well formed FLL for each data point, the sample is field cooled from above T_c. At each temperature, the field dependence of σ_{sc} is

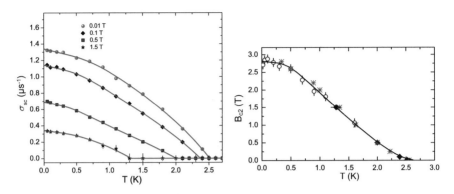

Fig. 6.34 Left: temperature dependence of the depolarization rate σ_{sc} (single Gaussian fit) for different applied fields (lines are guides to the eyes). Right: upper critical field for RbFe$_2$As$_2$. The open circles are obtained by analyzing the field dependence of the depolarization rate at different temperatures using Eq. 6.65. The diamonds are values obtained by analyzing the temperature dependence of the depolarization rate. The stars are from magnetoresistance measurements and correspond to the complete disappearance of the resistivity in field. The line is to guide the eyes. Modified from Shermadini et al. (2010), © American Physical Society. Reproduced with permission. All rights reserved

analyzed with Eq. 6.65 by leaving λ and B$_{c2}$ as free fitting parameters. This yields, under the implicit assumption that λ is field independent, $\lambda(T)$ and $B_{c2}(T)$.

As demonstrated by Fig. 6.34, right panel, the values of B_{c2} obtained by fitting the field dependence of σ_{sc} with a field independent penetration depth agree very well with the values of B_{c2} obtained from magnetoresistance measurements and those directly deduced from the temperature dependence of σ_{sc}, see Fig. 6.34 left panel. This fact can be taken as a strong indication that the theoretical field dependence of $\langle \Delta B_z^2 \rangle$ calculated in the framework of the GL theory, Eq. 6.65, can indeed be used for a quantitative analysis of isothermal experimental data even at temperatures $T \ll T_c$. It also supports the assumption of a field independent penetration depth in this s-wave superconductor.[56] From the $T \to 0$ values of σ_{sc} and B_{c2} the characteristic lengths of this superconductor are $\xi_{GL}(0) = 10.8(4)$ and $\lambda(0) = 267(5)$ nm.

Multiple Gaps

Figure 6.35 shows the temperature dependence of λ^{-2} in the system RbFe$_2$As$_2$. Information about the superconducting gap and its structure can be obtained from the data. A fit assuming a single gap with s-wave symmetry does not reproduce the data in a satisfactory manner. The indication of an inflection point around

---

[56] This result also rules out the possibility that RbFe$_2$As$_2$ is a nodal superconductor, since a field would induce excitations at the gap nodes due to nonlocal and nonlinear effects, thus reducing the superconducting carrier concentration n_{sc} and therefore affecting the effective penetration depth as measured by μSR (Sonier et al. 1999; Amin et al. 2000).

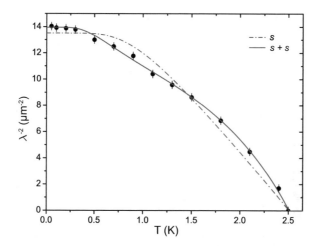

Fig. 6.35 Temperature dependence of λ^{-2}. The red dashed line is for a single s-wave symmetry gap, while the solid line represents the fit using a two-gap $s + s$ model. Modified from Shermadini et al. (2010), © American Physical Society. Reproduced with permission. All rights reserved

1 K suggests the existence of two superconducting gaps. Additional evidence from angular resolved electron spectroscopy (ARPES) suggest that disconnected Fermi surface sheets contribute to the superconductivity, resulting in two distinct values of the gaps.

Figure 6.35 shows that the model of two s-wave gaps $(s + s)$ fits the experimental data rather well[57] with a small gap of value $\Delta_1(0) = 0.15(2)$ meV contributing $w = 36\%$ to the total n_{sc} and a second larger gap $\Delta_2(0) = 0.49(4)$ meV. Note that the low-energy excitations at low temperatures are determined by the smaller gap, while T_c is determined by the larger gap.

Uemura Relation
The discovery of the high-T_c cuprate superconductors in the late '80s led to considerable efforts to find universal trends and correlations among the physical properties in order to identify the relevant parameters and to understand the microscopic mechanism of high-T_c superconductivity.
Already in the early stages of these studies, μSR TF-measurements of polycrystalline samples in the vortex state proved to be a powerful method for directly measuring the magnetic penetration depth with sufficient accuracy. The small amount of material necessary to collect data (less than 10 mg) and the direct relationship between the Gaussian damping of the spectra and the magnetic

---

[57] Note that according to Eq. 6.28 λ^{-2} is insensitive to the sign (or phase) of the superconducting gap. So it cannot distinguish between two s-wave gaps of opposite signs, or the more trivial case of the same sign.

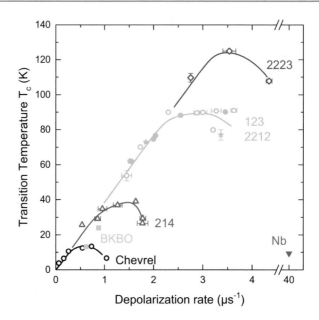

Fig. 6.36 Plot of T_c as a function of $\sigma_{sc}(0) \propto 1/\lambda^2 \propto n_{sc}/m_e^*$ for different families of superconductors: cuprates $YBa_2Cu_3O_{7-\delta}$ (123), $La_{2-x}Sr_xCuO_4$ (214),
$Bi_2Sr_2CaCu_2O_8$ and $Tl_{0.5}Pb_{0.5}Sr_2CaCu_2O_7$ (2212), $Bi_{2-x}Pb_xSr_2Ca_2Cu_3O_{10}$, $Tl_2Ba_2Ca_2Cu_3O_{10}$,
and $Tl_{0.5}Pb_{0.5}Sr_2Ca_2Cu_3O_9$ (2223); bismuthates $Ba_{1-x}K_xBiO_3$ (BKBO); chevrel phase systems
$LaMo_6Se_8$, $LaMo_6S_8$ and $PbMo_6S_8$; organic superconductors $(BEDT-TTF)_2Cu(SCN)_2$; conventional superconductor Nb. Modified from Uemura et al. (1991), © American Physical Society.
Reproduced with permission. All rights reserved

penetration depth expressed by Eq. 6.70 allowed systematic studies of high-T_c
superconductors and other exotic systems. One of the first patterns to emerge for the
high-temperature cuprate superconductors in the underdoped regime, was a linear
increase of the superconducting transition temperature with the Gaussian relaxation
rate σ_{sc}. Remarkably, other superconductors such as heavy fermions, organics,
Chevrel phases and fullerene compounds were also found to follow a similar scaling
behavior. Figure 6.36 shows the original plot, now referred to as "Uemura plot" or
relationship (Uemura et al. 1989, 1991).

The linear relation between T_c and σ_{sc} implies a direct correlation between T_c
and n_{sc}/m_e^*, since $\sigma_{sc} \propto 1/\lambda^2 \propto n_{sc}/m_e^*$. Such a scaling is not expected for phonon
coupled-superconductors, which are well described by the weak-coupling BCS
theory, where $T_c \cong (2\hbar\omega_D)/k_B \exp(-2/VD(E_F))$, with ω_D the Debye frequency,
V the effective attractive Cooper pair potential, and $D(E_F)$ the density of electronic
states at the Fermi level. The Uemura plot is frequently used to classify superconductors. Typical BCS-like superconductors such as the elemental Al, Sn, Zn, Nb are
found in the lower right part of the Uemura plot (see Fig. 6.36). At equivalent T_c they
have a much higher superfluid density compared to the other compounds, which are

loosely classified as "exotic" or "unconventional" superconductors.[58] From $\sigma_{sc}(0)$ one can estimate the Fermi temperature T_F. The data indicate a correlation between T_c and T_F. Within this scheme, exotic superconductors such as cuprates, heavy fermions, organic, fullerenes and Chevrel phase superconductors show a more or less similar linear trend with $1/100 < T_c/T_F \ll 1/10$. This is not observed in conventional BCS superconductors for which $T_c/T_F < 1/1000$.

6.3.1.2 Multi-Gaussian Analysis

We have already mentioned that for polycrystalline samples consisting of randomly oriented grains the field distribution is nearly symmetric and a single Gaussian fit of the μSR spectra is usually sufficient.[59] This is not the case for high-purity crystals.

Figure 6.37 (right panel) exhibits an asymmetry signal and the FFT obtained in the vortex state of a single crystal NbSe$_2$ sample by field-cooling in 0.35 T (Sonier et al. 2000). The quality of the asymmetry fit with a single Gaussian (red line) is clearly poor. The reason is that a symmetric function cannot reproduce the long tail in the high-field region of the field distribution, Fig. 6.22.

An empirical approach to handle asymmetric shapes is to fit the μSR data with a sum of Gaussian functions. From the fits the variance of the field distribution can be obtained and then fitted with Eq. 6.59 or 6.65 or 6.64 to obtain the magnetic penetration depth λ and possibly the coherence length ξ_{GL} as a function of parameters such as temperature, applied field or pressure.

This means that the asymmetry spectra are fitted by following expression

$$A(t) = \sum_{i=1}^{N} A_i \, \exp(-\sigma_i^2 t^2/2) \cos(\gamma_\mu B_i t + \varphi) \ . \tag{6.72}$$

The first moment (average field) of the multi-Gaussian internal field distribution is given by

$$\langle B_z \rangle = \sum_{i=1}^{N} c_i \, B_i \ , \tag{6.73}$$

---

[58] Note that the terms unconventional or exotic superconductor are not uniquely used in the literature. More precisely, an unconventional superconductor has a superconducting pairing state that is different from the usual s-wave spin singlet BCS-type pairing; it can be classified by analyzing the symmetries broken by the superconducting state. For example, the crystals of tetragonal high-T_c superconductors have a fourfold symmetry axis perpendicular to the ab-plane but the superconducting order parameter with $d_{x^2-y^2}$ symmetry has only a twofold symmetry. See Stewart (2017); Sigrist and Ueda (1991) for a detailed discussion and classification of unconventional superconductors.

[59] A simple single Gaussian fit can still be useful for a first analysis during the experiment, since it yields a fair qualitative picture of the temperature variation of σ_{sc} and hence of the magnetic penetration depth.

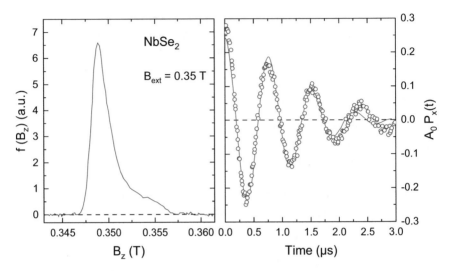

Fig. 6.37 Right: asymmetry spectrum obtained in the vortex state of a NbSe$_2$ crystal ($B_{ext} = 0.35$ T, field cooled applied parallel to the c-axis). The solid line is a fit to a Gaussian function, which clearly fails to reproduce the data. Left: Corresponding magnetic field distribution obtained by a Fourier transform of the μSR signal, background corrected. Modified from Sonier et al. (2000), © American Physical Society. Reproduced with permission. All rights reserved

and the variance by

$$\langle \Delta B_z^2 \rangle = \frac{\sigma_{sc}^2}{\gamma_\mu^2} = \sum_{i=1}^{N} c_i \left[\langle \Delta B_i^2 \rangle + (B_i - \langle B_z \rangle)^2 \right] \ . \tag{6.74}$$

Here φ is the initial muon spin phase, while $c_i = A_i / \sum_{i=1}^{N} A_i$ represents the weight of the i-th Gaussian with asymmetry amplitude A_i, B_i is the first moment and $\langle \Delta B_i^2 \rangle \equiv \sigma_i^2 / \gamma_\mu^2$ the variance of the i-th Gaussian component.[60] Such a fit in time domain corresponds to the assumption that the field distribution of the FLL can be described by a sum of N Gaussian functions[61]

$$f(B_z) = \frac{1}{\sqrt{2\pi}} \sum_{i=1}^{N} \frac{c_i}{\sqrt{\langle \Delta B_i^2 \rangle}} \exp \left[-\frac{(B_z - B_i)^2}{2\langle \Delta B_i^2 \rangle} \right] \ . \tag{6.75}$$

---

[60] Additional broadening of the field distribution due to nuclear moments or vortex disorder is accounted for by multiplying the right-hand side of Eq. 6.72 with a Gaussian function $\exp[-(\sigma_N^2 + \sigma_V^2) t^2/2]$.

[61] Generally, 2 or 3 Gaussians are sufficient to well represent a typical μSR lineshape, see Fig. 6.38.

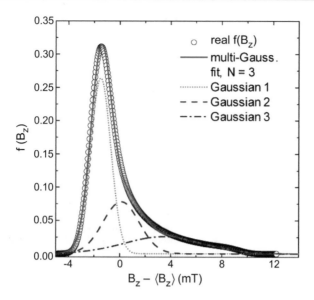

Fig. 6.38 Comparison of a realistic field distribution $f(B_z)$ (open circles) simulated using the numerical Ginzburg-Landau model with parameters $\lambda = 200$ nm, $\xi_{GL} = 4$ nm, $\langle B_z \rangle = 0.5$ T and broadening due to nuclear moments and vortex disorder $\sigma_c = 0.7$ mT with a fit using the superposition of 3 Gaussians (red solid line). The dotted, dashed, and dash–dotted lines represent the individual Gaussian components used in the fit. Clearly the "real" field distribution is well described by a multi-Gaussian fit. Modified from Maisuradze et al. (2009), © IOP Publishing. Reproduced with permission. All rights reserved

6.3.1.3 Full Model Analysis

In this case, the relation between the field distribution $f(B_z)$ and the polarization function $P_x(t)$, Eq. 6.41, is used to analyze the μSR spectra. The field distribution, which depends on λ and ξ_{GL}, is obtained from the theoretical model for the spatial variation of $B_z(\mathbf{r})$ via Eq. 6.60. From the distribution a theoretical muon polarization function is generated and then used to fit the μSR spectra. Before that, the ideal theoretical distribution $f(B_z)$ is convoluted with a Gaussian distribution, i.e. following replacement of $f(B_z)$ takes place[62]

$$f(B_z) \;\Rightarrow\; \frac{\gamma_\mu}{\sqrt{2\pi}\,\sigma_c} \int_{-\infty}^{\infty} f(B_z') \exp\left[-\frac{\gamma_\mu^2 (B_z - B_z')^2}{2\sigma_c^2} \right] dB_z' \qquad (6.76)$$

The parameter σ_c takes into account possible broadening effects such as vortex disorder, which causes random displacement of the vortex core from the ideal position, the nuclear moments contribution and, in powder samples, the distribution

---

[62] This is equivalent to multiplying $P_x(t)$ with a Gaussian function $\exp(-\sigma_c^2 t^2/2)$.

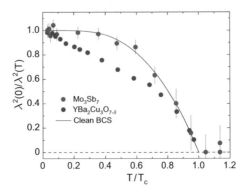

Fig. 6.39 Comparison between the normalized superfluid density $\lambda^2(0)/\lambda^2(T)$ in Mo_3Sb_7 (red points, s-wave) and $\lambda_{ab}^2(0)/\lambda^2(T)_{ab}$ in $YBa_2Cu_3O_{7-\delta}$ (blue, d-wave). In YBCO the field (0.5 T) is applied parallel to the c-axis, i.e. and supercurrents flow in the ab-planes. The solid line is a fit for a clean BCS s-wave superconductor. Note the different temperature dependence. Modified from Khasanov et al. (2008) and Sonier et al. (2000), © American Physical Society. Reproduced with permission. All rights reserved

of demagnetization factors (Brandt 1988b; Maisuradze et al. 2009). For the analysis, analytical solutions of the London or Ginzburg-Landau theories mentioned in Sect. 6.2.3 are used, sometimes even beyond their theoretical range of application. Figure 6.39 shows the temperature dependence of the normalized superfluid density for YBCO. The values were obtained by a full model fit of the μSR time spectra at 0.5 T with the analytical Ginzburg-Landau (AGL) model (Sonier et al. 2000, 1994).

The numerical iterative procedure for solving the GL equations introduced by Brandt (NGL) (Brandt 1997, 2003) has a wider range of applications than the analytical solutions. Figure 6.23 shows an example; the solid red lines are fits of the experimental field distribution obtained from the numerical solutions of the GL equations (Zhang et al. 2015).

Figure 6.39 compares the temperature dependence of the normalized superfluid densities for Mo_3Sb_7 and YBCO. The Mo_3Sb_7 data display the typical behavior of an s-wave superconductor with the almost flat low-temperature behavior of a gapped superconductor where thermal energy of the order of the gap must be provided to the system in order to break Cooper pairs, see Eqs. 6.34 and 6.35. A completely different behavior is observed in YBCO, Figs. 6.39 and 6.40, where the value of λ_{ab} versus temperature is plotted. At low applied field and low temperature, $\lambda_{ab}(T)$ exhibits the characteristic linear temperature dependence reflecting the presence of line nodes in the gap of a $d_{x^2-y^2}$-wave superconductor, see Eq. 6.37. The comparison demonstrates why TF-μSR measurements in the vortex state are routinely used to determine whether the superconducting gap has nodes or not, and to obtain information about the symmetry of the order parameter.

Another theoretical approach is to begin with a microscopic calculation of $\mathbf{B}(\mathbf{r})$ to obtain the field distribution and the characteristic parameters in the vortex state. The starting point is generally the Gor'kov formulation of the BCS theory, which can account for the formation of a FLL, see footnote 6. Specifically, the quasiclassical

Fig. 6.40 In plane penetration depth $\lambda_{ab}(T)$ for YBa$_2$Cu$_3$O$_{7-\delta}$. Black dots are μSR measurement. The line represents $\Delta\lambda_{ab}(T)$ measurements obtained with zero-field microwave absorption (Hardy et al. 1993) offset by the $\lambda_{ab}(0)$ value from μSR. Modified from Sonier et al. (1994), © American Physical Society. Reproduced with permission. All rights reserved

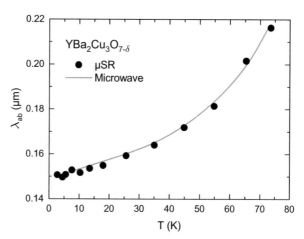

Eilenberger equations (Eilenberger 1968), which transform Gor'kov equations into a more tractable form, are solved numerically (Rammer 1991; Ichioka et al. 1999).[63] See Sect. 6.3.1.4 for some results from this approach.

6.3.1.4 Model Comparison

The validity of the different theoretical vortex state models and approaches used to analyze the μSR data has been tested with real data as well as with simulations (Maisuradze et al. 2009; Landau & Keller 2007; Laulajainen et al. 2006). A detailed comparison of simulated μSR spectra was performed in Maisuradze et al. (2009). For this, field distributions $f(B_z)$ for different values of λ, ξ_{GL} and $b \equiv \langle B_z \rangle / B_{c2}$ were first generated by using the exact numerical solution of the Ginzburg-Landau equation (NGL), (Brandt 1997, 2003) and then compared with different analytical models that we presented in Sect. 6.2.3: London model with Gaussian cutoff (LG), modified London model (ML), and analytical Ginzburg-Landau (AGL).

The comparison finds that for $\kappa > 5$ there is no advantage in using the AGL model instead of the simpler LG model. Among all models, the ML model best approximates the NGL results if the applied field is not too high $b < 0.1$ (a condition that is easier fulfilled in high-T_c superconductors, which have very high B_{c2}). There is a strong dependence of the μSR lineshape on λ as well as on ξ_{GL} with increasing B_{ext}. Care must be taken when trying to determine both length scales simultaneously, as they are strongly correlated. This is the case for all models and κ values.

For practical purposes, an important finding is that fitting the simulated experimental data[64] with the multi-Gaussian approach often used for μSR data analysis gives reliable results for λ (within a few percent) over the whole field range

---

[63] Note that one of the first μSR measurements of the vortex state in Nb was analyzed using the microscopic BCS-Gor'kov theory (Herlach et al. 1991).

[64] Adding statistical noise corresponding to 20 million events per histogram (a typical number of events in the spectra) and including typical broadening from nuclear moments and vortex disorder.

$0 < b < 1$, provided the vortex lattice is not too disordered. This shows that the field distribution of a FLL and its variance can be well approximated by a sum of a few Gaussian functions, see Fig. 6.38. The experimental values of σ_{sc} obtained in this way can be compared with the theoretical predictions for $\langle \Delta B^2 \rangle$ (e.g., Eqs. 6.64 or 6.65, derived from the Fourier decomposition of the spatial dependence of the local field.

In Landau and Keller (2007) experimental data have been reanalyzed. In particular the applied field dependence of the field variance, $\sigma_{sc}^2/\gamma_\mu^2$, was determined from the published data and compared with the NGL interpolation formula Eq. 6.65. It turns out that the results of several μSR measurements are very well described by this equation with a field independent λ, indicating that NGL calculations are a reliable and powerful tool to analyze μSR spectra. From the applied field dependence $\langle \Delta B_z^2(B_{ext}) \rangle$ not only λ but also $B_{c2}(T)$ can be extracted, see Sect. 6.3.1.1.

However, some remarks are in order. The theoretical calculations performed in the framework of London or GL theories and underlying Eqs. 6.59, 6.65 and 6.64 (Brandt 1988a, 1997, 2003) are for a superconductor with a single s-wave gap and are not generally valid.[65] When using these models outside their range of application (for instance for multigap or nodal superconductors, like many unconventional superconductors) special caution must be exercized.

Several experiments have reported an applied field dependence of the magnetic penetration depth (Sonier et al. 1999; Kadono et al. 2001; Kadono 2004). In this case, the extracted parameter $\lambda_{eff}(B_{ext})$, is not a true measure of the magnetic penetration depth and does not indicate a variation of the superfluid density with field. Rather, it points to shortcomings or omissions (e.g., anisotropy of the superconducting gap or vortex core, changes in the electronic structure of the vortex with respect to GL) of the employed theoretical model for $f(B_z)$, which is not able to describe all the relevant physics (Amin et al. 2000; Yaouanc et al. 1997). The true magnetic penetration depth corresponds to the value extrapolated to zero applied field $\lambda_{eff}(B_{ext} \to 0)$ (Laulajainen et al. 2006; Landau & Keller 2007; Sonier 2007).

As far as results from microscopic Gor'kov-like theories are concerned, microscopic calculations of the vortex structure in high-κ superconductors find that the "exact" Gor'kov-Eilenberger field distribution is very well approximated by the London model with Gaussian cutoff, provided that $B_{ext} \ll B_{c2}$ (Rammer 1991). Investigating the effect of an anisotropic gap, the local field B_z of a d-wave superconductor is found to have a fourfold symmetric shape around and in the vortex core. Of interest for the analysis of μSR spectra is the finding that at low fields the d-wave nature of the pairing hardly affects the outer core region. In particular, a comparison between an s- and a d-wave superconductor shows that the field distribution $f(B_z)$ is practically the same as long as $B_{ext} \ll B_{c2}$ (Ichioka et al. 1999). This indicates that, at not too high applied fields, the previously discussed results based on London and GL theory are still applicable to d-wave superconductors.

---

[65] These theories also assume a circularly symmetric vortex core and field decay away from it.

6.4   Interplay of Magnetism and Superconductivity

The phase diagrams of many unconventional superconductors show striking similarities, such as the presence of superconducting phases in the vicinity of magnetism.[66] Figure 6.41 shows as an example the phase diagram of some cuprates and Fe-based superconductor families. For both families, superconductivity emerges when doping the magnetic parent compound with electrons or holes, thereby suppressing the antiferromagnetic magnetic order. As Fig. 6.41 evidences, the disappearance of magnetism can be sudden, indicating a first order phase transition, or more gradual. In the latter case there are regions where both magnetism and superconductivity may coexist.

The exact role of magnetism with respect to superconductivity (phase separation, competitive or cooperative coexistence) is a matter of ongoing debate. μSR (and also neutron scattering) are useful probes for addressing these questions. With respect to neutron scattering, μSR has a better sensitivity to static magnetic order with randomly ordered (e.g., spinglass-like magnetism) or small moments and can clearly determine the volume fraction of magnetically ordered regions near the phase boundaries. Besides yielding information on the doping dependence of the transition temperatures and the respective order parameters, μSR can address the question of the phase coexistence at microscopic level.

The questions of whether and under what conditions magnetism occurs in high-T_c superconductors, and whether the two properties coexist under competing or cooperative conditions, are central to the understanding of unconventional superconductors. These questions can be addressed by a combination of ZF- and TF-μSR measurements.

As an example, we discuss the phase diagram of $YBa_2Cu_3O_{6+x}$ (Sanna et al. 2004), Fig. 6.41c.[67]

Polycrystalline samples with $0.20 \leq x \leq 0.42$[68] were investigated. The samples with low oxygen content $x \leq 0.32$ do not superconduct and show only antiferromagnetic behavior (Sanna et al. 2003). At low oxygen (or hole carrier) content two transitions of the magnetic ground state are clearly distinguishable. First, there is a long-range 3D-antiferromagnetic state at T_N. Furthermore, at a lower temperature T_f, there is a freezing of the spin degrees of freedom and the formation of an other antiferromagnetic state of spin glass cluster or stripe-like character, see Fig. 6.45. Relevant for a study of the coexistence of superconductivity and antiferromagnetism, are the samples with $x \geq 0.32$ (or $h \geq 0.055$), see Figs. 6.44 and 6.45.

---

[66] Proximity of magnetism, both anti- and ferromagnetism, to superconductivity is also found in heavy fermion systems.

[67] Immediately after the discovery of the cuprate superconductors, μSR measurements provided evidence of antiferromagnetic ordering in the insulating oxygen-deficient $YBa_2Cu_3O_{6+x}$, $x \approx 0.2$ (Nishida et al. 1987; Brewer et al. 1988).

[68] This corresponds to a hole content of $0.033 \leq h \leq 0.08$ per Cu in the CuO_2 planes, which are the layers where superconductivity takes place.

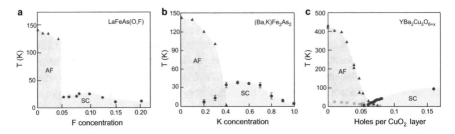

Fig. 6.41 Electronic phase diagrams as a function of composition in some high-T_c superconductors. (**a**) LaFeAs(O,F) (Luetkens et al. 2009); (**b**) (Ba,K)Fe$_2$As$_2$ (Chen et al. 2009); (**c**) YBa$_2$Cu$_3$O$_{6+x}$ (Sanna et al. 2004). All systems show an evolution from antiferromagnetism (AF) to superconductivity (SC). AF can abruptly disappear (**a**) or coexist with superconductivity near the phase boundary (**b, c**). (**a**) and (**c**) are based on μSR measurements, while (**b**) is based on neutron scattering. Modified from Uemura (2009), © Springer Nature. Reproduced with permission. All rights reserved

We first discuss the ZF data. In the paramagnetic phase, the μSR signal can be generically described by $A_z(t) = A_{ZF} P_{N,z}(t)$, where $P_{N,z}(t)$ describes the depolarization due to the presence of nuclear moments. Upon lowering the temperature, a superconducting or a magnetic transition may be crossed. For a superconducting transition, no specific signature is observed in the ZF measurement, which is insensitive to superconductivity.[69] Conversely, magnetic order implies that a well defined or a distribution of internal fields $\mathbf{B}_\mu$ appear, which are sensed by the muon. This gives rise to a signal in the ZF spectrum of the form

$$A_z(t) = A_L \, P_z(t) + A_T \, P_x(t) \;, \tag{6.77}$$

where $A_L + A_T = A_{ZF}$. This expression distinguishes the longitudinal component ($\mathbf{B}_\mu \parallel \mathbf{I}_\mu$, with amplitude A_L) and the transverse component ($\mathbf{B}_\mu \perp \mathbf{I}_\mu$, with amplitude A_T) with respect to the local field.

In a polycrystalline magnetic sample we expect for the weights $w_L \equiv A_L/A_{ZF} = 1/3$ and $w_T \equiv A_T/A_{ZF} = 2/3 = 2w_L$, see Eq. 4.11. If magnetism is present in only a part of the sample $w_L > 1/3$ and $w_T < 2/3$. The sample volume fraction where the muon experiences a local magnetic field is therefore given by

$$f_{AF} = \frac{3}{2} \frac{A_T}{A_{ZF}} = \frac{3}{2} \left(1 - \frac{A_L}{A_{ZF}} \right) \;. \tag{6.78}$$

---

[69] We do not consider here the case of TRSB, where superconductivity induces very small spontaneous magnetic fields, see Sect. 6.6.

Fig. 6.42 μSR results in ZF for a YBCO sample with 0.070 holes per planar Cu; **(a)** Normalized asymmetries at $T/T_f \approx 0.17, 0.96, 1.4, 3.2$; **(b)** Longitudinal and transverse weights $w_L(T)$, $w_T(T)$ with fit curves; **(c)** Longitudinal relaxation rate, λ_L, vs. temperature, peaking at T_f. Modified from Sanna et al. (2004), © American Physical Society. Reproduced with permission. All rights reserved

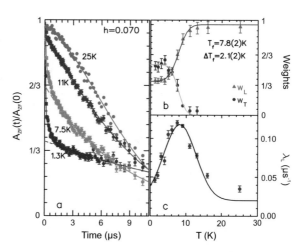

Depending on the type of order (long- or short-range), the function P_x can be represented by one (or more)[70] damped precession signal(s), if the local field is well defined (long-range order), or by an overdamped signal without clear oscillations if short-range order is present.

Figure 6.42 shows the ZF results for a superconducting sample ($h = 0.07$) with $T_c = 30$ K. The presence of static magnetism[71] is demonstrated by the appearance of a strong depolarization below ~ 11 K, Fig. 6.42a, reflecting a distribution of static local fields. Moreover, for $T \rightarrow 0$, w_L is very close to the $1/3$ value and w_T to $2/3$, which is expected for a fully magnetically ordered sample: see the dashed line in Fig. 6.42a for w_L and Fig. 6.42b for the temperature dependence of w_L and w_T.

The magnetic transition is also reflected in the peak of the longitudinal relaxation rate λ_L, obtained by fitting Eq. 6.77 with $P_z(t) = \exp(-\lambda_L t)$. All these features indicate that below a transition temperature $T_f \approx 8 K$, the entire sample is magnetic. Since, as mentioned above, a ZF measurement is insensitive to superconductivity, these results do not exclude the simultaneous presence of superconducting pairs in the sample.

Superconductivity can be investigated by TF measurements as a function of temperature. In the normal state the μSR spectrum shows a precession in the external field, weakly damped by the nuclear moments: $A_x(t) = A_0 \exp\left(-\sigma_N^2 t^2/2\right) \cos\left(\gamma_\mu B_{ext} t\right)$. Below T_c, when the flux-line lattice is formed,

---

[70] In the long-range AF phase for $x \lesssim 0.3$ two distinct local fields, corresponding to two muon sites, are detected (Sanna et al. 2003).

[71] It can be decoupled with longitudinal fields of some tens of mT.

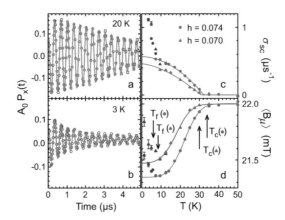

Fig. 6.43 Results of TF measurement with $B_{ext} = 22$ mT. (**a**) Asymmetry for $T_f < T = 20$ K $<$ T_c and (**b**) for $T = 3$ K $< T_f$ in a YBCO sample with $h = 0.070$. The solid curves are best fits to a Gaussian damped precession. (**c**) Relaxation rate σ_{sc} and (**d**) internal field $\langle B_\mu \rangle$ fitted with Eq. 6.79 for samples with $h = 0.074$ and $h = 0.070$. Modified from Sanna et al. (2004), © American Physical Society. Reproduced with permission. All rights reserved

the data can be fitted with[72]

$$A_x(t) = A_{TF} \exp\left[-\frac{(\sigma_{sc}^2 + \sigma_N^2)t^2}{2}\right] \cos\left(\gamma_\mu \langle B_{\mu,z} \rangle t\right) . \tag{6.79}$$

The Gaussian damping with variance σ_{sc}^2 reflects the Gaussian field distribution expected in a powder sample; $\langle B_{\mu,z} \rangle$ is the average field in the vortex state.

The results for a sample with hole content $h = 0.73$ are shown in Fig. 6.43. In the panels (c) and (d) the temperature dependence of σ_{sc} and $\langle B_{\mu,z} \rangle$ displays the characteristic increase of the variance at T_c and the concomitant diamagnetic shift of the local field $\langle B_{\mu,z} \rangle = B_{ext}(1 + \chi)$, with $\chi < 0$. Both parameters increase below T_f, revealing the onset of magnetic order. The signal amplitude in the normal phase A_0 is very close to the amplitude below T_c, see Fig. 6.43a, b. From this it follows that the volume fraction of the sample corresponding to the superconducting state, given by $f_{sc} \equiv A_{TF}/A_0$, is very close to one.

Combining the results from ZF- and TF-μSR leads to the conclusion that there is a region of the phase diagram (see in Fig. 6.44), where $f_{AF} \cong f_{sc} \cong 1$, meaning that practically all implanted muons detect a flux lattice for $T_f < T < T_c$ and at the same

---

[72] More precisely, Eq. 6.79 reflects the vortex state only in the interval $T_f < T < T_c$. Below T_f, σ_{sc} and $\langle B_{\mu,z} \rangle$ are affected by the presence of magnetism, see Fig. 6.43c, d. If the magnetism is strong it wipes out the signal from the FLL. In other cases, a detailed analysis of the spectra can disentangle the two contributions (Savici et al. 2002).

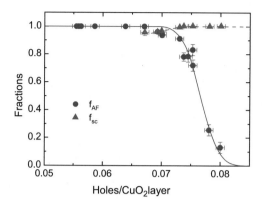

Fig. 6.44 Volume fractions of the antiferromagnetic and superconducting phases vs. hole content in $YBa_2Cu_3O_{6+x}$ from ZF and TF measurements. Blue circles: Antiferromagnetism at $T = 0$. Red triangles: Superconductivity for $T_f \leq T \leq T_c$ (SQUID measurements indicate that bulk superconductivity is present down to $T \approx 0$). Lines are guides to the eye. Modified from Sanna et al. (2004), © American Physical Society. Reproduced with permission. All rights reserved

time a local antiferromagnetic field for $T < T_f$. Furthermore, SQUID measurements (not shown) indicate that the bulk superconductivity persists also below T_f.

These findings can be understood in the following way. First, we remark that μSR probes superconductivity with a spatial resolution more than 100 times coarser than that for static magnetism. The magnetic field in the vortex structure varies on a length scale of the order of the magnetic penetration depth λ, typically a few hundred nm, so that any inhomogeneity of the superconducting properties on shorter length scales is averaged out. On the other hand, each muon site is subject to a local magnetic field (mainly of dipolar origin) if it is within a distance R from an electronic moment μ_j. Taking $\mu_j \approx 0.6\,\mu_B$, one estimates $R \approx 2$ nm. The magnetic ground state may still be inhomogeneous, but only on the length scale of the μSR experiment of a few nanometers. These nanoscopic magnetic clusters must be present within the superconducting volume.

The picture of a nanoscale coexistence of AF clusters (or stripes) and superconducting pairs with minimal reciprocal interference is supported by several μSR experiments on cuprates and other unconventional superconductors. The phase diagram of $YBa_2Cu_3O_{6+x}$ obtained from the experiment of (Sanna et al. 2004) is shown in Fig. 6.45 together with the similar diagram of $La_{2-x}Sr_xCuO_4$ (Niedermayer et al. 1998).

Fig. 6.45 Phase diagram of YBa$_2$Cu$_3$O$_{6+x}$ (solid lines) and of La$_{2-x}$Sr$_x$CuO$_4$, (dotted lines) from Niedermayer et al. (1998). Lines are guides to the eye. Modified from Sanna et al. (2004), © American Physical Society. Reproduced with permission. All rights reserved

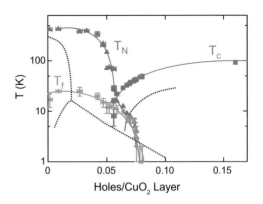

6.5   Study of Vortex Matter

The vortex state of a superconductor represents a unique state of the solid and can be compared to a crystal and its lattice. One speaks of vortex matter, where interestingly the "lattice constant" can be changed by an external field (see Eq. 6.49) and the temperature varied in a wide range, thus tuning the interaction between the vortices.

Differently to the rigid flux lines which are found in isotropic superconductors, in high-T_c and other unconventional superconductors, the combination of extreme anisotropy, thermal fluctuations (important since T_c is large), and pinning induces disorder and may lead to very flexible flux lines and complex vortex structures that can be described in the field-temperature diagram as solid-liquid or ordered-disordered transitions (Blatter et al. 1994).

The first microscopic investigations that demonstrated the melting of the flux-line-lattice of the highly anisotropic cuprate Bi$_{2.15}$Sr$_{1.85}$CaCu$_2$O$_{8+\delta}$[73] (BSCCO, $T_c \simeq 84$ K) were obtained with μSR (Lee et al. 1993) and confirmed by small-angle neutron diffraction (Cubitt et al. 1993b). Measurements as a function of magnetic field[74] and temperature show characteristic changes in the field distribution that can be associated with transitions in the flux vortex structure.

The changes in the field distribution (and in the associated vortex structure) can be quantified by the dimensionless skewness parameter α_{sk} derived from the third and second central moments of the field distribution and which represents a measure of the asymmetry of the distribution $f(B_{\mu,z})$. The skewness parameter is defined as

$$\alpha_{sk} = \frac{\langle \Delta B_{\mu,z}^3 \rangle^{1/3}}{\langle \Delta B_{\mu,z}^2 \rangle^{1/2}} \ , \tag{6.80}$$

---

[73] The sample consisted of a mosaic of single crystals.

[74] The magnetic field was applied parallel to the c-axis of the crystals.

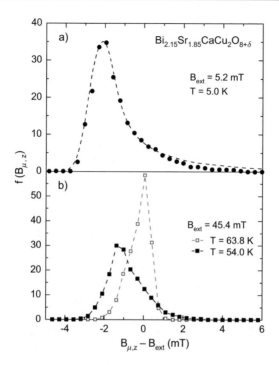

Fig. 6.46 Local field distribution in $Bi_{2.15}Sr_{1.85}CaCu_2O_{8+\delta}$. (**a**) $B_{ext} = 5.2$ mT and $T = 5.0$ K. The lineshape is asymmetric and exhibits features characteristic of a FLL structure. This is confirmed by the good agreement with the simulation (dashed line) of an ideal triangular lattice with magnetic penetration depth $\lambda = 160$ nm. The theoretical curve has been convoluted with a Gaussian of field width 0.24 mT to simulate the effects of lattice imperfection and instrumental resolution. (**b**) $B_{ext} = 45.4$ mT, for i) $T = 54.0$ K (solid squares) and ii) at $T = 63.8$ K (open red squares). The dashed lines are guides to the eye. The distribution at $T = 63.8$ K shows a significantly different lineshape. Modified from Lee et al. (1993), © American Physical Society. Reproduced with permission. All rights reserved

where

$$\langle \Delta B_{\mu,z}^n \rangle = \int_0^\infty (B_{\mu,z} - \langle B_{\mu,z} \rangle)^n \, f(B_{\mu,z}) dB_{\mu,z} \; . \tag{6.81}$$

Figure 6.46a displays $f(B_{\mu,z})$ obtained by Fourier transforming the μSR spectrum measured after cooling the sample to 5 K in an applied field of 5.2 mT. The measured local field distribution shows the typical features of a 3D flux-line lattice, as the high field tail, see Fig. 6.22, and can be well represented by a calculation of the field distribution in the lattice.[75]

---

[75] In an ideally arranged triangular vortex lattice $\alpha_{sk} = 1.2$.

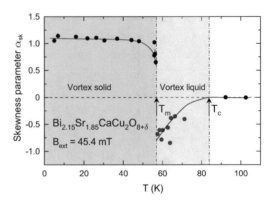

In Fig. 6.47 α_{sk} is plotted as a function of temperature for an applied field of 45.4 mT. At a temperature around 59.5 K, α_{sk} abruptly drops and changes within less than 2 K, indicating the existence of a phase transition in the vortex structure from solid to liquid, where the correlations between the vortex core positions are lost. Temperature scans at several fields from 5.2 to 401 mT allow to determine the transition line in the B_{ext} vs. T diagram. This line is found to be close to the so-called irreversibility line, below which hysteretic behavior is observable.[76]

Another change of the vortex structure is found in field scans at fixed temperature (after field-cooling in each field). This change is not as drastic as the previous one because it only involves a change of magnitude of α_{sk}, not of sign. This was first interpreted as a dimensional crossover from a 3D to a less ordered 2D solid state, later found to be better characterized not only by short-ranged correlations perpendicular to the conduction planes but also by nontrivial correlations within the planes, and named "vortex glass" (Menon et al. 2006; Heron et al. 2013).

The phase diagram of BSCCO, which can be considered a prototype vortex matter system, has been later reexamined with additional μSR data (Heron et al. 2013). The data were analyzed within a theoretical framework making use of liquid state and DFT calculations and relating the second and third central moments to two- and three-body spatial correlations of vortices (Menon et al. 2006). This, together with magnetization data, allows to refine the magnetic phase diagram of BSCCO, which is shown in Fig. 6.48. Besides the vortex lattice (region 1) and vortex liquid phases (region 4) inferred from the previous μSR investigations, an unusual glassy state can be identified at intermediate fields and below the irreversibility line. It appears to freeze continuously from the liquid state, but differs from the lattice and the conventional high-field vortex glass (region 3) and is termed pinned liquid state, region 2 in Fig. 6.48.

---

[76] The irreversibility line has crucial importance for high-current, high-field applications of super-conductors. This line separates the phase diagram into the regions of zero and of finite resistive dissipation and defines the boundary below which a superconductor can carry supercurrents without losses.

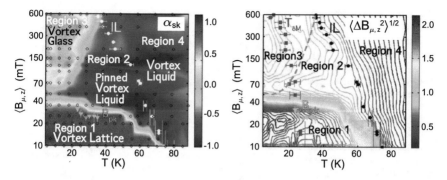

Fig. 6.48 Variations across the B_{ext}–T phase diagram of (**a**) Skewness parameter $\alpha_{sk} = \langle \Delta B_{\mu,z}^3 \rangle^{1/3} / \langle \Delta B_{\mu,z}^2 \rangle^{1/2}$ and (**b**) Field width $\langle \Delta B_{\mu,z}^2 \rangle^{1/2}$. The open black circles are the coordinates of the μSR data points. One distinguishes four regions. Region 1: $\alpha_{sk} > 0$ and close to one, high field width: vortex lattice. Region 2: $\alpha_{sk} < 0$, small field width: lines soften to a glassy arrangement of pancake vortices, with correlations maintaining a liquidlike character. Region 3: $\alpha_{sk} > 0$, small field width: glass of pancake vortices. Region 4: $\alpha_{sk} < 0$, low field width: liquid state above the macroscopic irreversibility line. Solid white or black circles: irreversibility line from bulk magnetization. This line is closely related to the contours of the width above which the gradient increases. Modified from Heron et al. (2013), © American Physical Society. Reproduced with permission. All rights reserved

6.5.1   Vortex Pinning

In order to extract characteristic lengths and microscopic information about a superconductor from the temperature and field dependence of the field width (by using Eqs. 6.59, 6.65, 6.64, or the full field distribution $f(B_{\mu,z})$), a periodic rigid lattice of flux lines must be established in the bulk of the superconductor. For this, in the experiment, the field is applied above T_c and then the sample cooled below T_c, so that the magnetic flux can be trapped in the sample by pinning. If the FLL signal does not change when the applied field is reduced after the field-cooling process, then the vortices are pinned.

Figure 6.49 compares the result of this procedure for YBa$_2$Cu$_3$O$_{6.95}$ Sonier et al. (2000). Clearly, only the narrow background peak corresponding to muons stopping outside the sample shifts from 1.5 to 1.48 T. In contrast, the signal corresponding to the FLL in the sample does not follow the change of B_{ext}, indicating that the vortex is rigidly pinned.[77]

---

[77] Generally, pinning in high-T_c superconductor is weaker than in conventional superconductors. However, it can be sizeable, due to point defects (e.g., oxygen vacancies in cuprates) or twin boundaries.

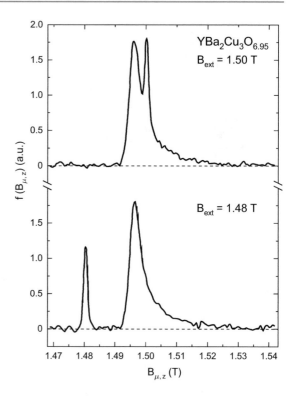

Fig. 6.49 Field distribution obtained by Fourier transforming the μSR spectra in detwinned $YBa_2Cu_3O_{6.95}$ crystals. Top panel: After field-cooling to 5 K in a field $B_{ext} = 1.5$ T. Bottom panel: After further lowering the field by 0.02 T while the sample is kept at 5 K. Modified from Sonier et al. (2000), © American Physical Society. Reproduced with permission. All rights reserved

6.6   Spontaneous Magnetic Field in Superconductors

Magnetism is detrimental to conventional superconductors, where magnetic impurities act as pair-breakers. Conversely, in several families of unconventional superconductors, superconductivity can coexist with magnetism and magnetic fluctuations are thought to mediate the pairing interaction similarly as phonons do in conventional superconductors, see Sect. 6.4.

In a limited but growing number of superconductors, weak magnetism is induced by the superconductivity itself. In these superconductors, at T_c, in addition to breaking the global gauge (or phase) symmetry,[78] an additional symmetry is broken, namely the time-reversal symmetry.[79]

Time-reversal symmetry breaking (TRSB) leads to the appearance of spontaneous weak magnetic fields below T_c. Most of the TRSB superconductors have been detected by μSR and in some cases by the observation of the Kerr effect (Xia et al. 2006) (for an overview of superconductors with time-reversal symmetry breaking

---

[78] The electrons, which have arbitrary phases in the normal state, condense into a single-phase macroscopic wave function below T_c.

[79] The time reversing operation transforms a state $|\mathbf{k}, \uparrow\rangle$ into $|-\mathbf{k}, \downarrow\rangle$.

Fig. 6.50 Muon spin relaxation rate $\sigma(T)$ in zero field of 1T-TaS$_2$4Hb-TaS$_2$. The increase at $T_c = 2.7$ K marks the onset of time-reversal symmetry breaking. Modified from Ribak et al. (2020), © American Association for the Advancement of Science. Reproduced with permission. All rights reserved

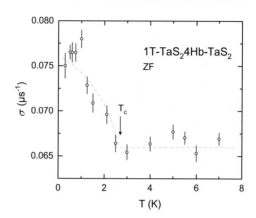

and their properties, see Ghosh et al. (2020b); for Re-based TRSB superconductors, see Shang and Shiroka (2021)).

Muon spin relaxation measurements are especially useful because they can detect internal magnetic fields as small as 10^{-5} T without applying an external magnetic field. The spontaneous field at the muon probe would be created by the spin moments of the Cooper pairs and/or by spontaneous supercurrents at inhomogeneities such as defects, edges or surfaces. The field is detected as an increase of the muon spin relaxation below T_c in zero field measurements. Special care must be taken to reduce stray and earth magnetic fields at the sample position as much as possible.[80]

In ZF, above T_c, the depolarization is mainly determined by the randomly oriented nuclear moments, leading to a Kubo-Toyabe polarization function with temperature independent σ_N. The increased relaxation in the superconducting state is found to be Gaussian or exponential in character. The Gaussian case corresponds to a dense distribution of TRSB sources. In this case, the asymmetry is fitted by the Kubo-Toyabe function, Eq. 4.19, with a Gaussian relaxation rate σ, which is allowed to vary with temperature

$$A(t) = \frac{1}{3} + \frac{2}{3}(1 - \sigma^2 t^2) \exp\left(-\frac{\sigma^2 t^2}{2}\right) \ . \tag{6.82}$$

Figure 6.50 shows the results of ZF measurements of 1T-TaS$_2$4Hb-TaS$_2$, (Ribak et al. 2020). This compound is a transition metal dichalcogenide which is naturally realized as an alternating stacking of two-dimensional monolayers consisting of the Mott insulator 1T-TaS$_2$ and the superconductor 1H-TaS$_2$. The plot of σ versus T in

---

[80] This can be done by pairs of Helmoltz coils actively compensating the three coordinates of the residual field. An additional procedure is to use a calibration sample in which muonium formation takes place. Its ~ 100 times larger effective gyromagnetic ratio, see Sect. 7.3.3, makes muonium correspondingly more sensitive to small residual fields than the muon.

Fig. 6.50 shows a constant value for $T \geq T_c$ and an increasing rate that sets in at $T_c = 2.7$ K . This reflects the appearance of weak magnetism in the superconducting state. From $B_{TRSB} = \sqrt{\sigma^2(T=0) - \sigma^2(T>T_c)}/\gamma_\mu$ we can estimate the field strength to be about 4.5×10^{-5} T.

In most observations of TRSB by μSR, the asymmetry is fitted with the product of a Gaussian Kubo-Toyabe function (with the temperature independent σ_N describing the nuclear moment contribution) with an exponential relation to account for the spontaneous time-reversal symmetry breaking field

$$P_t(t) = \left[\frac{1}{3} + \frac{2}{3} \left(1 - \sigma_N^2 t^2 \right) \exp\left(-\frac{\sigma_N^2 t^2}{2} \right) \right] \exp\left(-\lambda t\right) . \qquad (6.83)$$

Figures 6.51 and 6.52 show as an example ZF spectra and the temperature dependence of λ recorded in LaNiC$_2$ (Hillier et al. 2009). The non-centrosymmetric LaNiC$_2$ does not order magnetically, but exhibits superconductivity at 2.7 K. The onset of the spontaneous magnetic fields appears as an additional exponential relaxation $\lambda(T)$ below T_c. The right-hand side of Fig. 6.52 shows that the muon spin can be decoupled from its local environment by a small longitudinal field and indicates that the internal fields are static on the μSR time scale. The exponential relaxation indicates that the sources of the TRSB follow a dilute distribution. From the increase of the exponential relaxation by ≈ 0.01 μs^{-1} below T_c, as in the previous example, a characteristic field strength of about 1.2×10^{-5} T is deduced.

Spontaneous magnetic fields interpreted as signature of TRSB have been detected by ZF-μSR in a number of unconventional superconductors. It has been suggested that the perturbation of the muon charge on its environment may lead to a false TRSB signature (Willa et al. 2021). This question has been systematically

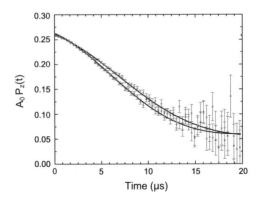

Fig. 6.51 ZF μSR spectra in LaNiC$_2$. The lines are least squares fits of Eq. 6.83 to the data. The red symbols are the data collected above T_c (at 3.0 K). The blue symbols are the data collected below T_c (at 54 mK) showing a small but significant increase of the muon spin depolarization. Modified from Hillier et al. (2009), © American Physical Society. Reproduced with permission. All rights reserved

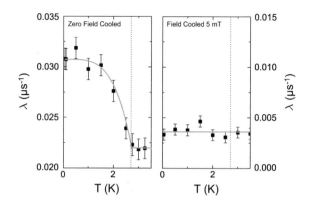

Fig. 6.52 Left: temperature dependence of the ZF relaxation rate λ in LaNiC$_2$. The data clearly show the increase of λ due to the spontaneous fields appearing at $T_c = 2.7$ K (dotted line). Right: Temperature dependence of λ for an applied longitudinal field of 5 mT. The flat temperature dependence indicates that the sources of the muon spin relaxation are static. The red lines are guides to the eye. Modified from Hillier et al. (2009), © American Physical Society. Reproduced with permission. All rights reserved

investigated using DFT calculations (Huddart et al. 2021). The calculations show that the positive muon induces only limited localized changes on the structural arrangement of nearby atoms or ions and, most importantly, only very small changes on the local electronic structure. These calculations imply that the detected spontaneous magnetic fields are of intrinsic nature.

The most prominent (but also intriguing) example of TRSB is probably strontium ruthenate Sr$_2$RuO$_4$ (SRO) (Luke et al. 1998), whose superconducting state's nature is still object of discussion (Pustogow et al. 2019).

In most superconductors, the electrons forming the Cooper pair are in a singlet spin and an even orbital state, see also Sect. 6.1.4. This is the case for conventional superconductors, which have $l = 0$ (s-wave), and for unconventional superconductors (such as cuprates and iron-based), which have $l = 2$ (d-wave). SRO, discovered in 1994 (Maeno et al. 1994), is a very special unconventional superconductor ($T_c \simeq 1.5$ K). Despite having a layered crystal structure practically identical to that of the cuprate LSCO family, it has very different electronic and superconducting properties. The detection of a spontaneous magnetic field by ZF-μSR (Luke et al. 1998) indicated that the pairing order of the superconducting state of SRO breaks the time-reversal symmetry. For a long time, the combined experimental evidence and

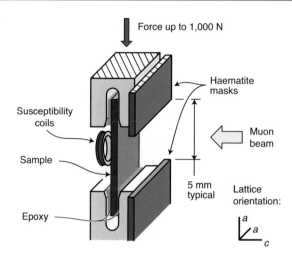

Fig. 6.53 Sample setup for ZF measurements of Sr_2RuO_4 under uniaxial stress of the ab-planes. The concentric coils behind the sample provide an in situ measurement of T_c. The haematite masks suppress the signal of muons from the holder sections in the beam. The strong antiferromagnetism of haematite quickly (50 ns) depolarizes the muon stopped in the masks. Modified from Grinenko et al. (2021b), © Springer Nature. Reproduced with permission. All rights reserved

theoretical considerations supported the view of SRO as a chiral[81] superconductor with spin triplet coupling and order parameter with $k_x \pm i\, k_y$ symmetry (odd p-wave parity) (Mackenzie and Maeno 2003; Kallin 2012).

However, recent experiments have cast doubt about the triplet pairing and p-wave assignment and therefore on the chiral character of SRO (Mackenzie et al. 2017). The lack of consensus on the superconducting properties of SRO has motivated a reconsideration of this fundamental issue.

The high sensitivity of ZF-μSR to weak magnetic fields has been used to measure SRO samples under uniaxial stress (Grinenko et al. 2021b) and hydrostatic pressure (Grinenko et al. 2021a). A piezoelectric-driven device (Ghosh et al. 2020a) was used to apply in-plane uniaxial stress to SRO crystals, thus lifting the tetragonal lattice symmetry of Sr_2RuO_4, see Fig. 6.53. In the presence of an order parameter with two components as is the case of $k_x \pm i\, k_y$, the degeneracy of the k_x and k_y components is expected to be lifted by the stress, which should become manifest

---

[81] A chiral superconductor is a superconductor where the phase of the complex order parameter $\Delta(\mathbf{k})$ winds up clockwise or counterclockwise as the wave vector $\mathbf{k}$ moves around an axis of the Fermi surface. $k_x \pm i\, k_y$, where the phase changes by $\pm 2\pi$ when $\mathbf{k}$ travels along a closed path around the k_z-axis, is the simplest example of chiral order. The combination of two or more order parameter components with complex coefficients (as is the case for $k_x \pm i\, k_y$) leads to the formation of a TRSB superconducting state.

Fig. 6.54 Temperature dependence of the ZF muon spin relaxation rate λ (data points with error, lines to guide the eyes) and of the diamagnetic susceptibility measured in situ (black solid line) for an SRO crystal at 0, −0.43, and −0.70 GPa . The temperature dependence of λ is obtained from an exponential fit of the asymmetry. The minus sign of the pressure values denotes the compression effect of the sample volume. Modified from Grinenko et al. (2021b), © Springer Nature. Reproduced with permission. All rights reserved

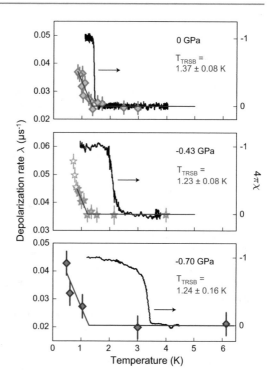

as a splitting of the phase transition temperature (Sigrist & Ueda 1991). The data, see Fig. 6.54, clearly show that the stress induces a splitting between the onset temperature of superconductivity T_c (obtained from the in situ susceptibility measurement) and the onset temperature of TRSB (obtained from the enhancement of the ZF muon spin relaxation). Transverse field measurements show that the sample is fully superconducting. No splitting is observed under hydrostatic pressure (Grinenko et al. 2021a). The results provide an important piece of evidence for a two-component superconducting order parameter of chiral character in SRO. While strongly supporting the chiral TRSB character of SRO, the μSR data do not allow to distinguish between spin singlet odd parity (such as $k_x \pm i\,k_y$) and spin triplet even parity order parameters (such as $d_{xz} \pm i\,d_{yz}$).

The μSR method has established itself as an efficient, reliable and sensitive probe to confirm or reject the presence of a TRSB state, which has important physical implications. For example, the observation of time reversal symmetry conservation in the topological superconductor[82] T_d-MoTe$_2$ (an orthorhombic two-dimensional

---

[82] A topological superconductor is a superconductor with nontrivial electronic states consisting of a full pairing gap in the bulk and gapless states at the surface composed of so-called Majorana fermions. These fermions (as opposed to the conventional Dirac fermions, see Appendix G.2) were originally proposed by E. Majorana as elementary particles that are their own antiparticles. In condensed matter they are the quasiparticle analogues.

transition metal dichalcogenide) has made this compound a singular candidate for a material that is both a topological superconductor and a time-reversal invariant (Guguchia et al. 2017).

6.7   Study of the Intermediate State

As a local probe, μSR is well suited to study inhomogeneous states such as the vortex state of a type-II superconductor. In fact, the vast majority of μSR experiments on superconductors have been devoted to this case. Less work has been dedicated to the study of the Meissner phase of type-I and type-II superconductors.[83] In particular, very little work has been done on the intermediate state (Egorov et al. 2001).[84]

The capabilities of the μSR technique are demonstrated in a recent study of the conventional type-I superconductor β-Sn (Karl et al. 2019) which shows that the full B_{ext} vs. T phase diagram can be reconstructed with a precise determination of the thermodynamic critical field as well as of the separation of the sample volume into normal and superconducting phases. Due to the local nature of the μSR technique, B_c values are determined directly from measurements of the internal field inside the normal state domains, see Sect. 6.1.3.

A very pure cylindrical sample of Sn (diameter 20 mm and height 100 mm) was used as a probe and a transverse field applied perpendicular to the cylindrical axis. In this geometry the demagnetization factor N is close to $1/2$.[85] Several series of measurements have been performed, B_{ext}-scans at fixed T, T-scans at fixed B_{ext}, and scans where applied field and temperature are simultaneously varied along a well-defined line in the middle of the phase diagram region of the intermediate state.

The asymmetry spectra are analyzed by including a sample (s) and a background (bg) contribution

$$A_0 P(t) = A_s P_s(t) + A_{bg} P_{bg}(t) \ . \tag{6.84}$$

---

[83] This remark does not apply to experiments with low-energy muons, which are ideally suited for Meissner state investigations, see Chap. 8.

[84] More recently μSR has been utilized to provide solid evidence for type-I superconductivity in binary compounds, which, when superconducting, generally exhibit type-II behavior, and whose properties hinted at the presence of unconventional superconductivity. Examples are $PdTe_2$ (Leng et al. 2019), a layered compound with unique and unconventional topological electronic states, and the non-centrosymmetric BeAu (Beare et al. 2019) .

[85] The μSR experiment also provides the exact value of N for the sample used, see below.

The relevant term is the sample contribution, which describes the possible separation between normal (n) and superconducting (sc) domains[86]

$$P_s(t) = f_n \exp(-\lambda_n t) \cos(\gamma_\mu B_n t + \varphi)$$

$$+ (1 - f_n) \left[\frac{1}{3} + \frac{2}{3}(1 - \sigma_N^2 t^2) \right] \exp\left(-\frac{\sigma_N^2 t^2}{2} \right). \tag{6.85}$$

The first term on the right-hand side of Eq. 6.85 describes the normal state part of the sample: f_n is the normal state volume fraction ($f_n = 1$ for $T > T_c$), λ_n accounts for an exponential depolarization, and B_n is the local field. $B_n = B_c$ for $T < T_c(B_{ext})$ and $B_n = B_{ext}$ for $T > T_c(B_{ext})$. The second term describes the fraction of the superconducting domains ($f_{sc} = 1 - f_n$) of the sample in the Meissner state. In these domains the local field is expelled, i.e., $B_{sc} = 0$. This contribution is approximated by a Gaussian Kubo-Toyabe function with a rate σ_N describing the relaxation due to the nuclear magnetic moments in zero field.

Figure 6.55 shows examples of asymmetries and field distributions from a temperature scan of the Meissner, intermediate, and normal states ($B_{ext} = 15$ mT). The data exhibit the expected signatures from a phase diagram such as that of Fig. 6.12. For instance, at $T = 0.90$ K, there is a pronounced sharp peak at $B_{\mu,z} = 0$ and only a weak peak at $B_{\mu,z} = B_c \simeq 29$ mT, Fig. 6.55d. This corresponds to the sample remaining in the Meissner state, where the field in most of the sample volume is zero, while only a small fraction of the sample is in the intermediate state.

At $T = 2.25$ K the Sn sample is clearly separated into normal state and superconducting domains. From the fit $f_n = 0.562(2)$, thus indicating that the normal state domains occupy more than half of the sample volume, and $B_n = B_c = 18.76(2)$ mT.

Finally, at $T = 2.70$ K $B_n = B_{ext}$. At $B_{\mu,z} = 0$ no peak is present. This indicates that at $T = 2.70$ K and $B_{ext} = 15$ mT the Sn sample is in the normal state. The temperature dependence of the thermodynamic critical field can be discerned from the two-dimensional plot of the Fourier transform of a T-scan, Fig. 6.56.

From the data a critical temperature value $T_c(B_{ext}=0) = 3.717(3)$ K and a zero-temperature value of the thermodynamic critical field $B_c(T=0) = 30.578(6)$ mT are extracted. Analyzing the temperature dependence $B_c(T)$ with the α-model (Johnston 2013), see page 244, one obtains various material parameters.[87]
Particularly interesting is the specific information provided by the local character of μSR. Thus, the field and temperature dependence of the sample volume fraction

---

[86] The background term corresponding to a damped precession in the applied field is not discussed here. For the fit, parameters which are common to the measured points are taken as global parameters (e.g., A_s, A_{bg}, or σ_N) and only those dependent on the experimental tuning parameters (T or B_{ext}) are taken as individual parameters for each data point (e.g., f_n, B_n). This is the preferred procedure for similar data sets.

[87] For the superconducting energy gap $\Delta(0) = 0.59(1)$ meV, for the electronic specific heat $\gamma_e = 1.781(3)$ mJ/(mol K^2), and for the jump in the heat capacity at the transition temperature $\Delta C(T_c)/\gamma_e T_c = 1.55(2)$. All these quantities are found in good agreement with literature values.

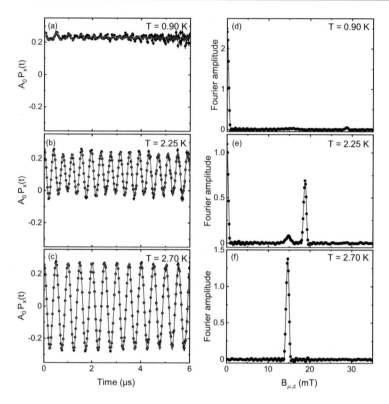

Fig. 6.55 (a)–(c): TF-μSR spectra in β-Sn taken at fixed $B_{ext} = 15$ mT and different temperatures (T-scans): panel (a) $T = 0.90$ K, panel (b) 2.25 K and panel (c) 2.70 K. The red lines are fits of Eq. 6.85 to the data (with background contribution). Panels (d)–(f): Magnetic field distribution obtained by Fourier transform of the time spectra (a)–(c). See text for details. Modified from Karl et al. (2019), © American Physical Society. Reproduced with permission. All rights reserved

in normal state f_n can be determined and scaling properties of this quantity can be established.

From Eq. 6.24, the value of the applied field at which $f_n \rightarrow 0$ (i.e., when, on lowering the field, the intermediate state disappears and a pure Meissner state sets in) corresponds to $B_{ext} = (1 - N)B_c$. This allows the experimental determination of the demagnetization factor N, which is found in very good agreement with the theoretical value $N_{th} = 0.469$.[88]

---

[88] According to Prozorov et al. (2018) N_{th} for a finite cylinder in a magnetic field applied perpendicularly to its axis is given by

$$\frac{1}{N_{th}} = 2 + \frac{D}{\sqrt{2}L} \,, \tag{6.86}$$

where D and L are the diameter and the length of the cylinder, respectively.

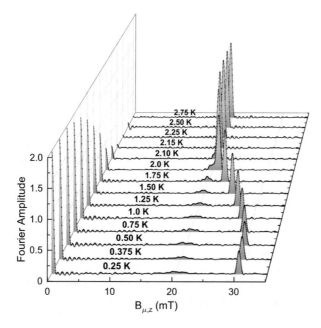

Fig. 6.56 Magnetic field distribution in the β-Sn sample at different temperatures at $B_{\text{ext}} = 20.0$ mT. The peak at the applied field position corresponds to the background contribution. Modified from Karl et al. (2019), © American Physical Society. Reproduced with permission. All rights reserved

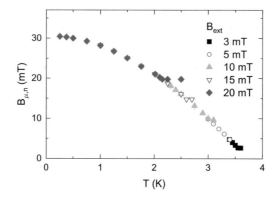

Fig. 6.57 Temperature dependence of the internal field in the normal state domains measured by μSR at different applied fields $B_{\text{ext}} = 3.0, 5.0, 10.0, 15.0,$ and 20.0 mT. The curve represents the temperature evolution of the thermodynamic critical field B_c. Modified from Karl et al. (2019), © American Physical Society. Reproduced with permission. All rights reserved

Measurements at different (B_{ext}, T) points of the phase diagram show that the local field within the normal state domains, Fig. 6.57, does not depend on f_n, i.e., it is independent of the relative sample volumes occupied by the normal state and superconducting domains.

Magnetic history effects caused by different paths to reach a point in the phase diagram have also been studied. ZF-cooling (where the sample is first cooled in zero field and then, while keeping the temperature constant, B_{ext} is continuously increased) and field-cooling warming path (where the sample is first cooled down to the lowest temperature of 0.25 K in a constant field and then, without changing the applied field, the temperature is increased up to the measuring point) do not lead to an observable change of the magnetic field distributions $f(B_{\mu,z})$ measured by μSR. This implies that even if the distribution and/or the shape of the domains depends on the magnetic history (Prozorov 2007), the internal field inside the normal state domains as well as the relative volumes between normal and Meissner state do not change.

Exercises

6.1. Superconducting gap
Compare the parametrizations of the normalized superconducting gap given in Carrington and Manzano (2003), Prozorov and Giannetta (2006), Gross et al. (1986), and Sheahen (1966). What do you conclude?

6.2. London model
To obtain the simple field independent relationship between field variance and magnetic penetration depth Eq. 6.59 the assumption $k^2\lambda^2 \gg 1$ is made. Justify this assumption.

References

Abrikosov, A. A. (1957). *Soviet Physics JETP, 5*, 1174.
Ager, C., Ogrin, F. Y., Lee, S. L., et al. (2000). *Physical Review B, 62*, 3528.
Amin, M. H. S., Franz, M., & Affleck, I. (2000). *Physical Review Letters, 84*, 5864.
Annett, J., (2004). *Superconductivity, Superfluids and Condensates*. Oxford University Press. ISBN: 978-0198507567.
Bardeen, J., Cooper, L. N., & Schrieffer, J. R. (1957). *Physical Review, 108*, 1175.
Barford, W., & Gunn, J. (1988). *Physica C: Superconductivity, 156*, 515.
Bauer, E., Rogl, G., Chen, X.-Q., et al. (2010). *Physical Review B, 82*, 064511.
Bauer, E., Sekine, C., Sai, U., et al. (2014). *Physical Review B, 90*, 054522.
Beare, J., Nugent, M., Wilson, M. N., et al. (2019). *Physical Review B, 99*, 134510.
Bednorz, J. G., & Müller, K. A. (1986). *Zeitschrift für Physik B Condensed Matter, 64*, 189.
Bennemann, K. H., & Ketterson, J. B. (Eds.). (2008). *Superconductivity* (vol. 1–2). Springer. ISBN: 978-3540732525.
Blatter, G., Feigel'man, M. V., Geshkenbein, V. B., et al. (1994). *Reviews of Modern Physics, 66*, 1125.
Blundell, S. J. (1999). *Contemporary Physics, 40*, 175.
Bouquet, F., Wang, Y., Fisher, R. A., et al. (2001). *Europhysics Letters, 56*, 856.
Brandt, E. H. (1972). *physica Status Solidi (b), 51*, 345.
Brandt, E. H. (1988a). *Physical Review B, 37*, 2349.
Brandt, E. H. (1988b). *Journal of Low Temperature Physics, 73*, 355.

Brandt, E. H. (1997). *Physical Review Letters, 78,* 2208.

Brandt, E. H. (2003). *Physical Review B, 68,* 054506.

Brandt, E. H., & Seeger, A. (1986). *Advances in Physics, 35,* 189.

Brewer, J. H., Ansaldo, E. J., Carolan, J. F., et al. (1988). *Physical Review Letters, 60,* 1073.

Carbotte, J. P. (1990). *Reviews of Modern Physics, 62,* 1027.

Carrington, A., & Manzano, F. (2003). *Physica C: Superconductivity, 385,* 205.

Chandrasekhar, B. S., & Einzel, D. (1993). *Annalen der Physik, 505,* 535.

Chen, H., Ren, Y., Qiu, Y., et al. (2009). *Europhysics Letters, 85,* 17006.

Clem, J. R. (1975). *Journal of Low Temperature Physics, 18,* 427.

Cubitt, R., Forgan, E. M., Warden, M., et al. (1993a). *Physica C: Superconductivity and its Applications, 213,* 126.

Cubitt, R., Forgan, E. M., Yang, G., et al. (1993b). *Nature, 365,* 407.

Dalmas de Réotier, P., & Yaouanc, A. (2011). *Physical Review B, 84,* 012503.

Dalmas de Réotier, P., & Yaouanc, A. (1997). *Journal of Physics: Condensed Matter, 9,* 9113.

de Gennes, P.-G. (1999). *Superconductivity of Metals and Alloys. Advanced books classics.* CRC Press. ISBN: 978-0429497032.

Dolgov, O. V., Kremer, R. K., Kortus, J., et al. (2005). *Physical Review B, 72,* 024504. Erratum, *Physical Review B, 72,* 059902.

Drozdov, A. P., Eremets, M. I., Troyan, I. A., et al. (2015). *Nature, 525,* 73.

Drozdov, A. P., Kong, P. P., Minkov, V. S., et al. (2019). *Nature, 569,* 528.

Egorov, V. S., Solt, G., Baines, C., et al. (2001). *Physical Review B, 64,* 024524.

Eilenberger, G. (1968). *Zeitschrift für Physik A Hadrons and Nuclei, 214,* 195.

Eliashberg, G. (1960). *Soviet Physics JETP, 11,* 696.

Fesenko, V., Gorbunov, V., & Smilga, V. (1991). *Physica C: Superconductivity, 176,* 551.

Forgan, E. M., Lee, S. L., Sutton, S., et al. (1991). *Hyperfine Interactions, 63,* 71.

Gasparovic, R., Taylor, B., & Eck, R. (1966). *Solid State Communications, 4,* 59.

Ghosh, S., Brückner, F., Nikitin, A., et al. (2020a). *Review of Scientific Instruments, 91,* 103902.

Ghosh, S. K., Smidman, M., Shang, T., et al. (2020b). *Journal of Physics: Condensed Matter, 33,* 033001.

Ginzburg, V., & Landau, L. (1950). *Zhurnal Eksperimental'noi i Teoreticheskoi Fiziki, 20,* 1064.

Gor'kov, L. P. (1959). *Soviet Physics JETP, 36,* 1364.

Gorter, C. J., & Casimir, H. B. G. (1934a). *Physica, 1,* 306.

Gorter, C. J., & Casimir, H. B. G. (1934b). *Physikalische Zeitschrift, 35,* 963.

Grinenko, V., Das, D., Gupta, R., et al. (2021a). *Nature Communications, 12,* 3920.

Grinenko, V., Ghosh, S., Sarkar, R., et al. (2021b). *Nature Physics, 17,* 748–754.

Gross, F., Chandrasekhar, B. S., Einzel, D., et al. (1986). *Zeitschrift für Physik B Condensed Matter, 64,* 175.

Guguchia, Z., von Rohr, F., Shermadini, Z., et al. (2017). *Nature Communications, 8,* 1082.

Hao, Z., Clem, J. R., McElfresh, M. W., et al. (1991). *Physical Review B, 43,* 2844.

Hardy, W. N., Bonn, D. A., Morgan, D. C., et al. (1993). *Physical Review Letters, 70,* 3999.

Herlach, D., Majer, G., Major, J., et al. (1991). *Hyperfine Interactions, 63,* 41.

Heron, D. O. G., Ray, S. J., Lister, S. J., et al. (2013). *Physical Review Letters, 110,* 107004.

Hillier, A. D., Quintanilla, J., & Cywinski, R. (2009). *Physical Review Letters, 102,* 117007.

Hirschfeld, P. J., & Goldenfeld, N. (1993). *Physical Review B, 48,* 4219.

Huddart, B. M., Onuorah, I. J., Isah, M. M., et al. (2021). *Physical Review Letters, 127,* 237002.

Ichioka, M., Hasegawa, A., & Machida, K. (1999). *Physical Review B, 59,* 8902.

Johnston, D. C. (2013). *Superconductor Science and Technology, 26,* 115011.

Kadono, R. (2004). *Journal of Physics: Condensed Matter, 16,* S4421.

Kadono, R., Higemoto, W., Koda, A., et al. (2001). *Physical Review B, 63,* 224520.

Kallin, C. (2012). *Reports on Progress in Physics, 75,* 042501.

Kamerlingh Onnes, H. (1911a). *Communications from the Physical Laboratory of the University of Leiden, Supplement, 29.*

Kamerlingh Onnes, H. (1911b). *Communications from the Physical Laboratory of the University of Leiden, 12,* 120.

Kamihara, Y., et al. (2008). *Journal of the American Chemical Society, 130*, 3296.

Kamihara, Y., Hiramatsu, H., Hirano, M., et al. (2006). *Journal of the American Chemical Society, 128*, 10012.

Karl, R., Burri, F., Amato, A., et al. (2019). *Physical Review B, 99*, 184515.

Ketterson, J. B., & Song, S. (1999). *Superconductivity*. Cambridge University Press. ISBN: 978-1107034266.

Khasanov, R., Klamut, P. W., Shengelaya, A., et al. (2008). *Physical Review B, 78*, 014502.

Kiefl, R. F., Hossain, M. D., Wojek, B. M., et al. (2010). *Physical Review B, 81*, 180502.

Kleiner, R., & Buckel, W. (2016). *Superconductivity: An Introduction* (3rd edn.). Wiley. ISBN: 978-3527411627.

Kogan, V. G. (1981). *Physical Review B, 24*, 1572.

Kogan, V. G., Martin, C., & Prozorov, R. (2009). *Physical Review B, 80*, 014507.

Landau, I., & Keller, H. (2007). *Physica C: Superconductivity, 466*, 131.

Laulajainen, M., Callaghan, F. D., Kaiser, C. V., et al. (2006). *Physical Review B, 74*, 054511.

Lee, S. L., Zimmermann, P., Keller, H., et al. (1993). *Physical Review Letters, 71*, 3862.

Leng, H., Orain, J.-C., Amato, A., et al. (2019). *Physical Review B, 100*, 224501.

London, F., & London, H. (1935). *Proceedings of the Royal Society of London A: Mathematical, Physical and Engineering Sciences, 149*, 71.

Luetkens, H., Klauss, H.-H., Kraken, M., et al. (2009). *Nature Materials, 8*, 305.

Luke, G. M., Fudamoto, Y., Kojima, K. M., et al. (1998). *Nature, 394*, 558.

Mackenzie, A. P., Scaffidi, T., Hicks, C. W., et al. (2017). *NPJ Quantum Materials, 2*, 40.

Mackenzie, A. P., & Maeno, Y. (2003). *Reviews of Modern Physics, 75*, 657–712.

Maeno, Y., Hashimoto, H., Yoshida, K., et al. (1994). *Nature, 372*, 532–534.

Maisuradze, A., Khasanov, R., Shengelaya, A., et al. (2009). *Journal of Physics: Condensed Matter, 21*, 075701.

Mangin, P., & Kahn, R. (2017). *Superconductivity: An Introduction*. ISBN: 978-3319505251.

Marsiglio, F. (2020). *Annals of Physics, 417*, 168102.

Meissner, W., & Ochsenfeld, R. (1933). *Naturwissenschaften, 44*, 787.

Menon, G. I., Drew, A., Divakar, U. K., et al. (2006). *Physical Review Letters, 97*, 177004.

Minervini, J. (2014). *Nature Materials, 13*, 326.

Mühlschlegel, B. (1959). *Zeitschrift für Physik, 155*, 313.

Niedermayer, C., Bernhard, C., Blasius, T., et al. (1998). *Physical Review Letters, 80*, 3843.

Nishida, N., Miyatake, H., Shimada, D., et al. (1987). *Japanese Journal of Applied Physics, 26*, L1856.

Padamsee, H., Neighbor, J. E., & Shiffman, C. A. (1973). *Journal of Low Temperature Physics, 12*, 387.

Parks, R. D. (Ed.). (1969). *Superconductivity* (1st edn.). CRC Press. ISBN: 978-0824715205.

Pincus, P., Gossard, A., Jaccarino, V., et al. (1964). *Physics Letters, 13*, 21.

Pippard, A. B. (1953). *Proceedings of the Royal Society of London. Series A. Mathematical and Physical Sciences, 216*, 547.

Poole, C., Farach, H., Creswick, R., et al. (2014). *Superconductivity* (3rd edn.). Elsevier. ISBN: 978-0124166103.

Prozorov, R., & Kogan, V. G. (2018). *Physical Review Applied, 10*, 014030.

Prozorov, R. (2007). *Physical Review Letters, 98*, 257001.

Prozorov, R., & Giannetta, R. W. (2006). *Superconductor Science and Technology, 19*, R41. Erratum: R. Prozorov, *Superconductor Science and Technology 21*, 082003.

Pümpin, B., Keller, H., Kündig, W., et al. (1990). *Physical Review B, 42*, 8019.

Pustogow, A., Luo, Y., Chronister, A., et al. (2019). *Nature, 574*, 72.

Rammer, J. (1991). *Physica C: Superconductivity, 177*, 421.

Ribak, A., Skiff, R. M., Mograbi, M., et al. (2020). *Science Advances, 6*, eaax9480.

Riseman, T. M., Brewer, J. H., Chow, K. H., et al. (1995). *Physical Review B, 52*, 10569.

Sanna, S., Allodi, G., Concas, G., et al. (2004). *Physical Review Letters, 93*, 207001.

Sanna, S., Allodi, G., & De Renzi, R. (2003). *Solid State Communications, 126*, 85.

Savici, A. T., Fudamoto, Y., Gat, I. M., et al. (2002). *Physical Review B, 66*, 014524.

Schrieffer, J. (1999). *Theory of Superconductivity. Advanced Books Classics.* CRC Press. ISBN: 978-0738201207.

Shang, T., & Shiroka, T. (2021). *Frontiers in Physics, 9,* 651163.

Shapiro, I., & Shapiro, B. Y. (2019). *Superconductor Science and Technology, 32,* 085011.

Sheahen, T. P. (1966). *Physical Review, 149,* 368.

Shermadini, Z., Kanter, J., Baines, C., et al. (2010). *Physical Review B, 82,* 144527.

Shubnikov, L., Khotkevich, V. I., Shepelev, Y. D., et al. (1937). *Journal of Experimental and Theoretical Physics, 7,* 221.

Sidorenko, A. D., Smilga, V. P., & Fesenko, V. I. (1991). *Hyperfine Interactions, 63,* 49.

Sigrist, M., & Ueda, K. (1991). *Reviews of Modern Physics, 63,* 239–311.

Sonier, J. E. (2007). *Reports on Progress in Physics, 70,* 1717.

Sonier, J. E., Brewer, J. H., Kiefl, R. F., et al. (1999). *Physical Review Letters, 83,* 4156.

Sonier, J. E., Kiefl, R. F., Brewer, J. H., et al. (1994). *Physical Review Letters, 72,* 744.

Sonier, J. E., Brewer, J. H., & Kiefl, R. F. (2000). *Reviews of Modern Physics, 72,* 769.

Stewart, G. R. (2017). *Advances in Physics, 66,* 75.

Thiemann, S. L., Radović, Z., & Kogan, V. G. (1989). *Physical Review B, 39,* 11406.

Tinkham, M. (1996). *Introduction to Superconductivity.* McGraw-Hill. ISBN: 978-0070648784.

Uemura, Y. J., Le, L. P., Luke, G. M., et al. (1991). *Physical Review Letters, 66,* 2665.

Uemura, Y. J., Luke, G. M., Sternlieb, B. J., et al. (1989). *Physical Review Letters, 62,* 2317.

Uemura, Y. J. (2009). *Nature Materials, 8,* 253.

Weber, M., Amato, A., Gygax, F. N., et al. (1993). *Physical Review B, 48,* 13022.

Willa, R., Hecker, M., Fernandes, R. M., et al. (2021). *Physical Review B, 104,* 024511.

Xia, J., Maeno, Y., Beyersdorf, P. T., et al. (2006). *Physical Review Letters, 97,* 167002.

Yaouanc, A., Dalmas de Réotier, P., & Brandt, E. H. (1997). *Physical Review B, 55,* 11107.

Zhang, J. L., Pang, G. M., Jiao, L., et al. (2015). *Physical Review B, 92,* 220503.

Muonium

7

7.1   Introduction

Muonium ($\mu^+ e^-$, chemical symbol Mu) is the atom consisting of a positive muon
as the nucleus and a bound electron. It was discovered in 1960 by stopping positive
muons in pure high pressure argon gas and observing one of its characteristics
Larmor precession frequency (Hughes et al. 1960), see Sect. 7.3. In this experiment,
muonium is formed in its ground state $^2S_{1/2}$ after capturing an electron from an argon
atom. For more details about the mechanisms of muonium formation in matter, see
Sect. 2.3.

Virtually, muonium is a light isotope of hydrogen ($m_{\mathrm{Mu}}/m_{\mathrm{H}} \approx 1/9$). Hydrogen is
a ubiquitous impurity in semiconductors and oxides, where it can be unintentionally
incorporated into the lattice during growth or subsequent processing. In crystals,
it represents the simplest impurity defect center and can significantly alter their
electronic and optical properties. It is therefore essential to obtain information
about the behavior of hydrogen and its states in materials that are fundamental for
electronic and optical applications.

Whereas isolated H is very difficult to investigate directly, spectroscopy of
muonium defect centers provides a unique opportunity to probe the electronic
structure of a hydrogen defect and its dopant properties in a wide range of materials.

Since $m_\mu \gg m_e$ muonium has practically equal reduced mass, size and
ionization energy as the hydrogen atom[1] and its electronic structure is expected to

---

[1] The muonium Bohr radius and ionization energy are within 0.5% of those of hydrogen,
deuterium, and tritium. The chemical properties of Mu are similar to those of the hydrogen isotopes
because they depend on the reduced mass. However, dynamical aspects (e.g., diffusion) can be
considerably different between Mu and H due to the light mass of Mu. Mu and H also have different
zero-point energies when localized.

© Springer Nature Switzerland AG 2024
A. Amato, E. Morenzoni, *Introduction to Muon Spin Spectroscopy*,
Lecture Notes in Physics 961, https://doi.org/10.1007/978-3-031-44959-8_7

Table 7.1 Main properties of free muonium, compared to hydrogen

Mass m_{Mu}	$0.1131\ m_{\text{H}} = 207.77\ m_e \cong m_\mu$
Reduced mass $\overline{m}_{\text{Mu}} = \dfrac{m_\mu m_e}{m_\mu + m_e}$	$0.9957\ \overline{m}_{\text{H}} = 0.9952\ m_e \cong m_e$
Radius ($n = 1$ shell) $a_{\text{Mu}} = \dfrac{4\pi\varepsilon_0\hbar^2}{\overline{m}_{\text{Mu}}e^2}$	$1.0048\ a_0 = 1.0043\ a_{\text{H}} = 0.05317$ nm
ionization energy $R_{\text{Mu}} = \dfrac{\overline{m}_{\text{Mu}}\ e^4}{(4\pi\varepsilon_0)^2 2\hbar^2}$	$0.9952\ R_y = 13.54$ eV
Hyperfine frequency ν_0	$3.1423\ \nu_{\text{H}} = 4463.30$ MHz
Gyromagnetic factor $\gamma_{\text{Mu}}^{\text{T}} = \dfrac{1}{2}(\gamma_e - \gamma_\mu)$ in the triplet state $\lvert F = 1, M = \pm 1\rangle$ and weak fields	$0.99668\ \gamma_{\text{H}}^{\text{T}} = 2\pi \times 1.3944706$ MHz/G $\cong 102.88\ \gamma_\mu$

be equivalent to that of hydrogen.[2] Muonium's most important properties compared to those of hydrogen are summarized in Table 7.1.

In solids, in principle, all the possible states of muonium can be found: besides the neutral state, the positively and the negatively charged states.[3]

In this chapter, we concentrate on the spectroscopy of muonium by μSR where its hyperfine interaction is measured. This provides information on the electronic structure of a nonionized hydrogen atom, but also on its position in the lattice and on reaction kinetics. Muonium states have been detected in various semiconductors, insulators, semimetals, and organic compounds (Patterson 1988; Cox 2009; Cox et al. 2013).

Muon spin rotation spectroscopy has played a pioneering role in the discovery and identification of intrinsic hydrogen-like states in semiconductors and oxides. In fact, a large amount of information on the structure and electrical activity of isolated interstitial hydrogen states in semiconductors and insulators has been obtained by muonium spectroscopy.[4]

In the beginning, the investigations focused on elemental semiconductors (such as Si and Ge) and binary III-V compound (such as GaAs) (Patterson 1988). Two distinct but coexisting, neutral states, an isotropic, Sect. 7.3, and an anisotropic with

---

[2] This has been directly verified in some cases, for example in rutile TiO_2, where it has been found that Mu and H form the same configuration with identical electronic structure (Vilão et al. 2015).

[3] To emphasize the charge state, in the literature about muonium spectroscopy often the notation Mu^0, Mu^+ and Mu^- ($\mu^+e^-e^-$) is used. Mu^+ underlines the fact that μ^+ in the solid may have some chemical association with the host atoms; the symbol μ^+ is then reserved for the free μ^+ impinging and thermalizing in the sample. Mu^+ and Mu^- are difficult to separate spectroscopically because they have essentially the same Larmor frequency of a diamagnetic state. See Ito et al. (2020) and references therein for methods of separation.

[4] Isolated hydrogen atoms in materials are difficult to investigate by other spectroscopic means, which generally require high hydrogen concentrations.

axial symmetry, Sect. 7.4, were found in these materials. Later, the measurements were extended to semiconductors of the II-VI groups as well as nonmagnetic oxides. A very weakly bound anisotropic muonium state was first found in CdS (Gil et al. 1999), see Sect. 7.5.

Unlike the muon in magnetic substances or superconductors, the positive muon in a semiconductor is not a passive probe, but represents a defect center carrying unit charge. The muon acts as an active probe, in the sense that μSR spectroscopy probes a state, which is created by the muon itself.

In addition to its use in condensed matter and materials science, muonium spectroscopy plays an important role in particle physics. Here muonium is of particular interest for high precision spectroscopy because it is the simplest purely leptonic system, sensitive only to electroweak and gravitational interactions. Moreover, in contrast to H, it has a "nucleus" that can be considered pointlike.[5] High precision muonium spectroscopy of the hyperfine structure, Lamb shift, and $1s$-$2s$ interval has provided tests of fundamental laws and symmetries and the precise determination of fundamental constants (for an overview see Jungmann 2016). New experiments are expected to improve the present limits and values.

7.2 Muonium Ground State and Hyperfine Interaction

7.2.1 Ionization Energy

In Sect. 5.1.1 we have already introduced the muon-electron interaction. Here we concentrate on the aspects relevant for muonium spectroscopy.

As with the hydrogen atom, we first consider the gross structure of the energy levels, i.e., we ignore all spin interactions. The Hamiltonian $\mathcal{H}_0$ then consists of the kinetic and potential energy terms of Eq. 5.3[6] and the energy levels only depend on the principal quantum number n

$$
\begin{aligned}
\mathcal{H}_0 &= \frac{\mathbf{p}^2}{2\overline{m}_{\mathrm{Mu}}} + V(r) \\
&= -\frac{\hbar^2}{2\overline{m}_{\mathrm{Mu}}}\nabla^2 - \frac{e^2}{4\pi\varepsilon_0 r} \quad,
\end{aligned}
\tag{7.1}
$$

---

[5] From scattering experiments the μ radius is estimated to be $< 10^{-18}$ m $\simeq 1/1000$ of the proton radius.

[6] The Hamiltonian $\mathcal{H}_0$ contains the reduced mass of the muon-electron system $\overline{m}_{\mathrm{Mu}} = m_\mu m_e/(m_\mu + m_e)$, which is the "effective" inertial mass moving in the potential field. In Eq. 5.3 the reduced mass was approximated by the electron mass, because in muonium it is practically equal to m_e, see Table 7.1.

where the momentum operator $\mathbf{p} = -i\hbar\nabla$, and the potential energy is determined by the Coulomb force $F = -e^2/(4\pi\varepsilon_0 r^2) = -dV(r)/dr$, hence $V(r) = -e^2/(4\pi\varepsilon_0 r)$. In spherical coordinates the time independent Schrödinger equation is

$$\left(-\frac{\hbar^2}{2\overline{m}_{Mu}}\nabla^2 - \frac{e^2}{4\pi\varepsilon_0 r}\right)\psi(r, \theta, \phi) = E\psi(r, \theta, \phi) \ . \tag{7.2}$$

The solution can be represented by the product of a radial part, expressed in terms of associated Laguerre functions, and an angular part given by spherical harmonic functions (Cohen-Tannoudji et al. 2005).

The wave function and energy of the ground state level $1^2S_{1/2}$, principal quantum number $n = 1$ and $L = 0$, are given by

$$\psi_{1s}(r) = \frac{1}{\sqrt{\pi a_{Mu}^3}}\exp\left(-\frac{r}{a_{Mu}}\right) \tag{7.3}$$

$$E_{1s} = -\frac{\overline{m}_{Mu}\,e^4}{(4\pi\varepsilon_0)^2 2\hbar^2} = -13.54 \text{ eV} \ , \tag{7.4}$$

which is the negative of the ionization energy R_{Mu}.

7.2.2 Isotropic Hyperfine Interaction

We first consider isotropic muonium in the ground state, Eq. 7.4. The general spin interaction, treated in Sect. 5.1.1, leads to the Hamiltonian in Eq. 5.11. Due to the spherical symmetry of the wave function only the Fermi contact term remains, since the probability to find an s-wave electron at the muon position is finite. Therefore, the spin part of the muonium Hamiltonian is given by Eq. 5.16 and is

$$\mathcal{H} = \mathcal{H}_{F.\,cont}$$

$$= \frac{2\mu_0}{3}g_e\,\mu_B\,g_\mu\,\mu_B^\mu\,|\psi_{1s}(0)|^2\,\mathbf{I}_\mu\cdot\mathbf{S}$$

$$= A_{Mu}\,\mathbf{I}_\mu\cdot\mathbf{S} \ , \tag{7.5}$$

where

$$|\psi_{1s}(0)|^2 = \frac{1}{\pi a_{Mu}^3} \ , \tag{7.6}$$

and for the hyperfine coupling constant A_{Mu} of free or vacuum muonium (also named hyperfine splitting or energy) we obtain

$$
\begin{aligned}
A_{Mu} &= \frac{2\mu_0}{3} g_e \, \mu_B \, g_\mu \, \mu_B^\mu \, |\psi_{1s}(0)|^2 \\
&= \frac{2\mu_0}{3} \hbar^2 \gamma_e \gamma_\mu \frac{1}{\pi a_{Mu}^3} \\
&= \hbar\omega_0 \quad \text{with} \quad \omega_0 = 2\pi \nu_0 = 2\pi \times 4463.3 \text{ MHz} \ .
\end{aligned}
\tag{7.7}
$$

The Hamiltonian Eq. 7.5 is a typical example of a two spin $1/2$ system (the degeneracy of the $1s$ level is fourfold). To determine eigenvalues and eigenstates, we can use as a basis the four states formed by the product of the two spins $|I_\mu = 1/2, S = 1/2, m_{I_\mu}, m_S\rangle$ where m_{I_μ} and m_S can be $\pm 1/2$. According to the direction of the two spins given by m_{I_μ} and m_S we denote the four levels as $|\uparrow_\mu \uparrow_e\rangle$, $|\downarrow_\mu \downarrow_e\rangle$, $|\downarrow_\mu \uparrow_e\rangle$ and $|\uparrow_\mu \downarrow_e\rangle$ (Zeeman states basis).

Another basis can be formed by taking the total spin of the system $\mathbf{F} = \mathbf{I}_\mu + \mathbf{S}$. The four states $|I_\mu = 1/2, S = 1/2, F, m_F\rangle$ are then labeled with the total spin quantum number F (with values 0 or 1) and its magnetic quantum number m_F as $|0, 0\rangle$ (singlet state) and $|1, 1\rangle$, $|1, 0\rangle$ and $|1, -1\rangle$ (triplet state).

The two bases are related to each other. For the triplet state with $F = 1$ and $m_F = 1, 0, -1$

$$
\begin{aligned}
|1, 1\rangle &= |\uparrow_\mu \uparrow_e\rangle \\
|1, 0\rangle &= \frac{1}{\sqrt{2}} \left(|\uparrow_\mu \downarrow_e\rangle + |\downarrow_\mu \uparrow_e\rangle \right) \\
|1, -1\rangle &= |\downarrow_\mu \downarrow_e\rangle \ ,
\end{aligned}
\tag{7.8}
$$

and for the singlet state with $F = 0, m_F = 0$

$$
|0, 0\rangle = \frac{1}{\sqrt{2}} \left(|\uparrow_\mu \downarrow_e\rangle - |\downarrow_\mu \uparrow_e\rangle \right) \ .
\tag{7.9}
$$

Using $\mathbf{F}^2 = \mathbf{I}_\mu^2 + \mathbf{S}^2 + 2\,\mathbf{I}_\mu \cdot \mathbf{S}$ and with the spin operators $\mathbf{I}_\mu$ and $\mathbf{S}$ commuting, we can express $\mathbf{I}_\mu \cdot \mathbf{S} = \frac{1}{2}[\mathbf{F}^2 - \mathbf{I}_\mu^2 - \mathbf{S}^2]$ from which it follows that $\mathbf{I}_\mu \cdot \mathbf{S}$ and the contact Hamiltonian are diagonal in the total spin basis

$$
A_{Mu} \, \mathbf{I}_\mu \cdot \mathbf{S} \, |F, m_F\rangle = \frac{A_{Mu}}{2} \left[F(F+1) - I_\mu(I_\mu + 1) - S(S+1) \right] |F, m_F\rangle \ .
\tag{7.10}
$$

Fig. 7.1 Schematic of the
energy levels of the muonium
ground state with the splitting
due to the hyperfine Fermi
contact interaction. For
clarity the splitting
$A_{\mathrm{Mu}} \simeq 18.5 \times 10^{-6}$ eV is
strongly expanded compared
to the ionization energy

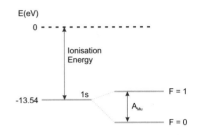

The energy eigenstates and eigenvalues of isotropic muonium in the ground state
and zero field are

$$\mathrm{Triplet} : \ E_{F=1} = \frac{1}{4} A_{\mathrm{Mu}}$$

$$\mathrm{Singlet} : \ E_{F=0} = -\frac{3}{4} A_{\mathrm{Mu}} \tag{7.11}$$

where A_{Mu} given by Eq. 7.7 contains the spatial contribution to the energy value.
Note that the hyperfine splitting of the ground state leaves the average energy of the
state unchanged, since $3E_{F=1} + 1E_{F=0} = 0$ (Fig. 7.1).

7.2.3 Hyperfine Splitting in an External Field

We now examine how the hyperfine levels of the muonium ground state are modified
by an applied magnetic field $\mathbf{B}_{\mathrm{ext}}$. Because of the Zeeman interaction with both
spins, the spin Hamiltonian acquires two additional terms[7]

$$\mathcal{H} = \mathcal{H}'_{\mathrm{F.\ cont}} + \mathcal{H}'_{\mathrm{Zeeman}} =$$
$$= A_{\mathrm{Mu}} \mathbf{I}_\mu \cdot \mathbf{S} + g_e\, \mu_{\mathrm{B}}\, \mathbf{S} \cdot \mathbf{B}_{\mathrm{ext}} - g_\mu\, \mu_{\mathrm{B}}^\mu\, \mathbf{I}_\mu \cdot \mathbf{B}_{\mathrm{ext}} \ , \tag{7.12}$$

where the positive sign for the electron Zeeman interaction reflects the opposite
direction of its spin with respect to the magnetic moment.[8] The total spin eigenstates
in zero field Eqs. 7.8 and 7.9, are no longer good eigenstates in field. To find
the eigenvalues of the Hamiltonian, we first express its matrix components in a
chosen basis. The easiest choice is to use the Zeeman states. With the z-axis as

---

[7] We recall that this is the Hamiltonian for free or vacuum muonium, i.e., there are no depolarizing
interactions with the environment.

[8] Note that, strictly speaking, the electron and muon g-factors in Eq. 7.12 are the electron and muon
g-factors in the muonium bound state. The difference from the free particle values is negligibly
small and can be expressed in powers of the fine structure constant α. For both particles the leading
term of the difference $\left(g_{\mu,e}(\mathrm{Mu\ bound}) - g_{\mu,e}(\mathrm{free})\right)$ is of order α^2 (Karshenboim and Ivanov
2002).

field direction and expressing the scalar product of the two spins with raising and lowering spin operators

$$\mathbf{I}_\mu \cdot \mathbf{S} = I_{\mu,z} S_z + \frac{1}{2} \left(I_{\mu,-} S_+ + I_{\mu,+} S_- \right) \ , \tag{7.13}$$

we obtain the following matrix elements

$$\mathcal{H}'_{ij} =$$

$$= \begin{pmatrix} \frac{1}{4}\hbar\omega_0 + \frac{1}{2}\hbar\omega_e - \frac{1}{2}\hbar\omega_\mu & 0 & 0 & 0 \\ 0 & -\frac{1}{4}\hbar\omega_0 + \frac{1}{2}\hbar\omega_e + \frac{1}{2}\hbar\omega_\mu & \frac{1}{2}\hbar\omega_0 & 0 \\ 0 & \frac{1}{2}\hbar\omega_0 & -\frac{1}{4}\hbar\omega_0 - \frac{1}{2}\hbar\omega_e - \frac{1}{2}\hbar\omega_\mu & 0 \\ 0 & 0 & 0 & \frac{1}{4}\hbar\omega_0 - \frac{1}{2}\hbar\omega_e + \frac{1}{2}\hbar\omega_\mu \end{pmatrix} , \tag{7.14}$$

where $\hbar\omega_0 = A_{\mathrm{Mu}}$, Eq. 7.7, and the Zeeman energies are written in terms of the angular Larmor velocities $\omega_e = \gamma_e B_{\mathrm{ext}}$ and $\omega_\mu = \gamma_\mu B_{\mathrm{ext}}$.

To find the eigenvalues and eigenstates, we have to solve the system of linear equations with unknown a_j

$$\sum_{j=1}^{4} \left[\mathcal{H}'_{ij} - E\, \delta_{ij} \right] a_j = 0 , \quad i = 1, 2, 3, 4 \ , \tag{7.15}$$

where E defines the eigenvalues and the a_j are the coefficients to express the spin eigenstates in the basis chosen to write $\mathcal{H}'_{ij}$ (i.e., the Zeeman basis). This system has a nontrivial solution only if the determinant of the matrix $\left| \mathcal{H}'_{ij} - E\, \mathbb{1}_4 \right|$ is zero (where $\mathbb{1}_4$ is the 4×4 unit matrix). The determinant equation defines the eigenvalues, which are then inserted into Eq. 7.15 to obtain the eigenstates.

The following eigenvalues are found, see Fig. 7.2[9]

$$E_1 = \frac{1}{4}\hbar\omega_0 + \frac{1}{2}\hbar\omega_e - \frac{1}{2}\hbar\omega_\mu$$

$$E_2 = -\frac{1}{4}\hbar\omega_0 + \frac{1}{2}\hbar \left[\omega_0^2 + (\omega_e + \omega_\mu)^2 \right]^{1/2} = -\frac{1}{4}\hbar\omega_0 + \frac{1}{2}\hbar\omega_0 \left[1 + x_B^2 \right]^{1/2}$$

$$E_3 = \frac{1}{4}\hbar\omega_0 - \frac{1}{2}\hbar\omega_e + \frac{1}{2}\hbar\omega_\mu$$

$$E_4 = -\frac{1}{4}\hbar\omega_0 - \frac{1}{2}\hbar \left[\omega_0^2 + (\omega_e + \omega_\mu)^2 \right]^{1/2} = -\frac{1}{4}\hbar\omega_0 - \frac{1}{2}\hbar\omega_0 \left[1 + x_B^2 \right]^{1/2} . \tag{7.16}$$

---

[9] We number 1 to 3 the eigenvalues and eigenstates of the triplet state and 4 of the singlet state.

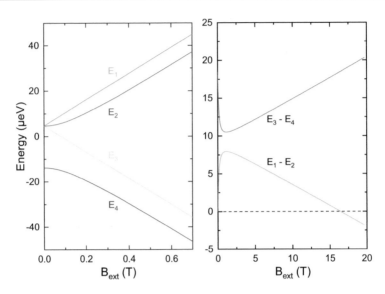

Fig. 7.2 Left: energy levels of isotropic vacuum muonium in a magnetic field (so-called Breit-Rabi diagram Breit and Rabi 1931). In the absence of a magnetic field, the triplet states are degenerate and have a higher energy than the singlet state. The degeneracy of the triplet states is lifted in the magnetic field. Right: difference between the energies of the $|1\rangle$ and $|2\rangle$ states, and of the $|3\rangle$ and $|4\rangle$ states. The energy levels E_1 and E_2 cross at 16.385 T, see Exercise 7.1

If $B_{ext} = 0$, $\omega_e = \omega_\mu = 0$ and we recover, as expected, Eq. 7.11

$$E_1 = E_2 = E_3 = \frac{1}{4}\hbar\omega_0 = E_{F=1}$$

$$E_4 = -\frac{3}{4}\hbar\omega_0 = E_{F=0} \ .$$

In Eq. 7.16 we have expressed the field dependence of the $|2\rangle$ and $|4\rangle$ states in terms of a normalized field x_B

$$x_B \equiv \frac{(\omega_e + \omega_\mu)}{\omega_0} = \frac{B_{ext}}{B_0} \ , \qquad (7.17)$$

where $B_0 = \omega_0/(\gamma_e + \gamma_\mu) = \hbar\omega_0/(g_e\mu_B + g_\mu\mu_B^\mu)$ is the field where the Zeeman energy splitting of the electron and muon is equal to the hyperfine splitting. For vacuum, isotropic muonium, which we have discussed so far[10] $B_0 = 0.1585$ T.

We can now put the eigenvalues from Eq. 7.16 in 7.15 to obtain the eigenstates. The states $|1\rangle$ and $|3\rangle$ are already eigenstates. The states $|2\rangle$ and $|4\rangle$ are obtained

---

[10] We will see in the next sections that in solids the muonium spectrum may differ, thus giving information about the configuration and electronic state of a hydrogen impurity.

by diagonalizing the 2×2 submatrix in Eq. 7.14 and can be expressed as a linear combination of two Zeeman states. We obtain

$$|1\rangle = |\uparrow_\mu \uparrow_e\rangle$$
$$|2\rangle = \sin \delta \, |\uparrow_\mu \downarrow_e\rangle + \cos \delta \, |\downarrow_\mu \uparrow_e\rangle$$
$$|3\rangle = |\downarrow_\mu \downarrow_e\rangle$$
$$|4\rangle = \cos \delta \, |\uparrow_\mu \downarrow_e\rangle - \sin \delta \, |\downarrow_\mu \uparrow_e\rangle \ , \qquad (7.18)$$

where the amplitudes are given by[11]

$$\cos \delta = \frac{1}{\sqrt{2}} \left(1 + \frac{\omega_e + \omega_\mu}{\left[\omega_0^2 + (\omega_e + \omega_\mu)^2\right]^{1/2}} \right)^{1/2} = \frac{1}{\sqrt{2}} \left(1 + \frac{x_B}{\left[1 + x_B^2\right]^{1/2}} \right)^{1/2}$$

$$(7.19)$$

and

$$\sin \delta = \frac{1}{\sqrt{2}} \left(1 - \frac{\omega_e + \omega_\mu}{\left[\omega_0^2 + (\omega_e + \omega_\mu)^2\right]^{1/2}} \right)^{1/2} = \frac{1}{\sqrt{2}} \left(1 - \frac{x_B}{\left[1 + x_B^2\right]^{1/2}} \right)^{1/2} .$$

$$(7.20)$$

In zero field with $\cos \delta = \sin \delta = 1/\sqrt{2}$ we recover the $|F, m_F\rangle$ eigenstates given by Eqs. 7.8 and 7.9. At high field $\sin \delta \to 0$ and $\cos \delta \to 1$, we retrieve the Zeeman states. This is understandable, because for $B_{ext} \to \infty$, the Zeeman term dominates the hyperfine interaction between muon and electron spin (Paschen-Back regime). We have

$$|i\rangle \text{ basis vectors} \xrightarrow[B_{ext}, \, x_B \to 0]{} \text{total spin basis}$$

$$|i\rangle \text{ basis vectors} \xrightarrow[B_{ext}, \, x_B \to \infty]{} \text{Zeeman basis}$$

7.3 Time Evolution of the Muon Polarization in the Muonium State

7.3.1 Introduction

In this section we consider the time evolution of the spin polarization of a muon bound in the ground state of free muonium in the presence of an external

---

[11] The amplitudes are expressed as sine and cosine, so that they naturally satisfy the normalization condition $\cos^2 \delta + \sin^2 \delta = 1$.

field applied parallel to the initial muon spin polarization (LF configuration) or perpendicular to it (TF configuration). The external field acts directly on the magnetic moment of the muon and the much larger moment of the electron via the Zeeman interaction. Additionally, the muon magnetic moment is coupled via the Fermi contact term of Eq. 7.12 to the electron magnetic moment.[12] The large electron moment makes muonium more sensitive to the local field than the "free" muon and, via coupling, strongly influences the time evolution of the muon spin polarization in muonium, which as we will see follows a rather complex time dependence.

Muonium is mainly formed during slowing down, when the muon, after a cycle of electron loss and capture from the medium, thermalizes as muonium (prompt or epithermal muonium), see Chap. 2 and particularly Sect. 2.3 for more details. In a solid, the thermalization time is very short, see Sect. 2.2.2, so that we can take $t = 0$,[13] as the time of muonium formation and onset of the muon spin precession.

Starting with a 100% polarized muon beam, (e.g., in the z direction) all the muons are in the $|\uparrow_\mu\rangle$ state. The captured electron is not polarized, so along the quantization z-axis, we have 50% of the electrons in the state $|\uparrow_e\rangle$ and 50% in the state $|\downarrow_e\rangle$. Thus, the muonium initial state is the incoherent superposition of two equally probable states

$$\text{Probability and states:} \quad \begin{cases} 50\% & |\uparrow_\mu\uparrow_e\rangle \\ 50\% & |\uparrow_\mu\downarrow_e\rangle \, . \end{cases} \tag{7.21}$$

In the zero and longitudinal field configurations, the state $|\uparrow_\mu\uparrow_e\rangle$ is an eigenstate of the Hamiltonian 7.12, see Eq. 7.18. As a consequence, this state will not show any time dependence when the external field is applied along the muon spin polarization. On the other hand, the state $|\uparrow_\mu\downarrow_e\rangle$, which is a linear combination of $|2\rangle$ and $|4\rangle$, see Eq. 7.18, is not an eigenstate and oscillates between the two states. The results for a longitudinally applied field are presented in Sect. 7.3.2.

In the transverse field configuration, both states of Eq. 7.21 are not eigenstates because the field defines a different quantization axis than the initial polarization. In this case we expect the full polarization to oscillate. The results for the transverse field are presented in Sect. 7.3.3.

---

[12] Remember that each particle feels the magnetic field produced by the other particle.

[13] Formation of a fraction of muonium via a potentially slower, several ns to μs, delayed formation has been discussed (Storchak et al. 2004), see Sect. 2.3. However, the observation of well-defined muonium precession lines with full or almost full amplitude in the GHz range (see for example Fig. 7.3) implies that, in this case, independent of the mechanism, the formation of the final muonium state occurs within a fraction of the typical hyperfine precession periods of some 100 ps.

According to Appendix F.4, the time evolution of the muon spin polarization in the Heisenberg picture can be calculated using the density matrix formalism,[14] Eq. F.18,

$$\mathbf{P}(t) = \langle \boldsymbol{\sigma}_\mu \rangle(t)$$
$$= \mathrm{Tr}\left(\rho_{\mathrm{Mu}}(0)\,\boldsymbol{\sigma}_\mu(t)\right) \ , \tag{7.22}$$

where ρ_{Mu} is the density matrix of muonium. As detailed in Appendix F.5, the density matrix of muonium at $t = 0$ is the product[15] of the density matrix of the electron with that of the muon. Taking into account that the captured electrons are unpolarized, we obtain

$$\rho_{\mathrm{Mu}}(0) = \frac{1}{4}\left[\mathbb{1}_2 + \mathbf{P}(0)\cdot\boldsymbol{\sigma}_\mu\right] \ . \tag{7.23}$$

Introducing this result into Eq. 7.22 we have

$$\mathbf{P}(t) = \frac{1}{4}\sum_k \langle k\,|(\mathbb{1}_2 + \mathbf{P}(0)\cdot\boldsymbol{\sigma}_\mu)\,\boldsymbol{\sigma}_\mu(t)|k\rangle$$
$$= \frac{1}{4}\sum_{k,n} \langle k\,|(\mathbb{1}_2 + \mathbf{P}(0)\cdot\boldsymbol{\sigma}_\mu)|n\rangle\,\langle n|\boldsymbol{\sigma}_\mu(t)|k\rangle \ , \tag{7.24}$$

where the $|k\rangle$ and the $|n\rangle$ are the eigenstates defined in Eq. 7.18. The matrix elements $\langle n|\boldsymbol{\sigma}_\mu(t)|k\rangle$ can be calculated using the equation defining the time evolution of the operator in the Heisenberg representation

$$-i\hbar\frac{\mathrm{d}}{\mathrm{d}t}\boldsymbol{\sigma}_\mu(t) = \left[\mathcal{H}, \boldsymbol{\sigma}_\mu(t)\right] \ , \tag{7.25}$$

which has the formal solution

$$\boldsymbol{\sigma}_\mu(t) = \exp(i\mathcal{H}t/\hbar)\,\boldsymbol{\sigma}_\mu\,\exp(-i\mathcal{H}t/\hbar) \ , \tag{7.26}$$

giving for the matrix elements

$$\langle n|\boldsymbol{\sigma}_\mu(t)|k\rangle = \langle n|\boldsymbol{\sigma}_\mu|k\rangle\,\exp(i\omega_{nk}t) \ , \tag{7.27}$$

where

$$\omega_{nk} = \frac{E_n - E_k}{\hbar} \ . \tag{7.28}$$

---

[14] We explicitly write here the index μ for the σ matrix to distinguish it from the matrix acting on the electron spin.

[15] Strictly speaking, it is the tensor product. See Appendix F.5.

This allows us to calculate the time dependence of the muon spin polarization. Taking into account that $\langle k|\mathbb{1}_2|n\rangle = \delta_{kn}$ and that $\sum_{k,n} \delta_{kn}\langle n|\sigma_{\mu,i}|k\rangle = 0$ (since $\mathrm{Tr}\,\sigma_{\mu,i} = 0$ for $i = x, y, z$), Eq. 7.24 simplifies to

$$\mathbf{P}(t) = \frac{1}{4}\sum_{k,n}\langle k|\,\mathbf{P}(0)\cdot\boldsymbol{\sigma}_\mu\,|n\rangle\,\langle n|\boldsymbol{\sigma}_\mu|k\rangle\cdot\exp(i\omega_{nk}\,t)\ . \tag{7.29}$$

From this expression we note that the precession frequencies in muonium are given by the transition frequencies between the muonium eigenstates of the Breit-Rabi diagram.

7.3.2 Longitudinal (and Zero) Field

With Eq. 7.29 we are now in position to calculate typical μSR configurations. In the longitudinal and zero field configurations, magnetic field and initial muon spin polarization are both along the z direction: $\mathbf{P}(0) = P_z(0)\hat{\mathbf{z}}$ (with $P_z(0) = 1$) and $\mathbf{B}_{\mathrm{ext}} = B_{\mathrm{ext}}\hat{\mathbf{z}}$.

The time evolution of the muon spin polarization is in this case (see Exercise 7.2 for a derivation of this expression)

$$P_z(t) = \frac{1}{2}\left[1 + \frac{1}{1 + x_B^2}\left[x_B^2 + \cos(\omega_{24}\,t)\right]\right]\ , \tag{7.30}$$

where the transition frequency between levels $|2\rangle$ and $|4\rangle$ is

$$\omega_{24} = \frac{E_2 - E_4}{\hbar}$$

$$= \sqrt{\omega_0^2 + (\omega_e + \omega_\mu)^2}$$

$$= \omega_0\sqrt{1 + x_B^2}\ . \tag{7.31}$$

In zero field $x_B = 0$ and $\omega_{24} = \omega_0 = 2\pi\nu_0$. Half of the muonium is in the state $|\uparrow_\mu\uparrow_e\rangle$, Eq. 7.21. This corresponds to the stationary state $|F = 1, m_F = 1\rangle$, which thus maintains the initial muon polarization. For the other half, the muon spin polarization oscillates at an angular frequency ω_0 corresponding to transitions between the triplet $|F = 1, m_F = 0\rangle$ and the singlet $|F = 0, m_F = 0\rangle$ states[16]

$$P_z(t, B_{\mathrm{ext}} = 0) = \frac{1}{2}[1 + \cos(\omega_0\,t)]\ . \tag{7.32}$$

---

[16] This corresponds to a $\Delta F = \pm 1\ \Delta m_F = 0$ selection rule.

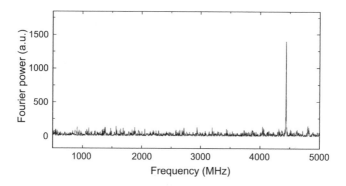

Fig. 7.3 Fourier transform of a zero field μSR spectrum of fused quartz at room temperature. The hyperfine frequency of muonium in quartz (4437 $\pm$1 MHz) is close to the vacuum value (4463 MHz). The effective time resolution of this experiment is 150 $\pm$ 10 ps (FWHM). Modified from Scheuermann et al. (2000), © Elsevier. Reproduced with permission. All rights reserved

This oscillation is due to the Fermi contact field at the "nucleus" created by the magnetic moment of the electron in the $1s$ state. To detect such high precession frequencies, a time resolution (FWHM) of the order of or better than $1/(2\nu_0) = 112$ ps is required,[17] which is beyond the capabilities of conventional μSR spectrometers with a typical time resolution of $\sim$ 1 ns. Special spectrometers developed for high-field measurements and based on solid state detectors[18] for the muon and positrons detection achieve such resolutions.[19]

Figure 7.3 shows the Fourier spectrum of a ZF-μSR signal in quartz taken with a high-resolution spectrometer at PSI (Scheuermann et al. 2000). A conventional spectrometer would show no Fourier peak because it averages out the oscillating part of the polarization and reduces the observable μSR to the non-oscillating part. By applying a longitudinal field, this apparent loss of muon spin polarization can be reduced ("repolarization" or "quenching" of the depolarization) according to

---

[17] To detect an oscillation with period $1/\nu$ one needs at least a signal in two time channels, see also Exercise 7.3.

[18] Note that very good resolution can also be achieved with the traditional photomultipliers in dedicated setups, as already demonstrated in the 1980s (Holzschuh et al. 1981; Holzschuh 1983).

[19] High-field μSR instruments must be able to detect muon spin precession frequencies in $\sim$ 10 T fields. At this field the spin precession of a diamagnetic muon has a frequency $\nu_\mu = 1.35$ GHz. This goal can be achieved with a compact detector system using silicon photomultipliers (SiPMs), sometimes also called Avalanche Photodiodes (APDs), solid state detectors (Stoykov et al. 2012; Scheuermann et al. 2000). High-resolution TF experiments are only possible at continuous muon beam facilities such as PSI and TRIUMF (Kreitzman & Morris 2018). The high-field μSR instrument at PSI has a time resolution of $\sim$ 60 ps (rms). This makes muonium and μ^+ spectroscopy in high fields possible. See Exercise 7.3 for the effect of finite time resolution on the observable asymmetry.

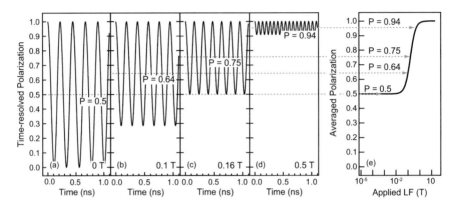

Fig. 7.4 (**a**) to (**d**) expected μSR signals for isotropic vacuum muonium in zero and longitudinally applied fields. Note the very short oscillation periods. The dotted lines indicate the polarization when the oscillating part is averaged out. (**e**) Repolarization curve according to Eq. 7.33. Adapted from Nuccio et al. (2014), © IOP Publishing. Reproduced with permission. All rights reserved. See also Blundell (2004)

the expression 7.30.[20] This effect is visible in Figs. 7.4 and 7.5. Increasing the longitudinal field increases the non-oscillating at the expense of the oscillating part, reflecting an external field reaching and exceeding the internal field due to the contact interaction. The observed polarization curve becomes in this case time independent

$$P_z^{\text{non-osc}}(t, x_B) = \frac{1}{2}\left[1 + \frac{x_B^2}{1 + x_B^2}\right] . \tag{7.33}$$

For $x_B = 0$, $P_z^{\text{non-osc}} = 1/2$, for $x_B \to \infty$, $P_z^{\text{non-osc}} = 1$.

From a fit of repolarization data with Eq. 7.33 and using the definition of x_B, Eq. 7.17, the hyperfine coupling constant or splitting ν_0 between the triplet and the singlet states can be determined, see Fig. 7.5. This quantity can also be deduced in an LF measurement from the field at which half of the polarization is recovered. For the field $B_{\text{ext},1/2}$ it holds

$$\frac{x_{B_{1/2}}^2}{1 + x_{B_{1/2}}^2} = \frac{1}{2} \Rightarrow x_{B_{1/2}} = 1,$$

$$B_{\text{ext},1/2} = B_0 = \frac{2\pi \nu_0}{\gamma_e + \gamma_\mu} . \tag{7.34}$$

---

[20] Note that in free (vacuum) muonium the polarization is not lost, but periodically returns to its full value, see Fig. 7.4. In a medium in which muonium more or less strongly interacts with its surrounding, irreversible processes can cause relaxation of the electron spin (and hence of the muon) so that the polarization disappears completely.

Fig. 7.5 (a) μSR asymmetry spectra and fits of a weakly bound muonium in BaSi$_2$ at longitudinal fields B$_{ext}$ = 0, 5, and 35 mT measured at 50 K. (**b**) Repolarization curve. In this weakly bound isotropic muonium, see Sect. 7.5, the hyperfine coupling constant and hence B_0 are much smaller than in vacuum muonium and only small fields are necessary to repolarize muonium. The plot of the initial asymmetry at 50 K versus longitudinal field shows the increase of the non-oscillating part with field. Fitting the data to Eq. 7.33 yields B$_{ext,1/2}$ = 8.8 mT corresponding to an hyperfine coupling constant A/h = 248 MHz. Modified from Xu et al. (2020), © IOP Publishing. Reproduced with permission

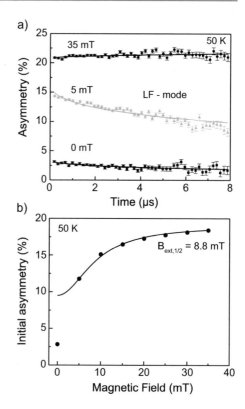

7.3.3   Transverse Field

In a transverse field geometry we take as before the external field along the z direction and the initial muon spin polarization along the x direction. We have therefore $\mathbf{P}(0) = P_x(0)\,\hat{\mathbf{x}}$ (with $P(0) = 1$) and $\mathbf{B}_{ext} = B_{ext}\,\hat{\mathbf{z}}$. Therefore, $\mathbf{P}(0)\cdot\boldsymbol{\sigma}_\mu = \sigma_{\mu,x}$. With the magnetic field perpendicular to the initial muon spin polarization, neither of the states $|\uparrow_\mu\uparrow_e\rangle_\perp$ and $|\uparrow_\mu\downarrow_e\rangle_\perp$ is an eigenstate, since their quantization axis is no longer along the field.[21] Therefore more frequencies appear in the TF case. Using again Eq. 7.29 and the eigenstates defined in Eq. 7.18 one obtains for the time evolution of the TF polarization in vacuum muonium

$$P_x(t) = \frac{1}{2}\left\{\cos^2\delta\,[\cos(\omega_{12}\,t) + \cos(\omega_{34}\,t)] + \sin^2\delta\,[\cos(\omega_{14}\,t) + \cos(\omega_{23}\,t)]\right\}\,. \tag{7.35}$$

---

[21] This is indicated with the index $\perp$.

The frequencies correspond to transitions between eigenstates. Note that the $|1\rangle \rightarrow |3\rangle$ and $|2\rangle \rightarrow |4\rangle$ transitions do not contribute to $P_x(t)$. Similar to the LF/ZF case, too high frequencies cannot be resolved with a conventional μSR spectrometer with a time resolution of about 1 ns.

We distinguish some typical field ranges:

- **Very low fields,** $x_B \ll 1$ For very low fields ω_{14} and $\omega_{34} \rightarrow \omega_0$. As for ω_0 these transition frequencies are too high to be observed, i.e., $\langle \cos(\omega_{14} t)\rangle = 0$ and $\langle \cos(\omega_{34} t)\rangle = 0$ and only ω_{12} and ω_{23} are measurable.[22]
  Moreover, for fields $\lesssim 1$ mT the two remaining frequencies collapse into a single one

$$\omega_{12} \simeq \omega_{23} \equiv \omega_{Mu}^T$$

$$= \frac{1}{2}(\omega_e - \omega_\mu) = \frac{1}{2}(\gamma_e - \gamma_\mu)B_{ext}$$

$$= \gamma_{Mu}^T B_{ext} \simeq \frac{1}{2}\gamma_e B_{ext} \ , \tag{7.36}$$

and $\sin\delta = \cos\delta = 1/\sqrt{2}$. The time dependence of the polarization simplifies to

$$P_x(t, x_B) \ll 1 = \frac{1}{2}\cos(\omega_{Mu}^T t) \ . \tag{7.37}$$

This means that in very small transverse fields, muonium shows only half of the polarization amplitude (corresponding to the precession of the $m_F = \pm 1$ components in the triplet state[23] at zero field, where muon and electron are locked together) and precesses with an effective gyromagnetic ratio γ_{Mu}^T about 100 times higher than γ_μ, giving a Larmor frequency $\omega_{Mu}^T \simeq 102.883 \times \omega_\mu$, see Table 7.1. Since $\gamma_e \gg \gamma_\mu$ the precession is dominated by the electron magnetic moment and therefore the triplet state precesses in the opposite sense to that of the free muon.[24]

In an experiment, to detect possible muonium formation, one performs a TF measurement at very low fields ($\lesssim 1$ mT). No information about the hyperfine coupling constant is found this way, but the detection of an oscillation with a frequency about two orders of magnitude higher than expected for a "free"

---

[22] The averaging here is over a time interval of the order of the time resolution of the spectrometer.

[23] This fact is underlined by the superscript T.

[24] Note that the angular frequency vectors ω_μ and ω_e are antiparallel. In this book we choose the g-factors and gyromagnetic ratios γ's positive for muon as well as for electron. Therefore the related quantities ω_μ and ω_e (and ω_{Mu}^T) are also positive and express the absolute values of the frequencies. One has to remember the different senses of precession when muon and muonium precession both contribute to the μSR spectrum.

muon, together with a missing fraction of equal amplitude,[25] is a clear signature of muonium formation.[26] In Sect. 8.4, an example is shown where the different magnitude of the precession frequencies has been used to distinguish muonium from the μ^+ state.

- **Low fields** Again, x_B is much less than 1, but higher than in the case of very low fields. Typical low fields for vacuum muonium are 10–20 mT. The splitting of the muonium precession and the hyperfine oscillation frequencies cannot be ignored. The four frequencies ω_{12}, ω_{34}, ω_{14} and ω_{23} have practically the same amplitude, Eq. 7.35. However, with conventional spectrometers ω_{14} and ω_{34} are too high to be observed and only the frequencies ω_{12} and ω_{23} are measurable. Since the approximation $\omega_{12} \simeq \omega_{23}$ can no longer be used

$$P_x(t, x_B \ll 1) = \frac{1}{2}\left\{\cos^2\delta \cos(\omega_{12}\,t) + \sin^2\delta \cos(\omega_{23}\,t)\right\} \;, \qquad (7.38)$$

which, using common trigonometric identities, can be rewritten as

$$P_x(t, x_B) = \tfrac{1}{2}\cos\left(\frac{\omega_{23} - \omega_{12}}{2}\,t\right)\cos\left(\frac{\omega_{23} + \omega_{12}}{2}\,t\right) +$$

$$+\; \tfrac{1}{2}(\cos^2\delta - \sin^2\delta)\sin\left(\frac{\omega_{23} - \omega_{12}}{2}\,t\right)\sin\left(\frac{\omega_{23} + \omega_{12}}{2}\,t\right) \;. $$

$$(7.39)$$

It is useful to introduce the variables $\omega_\pm = (\omega_{23} \pm \omega_{12})/2$. With these, and using the normalized field x_B

$$P_x(t, x_B \ll 1) = \frac{1}{2}\cos\left(\omega_+ t\right)\cos\left(\omega_- t\right) + \frac{1}{2}\frac{x_B}{\sqrt{1 + x_B^2}}\sin\left(\omega_+ t\right)\sin\left(\omega_- t\right) \;. $$

$$(7.40)$$

The contribution of the second term of Eq. 7.40 is negligible. Therefore, the signal corresponds to a muonium precession at frequency ω_+, modulated at the "beating frequency" ω_-, with half the initial muon polarization amplitude. This behavior can be clearly seen in the quartz measurement shown in Fig. 7.6.

---

[25] One speaks of missing fraction if the total initial asymmetry is smaller than the experimental asymmetry of the setup. A missing fraction is generally due to very fast relaxing processes, e.g., in the case of muonium interacting with a medium or of μ^+ in a magnetic material.

[26] As already pointed out the expressions in this section for the time evolution of a muon bound in Mu have been derived under assumption of a muonium not interacting with a medium. In a medium, due to its large γ_e, the electron of muonium is very sensitive to dynamic or static fields that depolarize the muonium and can suppress the precessing signal. Muonium is generally observed when the host medium does not have strong nuclear moments or when the muonium diffuses fast enough to be in the motionally narrowing regime, see Sect. 4.3.1.1. If muonium completely depolarizes, its "presence" appears as a missing fraction.

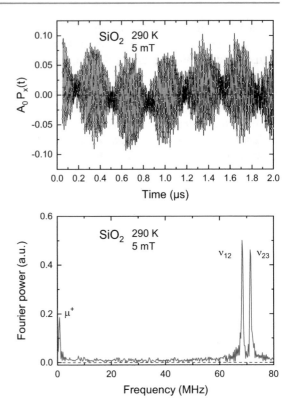

Fig. 7.6 Top: high statistics transverse field μSR spectrum in quartz at 3 mT and room temperature. Note the beating between the ν_{12} and ν_{23} frequencies superposed to a diamagnetic signal (dashed blue curve) due to the diamagnetic fraction in quartz. Bottom: frequency spectrum. One can identify the precession frequency of the diamagnetic muon (at ~ 0.68 MHz) and, at about 100 times higher frequency, the pair from isotropic muonium ν_{12} and ν_{23} centered at ~ 70 MHz

The observation of this two-frequency precession pattern allows to determine the hyperfine splitting frequency of the muonium atom (Gurevitch et al. 1971), because in the low-field approximation the hyperfine frequency is inversely proportional to the splitting of the two intratriplet precession frequencies $\omega_{23} - \omega_{12}$, see Exercise 7.4. The determination of the hyperfine frequency in this way is more precise than from the longitudinal repolarization measurement described above.

- **Very high fields,** $x_B \gg 1$ From Eq. 7.35 for $x_B \gg 1$ it follows $\cos\delta \to 1$ and $\sin\delta \to 0$, so that the amplitudes for the $|1\rangle \to |4\rangle$ and $|2\rangle \to |3\rangle$ transitions are practically zero and only the two lower frequencies are observable

$$P_x(t, x_B \gg 1) = \frac{1}{2}\{\cos(\omega_{12}\,t) + \cos(\omega_{34}\,t)\} \quad . \tag{7.41}$$

The limits of the transitions are

$$\nu_{12} \to \frac{\nu_0}{2} - \nu_\mu$$

$$\nu_{14} \to \nu_e + \frac{\nu_0}{2}$$

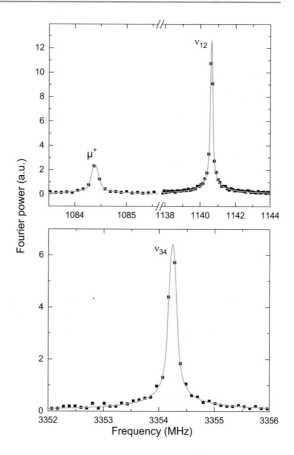

$$\nu_{23} \rightarrow \nu_e - \frac{\nu_0}{2}$$

$$\nu_{34} \rightarrow \nu_\mu + \frac{\nu_0}{2} \,. \qquad\qquad (7.42)$$

We have to distinguish two cases here, depending on whether ν_{12} is positive or
negative, i.e., whether ν_μ is smaller or larger than $\nu_0/2$.[27]
If $\nu_\mu < \nu_0/2$, the sum of the two observable frequencies gives us the hyperfine
constant $\nu_{12} + \nu_{34} = \nu_0$. Figure 7.7 shows that a measurement at 8 T of syntetic
quartz, which has an hyperfine coupling constant close to the vacuum value,
represents such a case: $\nu_\mu = 1084.4$ MHz $< \nu_0/2 \simeq 2240$ MHz.
If $\nu_\mu > \nu_0/2$, the fit of the μSR spectra in time domain or their Fourier transform
gives us the absolute value of the transition frequency, $|\nu_{12}|$ which corresponds to

---

[27] Remember Fig. 7.2 showing how the energy difference of the two levels evolves with the applied
field.

Fig. 7.8 Breit-Rabi diagram of isotropic muonium in a magnetic field. The vertical black lines indicate the transitions which are observed in zero, longitudinal (solid line) and in transverse field experiments (dashed lines). The eigenstates in the limiting cases $x_B \ll 1$ and $x_B \gg 1$ are also shown, as well as the directions of the muon and electron magnetic moments. Note the energy scale and the hyperfine splitting energy of 18.477 μeV

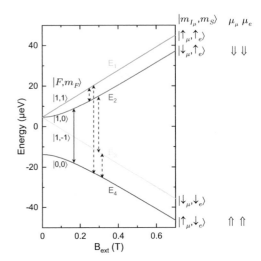

-v_{12}. In this case, the difference between the two transition frequencies is a direct measure of the hyperfine coupling constant, $v_{34} - |v_{12}| = v_0$ and the v_{12} and v_{34} lines are symmetrically placed around the diamagnetic line at $\gamma_\mu B_{\text{ext}}/2\pi$.

As we will see in Sect. 7.5, the very high field condition is realized in shallow muonium, where the hyperfine coupling constant is much smaller than the vacuum value $v_0 = 4463$ MHz, so that the high field limit is reached at fields of a few tens of mT.

Figure 7.8 summarizes in the Breit-Rabi diagram of isotropic muonium the transitions observed in zero, longitudinal, or transverse field experiments and discussed up to now.

The selection rules determining which frequencies are actually observed are a function of x_B, i.e., of the relative strength of B_{ext} with respect to B_0. If the applied field is small enough, in the total spin basis, the possible transitions correspond to $\Delta F = 0, \pm 1$ and $\Delta m_F = \pm 1$ but usually only the transitions with $\Delta F = 0$ and $\Delta m_F = \pm 1$ are observed corresponding to the frequencies ω_{12} and ω_{23}. If the applied field is high enough, the eigenstates tend to the Zeeman states and the possible transitions obey the selection rules $\Delta m_S = 0, \pm 1$ and $\Delta m_{I_\mu} = 0, \pm 1$. The limiting case $x_B \gg 1$ corresponds to $\Delta m_S = 0$ and $\Delta m_{I_\mu} = \pm 1$ and practically only ω_{12} and ω_{34} appear.

7.3.4   Nuclear Hyperfine Interaction

With a spectrometer of very good time resolution one can measure the muonium hyperfine frequency from the energy splitting between triplet and singlet states in zero field. However, this can only be realized in systems or conditions where the line broadening (or the muonium spin relaxation) is small. For this, the nuclear moments must not be too strong. An important source of line broadening is the contact interaction with the surrounding host nuclear spins, so-called nuclear hyperfine (NHFI) or superhyperfine interaction, since the muonium electron may have a significant density probability at these nuclei. This nuclear contact term and the Zeeman splitting of the nuclear levels add two terms to the Hamiltonian 7.12

$$\mathcal{H}_{\text{NHFI}} = \sum_{J=1}^{n} \mathbf{I}_{\text{N},j} \cdot \tilde{\mathbf{A}}_{\text{N},j} \cdot \mathbf{S} - \sum_{j=1}^{n} g_{\text{N},j}\, \mu_{\text{NM}} \mathbf{I}_{\text{N},j} \cdot \mathbf{B}_{\text{ext}} \ , \qquad (7.43)$$

where $\tilde{\mathbf{A}}_{\text{N}}$ is the nuclear hyperfine tensor. The nuclear hyperfine interaction produces a splitting of the muonium energy levels leading to a complex frequency spectrum with many weak lines and lines that cannot be resolved.[28] In special cases, e.g., with isotropic hyperfine and isotropic nuclear hyperfine interactions, analytical expressions for the precession frequencies and amplitudes can be obtained (Roduner & Fischer 1981).

A practical way to determine the hyperfine coupling constant A is to use the relation

$$h\nu_{12} + h\nu_{34} = A \ , \qquad (7.44)$$

which is valid in any field, see Eq. 7.16, and measure at sufficiently high fields (of the order of T) to quench the nuclear hyperfine interactions.

7.3.5   Isotropic Muonium in Solids

The Hamiltonian 7.12 describes an isotropic isolated muonium,[29] i.e., without interaction with its environment. In inert, nonconducting materials, muonium localized at interstitial lattice positions can exist in a state very similar to the atomic state. Favorable conditions for this to occur are (i) a free electron density at the muon site in the material small enough to weaken the screening of the Coulomb interaction and (ii) a high point symmetry of the muonium site. This condition is met in a number of semiconductors, dielectrics and organic compounds.

Figure 7.9 shows the hyperfine coupling constant (normalized to the vacuum value) versus the bandgap energy of various host materials (Cox 2003). In alkali fluorides the hyperfine coupling is slightly higher than in vacuum muonium. This

---

[28] Nuclear hyperfine interactions play an important role in the study of muoniated organic radicals, because the paramagnetic electron may interact with one or more hydrogen nuclei of the compound, see references in footnote 33, Chap. 5.

[29] This is the case in vacuum, but also in inert gases at not too high density. In fact, the most precise measurement of the muonium hyperfine splitting frequency ($\nu_0 = 4,463,302,765(53)$ Hz (12 ppb), Liu et al. 1999) has been obtained in krypton making use of Eq. 7.44.

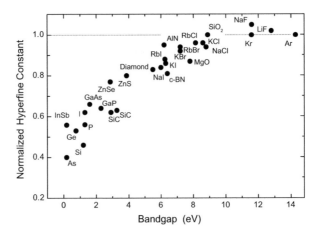

Fig. 7.9 Hyperfine coupling constant (normalized to the vacuum value) for isotropic muonium in semiconductors and dielectrics. The graph shows a broad correlation between the hyperfine coupling constant (which represents the spin density at the muon site) and the bandgap of the host material. Modified from Cox (2003), © IOP Publishing. Reproduced with permission. All rights reserved

corresponds to a slightly compressed wave function with a shorter radius of the $1s$ state, see Eq 7.6. In general, the hyperfine coupling is smaller than the vacuum value, corresponding to an expanded Mu $1s$ state. In the elementary semiconductors Si, Ge (group IV) it is only half of the free muonium value reflecting a lower electron spin density at the muon site and an increased Mu radius. The broad correlation between the hyperfine coupling constant and the bandgap value reflects the fact that muonium retains its atomic character when there is a large separation between the bonding and anti-bonding orbitals of the host material. When these energy levels are close (narrow gap), greater mixing with the host states is possible and the free atomic character of muonium is lost.

Example of Muonium Spectroscopy An example of a geophysical application of muonium spectroscopy to understand the behavior of hydrogen is the observation of isotropic muonium with a large hyperfine constant ($(A/A_{Mu}) > 1$) in stishovite (Funamori et al. 2015). Stishovite is a hard and dense tetragonal form of silicon dioxide. It is a rutile-type high-pressure phase whereas quartz is an ambient-pressure phase of SiO_2. It is very rare on the Earth's surface, but it appears to be a predominant form of silicon dioxide especially in the lower Earth's mantle.

The aim of the experiment was to address the question in which form hydrogen, which has been thought to exist as a OH group in high-pressure minerals, exists in the Earth's deep mantle. In stishovite, as in quartz, a large muonium fraction is found ($\approx 56\%$). In stishovite the hyperfine coupling parameter and the relaxation rate of the spin polarization of muonium are very large: the hyperfine frequency, A/h, at 300 K is found to be 4.67(3) GHz and at 2.5 K 5.17(14) GHz. These values

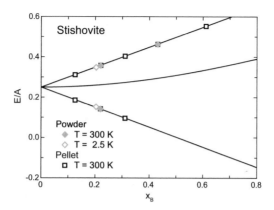

Fig. 7.10 Energy (normalized to the hyperfine coupling constant) diagram of the muonium triplet state in stishovite as a function of the normalized transverse field. The three solid lines from top to bottom, represent theoretical values of E_1/A, E_2/A and E_3/A. The data of E_1/A and E_3/A, from left to right, are measurements at 20, 35, 50, 70, and 100 mT. Because A is larger at 2.5 K than at 300 K, the data for 2.5 K lie to the left of the corresponding data at 300 K. Adapted from Funamori et al. (2015), under CC-BY-4.0 license. © The Authors

are significantly larger than 4.463 GHz for muonium in vacuum and 4.49(2) GHz for muonium in quartz, see Fig. 7.10. This implies that muonium is squeezed in small interstitial voids without binding to silicon or oxygen. In fact, the interstitial voids of stishovite, which consists of SiO_6 octahedra, are much smaller than those of quartz, which consists of SiO_4 tetrahedra. The results suggest that the formation of muonium is not controlled by the size of interstitial voids and that hydrogen may also exist in the form of neutral atomic hydrogen in the deep Earth's mantle.

7.4 Anisotropic Muonium

In general, we can write the Hamiltonian for a nonisotropic muonium by replacing the scalar hyperfine coupling constant A_{Mu} in Eq. 7.12 with a tensor $\tilde{\mathbf{A}}$

$$\mathcal{H}_{anis} = \mathbf{I}_\mu \cdot \tilde{\mathbf{A}} \cdot \mathbf{S} + g_e \mu_B \, \mathbf{S} \cdot \mathbf{B}_{ext} - g_\mu \mu_B^\mu \, \mathbf{I}_\mu \cdot \mathbf{B}_{ext} \ . \tag{7.45}$$

In several semiconductors, including Si, Ge, and GaAs, an anisotropic muonium state has been found[30] with a hyperfine constant that is axially symmetric around any of the [111] crystal axes. In the early literature, this anisotropic state is often called "anomalous muonium" or denoted as Mu*. It is described by the Hamiltonian

$$\mathcal{H}_{anis} = A_{||} I_{\mu,z} S_z + A_\perp \left(I_{\mu,x} S_x + I_{\mu,y} S_y \right) + g_e \mu_B \, \mathbf{S} \cdot \mathbf{B}_{ext} - g_\mu \mu_B^\mu \, \mathbf{I}_\mu \cdot \mathbf{B}_{ext} \ , \tag{7.46}$$

---

[30] This state was first clearly observed in silicon (Brewer et al. 1973) and later in many other semiconductors, see Patterson (1988); Cox (2009); Cox et al. (2013) for a review.

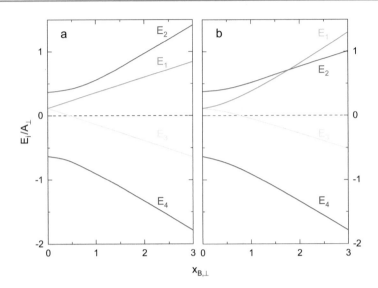

Fig. 7.11 Schematic of the hyperfine energy level diagram for anisotropic Mu (with $A_{\parallel}/A_{\perp} = 1/2$) when the field is applied along a [111] symmetry axis (a panel) or perpendicular to it (b panel). The field is expressed in the scaled variable $x_{B,\perp}$. The lower symmetry of the hyperfine interaction lifts part of the triplet state degeneracy of vacuum muonium at zero field. A nonphysical value of the gyromagnetic ratio $\gamma_{\mu}/\gamma_e = 1/3$ has been chosen to expose the effects of the anisotropic coupling. The numbering of the levels is such that, in zero field, the energy is the same in both cases

where $A_{\parallel}$ and $A_{\perp}$ are the tensor components parallel and perpendicular to the symmetry axis.[31] The energy eigenvalues and eigenvectors of Mu* depend not only on the strength of the magnetic field but also on its orientation with respect to the symmetry axis, see Fig. 7.15. Generally, their determination requires the solution of a fourth-order algebraic equation, which must be performed numerically.

Simple analytical expressions can be obtained for special cases.[32] Figure 7.11 shows the energy level for $\mathbf{B}_{ext}$ applied parallel and perpendicular to a [111] symmetry axis (z direction).

For the parallel case we have

$$\mathcal{H}'_{anis} = A_{\parallel} I_{\mu,z} S_z + A_{\perp} \left(I_{\mu,x} S_x + I_{\mu,y} S_y \right) + g_e \mu_B \, S_z B_z - g_\mu \mu_B^\mu \, I_{\mu,z} B_z \ . \tag{7.47}$$

---

[31] In solids, the g-factor for a bound electron in muonium may be more different from the free particle value than for vacuum muonium, see footnote 8, and may even be anisotropic. For an example in diamond see (Holzschuh et al. 1982). In addition, the value of g_e can be modified by an orbital contribution to the electron moment. We do not consider these points here, as they do not change the essence of the discussion.

[32] See for instance Hintermann et al. (1980), where Mu* is presented as an exemplary quantum mechanical system for which analytical solutions can be found, and also Patterson (1988); Senba (2000).

The eigenvalues can be calculated as in the isotropic case and one obtains

$$E_1 = \frac{1}{4}\hbar\omega_{||} + \frac{1}{2}\hbar\omega_e - \frac{1}{2}\hbar\omega_\mu$$

$$E_2 = -\frac{1}{4}\hbar\omega_{||} + \frac{1}{2}\hbar\left[\omega_\perp^2 + (\omega_e + \omega_\mu)^2\right]^{1/2} = -\frac{1}{4}\hbar\omega_{||} + \frac{1}{2}\hbar\omega_\perp\left[1 + x_{B,\perp}^2\right]^{1/2}$$

$$E_3 = \frac{1}{4}\hbar\omega_{||} - \frac{1}{2}\hbar\omega_e + \frac{1}{2}\hbar\omega_\mu$$

$$E_4 = -\frac{1}{4}\hbar\omega_{||} - \frac{1}{2}\hbar\left[\omega_\perp^2 + (\omega_e + \omega_\mu)^2\right]^{1/2} = -\frac{1}{4}\hbar\omega_{||} - \frac{1}{2}\hbar\omega_\perp\left[1 + x_{B,\perp}^2\right]^{1/2} ,$$

$$(7.48)$$

where $A_{||} = \hbar\omega_{||}$, $A_\perp = \hbar\omega_\perp$ and $x_{B,\perp} = (\omega_e + \omega_\mu)/\omega_\perp$. The corresponding eigenstates can be obtained from Eq. 7.18 by substituting ω_0 with $\omega_\perp$ and x_B with $x_{B,\perp}$.

If $\mathbf{B}_{ext}$ is perpendicular to a [111] symmetry axis the eigenvalues read

$$E_1 = -\frac{1}{4}\hbar\omega_\perp + \frac{1}{2}\hbar\left[\left(\frac{\omega_{||} + \omega_\perp}{2}\right)^2 + (\omega_e + \omega_\mu)^2\right]^{1/2}$$

$$= -\frac{1}{4}\hbar\omega_\perp + \frac{1}{4}\hbar(\omega_{||} + \omega_\perp)\left[1 + x_{B,+}^2\right]^{1/2}$$

$$E_2 = \frac{1}{4}\hbar\omega_\perp + \frac{1}{2}\hbar\left[\left(\frac{\omega_{||} - \omega_\perp}{2}\right)^2 + (\omega_e - \omega_\mu)^2\right]^{1/2}$$

$$= \frac{1}{4}\hbar\omega_\perp + \frac{1}{4}\hbar(\omega_{||} - \omega_\perp)\left[1 + x_{B,-}^2\right]^{1/2}$$

$$E_3 = \frac{1}{4}\hbar\omega_\perp - \frac{1}{2}\hbar\left[\left(\frac{\omega_{||} - \omega_\perp}{2}\right)^2 + (\omega_e - \omega_\mu)^2\right]^{1/2}$$

$$= \frac{1}{4}\hbar\omega_\perp - \frac{1}{4}\hbar(\omega_{||} - \omega_\perp)\left[1 + x_{B,-}^2\right]^{1/2}$$

$$E_4 = -\frac{1}{4}\hbar\omega_\perp - \frac{1}{2}\hbar\left[\left(\frac{\omega_{||} + \omega_\perp}{2}\right)^2 + (\omega_e + \omega_\mu)^2\right]^{1/2}$$

$$= -\frac{1}{4}\hbar\omega_\perp - \frac{1}{4}\hbar(\omega_{||} + \omega_\perp)\left[1 + x_{B,+}^2\right]^{1/2} ,$$

$$(7.49)$$

where we have introduced the two scaled fields

$$x_{B,+} = \frac{\omega_e + \omega_\mu}{\frac{1}{2}(\omega_{||} + \omega_\perp)} \quad \text{and} \quad x_{B,-} = \frac{\omega_e - \omega_\mu}{\frac{1}{2}(\omega_{||} - \omega_\perp)} . \tag{7.50}$$

The corresponding eigenstates are given by

$$|1\rangle = \sin \zeta \, |\uparrow_\mu \downarrow_e\rangle + \cos \zeta \, |\downarrow_\mu \uparrow_e\rangle$$
$$|2\rangle = \cos \eta \, |\uparrow_\mu \uparrow_e\rangle + \sin \eta \, |\downarrow_\mu \downarrow_e\rangle$$
$$|3\rangle = -\sin \eta \, |\uparrow_\mu \uparrow_e\rangle + \cos \eta \, |\downarrow_\mu \downarrow_e\rangle$$
$$|4\rangle = \cos \zeta \, |\uparrow_\mu \downarrow_e\rangle - \sin \zeta \, |\downarrow_\mu \uparrow_e\rangle \ , \tag{7.51}$$

where

$$\cos \zeta = \frac{1}{\sqrt{2}} \left(1 + \frac{\omega_e + \omega_\mu}{\left[\left(\frac{\omega_\parallel + \omega_\perp}{2} \right)^2 + (\omega_e + \omega_\mu)^2 \right]^{1/2}} \right)^{1/2}$$

$$= \frac{1}{\sqrt{2}} \left(1 + \frac{x_{B,+}}{\left[1 + x_{B,+}^2 \right]^{1/2}} \right)^{1/2}$$

$$\sin \zeta = \frac{1}{\sqrt{2}} \left(1 - \frac{\omega_e + \omega_\mu}{\left[\left(\frac{\omega_\parallel + \omega_\perp}{2} \right)^2 + (\omega_e + \omega_\mu)^2 \right]^{1/2}} \right)^{1/2}$$

$$= \frac{1}{\sqrt{2}} \left(1 - \frac{x_{B,+}}{\left[1 + x_{B,+}^2 \right]^{1/2}} \right)^{1/2} \tag{7.52}$$

$$\cos \eta = \frac{1}{\sqrt{2}} \left(1 + \frac{\omega_e - \omega_\mu}{\left[\left(\frac{\omega_\parallel - \omega_\perp}{2} \right)^2 + (\omega_e - \omega_\mu)^2 \right]^{1/2}} \right)^{1/2}$$

$$= \frac{1}{\sqrt{2}} \left(1 + \frac{x_{B,-}}{\left[1 + x_{B,-}^2 \right]^{1/2}} \right)^{1/2}$$

$$\sin \eta = \frac{1}{\sqrt{2}} \left(1 - \frac{\omega_e - \omega_\mu}{\left[\left(\frac{\omega_\| - \omega_\perp}{2} \right)^2 + (\omega_e - \omega_\mu)^2 \right]^{1/2}} \right)^{1/2}$$

$$= \frac{1}{\sqrt{2}} \left(1 - \frac{x_{B,-}}{\left[1 + x_{B,-}^2 \right]^{1/2}} \right)^{1/2} . \tag{7.53}$$

The principal values of the coupling tensor $A_\|$ and $A_\perp$ can be decomposed into an isotropic (contact) term A_{iso} and an anisotropic (dipolar) term D

$$A_{iso} = \frac{1}{3}(A_\| + 2A_\perp) \tag{7.54}$$

and

$$D = (A_\| - A_\perp) . \tag{7.55}$$

With this, the $\mathbf{I}_\mu \cdot \mathbf{S}$ term in Eq. 7.46 can be written as

$$A_{iso}\mathbf{I}_\mu \cdot \mathbf{S} + D(I_{\mu,z}S_z - \frac{1}{3}\mathbf{I}_\mu \cdot \mathbf{S}) . \tag{7.56}$$

Anisotropic axially symmetric muonium is found to coexist with isotropic muonium in various semiconductors. In archetypal semiconductors such as Si, Ge, and GaAs, muonium with an isotropic hyperfine coupling is located at the tetrahedral interstitial site (T) and labeled as Mu_T. The anomalous species Mu_{BC} with an anisotropic hyperfine coupling constant is located at the center of a stretched bond[33] between lattice atoms, so-called bond center site (BC) labeled Mu_{BC}, see Figs. 7.12 and 7.13.

Figure 7.14 compares the Fourier spectra of measurements in quartz, where only isotropic muonium is formed, with those in silicon. In the latter, besides an isotropic muonium with weaker hyperfine coupling, additional lines due to transitions in Mu_{BC} (Mu^*) are observed. The hyperfine frequency of isotropic muonium in Si is 2006 MHz, indicating that the electron density at the μ^+ "nucleus" is only half the value[34] for vacuum muonium. The hyperfine interaction of Mu_{BC} is generally smaller than that of isotropic muonium. For example, in silicon $A_\perp(Mu_{BC})/A(Mu_T) \simeq 0.05$ and the anisotropy is $A_\|/A_\perp \simeq 0.2$.

---

[33] The insertion of Mu_{BC} leads to an elongation of the Si-Si bond length from 0.24 nm in the undistorted crystal to 0.32 nm.

[34] Remember $A \propto |\psi(\mathbf{r}_\mu)|^2$, see Eq. 7.6.

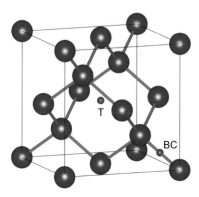

Fig. 7.12 A tetrahedral (T) isotropic and a bond-centered (BC) anisotropic muonium site in a crystal with the zinc blende structure, which is adopted for example by Si, Ge, diamond, and GaAs. Structure drawn with VESTA, © 2006–2022 Koichi Momma and Fujio Izumi (Momma and Izumi 2011)

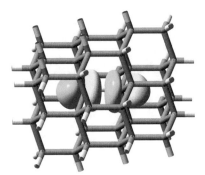

Fig. 7.13 Computational simulation of the H_{BC} wave function in silicon, which is expected to well represent Mu_{BC}. The singly occupied orbital has a node at the bond-centered site indicated by the white segment in the middle. The colors show the sign change of the wave function, with a spin density preferentially localized on the neighboring Si atoms. Modified from Cox (2009), © IOP Publishing. Reproduced with permission. All rights reserved

Figures 7.15 and 7.16 present the results from the first observation of normal (isotropic) and anisotropic muonium centers in the GaAs and GaP compound semiconductors (Kiefl et al. 1985). The results are well explained by the anisotropic spin Hamiltonian, axially symmetric about the ⟨111⟩ direction, Eqs. 7.46 and 7.47.[35]

---

[35] The z-axis corresponds to one of the four [111] axes of the crystal. Note that Mu_{BC} forms at each of the four [111] axes of the crystal.

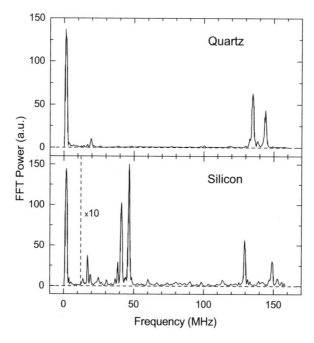

Fig. 7.14 Top: frequency spectra from TF measurements with an external field of 10 mT in fused quartz at room temperature. Bottom: spectrum in p-type silicon at 77 K (field along [111]) showing the precession components of the free muon ($\simeq$ 1.36 MHz) and isotropic Mu (the pair ν_{12} and ν_{23} centered around $\sim$ 140 MHz). Note the larger splitting of the Mu lines in Si, indicating a weaker hyperfine interaction, see Exercise 7.4. Note also the presence in Si, but not in fused quartz, of Mu* precession lines (ν_{12} and ν_{34}, with an angle of 70.5° between the direction of the field and the [111] axis) at 41 and 46 MHz. Modified from Brewer et al. (1973), © American Physical Society. Reproduced with permission. All rights reserved

The measurement was performed at a high field of 1.15 T to quench the line broadening due to the nuclear moments.[36] In high transverse magnetic fields ($\mu_B B_{\text{ext}} \gg |A_{||}|$ and $|A_{\perp}|$) for each angle θ between the $\langle 111 \rangle$ symmetry axis and the magnetic field direction there are two precession frequencies ν_{12}^* and ν_{34}^*, see Fig. 7.16. With the magnetic field applied parallel to a $\langle 111 \rangle$ direction, this leads to a total of four precession frequencies: two from centers with symmetry axis at an angle $\theta = 0°$ and the other two from centers with $\theta = 70.5°$, see Fig. 7.15. When the field is applied along a $\langle 110 \rangle$ axis the centers are at $\theta = 90°$ and $\theta = 35°$, respectively. Figure 7.16 shows the field dependence of the corresponding frequencies.

---

[36] GaAs (as well as GaP) contains a high percentage of nuclei with magnetic dipole moments. Muonium precession frequencies cannot be detected in low magnetic fields because of the line broadening due to the nuclear hyperfine interaction (NHFI), see Sect. 7.3.4.

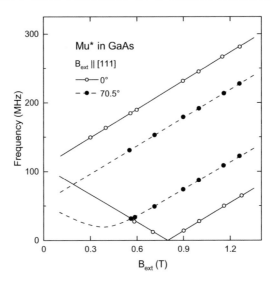

Fig. 7.15 Magnetic field dependence of the anisotropic muonium frequencies in GaAs at 23 K. The frequencies depend on the angle θ between a [111] axis and the applied field. The solid and dashed curves are a fit to the theoretical spectrum from Eq. 7.46 for $\theta = 0°$ and $\theta = 70.5°$, respectively. Modified from Kiefl et al. (1985), © American Physical Society. Reproduced with permission. All rights reserved

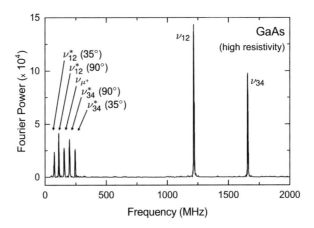

Fig. 7.16 Fourier transforms of TF-μSR spectra measured at 10 K with a high time resolution apparatus in high-resistivity GaAs with $B_{ext} = 1.15$ T applied along a $\langle 110 \rangle$ direction. Note the diamagnetic line (ν_{μ^+}), the two isotropic Mu lines ν_{12} and ν_{34}, and the Mu* lines $\nu_{ij}^*(\theta)$ (where θ is the angle between the $\langle 111 \rangle$ Mu* symmetry axis and the applied field direction). Modified from Kiefl et al. (1985), © American Physical Society. Reproduced with permission. All rights reserved

7.5   Shallow Muonium

In the group-IV elementary semiconductors and the III-V compounds described so far, one usually finds a diamagnetic state with a muon Larmor frequency corresponding to the external magnetic field, together with one or two paramagnetic Mu states, Mu_T and Mu_{BC}. These muonium states correspond to deep-level centers (or defects) in the material. The states are well separated from the conduction or valence bands, so that the energy required to remove an electron or hole from the center to the valence or conduction band is much larger than the characteristic thermal energy $k_B T$. Their wave function is quite compact with a hyperfine constant of the order of that of vacuum muonium, see Fig. 7.9. The wave function of Mu_{BC} is also rather compact and extends over only a few neighboring atoms. Its small hyperfine coupling constant ($A_{iso}/h = -67$ MHz in Si, -96 MHz in Ge and -206 MHz in diamond)[37] is due to the fact that the muon sits at a node of the wave function, see Fig. 7.13.

Studies of muonium in CdS and later in other II-VI semiconductors surprisingly revealed the existence of a third form of neutral anisotropic muonium (Gil et al. 1999, 2001). Its hyperfine interaction is very weak, about 10^{-4} of the vacuum value, indicating that the electron is only very weakly bound to the positive muon. The binding energy is characteristic of so-called shallow-level donor centers, most probably located at an antibonding site close to the anion, see Fig. 7.19. The site determination was made by comparing the width of the diamagnetic signal with that calculated from the magnetic field distribution created by the nuclear magnetic dipoles around the muon (Gil et al. 2001), see a similar example in Sect. 5.7.

Figure 7.17 shows the TF-μSR signal in CdS (10 mT), taken over a period of eight muon lifetimes. The Fourier transform of the signal shows five distinct frequencies: in addition to the μ^+ precession signal at 1.38 MHz, two pairs of lines symmetric around the central line are observed. The data can be explained by assuming an axially symmetric hyperfine interaction with the symmetry axis along the Cd-S bond direction.[38] In the geometry of Fig. 7.17, where the magnetic field is along the $\langle 0001 \rangle$ axis, there is one Cd-S bond, with direction at $\theta = 0°$ and three bonds at $\theta = 70.6°$ to the field direction (θ angle between the bond direction and the applied field). The intensity ratio of the muonium lines suggests that the outer pair of lines corresponds to $\theta = 0°$ and the inner pair to $\theta = 70.6°$.

The very low value of A ensures that even with the small applied field of 10 mT we are in the high field limit ($\mu_B B_{ext} \gg A$). In this case, with axial symmetry, a

---

[37] Only the relative sign between the Mu_{BC} hyperfine parameters ($A_\parallel$ and $A_\perp$) is known experimentally. How to determine the sign is shown in Odermatt et al. (1988).

[38] An anisotropic hyperfine constant with axial symmetry has been found in many systems. However, there are also examples of fully anisotropic shallow states, for example in $SrTiO_3$ (Salman et al. 2014).

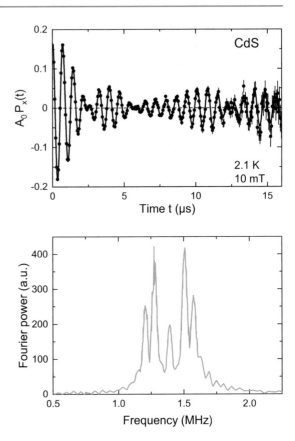

simple relation between the measured frequencies and the hyperfine tensor holds (Hintermann et al. 1980)

$$h\Delta v(\theta) = A(\theta) \cong A_{\parallel} \cos^2\theta + A_{\perp} \sin^2\theta = A_{\text{iso}} + \frac{D}{3}(3\cos^2\theta - 1) \ , \qquad (7.57)$$

where Δv is the separation of two lines symmetric around the central line, $A(\theta)$ is the hyperfine interaction for the angle θ, and $A_{\parallel}$ and $A_{\perp}$ are the hyperfine couplings constant parallel and perpendicular to the symmetry axis (Cd-S bond direction). The analysis of the results of Fig. 7.17 yields $A_{\parallel}/h = 335(7)$ kHz and $A_{\perp}/h = 199(6)$ kHz. This gives for the isotropic (contact) part of the interaction $A_{\text{iso}}/h = (A_{\parallel} + 2A_{\perp})/3h = 244$ kHz (Eq. 7.54) a value which is $1.8 \cdot 10^{-4}$ of the free muonium value ($A/h = 4463$ MHz). Remembering that $A \propto 1/a^3$, where a is the atom radius, the small A_{iso} indicates a shallow muonium (sMu) state with radius $a_{\text{sMu}} = 26 \, a_0 \approx 1.4$ nm. This means that its wave function extends over several lattice units.

An experimental estimate of the binding energy can be obtained from the temperature dependence of the asymmetry amplitudes of diamagnetic μ^+ and

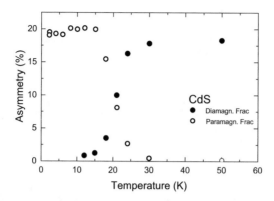

Fig. 7.18 Amplitudes of the diamagnetic (closed circles) and paramagnetic (open circles) fractions as a function of temperature in CdS, derived from fits in time domain of the spectra. The rapid decrease of the paramagnetic amplitude and the corresponding increase of the diamagnetic amplitude indicate the ionization of the paramagnetic muonium center. Modified from Gil et al. (1999), © American Physical Society. Reproduced with permission. All rights reserved

shallow muonium, Fig. 7.18, which shows that the "free" muon signal grows at the expense of the muonium signal. This result implies that with increasing temperature, the muonium center becomes ionized, so that the electron is no longer bound to the muon.

The binding energy of the electron obtained from the activation energy is $E = 18$ meV, i.e., much smaller than the bandgap of about 2.5 eV, see Fig. 7.19. The theoretical radius and binding energy of this state can be estimated using a hydrogenic model with effective electron mass and dielectric constant of the host material. With the electron distribution of shallow muonium simply described by a dilated hydrogen-like wave function in a dielectric medium (of dielectric constant ε) one can roughly estimate from the Bohr atom model $a_{\mathrm{sMu}} = a_0 \, \varepsilon / (m^*/m_e)$. Taking for CdS $\varepsilon = 9$ and an effective electron mass $m^*/m_e = 0.2$ we obtain a value of $a_{\mathrm{sMu}} = 45 \, a_0$ in reasonable agreement with the estimate from the hyperfine coupling constant. Within the same model a binding energy reduction from the free muonium value is expected to be $E_{\mathrm{sMU}} = R_y \, (m^*/m_e) \, \varepsilon^2 \approx 30$ meV, again in reasonable agreement with the ionization energy.[39]

In summary, the extremely weak hyperfine interaction and the disappearance of the muonium center at about 20 K by ionization clearly indicate that muonium forms a weakly bound shallow level close to the conduction band, acting as a donor. By analogy, the result shows that hydrogen atom impurities in materials with a shallow muonium form an electrically active shallow center, in contrast to the deep-level

---

[39] A later systematic investigation of shallow muonium states in the II-VI semiconductor compounds CdS, CdSe, CdTe, and in ZnO finds a better agreement between theory and experiment (Gil et al. 2001).

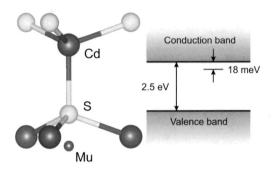

Fig. 7.19 Position of shallow muonium in CdS at the antibonding site. The site shown corresponds to the 0° site of Fig. 7.17. Note also that the three other sites close to sulphur are at 70.6°. The right side shows the energy level of the shallow state in the semiconducting gap, very close to the conduction band. Modified from Cox et al. (2000). © Science and Technology Facilities Council. Reproduced with permission. All rights reserved. Structure drawn with VESTA (Momma & Izumi 2011)

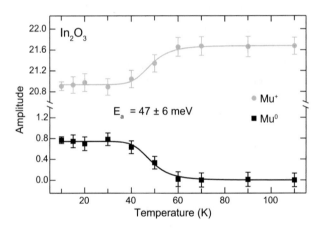

Fig. 7.20 Asymmetry amplitude of the paramagnetic shallow muonium (Mu^0, square symbol) and of the diamagnetic fraction (Mu^+ circle) in In_2O_3 as a function of temperature. Fitting the data to an ionization model yields an activation energy of 47 ± 6 meV. Modified from King et al. (2009), © American Physical Society. Reproduced with permission. All rights reserved

centers such as Mu^T and Mu_{BC} formed in elemental semiconductors and other dielectric materials.

Following the discovery of shallow muonium in CdS and related II-VI semiconductors this state has been found in many other compounds. Without claiming completeness we mention compounds belonging to the so-called transparent conducting oxides (TCO)[40] such as ZnO, CdO, In_2O_3 (Fig. 7.20), SnO_2 (King et al.

---

[40] Transparent conducting oxides are a class of oxide semiconductors, which can conduct and be transparent at visible wavelengths. They find applications in several devices including solar cells.

2009), and Ga_2O_3 [see King et al. 2010 and references therein], $SrTiO_3$ [a material with a very large dielectric constant at room temperature and low electric field ($\varepsilon \approx 300$) Salman et al. 2014], and FeS_2 (Okabe et al. 2018). The observation of shallow muonium donor states suggests that hydrogen is a shallow donor in these materials and that, in contrast to its behavior in conventional semiconductors, it may be an important source of conductivity when the weakly bound electron at low temperature becomes a free carrier at higher temperature.

A relevant question to be answered is which parameter governs the formation of either deep-level muonium or shallow-donor muonium. Of all the relevant properties, none of the known ones (e.g., bandgap, dielectric constant, bond-length, crystal structure etc.) shows a threshold at the transition from deep to shallow behavior.

The data suggest that the electron affinity (i.e., the position of the conduction-band minimum below the vacuum continuum) is a relevant parameter. The plot of the muonium hyperfine constant of a number of semiconductors (normalized to the free muonium value) versus the electron affinity, Fig. 7.21, shows a smooth decrease from a value essentially equal to one for SiO_2 to one half for silicon and a sharp transition to the weakly-bound shallow states once the electron affinity value of $\sim$ 3.5 eV is exceeded (Cox et al. 2013).

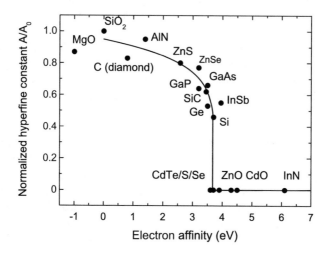

Fig. 7.21 Deep- to shallow-state transition in muonium. The figure shows that there is a correlation between the muonium hyperfine constant (normalized to that of free muonium, A_0) and the host electron affinity. The values for the shallow-donor states are not exactly zero, but very small $\lesssim 10^{-4}$. Modified from Cox et al. (2013), © IOP Publishing. Reproduced with permission. All rights reserved

7.6   Muon-Polaron Complexes

Nonmagnetic Oxides In oxides, H^+ and hence μ^+ tend to be attracted by the negatively charged oxygen ion, forming an $(OH)^-$ state, without breaking a nearby cation-O bond. Muonium spectroscopy experiments have found that in nonmagnetic transition metal oxides such as $SrTiO_3$ (Salman et al. 2014) and TiO_2 (Shimomura et al. 2015; Vilão et al. 2015) polaronic muonium centers are formed. In these centers, an oxygen-bound positive muon and an unpaired bound electron, which is predominantly located on a neighboring transition metal ion to form a small polaron, build up an overall charge-neutral complex.[41]

In rutile TiO_2, the hyperfine coupling constants indicate that at the lowest temperature (1.2 K), Mu is bound to one of the six oxygens surrounding Ti in the TiO_6 octahedron, with the electron located mainly at the Ti cation reducing it from Ti^{4+} to Ti^{3+}. The six O-Mu bonds are not equivalent. The "in plane" bond lies in the same ab-plane as Ti^{3+} and corresponds to the ground state configuration. The "out of plane" configuration (Mu bond to O atoms below and above the ab-plane) is slightly different and corresponds to an excited state position that becomes populated with increasing temperature. An example of the two configurations is sketched in Fig. 7.22. Comparison with hydrogen data (Brant et al. 2011; Vilão et al. 2015) shows that these configurations are the same for Mu and H.

Magnetic Oxides Muonium spectroscopy has been mostly limited to nonmagnetic materials. In magnetic materials it is generally assumed that neutral muonium states are not observable, because the muon spin in muonium, due to the large magnetic moment of the coupled electron, is expected to undergo a strong depolarization by the static or dynamic fields created by the magnetic moments of the host material.[42]

Recently, with the support of DFT calculations, neutral muon-polaron complexes have been shown to form in antiferromagnetic Cr_2O_3 (Dehn et al. 2020) and Fe_2O_3 (Dehn et al. 2021). These complexes are similar to those just described, where the positive muon is bound to the oxygen and the excess electron localizes on the nearby transition metal oxide, changing its valence state, Fig. 7.23. There is, however, a distinct difference to the state in the nonmagnetic transition metal oxides; the excess electron spin couples strongly to the unpaired d electrons of the Cr or Fe host ions, respectively. As a consequence, the electron spin degree of freedom of the paramagnetic muonium is suppressed and the μSR signal of this charge-neutral complex is practically indistinguishable from the diamagnetic signal of the bare μ^+.

---

[41] Charge-neutral is relative to the original cation O^{2-} and μ^+ system, i.e., the additional positive charge of the muon is compensated.

[42] MnF_2, an antiferromagnet with $T_N = 67.4$ K, is an exception (Uemura et al. 1986).

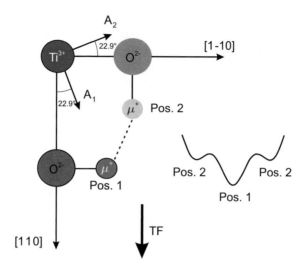

Fig. 7.22 The two possible muon configurations in rutile TiO_2 projected onto the (001) plane. The Ti^{3+}, O^{2-} and μ^+ particles with the darker colors and the solid borders lie in the same *ab*-plane as Ti^{3+} and form the ground state configuration (position 1). The excited state configuration is formed by particles below and above this plane (lighter color, position 2). Site changes (dashed line) are possible between the two muon positions in the same oxygen channel. A sketch of the potential between the two sites is shown. A_1 and A_2 are the principal axes of the hyperfine tensor for position 1. Modified from Vilão et al. (2015), © American Physical Society. Reproduced with permission. All rights reserved. See also Brant et al. (2011)

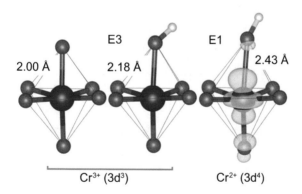

Fig. 7.23 Cr atom (blue) and oxygen octahedron (red) in Cr_2O_3. Left: Without the muon. Middle: With the positive muon. Right: Showing the neutral-charge state formed by the muon and an extra electron localized on the Cr, changing its valence state from Cr^{3+} to Cr^{2+}. The presence of the positive muon causes an elongation of the associated Cr-O bond from 2.00 to 2.18 Å. The localization of the extra electron at the Cr site, due to the Coulomb attraction of the muon and the energy gain from the lattice distortion, further elongates the Cr-O bond to 2.43 Å and leads to the formation of a charge-neutral muon-polaron complex. Yellow shows the charge density isosurface of the topmost occupied band for the charge-neutral case. Adapted from Dehn et al. (2020), under CC-BY-4.0 license. © The Authors

However, the presence of the neutral-complex and its dynamics manifest themselves in unusual temperature dependences of the μSR frequencies and relaxation rates.

The results pave the way to the study of the dopant characteristics of interstitial hydrogen impurities in magnetic oxides. For instance, the finding in Fe_2O_3, indicating that above about 200 K the neutral complex undergoes dissociation and polaron hopping takes place, suggests that at room temperature interstitial H represents a source of "free" polarons thus increasing the carrier density, while at the same time acting as a trap and therefore decreasing the carrier mobility.

Exercises

7.1. Level crossing in vacuum muonium
Calculate the magnetic field where the isotropic muonium levels $|1\rangle$ and $|2\rangle$ cross, Fig. 7.2. Argue why, by looking at the low field dependence in Fig. 7.2 left panel, you must expect a crossing at higher fields.

7.2. Muon spin polarization in muonium in a longitudinal field
Derive Eq. 7.30 using the density matrix formalism.

7.3. Time resolution of a spectrometer and detectable frequencies
Estimate the effect of the spectrometer's time resolution on the detectable frequency spectrum.

7.4. Frequency beating in muonium
Show that, for weak applied transverse fields in the intermediate regime, Sect. 7.3.3 and Eq. 7.40, the hyperfine frequency is inversely proportional to the splitting of the two intratriplet precession frequencies.

References

Blundell, S. J. (2004). *Chemical Reviews, 104,* 5717–5736.
Brant, A. T., Yang, S., Giles, N. C., et al. (2011). *Journal of Applied Physics, 110,* 053714.
Breit, G., & Rabi, I. I. (1931). *Physical Review, 38,* 2082.
Brewer, J., Crowe, K., Gygax, F., et al. (1973). *Physical Review Letters, 31,* 143.
Cohen-Tannoudji, C., Laloë, F., & Diu, B. (2005). *Quantum Mechanics.* Wiley. ISBN: 9780471569527.
Cox, S. F. J. (2003). *Journal of Physics: Condensed Matter, 15,* R1727.
Cox, S. F. J. (2009). *Reports on Progress in Physics, 72,* 116501.
Cox, S. F. J., Gil, J. M., Weidinger, A., et al. (2000). *ISIS Neutron and Muon Source, Annual Report,* 63.
Cox, S. F. J., Lichti, R. L., Lord, J. S., et al. (2013). *Physica Scripta, 88,* 068503.
Dehn, M. H., Shenton, J. K., Arseneau, D. J., et al. (2021). *Physical Review Letters, 126,* 037202.
Dehn, M. H., Shenton, J. K., Holenstein, S., et al. (2020). *Physical Review X, 10,* 011036.
Funamori, N., Kojima, K. M., Wakabayashi, D., et al. (2015). *Scientific Reports, 5,* 8437.
Gil, J. M., Alberto, H. V., Vilão, R. C., et al. (1999). *Physical Review Letters, 83,* 5294.

Gil, J. M., Alberto, H. V., Vilão, R. C., et al. (2001). *Physical Review B, 64,* 075205.

Gurevitch, I. I., et al. (1971). *Soviet Physics JETP, 33,* 253.

Hintermann, A., Meier, P. F., & Patterson, B. D. (1980). *American Journal of Physics, 48,* 956.

Holzschuh, E. (1983). *Physical Review B, 27,* 102.

Holzschuh, E., Kündig, W., Meier, P. F., et al. (1982). *Physical Review A, 25,* 1272.

Holzschuh, E., Kündig, W., & Patterson, B. D. (1981). *Helvetica Physica Acta, 54,* 552.

Hughes, V. W., McColm, D. W., Ziock, K., et al. (1960). *Physical Review Letters, 5,* 63.

Ito, T. U., Higemoto, W., & Shimomura, K. (2020). *Journal of the Physical Society of Japan, 89,* 051007.

Jungmann, K. P. (2016). *Journal of the Physical Society of Japan, 85,* 091004.

Karshenboim, S. G., & Ivanov, V. G. (2002). *Canadian Journal of Physics, 80,* 1305.

Kiefl, R. F., Schneider, J. W., Keller, H., et al. (1985). *Physical Review B, 32,* 530.

King, P. D. C., Lichti, R. L., Celebi, Y. G., et al. (2009). *Physical Review B, 80,* 081201.

King, P. D. C., McKenzie, I., & Veal, T. D. (2010). *Applied Physics Letters, 96,* 062110.

Kreitzman, S. R., & Morris, G. D. (2018). *JPS Conference Proceedings, 21,* 011056.

Liu, W., Boshier, M. G., Dhawan, S., et al. (1999). *Physical Review Letters, 82,* 711.

Momma, K., & Izumi, F. (2011). *Journal of Applied Crystallography, 44,* 1272–1276.

Nuccio, L., Schulz, L., & Drew, A. (2014). *Journal of Physics D: Applied Physics, 47,* 473001.

Odermatt, W., Baumeler, H., Keller, H., et al. (1988). *Physical Review B, 38,* 4388.

Okabe, H., Hiraishi, M., Takeshita, S., et al. (2018). *Physical Review B, 98,* 075210.

Patterson, B. D. (1988). *Reviews of Modern Physics, 60,* 69.

Roduner, E., & Fischer, H. (1981). *Chemical Physics, 54,* 261.

Salman, Z., Prokscha, T., Amato, A., et al. (2014). *Physical Review Letters, 113,* 156801.

Scheuermann, R., Dilger, H., Roduner, E., et al. (2000). *Physica B: Condensed Matter, 289,* 698.

Senba, M. (2000). *Physical Review A, 62,* 042505.

Shimomura, K., Kadono, R., Koda, A., et al. (2015). *Physical Review B, 92,* 075203.

Storchak, V. G., Eshchenko, D. G., & Brewer, J. H. (2004). *Journal of Physics: Condensed Matter, 16,* S4761.

Stoykov, A., Scheuermann, R., Sedlak, K., et al. (2012). *Physics Procedia, 30,* 7.

Uemura, Y. J., Keitel, R., Senba, M., et al. (1986). *Hyperfine Interactions, 31,* 313.

Vilão, R. C., Vieira, R. B. L., Alberto, H. V., et al. (2015). *Physical Review B, 92,* 081202.

Xu, Z., Sato, T., Nakamura, J., et al. (2020). *Japanese Journal of Applied Physics, 59,* 071004.

Investigations of Thin Films and Heterostructures with Low-Energy Muons

8

8.1 Introduction

Experiments making use of surface muons do not provide depth-selective information or allow the study of thin samples in the nanometer range. With the initial
implantation energy of 4.1 MeV typical for surface muons, the mean stopping range
in a solid varies from 0.1 mm to 1 mm with a range distribution of about 20% of the
mean value, see Fig. 8.1 and Sect. 2.2.1. This allows to study the bulk properties of
materials.

To extend the scope of the μSR technique to (possibly depth dependent)
investigations of thin films, heterostructures and, in general, objects of restricted
dimensionality, spin polarized muon beams with energies tunable from several eV
to several keV and a narrow energy distribution are required. These positive muons
can be implanted at relatively well-defined depths ranging from a few nanometers
to a few hundred nanometers. Muons with kinetic energies typically $\lesssim 100$ keV are
called "low-energy muons" (LEM) or sometimes "(very) slow muons".[1] Details on
their properties and use can be found in Morenzoni (1999), Bakule and Morenzoni
(2004); Morenzoni et al. (2004), and Prokscha et al. (2008).

8.2 Generation of Low-Energy Muons

Compared to stable particles, the production of low-energetic μ^+ faces more
difficult contraints inherent to their properties.[2] The muon originates from the pion
decay as an energetic (MeV) particle and therefore one of the first tasks towards a

---

[1] We prefer to avoid the term ultracold muons found sometimes in the literature: the kinetic energy
of 1 eV muon corresponds to $\approx 10^4$ K.

[2] Standard beam cooling techniques, such as stochastic (van der Meer 1985) or electron cooling
(Budker & Skrinskiĭ 1978), are not suitable for muon beams, due to the short muon lifetime.

© Springer Nature Switzerland AG 2024
A. Amato, E. Morenzoni, *Introduction to Muon Spin Spectroscopy*,
Lecture Notes in Physics 961, https://doi.org/10.1007/978-3-031-44959-8_8

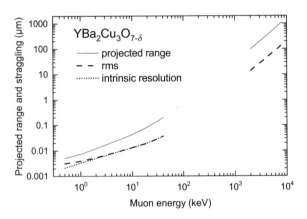

Fig. 8.1 Mean (red curve) and rms (dashed black curve) projected range of positive muons implanted in the high-T_c superconductor $YBa_2Cu_3O_{7-\delta}$ as a function of kinetic energy. An absolute energy uncertainty of 400 eV is assumed for the low energy and a relative uncertainty of 6% for the energetic (MeV) muons. The dotted black curve at low energies displays the intrinsic resolution for a monoenergetic beam of low-energy muons. Modified from Morenzoni et al. (2004), © IOP Publishing. Reproduced with permission. All rights reserved

low-energy beam is to slow down muons with some MeV kinetic energy. The phase space properties (beam size, divergence, and density) of a conventional μ^+ beam, as described in Sect. 1.8, are very poor compared to those of electron, proton, or ion beams. Typical muon beams have spots at the sample position of the order of few cm^2 and divergences on the order of 100 mrad. This is a consequence of the spatial extent of the muon source. In addition, the beam intensity (typically between 10^5 and 10^8 μ^+/s for surface muons) is many orders of magnitude lower than that of stable particles.

We present here only methods that are primarily designed or can reasonably be conceived for μSR studies. These experiments require great flexibility in the sample environment and in the choice of experimental parameters such as temperature, magnetic field, pressure, ... These conditions can be practically achieved only with a spatial separation between the low-energy muon source and the experiment, allowing the muons to be transported to a magnetic field-free region, where the μSR instrument with sample is located. For other experiments, such as particle physics precision experiments with muons or investigations of exotic atoms formed by the capture of negative muons in gases, see Chap. 9, it is often possible to integrate the slowdown section into the experiment. In this case, various methods have been developed to reduce the energy of the muons and thus increase the stopping density (for an overview of early attempts see Simons et al. 1992).[3]

8.2.1 Use of Degraders

In principle, a straightforward method to obtain low-energy muons is to tune a conventional surface muon beamline to momenta much lower than the surface muon momentum of 29.79 MeV/c, see Sect. 1.8.3. This procedure corresponds to the collection and transmission of muons originating from the pions stopped at some depth from the surface of the production target, see Sect. 1.8.2.

While passing through this amount of target material, the muon loses energy due to inelastic and elastic collisions with the target constituents. Energy dissipation, absorption and scattering also take place. The muon intensity N_μ at the end of a surface muon beamline has the following momentum dependence (Badertscher et al. 1985), see Sect. 2.2.1

$$N_\mu(p) \simeq N_S p^{3.5} \frac{\exp\left(-\dfrac{m_\mu L}{p\tau_\mu}\right)}{29^{3.5}} \qquad 5 \lesssim p \lesssim 29 \,\text{MeV/c} \;, \qquad (8.1)$$

where p is expressed in MeV/c, L is the length of the beamline and N_S is the intensity near the surface momentum edge, $p \approx 29$ MeV/c. This method provides reasonable intensities and manageable background rates only down to a few hundred keV kinetic energy (5 MeV/c corresponds to a kinetic energy of 125 keV). Below this energy, the intensity decrease is more pronounced than predicted by Eq. 8.1. Multiple scattering and energy straggling in the target strongly reduce the phase space density of the emerging particles. This makes their transport, focusing and momentum selection very inefficient so that no usable flux intensity can be achieved at the experiment.

Moving the degrader closer to the experiment and degrading a surface muon beam in a low Z metallic material would in principle reduce the transport and lifetime losses. This procedure corresponds to moving the outer part of the

---

[3] A novel device promising an intense (2×10^5 s^{-1}) almost fully polarized low-energy muon beam with sub-mm size and eV energy spread is under study at PSI for fundamental precision experiments (Belosevic et al. 2019). The compression of the phase space volume (i.e., momentum spread around its mean value times the spatial spread) would be achieved in several steps. First by stopping a standard μ^+ beam in a cryogenic helium gas and then by manipulating the stopped μ^+ into a small spot in successive transverse and longitudinal compression stages using electric and strong magnetic fields (5 T) in combination with gas density gradients. Compression of ~ 10 MeV/c muons in the 5 T field has been demonstrated separately for the first two stages (Bao et al. 2014; Antognini et al. 2020). Efficient extraction of the very slow muons from 5 T into a field-free region, while maintaining good phase space properties, is the most difficult step. It is an extraordinary challenge that remains to be demonstrated. The polarization of these muons, which is not relevant for most precision particle physics experiments, but crucial for μSR applications, has not yet been measured.

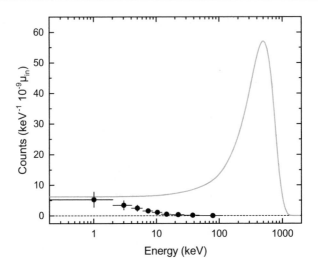

Fig. 8.2 Energy distribution of 27.5 MeV/c muons emerging from an Al foil degrader. The thickness of the foil (250 μm) has been chosen to optimize the intensity of muons with energy $\lesssim$ 30 keV. The full circles indicate the muonium component. Modified from Prokscha et al. (1998), © American Physical Society. Reproduced with permission. All rights reserved

production target close to the experiment. However, also this method does not produce useful low-energy muon intensities. For example, Fig. 8.2 shows the energy distribution of an impinging surface muon beam of 27.5 MeV/c after an Al degrader whose thickness was optimized for maximum low-energy muon intensity (Prokscha et al. 1998). The exiting muons have lost most of their initial kinetic energy of 3.6 MeV. As a consequence of the statistical nature of the slowing down process, the energy distribution of the outgoing particles is broad: the mean energy amounts to about 500 keV with a FWHM of the same order of magnitude. Energy selection of such a beam by electrostatic or magnetic analyzers would lead to unacceptable intensity losses. Below 30 keV, the yield of μ^+ exiting the degrader is very low. Moreover, at this energy, muons exiting the degrader preferably capture an electron and form muonium. Figure 8.2 shows that the probability of muonium formation increases with decreasing energy reaching almost 100% at a few keV. There is no threshold for electronic interactions of very slow muons in a metal. Because of this, as we will see in Sect. 8.2.3, a simple metallic degrader is not capable of producing a reasonable flux of outgoing particles with a few tens of eV kinetic energy.

8.2.2 Laser Resonant Ionization of Muonium

One method to obtain a slow muon source with small momentum and energy spreads is the laser ionization of thermal muonium. In selected materials, such as hot noble metals (W Mills et al. 1986, Pt, Ir or Re), low-density silica powder, or SiO_2 (Beer et al. 1986), a fraction of the implanted surface muons after reaching thermal energies diffuses to the surface and is reemitted as thermal muonium. The exiting muonium atoms are ionized by multiphoton excitation thus resulting in a source of

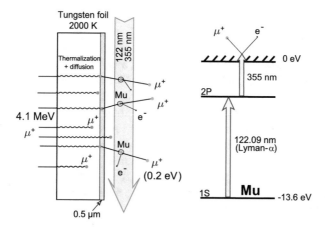

Fig. 8.3 Principle of thermal muonium generation from a tungsten foil and two-photon resonant ionization resulting in the production of very low-energy positive muons. Modified from Bakule and Morenzoni (2004), © Taylor & Francis. Reproduced with permission. All rights reserved

thermal positive muons, which are then extracted and accelerated toward the sample region, see Fig. 8.3.

The principle of this technique was originally demonstrated at a pulsed muon source of the KEK accelerator complex, Japan, using a hot tungsten target (2000 K) to produce a μ^+ source with a nominal energy of 0.2 eV (Nagamine et al. 1995). In the experiment, the muonium atoms evaporating from the back surface of the W-target are resonantly ionized by lasers with wavelengths of 122 nm (Lyman-α) and 355 nm. The Lyman-α source excites the muonium electron from the $1s$ state to the $2p$ state and the 355 nm laser releases this electron, see Fig. 8.3. A beam based on this scheme has been under development for a long time. It is presently implemented and further developed at the μSR facility MUSE at J-PARC (Japan), where the highest pulsed beam intensities are expected (Miyake et al. 2014) and where the repetition rate of the pulsed laser system matches and is synchronized with that of the muon beam (25 Hz).

Due to the challenges posed by this complex scheme and various unfortunate circumstances (such as the 2011 earthquake), progress has been slow and the current results are well below the original expectations. Very recently, intensities of ~ 300 slow muons per second (Kanda et al. 2023) (compared to the 10^5-10^6/s design values, Miyake et al. 2012) with a beam spot of diameter $\lesssim 5$ mm have been reported.[4] Crucial limiting parameters are, besides the thermal muonium production, the laser intensity and the overlap between the ionizing laser beam and the thermal muonium cloud. Improvement in the Mu emission rate could be obtained by using a different muonium emitting target. Silica powder is expected to

---

[4] With the present setup, the experimental asymmetry A_0 is about 0.08 and energies in the range ~ 0.5–30 keV with ~ 50 eV uncertainty can be achieved.

be one of the best target materials for this purpose but large variations of its long-term efficiency, probably related to surface and preparation conditions, have been reported. Silica in the more stable aerogel form appears to be a better candidate. A recent study has shown that an emitted thermal muonium to muon stop ratio up to 3% can be obtained under optimized conditions (Beer et al. 2014). The use of a 244 nm laser for the Mu $1s - 2s$ excitation instead of the Lyman-α laser for the first ionization step also promises higher ionization efficiency.

The muonium ionization method produces a low-energy muon beam with only 50% polarization, since, as shown in Sect. 7.3, the state $|\uparrow_\mu\downarrow_e\rangle$ is not an eigenstate of muonium, see Eq. 7.9. This reduces the maximum asymmetry that can be observed in a μSR spectrum by a factor of two and the figure of merit of the instrument (see Exercise 8.1) by a factor of four, compared to the methods preserving the initial full muon polarization, see also footnote 4.

8.2.3   Moderation in Thin Layers of Cryosolids

Implanting muons in some materials is an easy way to make them lose energy, see Chap. 2. Not all materials are suitable for generating a low-energy muon beam. As shown in Fig. 8.2 when an energetic muon beam passes through a thin metallic foil a few hundred μm thick, the energy spectrum of the transmitted particles is very broad, and only very few muons with energy in the ~ 10 eV range can be collected after the foil. This is not the case for some weakly-bound insulating materials with large bandgaps such as condensed van der Waals gases (cryosolids s-Ne, s-Ar, and s-N_2). In these wide bandgap insulators, the suppression of the electronic interactions of the implanted charged particles with an energy of the order of the bandgap leads to a preferential emission at epithermal energies of ~ 10 eV.[5]

The muon moderation technique in cryosolids[6] is presently the most successful method for generating muons with energies of the order of 15 eV. The moderation of surface muons to epithermal energies was first observed at TRIUMF, Canada (Harshman et al. 1986, 1987). The subsequent decisive observation at PSI, Switzerland, that the moderation process fully conserves the initial polarization of a surface beam (Morenzoni et al. 1994) paved the way for several systematic improvements of the technique and to the construction of a beam of polarized low-energy muons and associated μSR instrument and sample environment at PSI, Sect. 8.3. This apparatus is routinely used for nanoscale depth dependent μSR investigations (Morenzoni et al. 2004; Prokscha et al. 2008).

The different stages of the moderation process from surface muon (~ 4.1 MeV) to epithermal energy (~ 15 eV) can be summarized as follows, see Fig. 8.4:

---

[5] Epithermal, i.e., higher than thermal, because the particles are emitted before thermalization.

[6] We refer to the combination of a metallic substrate and a cryosolid as a moderator. A moderator can be thought of as a very special degrader.

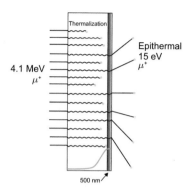

Fig. 8.4 Schematic of the low-energy muon production by moderation (not to scale). Surface muons impinge on a thin metallic foil, typically cooled to 10 K or less, on which a thin (∼500 nm) layer of a cryosolid (blue layer) has been deposited. The muons lose a large fraction of their kinetic energy in the foil. About half of them stop in the foil (stopping profile roughly represented by the green curve); the rest passes through and reach the thin cryosolid layer. From this layer, the small fraction of muons whose energy has decreased down to values of the order of the bandgap energy of the cryosolid is emitted as epithermal muons with ∼ 15 eV kinetic energy

- Surface muons with kinetic energy 4.1 MeV are focused to a thin-foil moderator substrate.[7] They rapidly lose energy by Coulomb collisions with electrons, ionizing and exciting the target atoms (electron-hole creation in a solid). This stage corresponds to the $1/\beta^2$ region of the Bethe energy loss formula, see Eq. 2.17 and Fig. 2.4.[8]
- When a muon has lost most of its kinetic energy, at energies below a few tens of keV, the charge-changing cycles described in Sect. 2.3, involving muonium formation in one collision (where the positive muon captures an electron) and muonium break-up in one of the subsequent collisions, also acquire importance as energy-dissipating mechanisms. This corresponds to the region on the left-hand side of the maximum of Fig. 2.4.
- In wide bandgap perfect weakly-bound insulators, such as the krypton, argon, nitrogen, and neon cryosolids (bandgap energy between 11 and 22 eV), these electronic processes have high threshold energies.[9] Therefore, once a muon has reached a kinetic energy of the order of the bandgap,

---

[7] The substrate is a 125 μm thick Ag foil. Its surface has a micro-grating structure, which enlarges the effective surface area by 50% with respect to a flat substrate (Prokscha et al. 2001).

[8] The total thickness of the material traversed by the surface μ^+ and the exact momentum of the incident beam are chosen so that the stopping density distribution of μ^+ is centered at the downstream surface of the moderator target, i.e., in the condensed gas layer. This optimizes the very slow muon yield and means that about 50% of the beam is stopped in the substrate.

[9] A muon must at least possess a kinetic energy comparable to the bandgap in order to create an electron-hole pair in a solid or, in an atomic picture which is justified for these weakly bound solids, to excite a target atom or molecule. In the cryosolids Ne, Ar, and N_2 (bandgaps 21.58, 14.16, and 15.1 eV, respectively) muonium formation has even positive threshold energies, see Table 2.1.

the efficient electronic energy loss mechanisms are strongly suppressed or become energetically impossible. As a consequence, the energy loss rate in the moderator layer drops considerably, since only the inefficient elastic scattering and phonon excitation processes remain as energy loss mechanisms, see Chap. 2.

Those muons, which in their statistical slowing down cascade have reached an energy of the order of a few tens of eV, can basically move unperturbed and be emitted as very slow μ^+ from the downstream side of the moderator. Their energy distribution is shown in Fig. 8.5. It has a ~ 20 eV broad maximum centered around 15 eV, with a tail extending to higher energies. The emission mechanism can be classified as hot emission: the emitted very slow μ^+ are particles that have not completely thermalized in the cryosolid layer. This mechanism results in a large effective depth of escape for epithermal μ^+, see Fig. 8.6, giving rise to a particularly efficient moderation to epithermal energies in these materials.[10] Within the cryosolids, the escape depth as well as the moderation efficiency increase with increasing bandgap energy.[11,12]

The moderation efficiency to obtain epithermal muons from surface muons can be defined as

$$\varepsilon_\mu = \frac{N_{\text{epith}}}{N_{\text{surf}}} \simeq \frac{\Delta\Omega(1 - f_{\text{Mu}})L}{\Delta R_p} \, , \tag{8.2}$$

where $\Delta\Omega$ is the probability to escape into vacuum,[13] $(1 - f_{\text{Mu}})$ is the fraction of muons avoiding Mu formation (f_{Mu} muonium formation probability), and ΔR_p is the spread of the projected range values of the surface muons, Eq. 2.27. Since the effective escape layer L is much shorter than the range straggling, the moderation efficiency is not very high. It is between $\sim 1.5 \times 10^{-4}$ for solid neon and $\sim 5 \times 10^{-5}$

---

[10] From Fig. 8.6 we can estimate the length of the layer thickness beyond which no increase of efficiency is observed. It is ~ 50 nm for s-Kr, 110 nm for s-Ar, 60 nm s-N_2 and ~ 390 nm for s-Ne.

[11] In metals, an effective threshold for conduction electron excitation is absent. Energy can always be transferred from the μ^+ to electrons near the Fermi level exciting them to empty states in the conduction band. This causes a large rate of μ^+ energy dissipation and consequently strongly suppresses the escape probability of the epithermal μ^+, which have been generated inside the metal. This makes metals very poor moderators, with moderator efficiencies between 2 and 3 orders of magnitude lower than in the cryosolids (Morenzoni 1999).

[12] A wide bandgap appears to be a necessary but not sufficient condition to make an efficient muon moderator. Very low efficiencies, comparable to those of metals, have been observed in insulators such as LiF and SiO_2 (Harshman et al. 1986; Morenzoni 1999), see Fig. 8.5. This finding points to different energy loss mechanisms of epithermal muons in insulators with ionic character compared to those in the weakly bound simple cryosolids, which to good approximation can be represented as dense gases.

[13] Epithermal muons are emitted with a $\cos\theta$ distribution (Lambert distribution), where θ is the angle with respect to the normal to the surface (Prokscha et al. 2001).

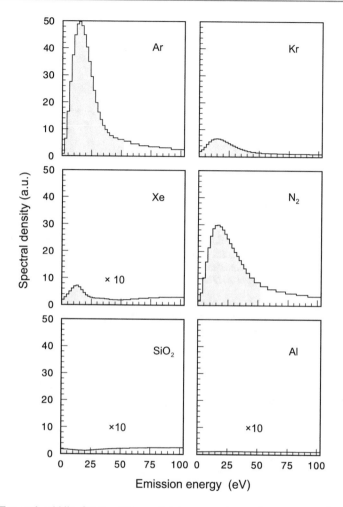

Fig. 8.5 Top and middle figures: Measured energy spectrum of muons escaping different cryosolids, normalized to the same number of incoming surface muons. The useful energy interval of epithermal muons is highlighted. For a comparison the bottom figures show the much smaller yield of epithermal muons from a typical wide bandgap insulator SiO_2 and a metal Al. The distributions, as well as yield measurements, clearly indicate that the emission of very slow muons is a mechanism peculiar to van der Waals condensed gases and that only these materials seem to possess true moderating properties for muons. Modified from Morenzoni (1999), © Taylor and Francis Group. Reproduced with permission. All rights reserved

for solid nitrogen or argon. Therefore, to obtain a useful intensity of low-energy muons, the availability of an intense surface muon beam is essential.

Another, key parameter is the polarization. Figure 8.7 plots the asymmetry signal in a 5 mT transverse field of epithermal muons moderated in solid argon, see also Morenzoni et al. (1994). The result shows that moderation conserves practically the

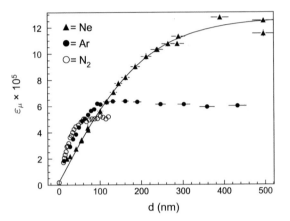

Fig. 8.6 Moderation efficiency ε_μ, defined as the number of epithermal muons divided by the number of incoming surface muons, for various moderating materials as a function of the thickness of the solid van der Waals layer condensed onto a patterned Ag substrate, held at a temperature of 6 K. Modified from Morenzoni et al. (2004), © IOP Publishing. Reproduced with permission. All rights reserved

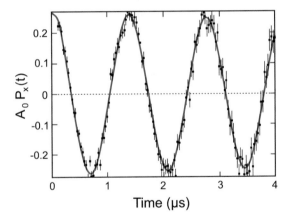

Fig. 8.7 TF-μSR signal of epithermal muons emitted from a solid argon layer, reaccelerated to 8 keV and stopped in a silver sample. The muon spin precesses in a 5 mT transverse magnetic field. The amplitude corresponds to a practically full polarization

full initial polarization of the surface muon beam since depolarization via electronic processes and Coulomb scattering is negligible. Furthermore, the overall time for moderation from 4.1 MeV to $\sim$ 10 eV is very short (of the order of 10 ps, see Sect. 2.2.2) much smaller than the muonium hyperfine period of $1/\nu_0 = 224$ ps, so that even transient muonium formation during moderation does not lead to loss of polarization (for a discussion of the time evolution of the muon spin polarization in muonium see Sect. 7.3.2).

8.3   The Low-Energy Muon Apparatus at PSI

Epithermal muons emitted from a moderator are a simple and well-suited source of a low-energy beam of fully polarized positive muons (LEM) with tunable energy in the sub keV to keV range, used for so-called low-energy μSR (LE-μSR) experiments.[14]

The practical realization of this scheme at PSI is shown in Fig. 8.8.[15] Schematically, the operation is as follows (Prokscha et al. 2008). Fully polarized muons from a high intensity surface beam with a utilizable continuous rate of presently $\sim 3 \times 10^8$ muons/s[16] (T. Prokscha, private communication 2023) enter the ultra-high vacuum (UHV) LEM apparatus through a 30 μm titanium window[17] and are sent into the cryogenic moderator.[18] Epithermal muons emerging at a rate of $\sim 1.8 \times 10^4$ μ^+/s,[19] from the moderator held at a positive potential between 12 and 20 kV, are extracted and accelerated in the potential gradient, transported and focused by electrostatic lenses and an electrostatic mirror to the sample, where they arrive at a rate of $\sim 7 \times 10^3$ μ^+/s. The beam diameter at the sample position depends on the low-energy beam parameters; it typically has a size of 10–15 mm (FWHM).[20]

An electrostatic mirror separates the slow from the fast muons exiting the moderator. For a μSR experiment with a continuous beam, the implantation time ($t = 0$ time of the μSR spectrum) for each low-energy muon has to be available, see Chap. 3. The low-energy muons are detected when they pass through a ~ 10 nm thick carbon foil (corresponding to only about 50 atomic layers) placed at an

---

[14] A related technique available at the TRIUMF ISAC facility, is β-detected NMR (β-NMR). It uses a low-energy (< 30 keV) beam of radioactive probe ions that can be introduced into solids in a depth-controlled manner. Typically, a beam of optically spin polarized ($\approx 70\%$) ^{8}Li$^+$ with a flux of $\sim 10^6$ ions/s is used. The ^{8}Li$^+$ probe has spin = 2, nuclear gyromagnetic ratio = 6.315 MHz/T, lifetime = 1.21 s, and quadrupole moment Q = +31.7(4) mb (Stone 2005). The high energy β-electrons from the asymmetric decay of ^{8}Li$^+$ are used to monitor the polarization (Kiefl et al. 2003; Morris 2014; MacFarlane 2022). To avoid quadrupolar effects, one has to use pure magnetic nuclei with spin ½, e.g., ^{31}Mg.

[15] The LEM beam is an example of a tertiary beam (the primary being the proton beam generating pions and the secondary the surface muon beam originating from the pion decay). At the time of this writing, this is worldwide the only low-energy muon beam routinely in operation and available to a wide user community.

[16] This corresponds to $\sim 40\%$ of the full intensity.

[17] The window separates the better than $\sim 10^{-10}$ mbar UHV of the LEM apparatus from the poorer vacuum ($\sim 10^{-5}$ mbar) of the surface muon beamline. The UHV conditions are needed to (i) avoid the deposition of residual gas molecules on the cryogenic moderator and (ii) to provide clean conditions for the surfaces of the investigated samples (Morenzoni et al. 1997).

[18] The routinely used cryogenic layer consists of a few 100 nm s-Ar layer covered with a protecting ~ 10 nm thin s-N$_2$ layer. This has proven to provide the best long term stable conditions.

[19] About a factor 2 to 3 higher yield is obtained with a s-Ne moderator, which, however, sustains a lower extraction potential and thus a lower maximum kinetic energy.

[20] This size can be reduced by a factor of 2 or better, with an intensity reduction of about a factor of 2, by using upstream collimators (Ni et al. 2023).

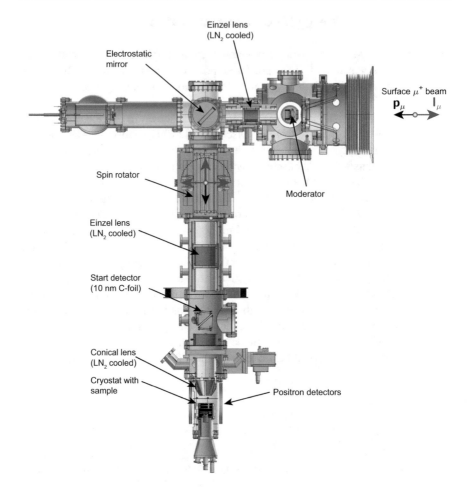

Fig. 8.8 Layout of the low-energy polarized muon beam and μSR spectrometer at the PSI for depth dependent experiments in the nm range

intermediate focus of the beam transport system (start detector in Fig. 8.8). By traversing the foil the muon ejects a few electrons, which are directed by a grid system to a microchannel plate detector where they are detected. This provides the implantation time of the muon in the sample and starts the time differential measurement.[21] This scheme keeps the amount of material interacting with the muon and the consequent effects on the trajectory minimal, while allowing for an efficient (> 80%) and fast detection. On passing through the foil, the muons lose about 1 keV and acquire an energy spread (rms) of ~ 0.4 keV.

---

[21] The total time resolution is 5-10 ns, depending on energy.

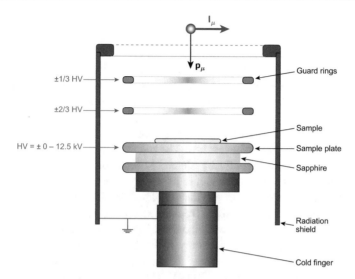

Fig. 8.9 Sample region of the LE-μSR spectrometer, with sample mounted on the cold finger of a cryostat. The sample is electrically insulated and can be biased to $\pm$ 12.5 kV. This, together with the variable extraction voltage at the moderator, allows to tune the implantation energy between $\sim$ 0.5 and 30 keV. The muon spin direction for a TF measurement is shown

The final kinetic energy of the muons implanted into the sample can be varied over the range $\sim$ 0.5–30 keV[22] by applying, in combination with the extraction potential, an accelerating or decelerating potential of up to $\pm$12.5 kV to the sample, which is mounted, in good thermal but electrically insulating contact, on a cryostat for low temperature experiments or on other types of sample holder, Fig. 8.9. The 90° deflection at the electrostatic mirror has also the practical effect of transforming the initially longitudinally polarized muon beam into a transversely polarized one for TF measurements. A small spin rotator with an $\mathbf{E} \times \mathbf{B}$ field can rotate the spin by 90° to have the spin parallel to the momentum and perform LF measurements (Salman et al. 2012). The decay positrons from the muons implanted in the sample are detected by a set of scintillator segments surrounding the vacuum tube and placed left, right, above and below the beam axis. The readout of the signal is performed via silicon photomultipliers (avalanche photodiodes) detectors.

8.4   Stopping Profiles of Low-Energy Muons in Thin Films

In μSR experiments making use of surface or decay muon beams the exact stopping depth in the sample is not relevant, as long as the sample is homogeneous. To obtain information about the bulk properties of the material it is enough to ensure that the

---

[22] This corresponds to typical depths between a few to a few hundred nanometers.

thickness of the sample is sufficient so that all the muons stop inside the sample. In contrast to bulk μSR experiments, knowledge of the implantation profile is essential for designing and analyzing experiments on thin films or near surface regions with a sub keV-keV muon beam. In particular, it is desirable to have simple prescriptions and computer codes giving reliable predictions of energy dependent quantities such as implantation profiles, backscattering, reflection and neutralization probability of positive muons in a large variety of materials.

As shown in Chap. 2, when a muon enters a solid sample the initial kinetic energy is dissipated within a few picoseconds. The muon continuously loses energy and changes direction predominantly by electronic collisions and scattering with the target nuclei. Due to the random nature of the collision processes, a more or less broad stopping or implantation profile $n(x, E_{kin})$ is obtained, where E_{kin} is the muon kinetic energy and x the depth from the sample surface.

In Fig. 8.1 we have shown calculations of mean range and rms of stopping profiles in $YBa_2Cu_3O_{7-\delta}$ for an energy interval ranging from low (LEM beam) to high energies (decay beam). Figure 8.10 shows, for the same compound, the full implantation profile for energies < 30 keV. The calculations have been performed adapting the Monte Carlo code TRIM.SP (Transport and Range of Ions in Matter) (Eckstein 1991) to the muon case. In the simulation the muon is treated as a proton like particle of mass $m_\mu \approx 1/9\ m_p$. To be applicable to muons, velocity-scaled electronic stopping powers and energy-scaled elastic energy losses of proton are used, see Chap. 2 (Morenzoni et al. 2002). The trajectory of the implanted particle is calculated step-by-step taking into account energy loss, scattering, and thermalization processes.

At low energies the profile rms is typically 5–10 nm. Note that even for perfectly monoenergetic particles there is an inherent limit to the depth resolution due to

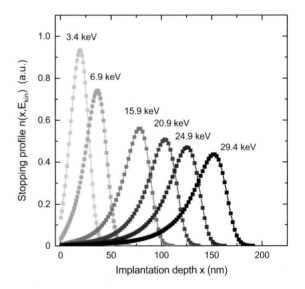

Fig. 8.10 Monte Carlo simulation of stopping profiles of low-energy muons in $YBa_2Cu_3O_{7-\delta}$, as a function of the implantation energy. An energy uncertainty of 400 eV (rms) is assumed, (Eckstein 1991; Morenzoni et al. 2002)

the broadening of the implantation profile caused by the statistical nature of the collision processes. This intrinsic broadening is the dominant effect for muons of energy $\gtrsim 2$ keV, as the comparison of the dotted with the dashed curves in Fig. 8.1 shows.

To test the reliability of the simulation in predicting the implantation profiles of low-energy muons in matter one can exploit the property that polarized muons thermalized in metals behave as a "free" diamagnetic particles whereas, in most insulators, the large majority of muons is bound to an electron and forms muonium. Because of the different effective gyromagnetic ratios, the two states can be easily distinguished from the different muon spin precession frequency in a static low magnetic field B_{ext} transverse to the initial spin direction; for a free muon $\omega_\mu = \gamma_\mu B_{ext}$ and for muonium in the triplet state ($m_F = \pm 1$) $\omega_{Mu}^{T} = \gamma_{Mu}^{T} B_{ext}$, see Eq. 7.36 and Chap. 7.

The principle of the experiment is depicted in Fig. 8.11 (Morenzoni et al. 2002). In a sample composed of a thin metallic layer (thickness d from 40 to 100 nm) deposited on an insulator, if the implantation profile extend over both layers, two oscillating signals contribute to the measured asymmetry with a time dependence of the form

$$A_0\,P(t) = A_\mu(E_{kin})\exp\left(-\frac{\sigma_\mu^2 t^2}{2}\right)\cos(\omega_\mu t + \varphi_\mu)+$$

$$+ \frac{A_{Mu}(E_{kin})}{2}\exp(-\lambda_{Mu} t)\cos(\omega_{Mu}^{T} t + \varphi_{Mu})\ . \tag{8.3}$$

Here, A_μ and A_{Mu} are the initial amplitudes of the diamagnetic and paramagnetic fraction with initial phases φ_μ and φ_{Mu}, respectively. The factor $1/2$ in A_{Mu} takes into account the unobserved high-frequency component due to the hyperfine coupling (see Sect. 7.3.2), the damping parameters σ_μ and λ_{Mu} describe possible depolarization processes due to spatial or temporal variations of the local magnetic fields.

After correcting for small diamagnetic contributions from muons stopped in the insulating layer but not forming muonium[23] and taking into account the probability that, at low energies, muons do not stop in the sample but are reflected at the surface or backscattered from the metallic layer,[24] the energy dependence of the diamagnetic and paramagnetic amplitudes can be compared with the prediction of the Monte Carlo simulation of the stopping profiles. Theoretically one expects that the two amplitudes are directly determined by the partial integral of the

---

[23] Muonium formation was found to be energy dependent in the keV energy range (Prokscha et al. 2007).

[24] These processes, which give no contribution to A_μ, must be taken in consideration at energies below ~ 5 keV. Their fraction was determined in an independent measurement, see Morenzoni et al. (2002) for details.

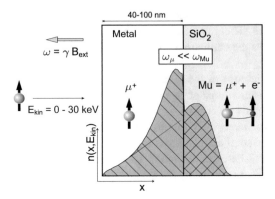

Fig. 8.11 Principle of an experiment to determine the stopping fraction in different layers. Positive polarized muons with energies between 0.5 and 30 keV are implanted into thin metallic layers deposited on quartz glass. A magnetic field is applied perpendicular to the muon spin. Muons stopped in the metal layer precess with a different Larmor frequency than those forming muonium in the insulating substrate. The amplitudes of the frequencies measured by the muon spin rotation technique give the energy dependent fractions of the particles stopped in the two layers. The results are compared with Monte Carlo simulations of $n(x, E_{kin})$. Modified from Morenzoni et al. (2002), © IOP Publishing. Reproduced with permission. All rights reserved

implantation distribution in the metallic $\int_0^d n(x, E_{kin})dx$ and in the insulating layer $\int_d^\infty n(x, E_{kin})\,dx$, respectively.

In the experiment thin films of Al, Cu, Ag and Au were investigated thus covering a range from low to high atomic number Z. The measurements were performed for different energies, enclosing different stopping ranges. Figure 8.12 shows as a representative example the results for a bilayer consisting of 98 nm Al deposited on quartz SiO_2.[25] After an increase at low energies, the diamagnetic asymmetry saturates at a value of about 27%, when essentially the totality of the muons thermalizes in the metallic layer. The observed initial increase of the "free" muon fraction with energy at a few keV is a consequence of reflection / backscattering and simultaneous neutralization of muons scattered at the metallic surface or reemerging from the bulk. This effect is especially pronounced in the samples containing heavy elements. When the muon energy is raised above ~ 12 keV, A_μ decreases, reflecting the fact that an increasing fraction of muons traverse the metallic layer and reach the insulating layer, where they predominantly form muonium. A measurement at $B_{ext} = 0.5$ mT, a field sufficiently low to observe the transverse spin rotation in muonium, confirms that the decrease of the diamagnetic muon fraction is accompanied by a corresponding increase of the muonium fraction $A_{Mu}(E_{kin})$ and that, as expected, $A_\mu(E_{kin}) + A_{Mu}(E_{kin}) = A_0$.

---

[25] Before decaying, the thermalized particles could in principle diffuse, thus modifying the implantation profile. To suppress this effect the samples were cooled to 20 K, where diffusion is negligible.

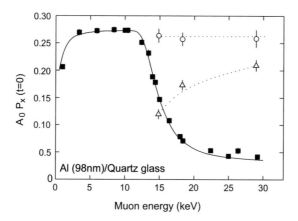

Fig. 8.12 Energy dependence of the free muon asymmetry, A_μ, in an Al layer ($d = 98$ nm) on quartz glass (squared symbols). The data were taken in a TF of 5 mT. The decrease of the diamagnetic asymmetry for energies above ~ 12 keV reflects the increasing fraction of muons passing the metal layer. The solid line shows the expected energy dependence of A_μ calculated with the modified TRIM.SP code. The open triangles display the muonium asymmetry A_{Mu}, which is proportional to the fraction of particles stopping in the insulating layer. The open circles represent the total asymmetry $A_\mu + A_{\text{Mu}}$. Modified from Morenzoni et al. (2002), © IOP Publishing. Reproduced with permission. All rights reserved

The results from the investigated samples are well reproduced by a simulation of the implantation profiles and reflection probabilities based on the modified TRIM.SP Monte Carlo program (Eckstein 1991; Morenzoni et al. 2002).

An additional experiment, able to determine the full differential implantation profile $n(x, E_{\text{kin}})$ in a single implantation and imaging measurement, was performed to verify the simulated stopping profiles (Morenzoni et al. 2004). In analogy with magnetic resonance imaging technique, the implantation profile can be obtained by analyzing the μSR signal when an inhomogeneous, transverse magnetic field $\mathbf{B}_{\text{ext}}(x)$ of known gradient is applied to the sample. The field is applied along the surface of a thin film sample (z direction) and muons of different energies are implanted in the sample.

The local magnetic field at position x causes a precession of the muons stopped at x with frequency $\gamma_\mu B_{\mu,z}(x)$ and probability $n(x, E_{\text{kin}})$. The time evolution of the polarization $P_x(t)$ measured at a well-defined energy E_{kin} can be written as

$$P_x(t) = \int_0^\infty f(B_{\mu,z}, E_{\text{kin})} \cos(\gamma_\mu B_{\mu,z}\, t + \varphi) dB_{\mu,z} \ , \tag{8.4}$$

where $f(B_{\mu,z}, E_{\text{kin}})$ is the field distribution sensed by the muons. The differential muon stopping distribution $n(x, E_{\text{kin}})$ is related to the field distribution by the relationship

$$n(x, E_{\text{kin}})\, dx = f(B_{\mu,z}, E_{\text{kin}})\, dB_{\mu,z} \ , \tag{8.5}$$

which expresses the fact that the probability for a muon to experience a field in the interval $[B_{\mu,z}, B_{\mu,z} + dB_{\mu,z}]$ is given by the probability that it will stop at a depth in the range $[x, x + dx]$.

From Eq. 8.5, we see that, if we apply a known field gradient to the sample and determine the field distribution probed by the muon ensemble, by performing a Fourier transform of $P_x(t)$, we obtain the differential stopping distribution by·use of the relationship

$$n(x, E_{\text{kin}}) = f(B_{\mu,z}, E_{\text{kin}}) \frac{dB_{\mu,z}}{dx} \quad . \tag{8.6}$$

For low-energy muons, the mean range and straggling are on nanometer scale, therefore a large field gradient is necessary to extract the implantation profile. For this, we make use of the magnetic field penetrating below the surface of an extreme type-II superconductor, $YBa_2Cu_3O_{7-\delta}$, in the Meissner state. With the field applied parallel to the c-axis, $B_{\mu,z}(x) = B_{\text{ext}} \exp(-x/\lambda_{ab})$ (where λ_{ab} is the magnetic penetration depth, see Sects. 6.1.1 and 6.2.4). With typical values of $B_{\text{ext}} \simeq 10$ mT and $\lambda_{ab} \simeq 100$ nm, field gradients of the order of 10^5 T/m can be generated within the range distribution of low-energy muons.[26]

Figure 8.13 shows, that the muon implantation profile obtained by the described procedure compares quite well with the predictions from Monte Carlo simulations. This finding together with the results from the bilayer samples indicate that the

Fig. 8.13 Implantation profile of 3.4 keV muons in a thin film of $YBa_2Cu_3O_{7-\delta}$ obtained by the direct imaging technique (circles). The profile is compared to predictions of Monte Carlo calculations using the code TRIM.SP with different assumptions about the scattering potential (solid, dotted and dash-dotted curves). Modified from Morenzoni et al. (2004), © IOP Publishing. Reproduced with permission. All rights reserved

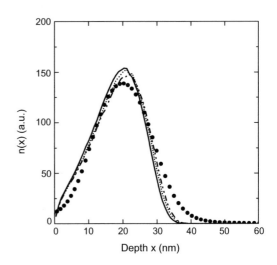

Depth x (nm)

[26] Note that in the next section we will assume the knowledge of the implantation profile $n(x, E_{\text{kin}})$ to microscopically prove that the field penetration below the surface of a superconductor in the Meissner state decays exponentially and to make an absolute measurement of the London penetration depth and of its temperature dependence. Here, by contrast, to measure the depth profile we assume an exponentially decaying magnetic field profile with known penetration depth. The argument is noncircular since for the present analysis we use the value of λ_{ab} obtained from an independent measurement of the same sample in the vortex state (Niedermayer et al. 1999).

stopping of low-energy muons in thin films and heterostructures can be reliably predicted with the Monte Carlo simulation program TRIM.SP adapted to muon implantation (Eckstein 1991; Morenzoni et al. 2004).

8.5 Examples

8.5.1 Magnetic Field Profiling at the Surface of Superconductors

The depth sensitivity in nanometer range of low-energy muons and the local character of the muon probe allow to directly measure single values of magnetic fields as a function of depth, and thus to image magnetic field profiles beneath the surface of materials on a nanometer length scale.[27] At the moment, no other technique is able to provide this information.[28]

For illustration, we consider first typical measurements of $B_z(x)$ in different superconducting samples in the Meissner state. The measurements yield a direct determination of otherwise not easily accessible quantities such as the magnetic penetration depth and, if not too small, the coherence length.

As discussed in Sect. 6.1.1, for a superconductor in the Meissner state the applied magnetic field is screened from the bulk of the sample by superficial diamagnetic shielding currents and penetrates only in a near surface region.

In the so-called London limit ($\lambda_L \gg \xi_0$), a magnetic field applied parallel to the surface of a semi-infinite slab of a clean superconductor ($\ell \gg \xi_{BCS}$) decays exponentially with depth x, Eq. 6.13, with the decay length determined by a single parameter, the London penetration depth λ_L.[29]

It is interesting to remark that Eq. 6.13 was predicted already in 1935 by the London brothers (London & London 1935), but never experimentally tested at microscopic level. Low-energy μSR provided the first experimental proof of it, see Sect. 8.5.1.1 and Jackson et al. (2000); Brown (2000). Differently from a measurement in the vortex state, a measurement in the Meissner state by LE-μ^+

---

[27] In addition to single values, magnetic field distributions can be measured as a function of depth. By covering a sample with a nonmagnetic, inactive capping layer (e.g. Ag) and stopping the muons in this layer, it is even possible to measure the field distribution above the sample and use this information to characterize the sample. This method has been applied, for example, to monitor the spatial evolution of the magnetic field distribution as the vortex flux lines emerge through the surface of a superconducting YBCO film (Niedermayer et al. 1999), or to investigate the superconducting and magnetic properties of a FeSe monolayer epitaxially grown on SrTiO$_3$ (Biswas et al. 2018).

[28] The profiling capabilities are not restricted to magnetic field profiling. By measuring the effect of the interaction between muon states and free carriers on the muon spin polarization in p- and n-type Ge wafers, it has been shown that it is possible to determine the profiles of free charge carriers in the near surface accumulation-depletion region between 10 and 160 nm depth (Prokscha et al. 2020).

[29] Note that also for dirty ($\ell \ll \xi_{BCS}$) local superconductors ($\lambda_L \gg \xi_0$), an exponential decay is predicted with decay length λ given by Eq. 6.17.

provides an absolute and model independent determination of the magnetic penetration depth. The determination of the magnetic penetration depth (and possibly of the coherence length) in the vortex state, see Sect. 6.3.1.1, is a very dependable, efficient, and widely used method but has to rely on (i) a theory describing the vortex state (Ginzburg-Landau, London, or microscopic theory, ...) relating the measured field distribution $f(B_{\mu,z})$ (or its moments) to the characteristic lengths of the superconductor, (ii) a regular vortex lattice of known symmetry and possibly (iii) take into account the effects of field dependence, nonlocal and nonlinear effects, and the influence of disorder (Sonier et al. 2000).

8.5.1.1 Strong Type-II Superconductors

The experiment is performed by applying a field smaller than B_{c1} and parallel to the surface of a superconductor of thickness d. With an exponential field decay away from the surface of the superconductor, Eq. 6.13, one expects a field profile in the sample given by

$$B_z(x) = B_{\text{ext}} \frac{\cosh\left(\dfrac{\frac{d}{2} - x}{\lambda_{\text{L}}}\right)}{\cosh\left(\dfrac{d}{2\lambda_{\text{L}}}\right)}. \tag{8.7}$$

This is the form taken by the exponential dependence with magnetic flux penetrating from both surfaces, see Fig. 8.15 left panel. To determine the penetration depth at a well defined temperature, one performs several TF measurements by varying the kinetic energy (and hence implantation depth) of the muons. Figure 8.15 shows the results from measurements performed on a 700 nm high-quality thin film of the high-T_c superconductor $YBa_2Cu_3O_{7-\delta}$,[30] which is in the London limit[31] (Jackson et al. 2000). The magnetic field of 9.5 mT is parallel to the ab-planes and perpendicular to the incoming muon momentum. The muon spin perpendicular to the momentum and to the field. In this configuration $\lambda_{\text{L}} = \lambda_{ab}$. To avoid flux trapping at the surface, the measurements in the superconducting state are performed under ZF-cooled conditions.

---

[30] For the analysis of the YBCO experiment the value of x is corrected by a small quantity x_0 corresponding to a "dead layer", which allows for a nonsuperconducting thin layer arising mainly from surface imperfections such as roughness or oxide layers (Lindstrom et al. 2016).

[31] Most of the high-T_c superconductors are in or close to the London limit.

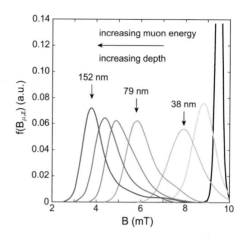

Fig. 8.14 Magnetic field distributions observed in YBa$_2$Cu$_3$O$_{7-\delta}$ at 20 K in the Meissner state with muons implanted at (from left to right) 29.4, 24.9, 20.9, 15.9, 6.9, and 3.4 keV. The field distribution centered at 9.5 mT (black line) is taken in the normal state. Modified from Jackson et al. (2000), © American Physical Society. Reproduced with permission. All rights reserved

To properly take into account the depth information provided by the implantation profile there are different approaches, which give very close results. In this experiment, first the distribution $f(B_{\mu,z}(x), E_{kin})$[32] of the values of magnetic field experienced by the implanted muons is derived by a Fourier analysis of the spin precession spectra.[33]

Distributions obtained at 20 K for various implantation energies are shown in Fig. 8.14. The muon stopping distribution $n(x, E_{kin})$ corresponding to each implantation energy of Fig. 8.14 was calculated using the modified Monte Carlo code TRIM.SP and is shown in Fig. 8.10.[34] To an excellent first approximation, to plot the local field as a function of depth, one can simply read off the peak fields in Fig. 8.14 and plot the values versus the peak depths in Fig. 8.10. This allows to iterate rapidly to the correct relationship between implantation profile and field distribution which is given by Eq. 8.6.[35] The values of $B_{\mu,z}^{peak}$ versus x_{peak}, obtained after this small correction of the order ~ 1 nm, are plotted for several sample temperatures in the right panel of Fig. 8.15. From the data λ_{ab} and its temperature dependence (not shown) can be determined. The values are in good agreement with mixed state measurements on the same film (Niedermayer et al. 1999) as well as with values obtained from measurement of bulk samples in the vortex state.

---

[32] Muons of a given energy stop over a certain range of depths, giving a corresponding range of fields with distribution $f(B_{\mu,z}(x), E_{kin})$.

[33] The Fourier analysis was performed by the so-called maximum entropy method, which is often more efficient than Fast Fourier Transform methods (Rainford & Daniell 1994; Riseman & Forgan 2003).

[34] A Gaussian spread of the muon kinetic energy of 500 eV, corresponding to the energy straggling introduced by the start detector, see Fig. 8.9, is included in the calculations.

[35] Equation 8.6 shows that the peak value of the field, $B_{\mu,z}^{peak}$, obtained from the maximum of $f(B_{\mu,z}, E_{kin})$ is reached at a depth x_{peak} where $n(x, E_{kin})/(dB_z/dx)$ is maximum rather than where $n(x, E_{kin})$ is maximum, i.e., $f(B_{\mu,z}^{peak}, E_{kin}) = n(x, E_{kin})/(dB_z/dx)|_{x_{peak}}$.

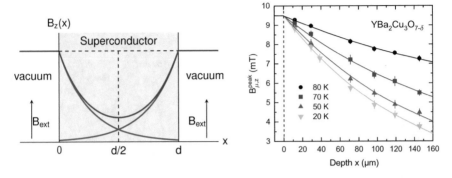

Fig. 8.15 Left panel: Red curve, field profile below the surface of a superconductor (thickness d) when the field penetrates only from one side of the superconductor (Eq. 6.13, valid for $d \gg \lambda_L$). The blue curve takes into account penetration from both sides, Eq. 8.7. The plot is for $\lambda_L = d/4$. Right panel: Values of magnetic field versus depth extracted from LE-μSR measurements on a thin film of $YBa_2Cu_3O_{7-\delta}$, see text for details. The different curves represent data at sample temperatures 20, 50, 70, and 80 K. The solid lines are fits of Eq. 8.7 to the data with λ_L as free parameter ($d = 700$ nm). The implantation depth has been corrected by a small "dead layer" $x_0 = 8$ nm. Right panel modified from Jackson et al. (2000), © American Physical Society. Reproduced with permission. All rights reserved

Another approach, which can be defined as an analytical model in the time domain, is to assume knowledge of the field profile $B_z(x)$ and thus to generate, at each implantation energy, the theoretical muon asymmetry signal by weighting the precession with the calculated implantation profile

$$A(t) = A_0 \exp\left(-\frac{\sigma^2 t^2}{2}\right) \int \cos(\gamma_\mu B_z(x)\, t + \varphi)\, n(x, E_{kin})\, dx \;, \qquad (8.8)$$

which is then used to fit the μSR spectra. Here σ takes into account any depolarizing contribution as nuclear moments, stray fields, ...

By assuming a London magnetic field profile, this approach has been used to analyze the profiles in the Meissner state of a mosaic of small detwinned single crystals of $YBa_2Cu_3O_{6.92}$.[36] The external magnetic field is applied parallel to the ab-surface. By orienting the field along the b- and a-axes respectively, it is possible to determine the magnetic penetration depths λ_a (shielding current $\parallel \hat{\mathbf{a}}$) and λ_b (shielding current $\parallel \hat{\mathbf{b}}$) and thus the planar anisotropy of this cuprate compound (Kiefl et al. 2010).[37] The mean value of the in-plane magnetic penetration depth $\lambda_{ab} = \sqrt{\lambda_a \lambda_b}$ obtained from the measured λ_a and λ_b in the Meissner state is in

---

[36] To suppress the background, the mosaic is mounted on a nickel-coated sample holder, which causes a very fast depolarization of the muons missing the small crystals (Saadaoui et al. 2012).

[37] The magnetic field profiles along the a- and b-axes are consistent with the London model, but there are deviations close to the surface where a dead layer of about 10 nm with suppressed supercurrents has to be included.

good agreement with values obtained from μSR bulk studies of the vortex state (Sonier et al. 2007). This result confirms the reliability of magnetic penetration depth measurements with μSR.

8.5.1.2 Nonlocal Superconductors

Experiments with low-energy muons have also provided the nonexponential field profiles of nonlocal superconductors (Suter et al. 2004, 2005).

In type-I and weak type-II superconductors the coherence length cannot be neglected. The electrodynamical response must be averaged over this length scale and becomes nonlocal. The local relationship between the supercurrent $\mathbf{j}_s$ and the vector potential $\mathbf{A}$ from London theory, Eq. 6.8,

$$\mathbf{j}_s(\mathbf{r}) = -\frac{1}{\mu_0 \lambda^2} \mathbf{A}(\mathbf{r}) \quad , \tag{8.9}$$

must be replaced by an expression that takes into account the fact that if the two electrons forming the Cooper pair are far apart at positions $\mathbf{r}$ and $\mathbf{r}'$, the supercurrent at point $\mathbf{r}$ depends on the values of $\mathbf{A}(\mathbf{r}')$ over a volume of radius $|\mathbf{r} - \mathbf{r}'|$ determined by the effective extent of a Cooper pair.[38]

Based on a earlier nonlocal generalization of Ohm's law to explain the anomalous skin effect in conductors where the mean free path ℓ is larger than the skin depth, Pippard (Pippard 1953) proposed a generalization of the London equation to

$$\mathbf{j}_s(\mathbf{r}) = -\frac{1}{\mu_0 \lambda_L^2} \frac{3}{4\pi \xi_0} \int \left(\frac{(\mathbf{r} - \mathbf{r}')[(\mathbf{r} - \mathbf{r}') \cdot \mathbf{A}(\mathbf{r}')]}{|\mathbf{r} - \mathbf{r}'|^4} \right) \exp\left(-\frac{|\mathbf{r} - \mathbf{r}'|}{\xi_P} \right) d\mathbf{r}' \quad . \tag{8.10}$$

Here[39]

$$\frac{1}{\xi_P} = \frac{1}{\xi_0} + \frac{1}{\ell} \quad , \tag{8.11}$$

---

[38] Nonlocal effects occur not only in superconductors where $\xi \gtrsim \lambda$, but also at nodes of a $\mathbf{k}$ dependent energy gap $\Delta(\mathbf{k})$ (see Eq. 6.27 for unconventional strong type-II cuprate superconductors). Since $\xi_{\mathbf{k}} \propto 1/\Delta(\mathbf{k})$, the coherence length becomes effectively infinite at the nodes. In this case, the effects are more subtle and not detectable in a field profile measurement. In YBa$_2$Cu$_3$O$_{6.95}$ evidence of nonlocal (and nonlinear) effects has been found in the field dependence of the effective magnetic penetration depth determined by μSR measurements in the bulk of the vortex lattice (Amin et al. 2000; Sonier et al. 2000)

[39] The weak temperature dependence of ξ_P, could be explained only later by the BCS theory, which replaces ξ_P with ξ, Eq. 6.16.

Fig. 8.16 Magnetic penetration profile in the Meissner state of aluminum ($\xi_{BCS} = 1600$ nm, $\lambda_L = 50$ nm) according to local (London) and nonlocal theories (Pippard and BCS). Note the virtually identical profiles predicted by the two nonlocal theories. The inset shows the same curves on a logarithmic scale. Modified from Suter et al. (2005), © American Physical Society. Reproduced with permission. All rights reserved

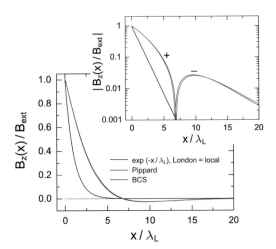

and ξ_0 can be identified with ξ_{BCS}, Eq. 6.16 (Tinkham 1996). The equivalent BCS expression can be written as Schrieffer (1999)

$$j_{s,\alpha}(\mathbf{r}) = \sum_{\beta=1}^{3} \int \left[P_{\alpha\beta}(\mathbf{r} - \mathbf{r}') - \frac{e^2 n_{sc}}{m^*} \delta(\mathbf{r} - \mathbf{r}') \delta_{\alpha\beta} \right] A_\beta(\mathbf{r}') \, d\mathbf{r}' \quad . \tag{8.12}$$

$P_{\alpha\beta}$ describes the nonlocal paramagnetic response, while the second term in parentheses represents the local diamagnetic contribution (of opposite sign). The presence of a paramagnetic contribution in the nonlocal theory strongly reduces the screening current $\mathbf{j}_s$ and leads to a deeper penetration of the magnetic field than predicted by the London theory.[40]

The shape of $B_z(x)$ is quite different from the exponential behavior of the local case: initially $B_z(x)$ decreases slowly and has a negative curvature, moreover a sign reversal of the magnetic field is predicted before it goes to zero, Fig. 8.16.

These properties can be understood qualitatively using hand-waving arguments. In the nonlocal case, the Cooper pairs are extended compared to the magnetic penetration profile. Therefore, the charges that make up a Cooper pair do not experience the same field, so the screening response is less effective, the magnetic penetration depth is greater and the slope is less steep compared to the local response. This has an additional consequence: as the field penetrates further below the surface, some Cooper pairs deep in the superconductor will still experience a nonzero average field and will "overcompensate", explaining the negative curvature as well as the field reversal of $B_z(x)$ before it approaches zero.

---

[40] In general, the magnetic penetration depth λ can be defined as $\lambda = 1/B_{ext} \int_0^\infty B(z) dz$ and depends on λ_L and ξ_0 and ℓ. For an exponentially decaying field, as in the local case, Eq. 6.13, this definition leads to $\lambda = \lambda_L$.

Note that if $\mathbf{A}$ does not change appreciably on the scale of the coherence length, a local London equation (predicting a pure diamagnetic response) is recovered in both expressions. Remarkably, the phenomenological Pippard expression, Eq. 8.10, gives very close results to the microscopically derived BCS expression, Eq. 8.12, see Fig. 8.16 for a comparison. To determine the field profile in the nonlocal case where deviations from a simple exponential are expected and no closed expression for $B_z(x)$ is available, a different approach than the one described in Sect. 8.5.1.1 may be used. The starting point is still Eq. 8.5, where $f(B_{\mu,z})$ is obtained from the Fourier transform of the $P_x(t)$ data, Eq. 8.4.

If one assume that $B_z(x)$ is a monotonically decaying function of x,[41] one can integrate Eq. 8.5 on both sides to obtain an integral equation for the field $B_{\mu,z}$ at a specified depth x, which can be numerically solved

$$\int_0^x n(\zeta, E_{kin})\, d\zeta = \int_{B_{\mu,z}(x)}^{\infty} f(\beta, E_{kin})\, d\beta \ . \tag{8.13}$$

The field distribution $f(B_{\mu,z}, E_{kin})$ is derived by maximum entropy Fourier analysis of the muon spin polarization spectra and $n(x, E_{kin})$ is the calculated stopping profile. With this method it is possible to determine almost the entire functional dependence of the local field from a single measurement made at a specific implantation energy. As expected, the determination of $B_{\mu,z}(x)$ at different energies results in a set of overlapping curves.[42]

This integral method has been used in Suter et al. (2004, 2005) to analyze the field profiles in Pb, Nb and Ta films. As an example we show in Fig. 8.17 the results for a Pb sample of thickness $d = 430(20)$ nm, $T_c = 7.1(1)$ K. The measurement was performed in a transverse field of 8.96 mT applied parallel to the surface after field-cooling the sample. The experimental curve shows clear deviations from the exponential behavior and has the characteristic curvature expected from BCS and Pippard theories, which fits the data quite well giving a low temperature value of $\lambda = 59(3)$ nm, $\xi_{BCS} = 90(5)$ nm, and a death layer 5.8(3) nm, with $\ell = 100$ nm kept fixed.[43] The deviations from the exponential behaviour of the London model (dashed line in Fig. 8.17) are more pronounced at low temperatures than near T_c, where the superconductor becomes increasingly local. This is a consequence of the fact that on approaching T_c the magnetic penetration length increases strongly, whereas the coherence length changes only weakly, thus satisfying the local condition. Figure 8.17 also shows the field profile obtained from the simplified analysis, where the field mean values $\langle B_{\mu,z} \rangle = \int B_{\mu,z}\, f(B_{\mu,z}, E_{kin}) dB_{\mu,z}$ are

---

[41] This is our case, the region of field reversal is outside the present range of low-energy muons.

[42] This self-consistency check further demonstrates the reliability of the Monte Carlo simulations of the muon implantation profiles (Eckstein 1991; Morenzoni et al. 2002).

[43] Note that BCS is a weak electron-phonon coupling theory ($\lambda_{e\text{-ph}} \ll 1$). The strong coupling character of Pb is taken into account by renormalizing the BCS parameters as $\lambda_L \to \lambda_L/\sqrt{Z}$ and $\xi_{BCS} \to \xi_{BCS} Z$, where $Z = 1 + \lambda_{e\text{-ph}} \approx 2.55$ (Carbotte 1990).

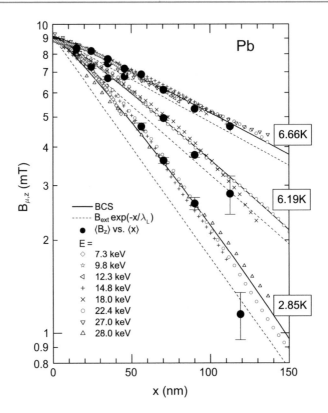

Fig. 8.17 Magnetic penetration profiles for a Pb film (thickness 430(20) nm) at low temperatures and close to $T_c = 7.1(1)$ K. The small open symbols represent the field profile obtained by the integral method at different implantation energies. The solid dots show $B_{\mu,z}(x)$ as determined from mean values (see text for details). The solid lines are fits of the BCS theory to the data, while the dashed line represents the exponential profile predicted by the London theory. Modified from Suter et al. (2005), © American Physical Society. Reproduced with permission. All rights reserved

plotted against the mean depth values $\langle x \rangle = \int x\, n(x, E_{kin})dx$. The points obtained are in good agreement with the $B_{\mu,z}(x)$ curves obtained from Eq. 8.13.

Field profile measurements on the two extreme type-I superconductors In and Sn by LE-μSR also quantitatively confirm the Pippard/BCS nonlocal electrodynamics (Kozhevnikov et al. 2013). In this experiment, the $B_{\mu,z}(x)$ data were used to determine, in addition to $\lambda_L(T)$ and the BCS coherence length, the renormalization factor Z resulting from the electron-phonon interaction, see footnote 43.[44]

---

[44] Magnetic field profiling has also been found to be a useful method for studying the quality of materials for superconducting radio frequency cavities, whose performance depends on surface treatment. A strong change in the Meissner screening in Nb films prepared for cavities has been reported for samples treated by low temperature baking (Romanenko et al. 2014).

8.5.2 Heterostructures

Low-energy muons can not only probe the near surface region of thin films, but also access buried layers of heterostructures. To illustrate the information that can be obtained by magnetic field profiling of heterostructures and how it is extracted from the data, we give some examples of experiments.

Cuprate Heterostructure The measurement of depth-resolved profiles of local magnetisation in heterostructures allows a variety of phenomena to be addressed that arise in multilayer structures due to the juxtaposition of materials with different electronic properties. In general, mutual influence and/or coupling between the layers is observed or predicted.

One example is the so-called proximity effect: if a thin normal metal layer is brought into close contact with a superconducting layer, Cooper pairs can enter the normal layer at the interface, making the normal layer superconducting and at the same time weakening the superconductivity in the superconducting layer. This phenomenon, schematically shown in Fig. 8.18, is well known for conventional superconductors, where it can generally be explained by the conventional proximity theory (Deutscher & de Gennes 1969).

The proximity effect, which is expected to show unusual features in heterostructures of unconventional superconductors, has been studied by LE-μSR. The field profiles along the c-axis of a heterostructure consisting of three cuprate layers, each 46 nm thick, have been measured at different temperatures (Morenzoni et al. 2011). The top and bottom layers consist of $La_{1.84}Sr_{0.16}CuO_4$ ($T_c \simeq 32$ K, optimally doped), while $La_{1.94}Sr_{0.06}CuO_4$ ($T_c' < 5$ K, underdoped) serve as a central "barrier"

Fig. 8.18
Normal-superconducting bilayer, showing qualitatively the order parameter and the proximity effect, the appearance of superconducting pairs in the normal layer and the suppression of superconductivity in the superconductor

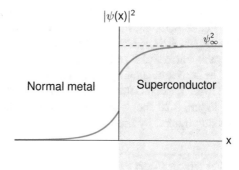

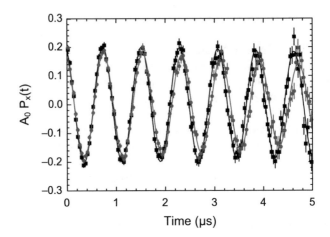

Fig. 8.19 Time dependence of the asymmetry for muons implanted at 12.5 keV in the central underdoped layer ($T_c' < 5$ K) of the $La_{1.84}Sr_{0.16}CuO_4$ (46 nm)/$La_{1.94}Sr_{0.06}CuO_4$ (46 nm)/$La_{1.84}Sr_{0.16}CuO_4$ (46 nm) heterostructure. After zero field-cooling an external transverse magnetic field of 9.5 mT is applied parallel to the ab-planes and to the interface (Meissner geometry). The signal at 40 K (black) represents the muon spin precession in the external field. At 9.5 K (red), i.e., well above the critical temperature of the underdoped layer, $T_c' < 5$ K, the lower precession frequency (proportional to the average local field) reflects the presence of diamagnetism in the underdoped layer. The curves are fits to the data (see text for details). Modified from Morenzoni et al. (2011), © Springer Nature. Reproduced with permission. All rights reserved

layer to search for proximity effects.[45] For the measurement, as usual for Meissner state experiments, the sample is cooled in the ZF from above T_c, a magnetic field of 9.5 mT is applied parallel to the ab-planes (z direction) and TF spectra are collected as a function of the muon implantation energy.

Figure 8.19, shows a μSR spectrum from muons stopped in the central underdoped $La_{1.94}Sr_{0.06}CuO_4$ layer. The figure directly displays the main result of the experiment: at 9.5 K, a temperature well above T_c', the spins of the muons implanted in the barrier coherently precess in a field that is diamagnetically shifted by 0.2–0.3 mT with respect to the applied field of 9.5 K.

Figure 8.20 shows the results of the field profile $\langle B_{\mu,z} \rangle$ versus $\langle x \rangle$ (or energy). As highlighted by the figure, up to 17 K, that is well above T_c', the local field is

---

[45] For experiments on thin films or heterostructures it is essential that the samples are of good quality and properly characterized. Several techniques are used to achieve this: Molecular Beam Epitaxy (MBE) for layer-by-layer growth on an oriented substrate, Reflection High Energy Electron Diffraction (RHEED) to check the quality of film growth and provide information on film thickness. Subsequent ex-situ characterization typically includes mutual inductance, X-ray diffraction (XRD), resistivity, atomic force microscopy (AFM) and Rutherford backscattering (RBS) to determine the thickness of each layer. In this experiment, the typical surface roughness determined by AFM was 0.5 nm, much less than a unit cell height (1.3 nm), and the interface roughness obtained from the X-ray reflectivity curve was about 1 nm (rms).

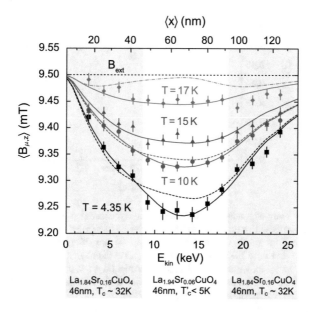

La$_{1.84}$Sr$_{0.16}$CuO$_4$ La$_{1.94}$Sr$_{0.06}$CuO$_4$ La$_{1.84}$Sr$_{0.16}$CuO$_4$
46nm, T$_c$~ 32K   46nm, T'$_c$< 5K   46nm, T$_c$~ 32K

Fig. 8.20 Depth profile of the local field in a cuprate heterostructure at different temperatures. The vertical lines indicate the position of the interfaces of the La$_{1.84}$Sr$_{0.16}$CuO$_4$ (46 nm) / La$_{1.94}$Sr$_{0.06}$CuO$_4$ (46 nm) / La$_{1.84}$Sr$_{0.16}$CuO$_4$ (46 nm) heterostructure. The horizontal dashed line shows the applied field of 9.5 mT. The points show the measured average field versus average implantation depth, shown on the upper scale. The whole heterostructure excludes the magnetic flux like a superconductor: it shows the Meissner effect with the central layer active in the shielding. The lines are obtained from fits using the London model of Eq. 8.15. The fit takes into account the energy dependent muon stopping profiles, which are also used to calculate the average stop depth $\langle x \rangle$. The grey dash-dotted line shows the field profile expected at $T = 4.35$ K when the shielding current flow is restricted to the upper and lower superconducting electrodes. The dashed lines are obtained by assuming that supercurrents in the barrier flow only in the c direction. Modified from Morenzoni et al. (2011), © Springer Nature. Reproduced with permission. All rights reserved

lower than the applied field in all layers and at all depths, meaning that the entire heterostructure excludes the magnetic flux like a conventional superconductor.

The observed field profile has the form of an exponential field decay in the Meissner state with the flux penetrating from both sides, Eq. 8.7, and looks like that for two superconductors with different magnetic penetration depths. The profile reflects a shielding supercurrent that runs along the c-axis as well as in the ab-planes of the barrier.[46] This is unexpected when one recalls that in this geometry the supercurrent must pass through the a 46 nm thick "barrier" La$_{1.94}$Sr$_{0.06}$CuO$_4$. At $T > T'_c$, as shown in a control measurement, Fig. 8.21, a single layer La$_{1.94}$Sr$_{0.06}$CuO$_4$ of same thickness does not show diamagnetism.

---

[46] Note that from London equation $\langle j_{ab} \rangle = (1/\mu_0)\langle dB_z/dx \rangle \neq 0$.

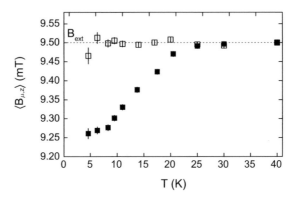

Fig. 8.21 Temperature dependence of the average field measured at the center of the underdoped layer acting as a barrier with a thickness of 46 nm in the trilayer (filled symbols). The open symbols are measurements on a single underdoped sample of 46 nm thickness, grown under the same conditions. In the first case the average local field has a diamagnetic shift up to $T_{eff} \simeq 22$ K. Above this temperature its value is, within the experimental error, equal to the applied field. No shift is observed for a single underdoped layer. Modified from Morenzoni et al. (2011), © Springer Nature. Reproduced with permission. All rights reserved

The analysis of this experiment is an example of a different approach to that described in Sects. 8.5.1.1 and 8.5.1.2.

The functional dependence of the magnetic field versus depth is assumed to be known a priori and in this case to be London-like. Since the field distribution $f(B_{\mu,z})$ can be well approximated by a Gaussian distribution, the TF-μSR spectra (Fig. 8.19) can be fitted by an asymmetry function of the form $A(t) = A_0 \exp(-\sigma^2 t^2/2) \cos(\gamma_\mu \langle B_{\mu,z} \rangle t + \varphi)$.[47] Parameters such as the magnetic penetration depth in the top, bottom, and central layers are obtained by modelling the measured field $\langle B_{\mu,z} \rangle$ by a solution of the London equations in the three layers. Specifically, the depth dependence of $\langle B_{\mu,z} \rangle$ is fitted with the model prediction for the mean field

$$\langle B_{\mu,z} \rangle = \int B_z^{mod}(x) \, n(x) \, dx \qquad (8.14)$$

where $n(x)$ is the Monte Carlo simulated energy dependent stopping distribution and B_z^{mod} is the London solution for three juxtaposed superconducting layers with magnetic penetration depth λ and λ' as fitting parameters and of the form

$$B_z^{mod} = \begin{cases} c_1 \exp(-x/\lambda) + c_2 \exp(x/\lambda) & x \text{ within upper layer} \\ c_3 \exp(-x/\lambda') + c_4 \exp(x/\lambda') & x \text{ within middle layer} \\ c_5 \exp(-x/\lambda) + c_6 \exp(x/\lambda) & x \text{ within bottom layer} \end{cases} \qquad (8.15)$$

---

[47] The spectra are fitted over a time interval where the fast relaxing parts, due to some magnetism, make a negligible contribution.

The coefficients c_1 to c_6 are determined from the boundary condition $B_z^{\text{mod}}(0) = B_z^{\text{mod}}(d) = B_{\text{ext}}$ and from the continuity of B_z^{mod} and of the vector potential at the interfaces $A_y = -\mu_0 \lambda_{ab}^2 j_y = -\lambda_{ab}^2 \, dB_z^{\text{mod}}/dx$, where y is the direction parallel to the ab-planes. The lines in Fig. 8.20 represents the fit results for $\langle B_{\mu,z}(E_{\text{kin}}) \rangle$ versus $\langle x \rangle$, which is calculated from the muon stopping profiles.

The local field profile in this cuprate heterostructure exhibits a Meissner effect in a thick underdoped barrier layer well above its intrinsic critical temperature T_c'. The superfluid density, which is induced by the juxtaposition of the bottom and top layers and which can be obtained from $\lambda(T)$ and $\lambda'(T)$, disappears at a temperature T_{eff} where $T_c' \ll T_{\text{eff}} < T_c$. In addition, the thickness of the layer is much larger than a few nanometers which is the expected depth of penetration of Cooper pairs into a normal metal in conventional proximity theory (Deutscher & de Gennes 1969). The experimental local findings by LE-μSR are not conventional and point to a giant proximity effect. Several models (existence of local superconducting clusters, quenching of phase fluctuations by the presence of neighboring layers with long-range phase order) have been proposed that are able to provide an enhanced proximity effect, see for example (Marchand et al. 2008).

Magnet-Superconductor Heterostructure A variety of physical phenomena can also occur at the interfaces between conventional superconductors and nonsuperconducting layers, especially when these are magnetic. One example is the so-called paramagnetic Meissner effect, as opposed to the conventional diamagnetic Meissner effect where the supercurrent is carried by Cooper pairs in a spin singlet state as in the previous examples. In the conventional Meissner effect the supercurrent $\mathbf{j}_s$, generated in response to a weak magnetic field, is linearly related to the vector potential $\mathbf{A}$ by the London equation, Eq. 6.8,

$$\mathbf{j}_s = -\frac{n_s e^2}{m_e^*} \mathbf{A} \ , \tag{8.16}$$

where here n_s is the density of the superconducting charges forming the spin singlet pairs.

If the superconducting layer is coupled to a magnetic layer, the exchange field of the magnetic layer can convert spin singlet into spin triplet states with a density n_t and Eq. 8.16 changes to

$$\mathbf{j}_s = -\frac{e^2}{m_e^*} \mathbf{A} \, (n_s - n_t) \ . \tag{8.17}$$

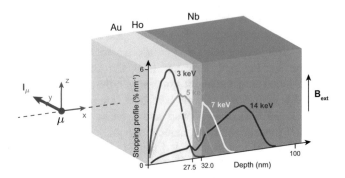

Fig. 8.22 Experimental transverse field setup to detect the paramagnetic Meissner effect in the Au / Ho / Nb multilayer. Normalized muon profiles simulated for a few representative implantation energies are shown. Modified from Di Bernardo et al. (2015), under CC-BY-3.0 license. © The Authors

The triplet and singlet components have different decay lengths. When n_t exceeds n_s an anomalous paramagnetic Meissner effect is predicted to occur (Yokoyama et al. 2011), a manifestation of the so-called odd-frequency pairing.[48]

The Meissner effect has been investigated by LE-μSR in a superconductor / magnet system, formed by a Au(27.5nm) / Ho(4.5nm) / Nb(150nm) trilayer (Di Bernardo et al. 2015). The conical magnetization of the Ho layer serves as a spin-active interface creating odd-frequency triplet Cooper pairs from singlet pairs leaking in from the superconducting Nb. Its small thickness ensures pair transmission into the Au layer, which is necessary since the presence of Cooper pairs cannot be probed directly in the magnetic Ho due to the rapid depolarization of the muon spin in a strong magnetic field. The depth profile is determined in the usual TF Meissner geometry, Fig. 8.22, by determining mean field values $\langle B_{\mu,z} \rangle$ from fits of the corresponding asymmetries $\propto \exp\left(-\lambda_m t\right) \cos\left(\gamma_\mu \langle B_{\mu,z}(E_{kin})\rangle\, t + \varphi\right)$ as a function of the implantation energy E_{kin}, Fig. 8.23a.

To obtain an accurate local field profile, similarly to the approach used in Sect. 8.5.2, a global fit to the asymmetry data of all implantation energies with a common model field profile is performed. In the Au layer it is modeled as $B_z^{mod}(x) = (B_{ext} + \mu_0 M_z(x))$, where the magnetization term is set to $\mu_0 M_z(x) = B_a \sin(x/\delta)$, which is a parametrization of the theoretical magnetization profile calculated for the Au / Ho / Nb heterostructure.

---

[48] In a general case pairing correlations in superconductors are described by a two-fermion correlation function, which depends not only on the spin and relative spatial coordinates but also on the relative times of the two particles. The superconducting correlation function must be odd under exchange of the two electrons. In this example the electrons are paired in an orbital s-state. If they are in a triplet spin state, they must be odd under time exchange to maintain the overall odd symmetry. If the correlation function is odd by time exchange, one speaks of odd-time (or odd-frequency) superconductivity (Berezinskii 1974; Linder & Balatsky 2019). Usually, for example in the BCS theory, the time coordinate is not taken into account.

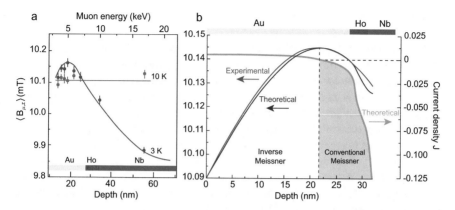

Fig. 8.23 (**a**) Average local magnetic field $\langle B_{\mu,z} \rangle$ in Au/Ho/Nb as a function of the muon implantation energy (top axis) and mean implantation depth (bottom axis). Magenta circles are measurements in the normal state, 10 K. Blue circles are in the superconducting state of Nb, 3 K. The solid lines are a guide to the eye. (**b**) Local field $\langle B_{\mu,z} \rangle$ and supercurrent profile at the Ho/Nb interface. Red curve, left y-axis: $\langle B_{\mu,z} \rangle$ determined from global fits of the LE-μSR measurement at different energies. Blue curve, left y-axis: Theoretical model. Gray curve, right y-axis: Calculated dimensionless screening current density flowing parallel to the y-axis, in the planes of the thin film heterostructure. The dashed line separates the region of the conventional from that of the anomalous Meissner effect. Modified from Di Bernardo et al. (2015), under CC-BY-3.0 license. © The Authors

The results of the global fit at 3 K is shown by the red curve in Fig. 8.23b. It illustrates the positive increase of $\langle B_{\mu,z} \rangle$ in Au over B_{ext}, where the (odd-frequency) triplet state dominates over the singlet state, indicating that the paramagnetic response is an intrinsic property of the odd-frequency superconducting state. By contrast, a conventional Meissner effect is measured in Nb up to the interface with Ho, where the contribution of spin singlet Cooper pairs to the screening current is larger than that due to the long-ranged spin triplet pairs.

Figure 8.23b shows that the experimental data can be well reproduced by the quasiclassical theory of superconductivity in proximity coupled systems (Chandrasekhar 2008), under the assumption that the time-reversal symmetry is spontaneously broken by the spatially dependent exchange field in Ho and by taking into account realistic boundary conditions for nonideal interfaces (Di Bernardo et al. 2015; Linder & Balatsky 2019).[49]

The paramagnetic Meissner effect is an example of propagation of superconducting pairs outside the superconductor, a "standard" effect reasonably well understood. Much less investigated experimentally as well as theoretically is another category of proximity effects, generally named inverse proximity effects, which involve the pairs inside the superconductor (Flokstra et al. 2014).

---

[49] LE-μSR has also provided experimental evidence of a similar proximity-induced odd-frequency superconductivity in a bilayer consisting of a topological insulator Bi_2Se_3 ($\sim$ 110 nm) on Nb ($\sim$ 80 nm) (Krieger et al. 2020).

Unusual long-range (of the order of the magnetic penetration depth) transfer of the magnetic field from a ferromagnet to a superconductors detected by LE-μSR spectroscopy[50] have motivated the theoretical formulation of the so-called electromagnetic proximity effect, which is basically given by the screening response of the superconductor to a vector potential at the superconductor/ferromagnet interface (Mironov et al. 2018). This gives rise to a previously overlooked long-range electromagnetic contribution to the inverse proximity effect.

A complete theoretical understanding of proximity coupled systems, especially those involving magnetic layers, is not yet available (Buzdin 2005). LE-μSR measurements, by providing a direct measurement of the spontaneous magnetic field in superconductor/magnet structures, are a very powerful tool for the characterization of proximity coupled systems and for the detection of new coupling mechanisms and have stimulated theoretical interpretations.

Spin Injection in an Organic Spin Valve The high sensitivity of μSR combined with depth dependent information has allowed the detection of induced electronic spin polarizations in different types of heterostructures. One of the first examples has been the observation of the RKKY-like spatially oscillating spin polarization of the conduction electrons in a Ag spacer of a Fe/Ag/Fe trilayer (Luetkens et al. 2003). Another example involves an heterostructure serving as a prototype of an electronic spin device based on an organic material. Besides the electric charge, the spin degree of freedom can be used to obtain new functionalities in electronic devices.[51] One of the most common method for exploiting the spin in devices is based on the alignment of the electron spin (up or down) relative to either a reference magnetic field or the magnetization orientation of a ferromagnetic layer. Device operation involves measuring a quantity such as the electrical current or resistivity that depend on the degree of spin alignment in the device.[52] The so-called "spin valve" is a prominent example of a spin dependent device. Spin valves consist essentially of two ferromagnetic layers, which can be magnetized parallel or antiparallel to each other, and a barrier layer.

---

[50] Field profiling in a Au/Nb/ferromagnet structure revealed a remote magnetic field in Au at a large distance ($\geq$ 50 nm) from the ferromagnet (Flokstra et al. 2016). In an other series of experiments a Cu/Nb/Co structure has displayed an anomalously large enhancement of the flux expulsion in Nb when a thin Co layer is added (Flokstra et al. 2018, 2019).

[51] To distinguish it from "conventional" electronics this field is named "spintronics". If superconductor–ferromagnet heterostructures are involved, as in the previous example, one speaks of "superconducting spintronics".

[52] The property of a material to change the value of its electrical resistance depending on the strength and orientation of an applied field is called magnetoresistance. If the electrical resistance notably changes depending on whether the magnetizations of adjacent ferromagnetic layers are parallel or antiparallel one speaks of giant magnetoresistance (GMR). The property of giant magnetoresistance in metallic multilayers has been discovered in 1988. Already in 1997, this effect has found applications in sensors and devices (e.g. in the read/write head of hard disks). In 2007 A. Fert and P. Grünberg won the Nobel prize for the discovery (Baibich et al. 1988; Binasch et al. 1989).

The use of organic materials in spintronics is attracting considerable interest, mainly due to their ease of processing, low cost, and electronic and structural flexibility. Another advantage is the long lifetime of electronic spin relaxation. In conventional inorganic semiconductors, the lifetime is mainly limited by the spin-orbit and hyperfine interactions. In contrast, organic semiconductors are composed of light elements, mainly carbon and hydrogen, where the spin-orbit and hyperfine interactions are weak (^{12}C has no nuclear moment), leading to long spin relaxation times.

One of the key parameters characterizing a spin valve is the spin diffusion length (or spin decoherence length), which is the length describing the spatial decay of the spin polarization of electrons injected across the barrier layer by applying a voltage difference between the top and bottom electrodes. It is an important parameter for understanding spin transport and relaxation in spintronic devices. Low-energy muon spin rotation has been used to measure this otherwise difficult to determine length in a working organic spin valve (Drew et al. 2009).

Figure 8.24 shows the device and the principle of the experiment. This heterostructure has an active area of 18×18 mm^2 and consists of NiFe(17 nm)/LiF(1.9

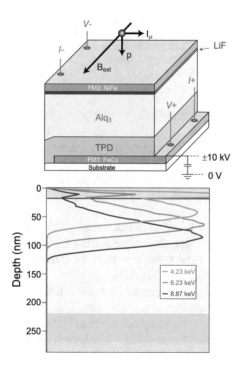

Fig. 8.24 Top: schematic diagram showing the structure of the organic spin valve and the experimental geometry. Muons with transverse spin I_μ are implanted with energy E_{kin} (momentum p). The external field B_{ext} is perpendicular to the spin and momentum of the muon. The layers FM1 and FM2 are the two ferromagnetic layers. The muons are stopped in the Alq3 barrier. They precess in a local field composed of the applied field and the field produced by the local electronic spin polarization. Bottom: Stopping profiles of muons with implantation energies of 4.23, 6.23 and 8.87 keV, respectively. Modified from Drew et al. (2009), © Springer Nature. Reproduced with permission. All rights reserved

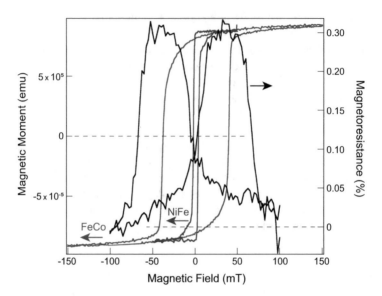

Fig. 8.25 Magnetoresistance and field hysteresis curves of the organic spin valve. The different coercive fields of the top (NiFe soft) and bottom (FeCo hard) ferromagnetic layers allow the spin valve to be switched into any of four possible states (two with parallel magnetization, two with antiparallel magnetization). Modified from Drew et al. (2009), © Springer Nature. Reproduced with permission. All rights reserved

nm)/Alq3(200 nm)/TPD(50 nm)/FeCo(17 nm) with Au contact pads at the edges. Spin polarized electrons are injected from the top NiFe electrode through a LiF tunnel barrier into Alq3,[53] in which the muons are implanted. The two lowest layers are the hole injection layer TPD[54] and the bottom electrode FeCo.

The principle for mapping the polarization profile of the injected spin and for determining the spin diffusion length by LE-μSR is as follows:

- Since the coercive field is higher in FeCo than in NiFe, by applying an external field of appropriate strength, the magnetization direction of the top and bottom ferromagnetic layers can be reversed separately, thus preparing the spin valve in one of the four different states, Fig. 8.25.
- Spin polarized electrons are injected from the top layer into the spin valve by applying a small voltage across the structure. Due to the small spin-orbit and hyperfine interactions in the organic semiconductor, these spins have a long spin coherence time, more than $\sim 10^{-5}$ s, which is much longer than the muon lifetime.

---

[53] Alq3, tris-(8-hydroxyquinoline) aluminium, is a common organic semiconductor widely used in organic light-emitting diodes (OLEDs), chemical formula $C_{27}H_{18}AlN_3O_3$.

[54] N,N'-Bis(3-methylphenyl)-N,N'-diphenylbenzidine, chemical formula $C_{38}H_{32}N_2$.

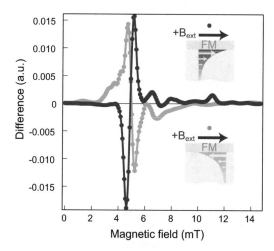

Fig. 8.26 Differences between the measured field distributions (TF 5 mT) with the injection current on and off for two spin valve states. The blue and green lines show the results for the two configurations where the direction of the external field with respect to the polarization of the injected carriers is either parallel or antiparallel. It is clear that in the parallel configuration, the current-on lineshape is skewed toward higher magnetic fields, while in the antiparallel configuration, the current-on lineshape is skewed toward lower magnetic fields. The insets schematically show the expected spatial distributions of the internal magnetic field in the organic layer for both configurations. When the spin moments of the injected carriers are aligned (anti-aligned) with the applied field, the μSR lineshape is skewed to the higher (lower) fields. Modified from Drew et al. (2009), © Springer Nature. Reproduced with permission. All rights reserved

- Therefore, from the μSR point of view, the injected spins give rise to a static electronic polarization $\langle \mathbf{S}_z(x) \rangle$, which produces a static magnetic field $\mathbf{B}_z^{spin}(x) \propto \langle \mathbf{S}_z(x) \rangle$. Depending on the spin valve state, this internal field adds to (or subtracts from) the external field $\mathbf{B}_{ext}$ used to select the spin valve state.
- The muons detect a local field $B_{\mu,z}(x) = B_{ext} \pm B_z^{spin}(x)$ at different depths in the barrier determined by the implantation energy.
- The field distribution $f(B_{\mu,z})$ is obtained from the Fourier transform of the TF-μSR signal.
- $B_z^{spin}(x)$ can be determined by comparing measurements when switching the injection on/off with current (voltage on or off) and by changing its sign with respect to the external field (this is obtained by reversing the magnetization of the NiFe electrode), Fig. 8.26.

From the global analysis of the field distributions in the different states of the spin valve, the spin diffusion length can be determined. An important finding is that its temperature dependence correlates with the temperature dependence of the magnetoresistance, Fig. 8.27. In another experiment, using the same methodology, it was shown that the inversion of the sign of the injected spin polarization can

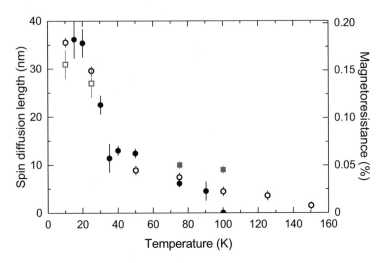

Fig. 8.27 Temperature dependence of the spin diffusion length extracted from the μSR measurements (two different samples indicated by open and filled red squares) plotted together with the temperature dependence of the magnetoresistance for two other samples (open and filled black circles). The figure evidences the qualitative agreement between the macroscopic and microscopic parameters. Modified from Drew et al. (2009), © Springer Nature. Reproduced with permission. All rights reserved

be achieved by introducing a very thin LiF layer at the interface between the top magnetic layer and Alq3 (Schulz et al. 2011), meaning that it is possible to control the polarization of the injected carriers.

8.5.3 Studies of Dynamics

Single Molecule Magnets As an example of the study of dynamics we present two experiments on soft matter materials.[55] LE-μSR has been used for a comparative study of the local spin dynamics of TbPc$_2$[56], a single molecule magnet compound (SMM)[57], deposited as a thin film (~ 100 nm), with the dynamics in a thick film (~ 1000 nm) and in the bulk (Hofmann et al. 2012).

---

[55] Soft matter comprises a very wide range of materials (from liquids to polymers to biological materials), which can be empirically characterized as soft to the touch. They are easily compressed, deformed by magnetic and electric fields, and modified by thermal stress or fluctuations at temperatures of the order of the room temperature.

[56] bis-phthalocyaninato terbium, Tb(C$_{32}$H$_{16}$N$_8$)$_2$ · CH$_2$Cl$_2$ in the bulk.

[57] Single molecule magnet compounds are metalo-organic nanosized clusters composed of transition metal or rare earth ions, which are coupled by exchange interaction, and associated ligands.

Fig. 8.28 Energy levels of the $J = 6$ ground state multiplet of TbPc$_2$. Two mechanisms can lead to spin flip: thermal excitation over the barrier, or quantum tunneling through the barrier (arrow)

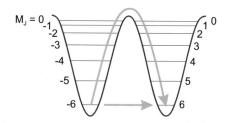

TbPc$_2$ consists of a Tb spin enclosed in an organic cage. Like bulk magnets, SMMs exhibit magnetic hysteresis at low temperatures (below the so-called blocking temperature). The hysteretic behavior results from a magnetic ground state, where the spin down and the spin up states are separated by an energy barrier, Fig. 8.28. This property and their smallness make SMMs interesting candidates for (dense) data storage and applications (as qubits) in quantum information processing and spintronics (Wernsdorfer et al. 2002; Gatteschi & Sessoli 2003; Leuenberger & Loss 2001). They are also used to study quantum tunneling of magnetization.

The ground spin state manifold $J = 6$ of TbPc$_2$ is split by a strong crystal field at the Tb^{3+} ion, resulting in a large separation (on the order of a few hundred K) between the ground state, $M_J = \pm 6$, and the first excited state, $M_J = \pm 5$.

For a useful operating temperature in applications, SMMs must have high energy barriers and blocking temperatures to prevent spin reorientation and fast fluctuations. Another important factor to understand is the behavior of SMMs when bonded to solid surfaces, as a change in their properties is often observed due to interaction with specific surfaces.

LE-μSR provides a direct observation of the spin dynamics of individual TbPc$_2$ molecules, studied as a function of depth in thin films deposited on an Au substrate and compared to results in thicker and bulk samples.

The ZF asymmetries measured in TbPc$_2$ are best fit by a Lorentz Kubo-Toyabe function multiplied by a square-root exponential relaxation

$$A(t) = A_0 \left[\frac{1}{3} + \frac{2}{3}(1 - at) \exp(-at) \right] \exp\left(-\sqrt{\lambda t}\right) , \qquad (8.18)$$

a/γ_μ is the HWHM of the static field distribution related to the Tb spins, while λ contains information about the dynamics of the local field. The square root exponential relaxation reflects the averaging of the muon spin relaxation over many inequivalent stopping sites. Such a behavior has also been observed in dilute spin glasses, see Eq. 4.75 and (Uemura et al. 1985).

The temperature dependences of $\delta \cong a/\gamma_\mu$ and of λ are plotted in Fig. 8.29. One can roughly distinguish two temperature regimes. In the high-temperature regime, the increase of λ with decreasing the temperature reflects the slowing down of the spin dynamics. It does not depend on depth and is practically the same as in the bulk. In this temperature range, $\delta \approx 0$ because the fast thermal fluctuations between the ground state $M_J = \pm 6$ and the first excited state $M_J = \pm 5$ average out the static

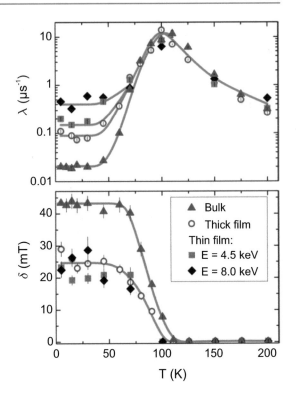

Fig. 8.29 Top: dynamic muon spin relaxation parameter, λ. Bottom: static field width parameter, $\delta = a/\gamma_\mu$ (bottom) as a function of temperature in bulk, thick, and thin film samples of TbPc$_2$. The lines are to guide the eye. Modified from Hofmann et al. (2012), © American Chemical Society. Reproduced with permission. All rights reserved

component. At low temperatures, $T \lesssim 70K$, the polarization curves (not shown) have a slowly relaxing tail, indicating that the muons sense a low spin dynamics in addition to the static component of the local field. The static field width is energy independent and saturates at low temperatures, see bottom panel of Fig. 8.29. Also λ saturates in this temperature range, but shows depth dependent values in the film that differ from the thick film and the bulk values.

The temperature dependence of the molecular spin fluctuation time τ_c, Fig. 8.30, can be obtained from the muon spin relaxation, by noting that in the fast fluctuation regime $\lambda = 4a^2\tau_c$, Eq. 4.75, and that in the slow fluctuations regime $A(t)$ is almost identical to the dynamic Lorentzian Kubo-Toyabe for slow fluctuations where only the 1/3-tail decays with $2\tau_c/3$, Eq. 4.62. At low temperatures, the correlation times in the film are much shorter than in the bulk, indicating much faster fluctuations (note the logarithmic scale).

The $\tau_c(T)$ data points can be well fitted by a model that considers two different fluctuation mechanisms (Branzoli et al. 2009). A high-temperature thermally activated one, mediated by spin-phonon coupling, and a low-temperature one ($T \lesssim$ 70 K) due to temperature independent quantum tunneling between the two (quasi-)degenerate $M_J = \pm 6$ ground states.

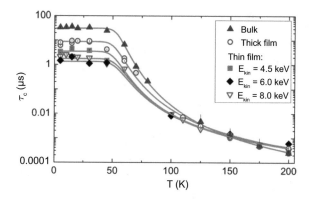

Fig. 8.30 Temperature dependence of the Tb spin correlation time in the bulk, thin, and thick films of TbPc$_2$. The lines are fits to the model described in the text. Modified from Hofmann et al. (2012), © American Chemical Society. Reproduced with permission. All rights reserved

From the data a quantum tunneling time τ_q can be derived. Noticing that with increasing implantation energy the muons stop closer and closer to the Au substrate, one finds that τ_q gradually increases from 1.4(1) μs about 10 nm close to the Au interface, to 6.6(2) μs a few 100 nm away (thick film data), to 31.2(1.6) μs in the bulk. This finding is explained by a change in the molecular arrangement (and hence the packing of the magnetic core) with distance from the substrate, from TbPc$_2$ molecules lying flat on the substrate (resulting in out-of-plane easy axes) to TbPc$_2$ molecules standing upright (in-plane easy axes), far from the interface (Hofmann et al. 2012).

Surface Dynamics of Polymer Films Muon spin rotation/relaxation techniques have been used to study various properties of polymers such as the structural changes taking place as a function of temperature or dynamical phenomena related to charge motion in conducting polymers (Pratt et al. 2003).

The properties of polymer thin films can differ significantly from those of the bulk. For example, the glass transition temperature T_g, a parameter characterizing the dynamic properties of polymers, is reduced in nanometer thick films (Forrest & Dalnoki-Veress 2001). LE-μSR has been used to probe the local dynamical properties and the associated spin glass temperature in various polymer compounds (Pratt et al. 2016; McKenzie et al. 2020). As an example, we present measurements of polystyrene, Fig. 8.31 (Pratt et al. 2005; Kanaya et al. 2015).[58]

In polystyrene (PS), about half of the implanted muons form a muonium state, where they add to the carbon atoms of the phenyl side rings of PS and form a so-

---

[58] Polystyrene, also called polystyrol, is one of the most widely used plastics. It is a long-chain hydrocarbon formed by carbon centers attached to phenyl groups (C_6H_5). The repeating chemical unit formula is C_8H_8.

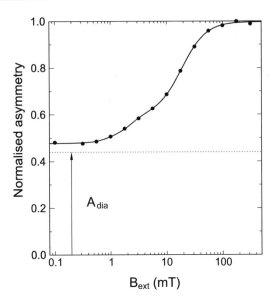

Fig. 8.31 Repolarization curve (initial asymmetry versus longitudinal field) of the muoniated radical state in PS at room temperature. The small repolarizing field compared to that of vacuum muonium indicates that the hyperfine coupling is much reduced compared to vacuum muonium. Most of the field independent polarization is due to a diamagnetic muon state. Note that the low field polarization in the Mu state, which accounts for half of the vacuum muonium repolarization curve is strongly suppressed in this case. The suppression is possibly related to the anisotropy of the hyperfine tensor or to the interaction with the nuclear spins adjacent to the muonium (superhyperfine interaction). Modified from Pratt et al. (2003), © Elsevier. Reproduced with permission. All rights reserved

called muoniated radical state, see Fig. 8.32.[59] This paramagnetic radical state has a small hyperfine coupling constant, as it can be inferred from the small repolarizing field of the repolarization curve Fig. 8.31.[60] The muon spin relaxation is related to the relaxation of the unpaired electron spin coupled to the muon spin via the hyperfine interaction tensor, which is modulated by the motion of the polymer and of the unpaired electron.

The muon spin relaxation provides information about the dynamics of the rings and is sensitive to the glass transition (McKenzie et al. 2020). At the glass transition temperature the dynamic correlation time of the polymer rapidly changes. The transition appears as a kink in the temperature dependence of the muon spin relaxation rate, Fig. 8.33.

---

[59] A radical is an atom or molecule with an unpaired electron. In the chemical formula a dot symbolizes the unpaired electron, which is placed, if possible, to indicate the atom with the highest spin density (International Union of PureApplied Chemistry 2009).

[60] The inflection point of the repolarization curve $B_{\text{ext},1/2}$ gives a measure of the hyperfine constant at the muon site $B_{\text{ext},1/2} \cong A/(\hbar\gamma_e)$, see Eqs. 7.17, 7.33 and Fig. 7.5.

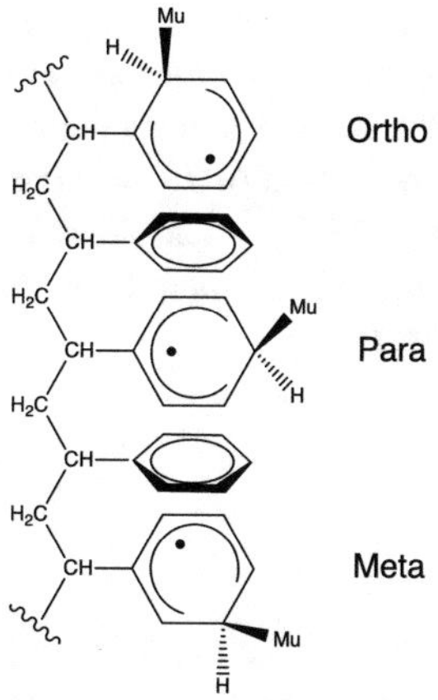

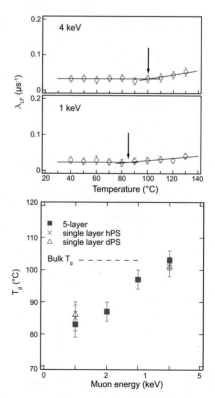

Fig. 8.32 Muoniated radicals in PS. Muonium can add at three different positions of the phenyl ring, which are characterized by different isotropic and dipolar hyperfine coupling constants. The three-dimensional ring is represented in the figure in the so-called dash-wedge convention. A wedged line represents a bond coming forwards, a dashed line represents a bond going backwards. Modified from McKenzie et al. (2020), © IOP Publishing. Reproduced with permission. All rights reserved

Fig. 8.33 Top panel: Temperature dependence of the longitudinal relaxation rate λ_{LF} for a 100 nm layer of polystyrene for two implantation energies. Bottom panel: Glass transition temperature T_g as a function of the muon energy (hPS: hydrogenated polystyrene, dPS: deuterated polystyrene). The results are in good agreement with neutron reflectivity measurements on the same samples. Modified from Kanaya et al. (2015), © American Physical Society. Reproduced with permission. All rights reserved

The results of the LF measurements (Kanaya et al. 2015) indicate that the polymer film has an heterogeneous dynamics as reflected by a depth dependence of T_g, which is lowest near the surface and increases with depth, Fig. 8.33. These measurements confirmed the reduction of T_g near the surface observed in previous ZF measurements (Pratt et al. 2005).[61] The T_g reduction can be explained by a model, where the local mobility of polymer segments is significantly enhanced at the free surface (Pratt et al. 2005).

8.5.4 Thin Films

Dilute Magnetic Semiconductors The application of LE-μSR is often motivated by the fact that the compound of interest can only be grown as a thin film. A typical example is the study of dilute magnetic semiconductors (DMS), materials that exhibit both semiconducting and ferromagnetic properties. This coexistence offers the possibility of developing novel devices in which information transfer (by charge) and storage (by spin) are performed in the same compound (spin electronics or spintronics).

Mn-doped GaAs, (Ga,Mn)As is a prototypical ferromagnetic semiconductor in which a Mn^{2+} atom replaces a Ga^{3+} atom, introducing a magnetic moment and a hole. It has received intensive attention not only for its potential applications in spintronics, but also for the fundamental interest in understanding the physics governing the evolution from a paramagnetic insulator to a ferromagnetic metal. Mn has a very low equilibrium chemical solubility limit of $< 0.1\%$. This limit can be overcome by epitaxial growth, which allows growth under conditions far from thermodynamic equilibrium and the fabrication of thin films with higher Mn concentrations in search of higher Curie temperatures (Ohno 1998). An important question is whether ferromagnetism develops throughout the sample volume. Intrinsic inhomogeneity of the material would have major practical limitations for device applications involving manipulation of the magnetization and injection of polarized electrons.

The issue of spatial inhomogeneity can be well addressed by μSR (Dunsiger et al. 2010), which with wTF measurements, Sect. 5.2, provides unique information about the volume fraction of regions with static magnetic order, as well as the size and distribution of the ordered moments with ZF measurements.

Low-energy muons with kinetic energy of 5 keV have been implanted in 60 nm samples with different Mn doping levels x. The chosen energy results in an average muon implantation depth of ~ 30 nm with a spread of ~ 10 nm. As an example,

---

[61] As discussed in Sect. 4.3, LF measurements provide a cleaner insight into dynamical effects than ZF or TF measurements. At the time of the (Pratt et al. 2005) experiment, the LF option was not available at the LE-μSR instrument.

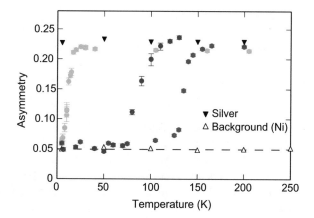

Fig. 8.34 Amplitude of the weak transverse field precession signal (10 mT) representing muons in a non- or paramagnetic environment. The maximum observable asymmetry is obtained by measuring a pure Ag target. The background level, shown by the dashed line, is obtained by replacing the sample with a Ni plate. Muons missing the sample and stopping in ferromagnetic Ni are quickly depolarized due to the presence of large internal magnetic fields and do not contribute to the precession. Modified from Dunsiger et al. (2010). © Springer Nature. Reproduced with permission. All rights reserved

Fig. 8.34 shows the amplitude of the wTF precession signal on high-quality thin film samples, which is caused by muons in a nonmagnetic or paramagnetic environment.

The results demonstrate that (Ga,Mn)As has a sharp onset of ferromagnetic order, which develops homogeneously in the full volume fraction in both insulating and metallic films.[62]

LE-μSR spectroscopy has also been applied to investigate the ferromagnetism of Co-doped TiO_2 thin films (Saadaoui et al. 2016), the first reported oxide-based DMS with a Curie temperature as high as 600 K (Matsumoto et al. 2001). Studies of epitaxially grown $\sim$ 50 nm thin films as a function of growth conditions show that $Co_{0.05}Ti_{0.95}O_{2-\delta}$ is an intrinsic dilute magnetic semiconductor with uniform structure and homogeneous ferromagnetism. These properties, combined with its high T_C, underline its position as a promising DMS candidate for spintronic applications.

Superlattices In the previous examples, the relevant layers were thick enough to allow the depth dependence to be studied by varying the kinetic energy of the muon. As Fig. 8.10 shows, the mean stop depth of LE-muons is of the order of tens of nanometer, with a relatively broad stopping range covering a region also on the order of tens of nanometers. Clearly, this depth resolution may be not sufficient to directly extract μSR parameters if the layers of interest are only a few unit cells

---

[62] Similar results have been obtained in investigations of other GaAs based ferromagnetic semiconductors (Levchenko et al. 2019).

thick (e.g., $\sim$ 1–2 nm). However, it is still possible to obtain information about the physical properties of these layers from a LE-μSR experiment, if the few nm thick relevant layers are repeated in a periodic structure (superlattice),[63] thus enhancing the μSR signal.

Not only can new phenomena occur by combining different materials (see, for example, the proximity effects discussed earlier), but by reducing the dimensionality of a compound, its electronic and physical properties can be significantly modified.

An example of a dimensionality effect is the detection of electronic phase transitions, which are controlled by the layer thickness in nickel oxide superlattices (Boris et al. 2011). The superlattices consist of atomically-precise layer-sequences of the metal lanthanum nickelate (LaNiO$_3$) and of the wide bandgap insulator lanthanum aluminate (LaAlO$_3$). They consist of N unit-cell-thick successive layers of the two materials (unit cell thickness $\sim$0.4 nm). Superlattices with $N = 2$ and $N = 4$ have been studied by LE-μSR and ellipsometry.[64] With a total thickness of 100 nm, the LE muon stop distribution is confined within the superlattice and the μSR signal as well as the response from ellipsometry measurements are enhanced. Figure 8.35 shows ZF and TF μSR spectra at selected temperatures and an implantation energy of 10 keV. At this energy, the mean stopping depth

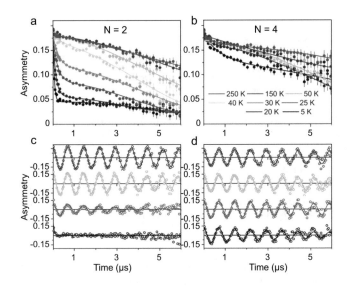

Fig. 8.35 Top panels: ZF μSR spectra at different temperatures for the (**a**) $N = 2$ and (**b**) $N = 4$ superlattices deposited on a LaSrAlO$_4$ substrate. Bottom panels: corresponding μSR spectra in a weak 10 mT transverse magnetic field. Modified from Boris et al. (2011). © American Association for the Advancement of Science. Reproduced with permission. All rights reserved

---

[63] The period consists of two (usually) or more different materials.

[64] Ellipsometry provides the frequency dependent complex dielectric function $\varepsilon(\omega) = \varepsilon_1(\omega) + i\varepsilon_2(\omega)$, which is related to the optical conductivity $\sigma(\omega)$ by $\varepsilon(\omega) = \varepsilon_0 + i\sigma(\omega)/\omega$.

is ~ 45 nm with a FWHM of ~ 20 nm. For the $N = 2$ superlattice, the stopping profile covers more than 100 layers of each constituent.

Figure 8.35a shows that, with the $N = 2$ superlattice, the asymmetry drops sharply below 50 K, indicating a magnetic transition. From the wTF measurements, see Sect. 5.2, at 10 mT the fraction of muons experiencing a local magnetic field larger than the applied field can be determined to be almost 100% in the $N = 2$ sample, Fig. 8.35c. The Figure manifests for the $N = 2$ superlattice a transition from a fully paramagnetic muon environment to a nearly full volume of static internal fields, with a sharp onset at $T_N \sim 50$ K.[65] In contrast, the ZF and wTF spectra with thicker $LaNiO_3$ layers ($N = 4$ superlattice) display no evidence of magnetism and show that $LaNiO_3$ remains paramagnetic down to the lowest temperatures, as in bulk. From the optical ellipsometry measurements it is additionally concluded that four-unit-cell $LaNiO_3$ superlattices remain metallic at all temperatures, whereas two-unit-cell $LaNiO_3$ exhibit clear evidence of a metal-insulator transition upon cooling, with a sharp onset at a temperature $T_{MI} > T_N$. The combination of the two techniques allows us to conclude that as the temperature decreases, superlattices with $LaNiO_3$ as thin as two unit cells undergo successive metal-insulator and antiferromagnetic transitions, whereas samples with thicker $LaNiO_3$ layers remain metallic and paramagnetic at all temperatures. The findings show that the electronic phase can be changed via $LaNiO_3$ thickness and demonstrate that the phase of metal-oxide correlated electron systems can be controlled by dimensionality.

Exercise

8.1. Figure of Merit
Derive the expression for the figure of merit of a μSR setup.

References

Amin, M. H. S., Franz, M., & Affleck, I. (2000). *Physical Review Letters, 84*, 5864.

Antognini, A., Ayres, N. J., Belosevic, I., et al. (2020). *Physical Review Letters, 125*, 164802.

Badertscher, A., Egan, P., Gladisch, M., et al. (1985). *Nuclear Instruments and Methods in Physics Research - Section A, 238*, 200.

Baibich, M. N., Broto, J. M., Fert, A., et al. (1988). *Physical Review Letters, 61*, 2472.

Bakule, P., & Morenzoni, E. (2004). *Contemporary Physics, 45*, 203.

Bao, Y., Antognini, A., Bertl, W., et al. (2014). *Physical Review Letters, 112*, 224801.

Beer, G. A., Fujiwara, Y., Hirota, S., et al. (2014). *Progress of Theoretical and Experimental Physics, 2014*, 091C01.

Beer, G. A., Marshall, G. M., Mason, G. R., et al. (1986). *Physical Review Letters, 57*, 671.

Belosevic, I., Antognini, A., Bao, Y., et al. (2019). *Hyperfine Interactions, 240*, 41.

---

[65] As a local probe, μSR does not allow definitive conclusions about the type of magnetic ordering. However, ferromagnetism can be ruled out based on an estimate of the Ni ordered moment from the width of the local field distribution experienced by the muons.

Berezinskii, V. L. (1974). *JETP Letters, 20,* 287.

Binasch, G., Grünberg, P., Saurenbach, F., et al. (1989). *Physical Review B, 39,* 4828.

Biswas, P. K., Salman, Z., Song, Q., et al. (2018). *Physical Review B, 97,* 174509.

Boris, A. V., Matiks, Y., Benckiser, E., et al., (2011). *Science, 332,* 937.

Branzoli, F., Filibian, M., Carretta, P., et al. (2009). *Physical Review B, 79,* 220404.

Brown, B. (2000). *Physical Review Focus, 5,* 22.

Budker, G. I., & Skrinskiĭ, A. N. (1978). *Soviet Physics Uspekhi, 21,* 277.

Buzdin, A. I. (2005). *Reviews of Modern Physics, 77,* 935.

Carbotte, J. P. (1990). *Reviews of Modern Physics, 62,* 1027.

Chandrasekhar, V. (2008). In K. H. Bennemann & J. B. Ketterson (Eds.), *Superconductivity* (p. 279). Springer. ISBN: 978-3540732525

Deutscher, G., & de Gennes, P. G. (1969). In R. D. Parks (Ed.), *Superconductivity* (vol. 2, p. 1005). CRC Press. ISBN: 978-0203737958

Di Bernardo, A., Salman, Z., Wang, X. L., et al. (2015). *Physical Review X, 5,* 041021.

Drew, A. J., Hoppler, J., Schulz, L., et al. (2009). *Nature Materials, 8,* 109.

Dunsiger, S. R., Carlo, J. P., Goko, T., et al. (2010). *Nature Materials, 9,* 299.

Eckstein, W. (1991). *Computer Simulation of Ion-Solid Interactions.* Springer. ISBN: 978-3642735134.

Flokstra, M. G., Ray, S. J., Lister, S. J., et al. (2014). *Physical Review B, 89,* 054510.

Flokstra, M. G., Satchell, N., Kim, J., et al. (2016). *Nature Physics, 12,* 57.

Flokstra, M. G., Stewart, R., Satchell, N., et al. (2018). *Physical Review Letters, 120,* 247001.

Flokstra, M. G., Stewart, R., Satchell, N., et al. (2019). *Applied Physics Letters, 115,* 072602.

Forrest, J., & Dalnoki-Veress, K. (2001). *Advances in Colloid and Interface Science, 94,* 167.

Gatteschi, D., & Sessoli, R. (2003). *Angewandte Chemie International Edition, 42,* 268.

Harshman, D. R., Mills, A. P., Beveridge, J. L., et al. (1987). *Physical Review B, 36,* 8850.

Harshman, D. R., Warren, J. B., Beveridge, J. L., et al. (1986). *Physical Review Letters, 56,* 2850.

Hofmann, A., Salman, Z., Mannini, M., et al. (2012). *ACS Nano, 6,* 8390.

International Union of Pure and Applied Chemistry. (2009). *IUPAC Compendium of Chemical Terminology (the "Gold Book").*

Jackson, T. J., Riseman, T. M., Forgan, E. M., et al. (2000). *Physical Review Letters, 84,* 4958, 434.

Kanaya, T., Ogawa, H., Kishimoto, M., et al. (2015). *Physical Review E, 92,* 022604.

Kanda, S., Teshima, N., Adachi, T., et al. (2023). *Journal of Physics: Conference Series, 2462,* 012030.

Kiefl, R. F., Hossain, M. D., Wojek, B. M., et al. (2010). *Physical Review B, 81,* 180502.

Kiefl, R., MacFarlane, W., Morris, G., et al. (2003). *Physica B: Condensed Matter, 326,* 189–195.

Kozhevnikov, V., Suter, A., Fritzsche, H., et al.(2013). *Physical Review B, 87,* 104508.

Krieger, J.A., Pertsova, A., Giblin, S. R., et al. (2020). *Physical Review Letters, 125,* 026802.

Leuenberger, M. N., & Loss, D. (2001). *Nature, 410,* 789.

Levchenko, K., Prokscha, T., Sadowski, J., et al. (2019). *Scientific Reports, 9,* 3394.

Linder, J., & Balatsky, A. V. (2019). *Reviews of Modern Physics, 91,* 045005.

Lindstrom, M., Fang, A. C. Y., & Kiefl, R. F. (2016). *Journal of Superconductivity and Novel Magnetism, 29,* 1499.

London, F., & London, H. (1935). *Proceedings of the Royal Society of London A: Mathematical, Physical and Engineering Sciences, 149,* 71.

Luetkens, H., Korecki, J., Morenzoni, E., et al. (2003). *Physical Review Letters, 91,* 017204.

MacFarlane, W. A. (2022). *Zeitschrift für Physikalische Chemie, 236,* 757–798.

Marchand, D., Covaci, L., Berciu, M., et al. (2008). *Physical Review Letters, 101,* 097004.

Matsumoto, Y., Murakami, M., Shono, T., et al. (2001). *Science, 291,* 854–856.

McKenzie, I., Cordoni-Jordan, D., Cannon, J., et al. (2020). *Journal of Physics: Condensed Matter, 33,* 065102.

Mills, A. P., Imazato, J., Saitoh, S., et al. (1986). *Physical Review Letters, 56,* 1463.

Mironov, S., Mel'nikov, A. S., & Buzdin, A. (2018). *Applied Physics Letters, 113,* 022601.

Miyake, Y., Shimomura, K., Kawamura, N., et al. (2012). *Physics Procedia, 30,* 46.

Miyake, Y., Shimomura, K., Kawamura, N., et al. (2014). *Journal of Physics: Conference Series, 551*, 012061.

Morenzoni, E. (1999). In S. Lee, S. Kilcoyne, & R. Cywinski (Eds.), *Muon Science: Muons in Physics, Chemistry, and Materials* (p. 343). CRC Press. ISBN: 978-0750306300.

Morenzoni, E., Kottmann, F., Maden, D., et al. (1994). *Physical Review Letters, 72*, 2793.

Morenzoni, E., Glückler, H., Prokscha, T., et al. (2002). *Nuclear Instruments and Methods in Physics Research - Section B, 192*, 254.

Morenzoni, E., Prokscha, T., Hofer, A., et al. (1997). *Journal of Applied Physics, 81*, 3340.

Morenzoni, E., Prokscha, T., Suter, A., et al. (2004). *Journal of Physics: Condensed Matter, 16*, S4583.

Morenzoni, E., Wojek, B. M., Suter, A., et al. (2011). *Nature Communications, 2*, 272.

Morris, G. D. (2014). *Hyperfine Interactions, 225*, 173–182.

Nagamine, K., Miyake, Y., Shimomura, K., et al. (1995). *Physical Review Letters, 74*, 4811.

Ni, X., Zhou, L., Martins, M. M., et al. (2023). arXiv:2302.06773.

Niedermayer, C., Forgan, E. M., Glückler, H., et al. (1999). *Physical Review Letters, 83*, 3932.

Ohno, H. (1998). *Science, 281*, 951.

Pippard, A. B. (1953). *Proceedings of the Royal Society of London. Series A. Mathematical and Physical Sciences, 216*, 547.

Pratt, F. L., Blundell, S. J., Marshall, I. M., et al. (2003). *Physica B: Condensed Matter, 326*, 34.

Pratt, F. L., Lancaster, T., Brooks, M. L., et al. (2005). *Physical Review B, 72*, 121401, 435.

Pratt, F. L., Lancaster, T., Baker, P. J., et al. (2016). *Polymer, 105*, 516.

Prokscha, T., Chow, K.-H., Salman, Z., et al. (2020). *Physical Review Applied, 14*, 014098.

Prokscha, T., Morenzoni, E., David, C., et al. (2001). *Applied Surface Science, 172*, 235.

Prokscha, T., Morenzoni, E., Deiters, K., et al. (2008). *Nuclear Instruments and Methods in Physics Research - Section A, 595*, 317.

Prokscha, T., Morenzoni, E., Eshchenko, D. G., et al. (2007). *Physical Review Letters, 98*, 227401.

Prokscha, T., Morenzoni, E., Meyberg, M., et al. (1998). *Physical Review A, 58*, 3739.

Rainford, B. D., & Daniell, G. J. (1994). *Hyperfine Interactions, 87*, 1129.

Riseman, T., & Forgan, E. (2003). *Physica B: Condensed Matter, 326*, 226.

Romanenko, A., Grassellino, A., Barkov, F., et al. (2014). *Applied Physics Letters, 104*, 072601.

Saadaoui, H., Luo, X., Salman, Z., et al. (2016). *Physical Review Letters, 117*, 227202.

Saadaoui, H., Salman, Z., Prokscha, T., et al. (2012). *Physics Procedia, 30*, 164–167. 12th International Conference on Muon Spin Rotation, Relaxation and Resonance (μSR2011).

Salman, Z., Prokscha, T., Keller, P., et al. (2012). *Physics Procedia, 30*, 55.

Schrieffer, J. (1999). *Theory of Superconductivity. Advanced Books Classics.* CRC Press. ISBN: 978-0738201207.

Schulz, L., Nuccio, L., Willis, M., et al. (2011). *Nature Materials, 10*, 39.

Simons, L., Morenzoni, E., & Kottmann, F. (1992). In G. Benedek & H. Schneuwly (Eds.), *Exotic Atoms in Condensed Matter. Proceedings in Physics* (vol. 59, p. 33). Springer. ISBN: 978-3642763700.

Sonier, J. E., Brewer, J. H., & Kiefl, R. F. (2000). *Reviews of Modern Physics, 72*, 769.

Sonier, J. E., Sabok-Sayr, S. A., Callaghan, F. D., et al. (2007). *Physical Review B, 76*, 134518.

Stone, N. (2005). *Atomic Data and Nuclear Data Tables, 90*, 75–176.

Suter, A., Morenzoni, E., Garifianov, N., et al. (2005). *Physical Review B, 72*, 024506.

Suter, A., Morenzoni, E., Khasanov, R., et al. (2004). *Physical Review Letters, 92*, 087001.

Tinkham, M. (1996). *Introduction to Superconductivity.* Mc Graw-Hill. ISBN: 978-0070648784.

Uemura, Y. J., Yamazaki, T., Harshman, D. R., et al. (1985). *Physical Review B, 31*, 546.

van der Meer, S. (1985). *Reviews of Modern Physics, 57*, 689.

Wernsdorfer, W., Aliaga-Alcalde, N., Hendrickson, D. N., et al. (2002). *Nature, 416*, 406.

Yokoyama, T., Tanaka, Y., & Nagaosa, N. (2011). *Physical Review Letters, 106*, 246601, 436.

Use of Negative Muons: μ^-SR and Elemental Analysis

9

In the previous chapters, we have discussed the use of positive muons and muonium to probe the microscopic properties of matter. The positive charge leads in solids to the localization, mainly at interstitial sites, of (diamagnetic) muon or muonium states, whose local probe properties have established the positive muon as a very versatile, sensitive probe that is used in the vast majority of μSR research.

For a variety of reasons discussed below, μSR with negative muons (μ^-SR) has a much more limited scope. In certain materials, however, it can still be used to extract specific information. When negative muons stop in matter muonic atoms are formed. Spectroscopy of muonic atoms can be used as a nondestructive, sensitive tool for depth-resolved elemental analysis, a technique that has experienced a revival in recent years. In this chapter we present these two main applications of negative muons in condensed matter and applied sciences.

9.1 Negative Muon Beams

Negative muons are produced by the decay of a negative pion, in a process which is the charge conjugated and parity symmetric (CP) process of the π^+ decay, see Fig. 1.12.

Negative pion (and hence negative muon) beams at meson factories are less intense than the corresponding positive beams. There are several reasons for this. Due to charge conservation, to create a π^- in a proton-nucleus reaction, at least two positive particles must be created thus reducing the number of possible reactions leading to the formation of a π^- and making the π^--production cross section smaller than that of π^+, see Sect. 1.4.2.

For protons with energies below $600\,\mathrm{MeV}$ practically only a single pion production reaction can be considered

$$p + n \rightarrow p + p + \pi^- \ . \tag{9.1}$$

© Springer Nature Switzerland AG 2024
A. Amato, E. Morenzoni, *Introduction to Muon Spin Spectroscopy*,
Lecture Notes in Physics 961, https://doi.org/10.1007/978-3-031-44959-8_9

Fig. 9.1 Intensity
comparison between positive
and negative muon beams as
a function of the muon
momentum. The data,
expressed in particles per
second and mA of 590 MeV
proton current, are for the
π E5 beamline at PSI

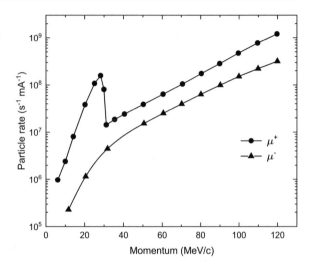

In addition to this, as Fig. 9.1[1] reveals, in the negative muon intensity versus
momentum curve, the typical enhancement at ~ 30 MeV/c reflecting the surface
muon production, Sect. 1.8.3, is absent. The intensity and full polarization of the
positive surface muon beam made it the preferred choice for most μ SR experiments
in the bulk[2] and its development fostered the evolution of the μ SR technique (Pifer
et al. 1976). Recall that surface muons are obtained from the decay of positive pions
that come at rest near the surface of the production target before decaying. This
process is possible because the positive pions can thermalize at an interstitial site
near the surface of the production target before decaying.

In contrast, the negatively charged pions that have slowed down in the production
target, have a very high probability of being captured into an orbit around a target
nucleus (formation of a pionic atom). After the atomic capture, they cascade down
to lower levels and, being hadrons, interact strongly with the nucleus where they
can be rapidly captured and annihilated. This process occurs at a rate many orders
of magnitude faster than the decay rate (π^- lifetime 26 ns) into a μ^-. Therefore, the
generation of a surface muon beam is only possible with positive muons.

Negative muon beams with momenta ~ 30 MeV originate mainly from pion
decays in flight inside or near the production target ("cloud" muons). These muons
can be created in forward or backward decay processes from pions with different
momenta at the time of decay: in contrast to the practically fully polarized surface
μ^+'s, cloud muons have a much lower initial polarization. The exact value depends
on beamline properties, such as acceptance and momentum spread; experiments, as
well as simulations, have found a polarization in the range 10–25% (van Dyck et al.
1979; Musser et al. 2005) (note that the polarization is in the opposite direction

---

[1] The muon curves from Fig. 1.26 are replicated here for convenience.

[2] Low-energy muon generation, Sect. 8.2, also relies on an intense surface muon beam.

to that of the surface μ^+). The low polarization and intensity make them not well suited for μ^- spin spectroscopy.

High-energy negative muons can be obtained from negative pions decaying in flight in a long solenoid after a kinematic selection in an identical manner as for positive high-energy muons, see Fig. 1.27 and, for the kinematics, Fig. 1.31. By choosing forward or backward decay products, these negative muons have a useful polarization similar to that of the positive counterpart.

9.2 Implantation of Negative Muons in Matter

When implanting μ^- into matter, the initial slowing down processes are similar to those for μ^+, which involve Coulomb excitation and ionization of the target electrons and are described by the Bethe formula 2.17.[3] This leads, for a large range of momenta, to the same μ^- stopping power $-dE/dx$ and related parameters as for μ^+, see also Fig. 2.4. The main difference appears at lower energies when the velocity of the negative muon becomes similar to that of the atomic electrons. Obviously, charge-changing cycles involving electron capture and loss, typical of μ^+, are not possible for μ^-; instead, μ^- strongly feels the attractive Coulomb force generated by a nearby nucleus so that it is captured forming a so-called "muonic atom" (or molecule).

9.2.1 Muonic Atoms

The capture process of the implanted negative muon into a bound level is the last step of the deceleration process in matter. At the same time, it is the first step in the formation of the muonic atom (or molecule) and is immediately followed by a very fast cascade process ($\sim 10^{-13}$ s) of the muon down to its $1s$ ground state, The gross features of a muonic atom can first be understood within a Bohr atomic model (Bohr 1913a,b) with μ^- playing the role of a heavy electron and where we introduce the reduced mass of the system formed by μ^- and the nucleus of an atom with mass number A and atomic number Z, $\overline{m}_{\mu A} = m_\mu m_A/(m_\mu + m_A) \simeq m_\mu$. The radius of an orbit with quantum number n is given by[4]

$$r_{n,\mu} = \frac{4\pi\varepsilon_0 \hbar^2 n^2}{\overline{m}_{\mu A} Z e^2} = \frac{\hbar\, n^2}{\overline{m}_{\mu A} Z c\alpha} = \frac{a_0}{Z}\frac{m_e}{\overline{m}_{\mu A}}n^2 \simeq \frac{m_e}{\overline{m}_{\mu A}}r_{n,e} \ . \tag{9.2}$$

---

[3] The Bethe formula predicts $\propto Z^2$ and therefore gives the same result for a positive or negative muon. There are higher-order corrections $\propto Z^3$ but they can be neglected for our purposes.

[4] As the electron and the negative muon are different particles they have their own quantum levels and numbers.

Similarly, for the energy we can write[5]

$$E_{n,\mu} = -\frac{\overline{m}_{\mu A}\, Z^2 e^4}{(4\pi\varepsilon_0)^2 2n^2\hbar^2} = -R_y \frac{\overline{m}_{\mu A}}{m_e}\frac{Z^2}{n^2} \simeq \frac{\overline{m}_{\mu A}}{m_e} E_{n,e} \ , \qquad (9.3)$$

which can also be expressed with the fine structure constant

$$E_{n,\mu} = -\frac{1}{2}\frac{(Z\alpha)^2 \overline{m}_{\mu A} c^2}{n^2} \ . \qquad (9.4)$$

Since the muon is ~ 207 times heavier than the electron, Table 1.2, its orbit in a muonic atom $r_{n,\mu}$ is a factor ~ 207 closer to the nucleus, and its energy $E_{n,\mu}$ a factor ~ 207 higher than the respective values of the electronic Bohr atom, $r_{n,e}$ and $E_{n,e}$.

For the muon capture, collisions with large impact parameters b are more likely; they correspond to collisions with atomic orbits with high L values.[6] A simple model predicts that the negative muon is captured in an excited state of similar energy and size as the 1s electronic level of the corresponding ordinary atom (where the Coulomb field of the nucleus is virtually unshielded by electrons), i.e., a state with principal quantum number n satisfying $r_{n,\mu} = r_{1s,e}$. This means $n = (\overline{m}_{\mu A}/m_e)^{1/2} \cong (m_\mu/m_e)^{1/2} \simeq 14$, see Fig. 9.2 for a schematic picture.

The muon cascade is characterized by different electronic processes. At the beginning of the cascade, the electron shells of the atom can be assumed to be filled and Auger transitions take place, ejecting electrons from the outer orbits. If, after the depletion of one shell, the remaining electrons are too tightly bound or if there are no electrons left, radiative transitions with emission of muonic X-rays dominate.[7] These X-rays, with energy ranging from a few keV to MeV, are characteristic of each element, making muonic X-ray spectroscopy a central technique in elemental analysis, see Sect. 9.4. For electric dipole transitions (E1)

---

[5] The Bohr model ignores the fine structure of the levels, that is the relativistic effects and the spin-orbit coupling. The level energy depends only on the principal quantum n. The Dirac equation includes the fine structure and gives the Coulomb potential energy for a pointlike infinite-mass nucleus (with J total angular momentum) (Adkins 2008).

$E_{n,\mu} = m_\mu c^2\left[\left[1 + (Z\alpha)^2\left[n - J - \tfrac{1}{2} + \sqrt{(J+1/2)^2 - (Z\alpha)^2}\right]^{-2}\right]^{-1/2} - 1\right]$. By expanding

one obtains $E_{n,\mu} = -\frac{1}{2}\frac{(Z\alpha)^2 m_\mu c^2}{n^2} - \frac{m_\mu c^2}{2n^4}\left(\frac{n}{J+\frac{1}{2}} - \frac{3}{4}\right)(Z\alpha)^4 + \dots$. The first term corresponds to Eq. 9.4 (for $m_A = \infty$) and the second term gives the first-order fine structure correction.

[6] The collision probability is $\propto b \propto L$, see Fig. 2.1, corresponding to the $2L + 1$ degeneracy of an n level.

[7] In this chapter, we use the generic term of X-rays, even though the energies involved sometimes correspond to gamma-rays. In heavy atoms, the energy of the muon transitions between low-lying states can be higher than the binding energy of a neutron leading to neutron release, called the nuclear Auger effect (Zaretsky & Novikov 1960).

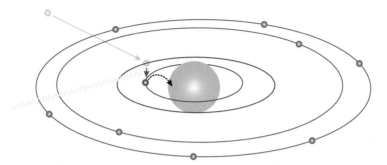

Fig. 9.2 Sketch of the capture of μ^- by an atom of the sample. The electron orbitals are in red and the muonic orbitals are in blue. For clarity, the ratio between the muon and electron orbital radii is fictitious (the muon orbitals are much closer to the nucleus than the electronic orbitals, see Eq. 9.2). The incoming negative muon (light violet) is first captured by the atom in an orbital with a radius close to that of the $1s$ electron orbital and cascades down to its $1s$ orbital (dark blue). During the cascade, characteristic X-rays are emitted. In a second step, the muon in the $1s$ orbital can be captured by the nucleus (black arrow), see Sect. 9.2.1

$\Delta l = \pm 1$ and the most intense transitions after capture in a $(n, L = n - 1)$ state are $(n, L = n - 1) \rightarrow (n - 1, L = n - 2) \rightarrow (n - 2, L = n - 3)\ldots.$ Transitions with $\Delta n > 1$ can also occur with reduced intensity.

The characteristic muonic X-rays play a central role in elemental analysis. As with the electronic transitions, the characteristic energy of emitted (or absorbed) X-rays is given by the energy difference between the initial and final states.

To first approximation, using Eq. 9.4, the energy of an X-ray emitted by the transition of an electron between atomic levels with principal quantum numbers n_i (initial) and n_f (final) is given by the empirical Moseley's law (Moseley 1913)

$$E_{i \rightarrow f, e} = \frac{\overline{m}_{Ae}}{m_e} R_y (Z - S_{scr,e})^2 \left(\frac{1}{n_f^2} - \frac{1}{n_i^2} \right) . \tag{9.5}$$

The atomic number Z is replaced by an effective parameter $(Z - S_{scr,e})$ to account for the fact that the charge of the nucleus seen by the electron involved in the transition is screened by electrons between itself and the nucleus. For a K_α transition,[8] $S_{scr,e} = 1$ since an L electron senses a nuclear charge shielded by the remaining electron in the K shell.

---

[8] A $2p \rightarrow 1s$ transition is denoted by K_α in the so-called Siegbahn notation and by $K - L$ in IUPAC notation. Taking into account the fine structure level splitting, the transition $2p_{3/2} \rightarrow 1s_{1/2}$ is denoted by K_{α_1} ($K - L_3$), and $2p_{1/2} \rightarrow 1s_{1/2}$ by K_{α_2} ($K - L_2$), respectively.

Analogously, for muonic X-ray transitions we expect

$$E_{i \to f, \mu} = \frac{\overline{m}_{A\mu}}{m_e} R_y (Z - S_{\text{scr}, \mu})^2 \left(\frac{1}{n_f^2} - \frac{1}{n_i^2} \right) \tag{9.6}$$

Since the radius of the $1s$ muonic orbital is much smaller than that of the corresponding electronic orbital, Eq. 9.2, for a muonic transition to the K shell the muon sees no nuclear charge screening, i.e., $S_{\text{scr}, \mu} = 0$, while $S_{\text{scr}, e} = 1$ for the electron.

Figure 9.3 shows experimental values of electronic and muonic transitions as a function of the atomic number $Z < 50$. The approximate Eqs. 9.5 and 9.6 predict that the square root of the transition energies should scale with the atomic number Z. As Fig. 9.3 shows, such a relationship is clearly observed for all electronic transitions and for muonic transitions if $Z \lesssim 20$. For large values of Z, the deviation from linearity observed for muonic transitions mainly reflects the fact that the size of the nucleus can no more be neglected with respect to the radius of the muonic

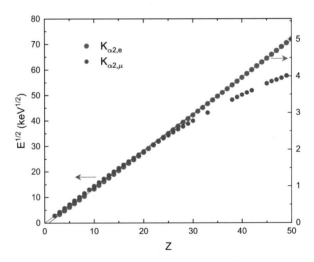

Fig. 9.3 Photon energy of a $K_{\alpha 2}$ transition in muonic and electronic atoms as a function of atomic number. The red dots and right scale are for electronic transitions in normal atoms, the blue dots and left scale are for muonic transitions. The vertical axes, which show the square root of the corresponding energies, scale with a factor $\sqrt{m_\mu / m_e}$, see text. Note that the abscissa intercept for the electronic transitions is $Z = 1$ (corresponding to $S_{\text{scr}, e} = 1$) and for the muonic transition is $Z = 0$ (corresponding to $S_{\text{scr}, \mu} = 0$). Note also the linearity (corresponding to $E_{K_{\alpha 2}} \propto (Z - S_{\text{scr}, X})^2$) for small values of Z, see Eqs. 9.5 and 9.6. Data for negative muons are taken from Engfer et al. (1974), Fricke et al. (1995) and for electrons from Bearden (1967)

level.[9] Note that such a deviation is less important for large values of n_f, where the radius of the muonic level is large enough that the point-charge approximation for the nucleus gives a reasonable estimate.

Polarization Until its capture by an atom, the negative muon retains its initial (beam) polarization (see Sect. 9.1) since, similarly to the positive muon, energy loss proceeds essentially by Coulomb processes. When the capture occurs and during the cascade down to the $1s$ state, strong depolarization occurs due to the spin-orbit coupling and, if the nucleus has nonzero spin, due to the hyperfine interaction with the nucleus (Winston 1963).[10] The details of the polarization loss process are complex and depend on the capture level, the atomic number Z, the effect of radiative or Auger transitions, and the nuclear spin.[11] Calculations as well as measurements indicate a remaining polarization of 5–20% of the original polarization, but there are also examples where no polarization has been detected (Evseev 1975; Grossheim et al. 2009).

Lifetime In vacuum, μ^- has the same lifetime as μ^+ and decays producing an electron, Eq. 1.7. The picture changes when it is implanted in matter. While the positive charge of μ^+ leads to a thermalization site in the lattice and to the avoidance of any interaction with the nucleus, the negative muon, cascading down to the $1s$ level, will be located very close to the nucleus with a substantial part of its wave function inside the nucleus, thus increasing the nuclear capture probability, see Fig. 9.4.

In the primary capture process, which is governed by the weak interaction, μ^- is captured by a bound proton of the nucleus (Measday 2001; Mukhopadhyay 1977)

$$\mu^- + p \rightarrow n + \nu_\mu \ . \tag{9.7}$$

The probability of this process depends directly on the overlap of the muonic $1s$ state density with the nucleus whose radial charge density can be reasonably

---

[9] The precise energy of the emitted X-rays depends on the size of the nuclei. In fact, precision muonic spectroscopy can be used to determine the charge distribution of nuclei. The most prominent example is the very accurate determination of the proton radius in muonic hydrogen by laser spectroscopy (Pohl et al. 2010), which reported a discrepancy with the results of other experimental investigations. This motivated a series of new experiments and theoretical calculations to understand the so-called proton radius puzzle, and led to a revision of the value adopted by CODATA (Karr & Marchand 2019).

[10] The strength of the hyperfine interaction in muonic atoms relative to ordinary atoms, can be illustrated by comparing the value of the $1s$ hyperfine splitting in atomic hydrogen ($e\,p$), A_H = $5.88 \cdot 10^{-6}$ eV with that of muonic hydrogen ($\mu^-\,p$), $A_{\mu p} = 0.183$ eV. This reflects the large probability of the negative muon to spend time in the nucleus, see Sect. 5.1.1 and Exercise 9.1.

[11] Simple models give an approximate expression for the residual polarization of μ^- in the $1s$ level of a muonic atom with nuclear spin I, to be about $1/6\,([1+2/(2I+1)]/3)$ of its initial (i.e., before the muonic atom is formed) polarization, see Nagamine (2003) for more details.

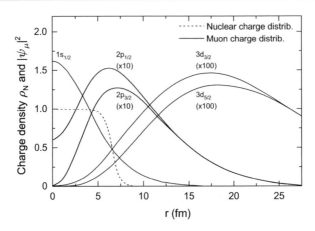

Fig. 9.4 Nuclear and muon charge distributions in lead ($Z = 82$). Note the large overlap in the $1s$ state, where the probability of finding the muon inside the nucleus is close to 50%. Modified from Devons and Duerdoth (1969), © Springer Nature. Reproduced with permission. All rights reserved

approximated by the Fermi function (Barrett 1974)

$$\rho_N(r) = \frac{\rho_0}{1 + \exp\left(\dfrac{r - R}{a_{st}}\right)} \ , \tag{9.8}$$

a_{st} is the nuclear surface thickness with a typical value of $a_{st} \simeq 0.5$ fm, R is the nuclear radius with $R = r_0 A^{1/3}$ ($r_0 \simeq 1.2$ fm) and A is the mass number. For a vanishing surface thickness, one obtains a homogeneous charge density $\rho_0 = 3 Ze/(4\pi R^3)$, over a radius R,

The capture probability increases with the nuclear charge Z, for $Z \gtrsim 7$, approximately as Z^4. This can be understood qualitatively by the following simple considerations. The probability for a μ^- in the $1s$ state to be in the volume of a nucleus with radius R is

$$\int_{\text{Nucl. Vol.}} |\psi_{1s,\mu}(r)|^2 d^3r \ \propto \ |\psi_{1s,\mu}(0)|^2 R^3 \ .$$

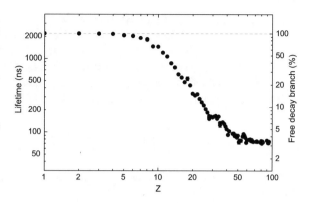

Fig. 9.5 Experimental mean lifetime of negative muons as a function of the atomic number of the element (note the logarithmic horizontal scale). The right-hand side axis indicates the fraction of the free decay branch (Eq. 1.7). Data are taken from Suzuki et al. (1987)

For the wave function of the muonic $1s$ state we can write (see Eq. 7.4)

$$\psi_{1s,\mu}(r) = \frac{1}{\sqrt{\pi a_{\mu A}^3}} \exp\left(-r/a_{\mu A}\right), \quad \text{with} \quad a_{\mu A} = \frac{a_0}{Z} \frac{m_e}{\overline{m}_{\mu A}}$$

$$|\psi_{1s,\mu}(0)|^2 = \frac{1}{\pi a_{\mu A}^3} \ . \tag{9.9}$$

The ratio A/Z does not change very much over the periodic table and we can write $R \propto Z^{1/3}$.

Summarizing, the capture rate is roughly proportional to $|\psi_{1s,\mu}(0)|^2 R^3 \propto Z^3 Z = Z^4$. Various theoretical and experimental studies have provided correction terms to the Z^4 law (Nagamine 2003; Suzuki et al. 1987).

The strong increase of the nuclear capture probability with increasing atomic number leads to a reduction of the bound μ^- lifetime. Whereas for light elements the μ^- lifetime is close to the value of the free particle, for heavy elements a drastic decrease is observed with $\tau_{\mu^-}(Z) < 200\,\text{ns}$ for $Z \gtrsim 30$. Figure 9.5 displays the Z dependence and shows that the muon lifetime in muonic atoms can vary from $\sim 70\,\text{ns}$ (for a heavy nucleus) to $\sim 2200\,\text{ns}$, close to the free μ^- lifetime, for hydrogen (Suzuki et al. 1987).

The capture process transforms the nucleus of charge Ze into an excited nucleus with charge $(Z - 1)e$.[12] For example, in iron this leads to the formation of a manganese isotope

$$\mu^- + {}^{56}_{26}\text{Fe} \rightarrow {}^{56}_{25}\text{Mn}^* + \nu_\mu \ . \tag{9.10}$$

---

[12] In general, a muonic atom looks like a pseudo-nucleus of charge $(Z - 1)$, when viewed from the atomic electron shells and the local environment.

The asterisk indicates the excited state of the Mn isotope, which releases energy in the form of characteristic gamma-rays.[13] The energies of these gamma-rays also depend on the isotope that captures the μ^-. Therefore, determining their energies provides additional confirmation of the muonic atoms involved, in addition to the information obtained by analyzing the muonic X-rays emitted during the cascade.

The negative muon lifetime in matter is given by[14]

$$\frac{1}{\tau_{\mu^-}(Z)} = \frac{1}{\tau_{\mu^-, \text{free}}} + \frac{1}{\tau_{\mu^-, \text{NC}}(Z)} \,, \tag{9.11}$$

which reflects the two possible decay channels, decay in orbit or capture by the nucleus.

9.3 μ^-SR

9.3.1 "Conventional" μ^-SR

When speaking of μSR studies it is implicitly assumed that the spin probe used is the positive muon. In fact, the overwhelming majority of experiments are and have been done with positive muons. Experiments with μ^- are much more scarce. In essence, the μ^-SR technique measures the precession and relaxation of the spin of a polarized negative muon bound in the ground state of a muonic atom (Yamazaki et al. 1975). From the previous discussion, it is clear that the negative muon spin rotation technique faces several challenges, primarily related to the strongly reduced lifetime in $Z \gtrsim 20$ elements, which sets an upper limit to the observation time window, and to the reduced polarization at the $t = 0$ time of the μSR measurement, due to the reduced initial beam polarization and the depolarization processes in the muonic atom.

Since there is no surface μ^- beam with nearly 100% polarization as for μ^+, for typical bulk μSR samples of $\sim 130\,\text{mg/cm}^2$ thickness, one has to resort to cloud muons with surface-like momentum $\sim 30\,\text{MeV/c}$, which is associated with

---

[13] Note that the excited state of the isotope is about 10–20 MeV above its ground state. This energy is usually larger than the binding energy of a nucleon. Therefore, one or more neutrons may be emitted in the nuclear deexcitation process. For example, in the case of iron, these reactions

$$\mu^- + {}^{56}_{26}\text{Fe} \rightarrow {}^{55}_{25}\text{Mn}^* + n + \nu_\mu$$

$$\mu^- + {}^{56}_{26}\text{Fe} \rightarrow {}^{54}_{25}\text{Mn}^* + n + n + \nu_\mu$$

can also take place. It is also possible that part of the deexcitation occurs through the emission of charged particles (protons or alpha particles) (Measday 2001).

[14] More precisely, there is a small correction Q (Huff factor), which takes into account that the free muon decay rate is reduced for a bound muon $[\tau_{\mu^-}(Z)]^{-1} = Q[\tau_{\mu^-, \text{free}}]^{-1} + [\tau_{\mu^-, \text{NC}}(Z)]^{-1}$. The value of Q decreases from 1 for light elements to 0.820 for uranium (Measday 2001).

a considerably reduced μ^- beam polarization. By contrast, the polarization of the higher energetic decay μ^- is similar to that of μ^+. However, the cascade processes after the capture and formation of the muonic atom lead to an additional polarization reduction of at least 80%. Since for a muon spin spectroscopy experiment the figure of merit is given by $A^2 N$, see Exercise 8.1, where A is the asymmetry amplitude of the μSR signal and N the number of muon decays recorded in a given time, to achieve a similar figure of merit μ^-SR measurements require measurement times typically a factor of 50 longer than standard μSR experiments. Additional limitations are related to the μ^- site, which in the $1s$ muonic shell is very close to the nucleus. The μ^-SR signal is straightforward only in muonic isotopes with zero nuclear spin. In this case the muon-nucleus system is a spin $1/2$ pure magnetic probe. In isotopes with nonzero nuclear spin, except for the lighter ones, the very strong contact hyperfine interaction, see Sect. 7.2.2, further reduces the polarization and locks the muon spin to that of the nucleus. This makes the use of μ^- as a local spin probe of condensed matter phenomena very challenging if not impossible.[15]

Note that Eq. 3.3, describing the μSR signal, is valid for positive muons, which have a single well-defined lifetime. In the usual case of samples composed of different elements, in which μ^- can be captured into, the different lifetimes $\tau_{\mu^-, i} = \tau_{\mu^-}(Z_i)$, Fig. 9.5, have to be taken into account.[16] In μ^-SR experiments, the histogram of the number of electrons detected as a function of time is then generally given by[17]

$$N_{e^-}(t) = \sum_i N_{0,i}\, e^{-\frac{t}{\tau_{\mu^-, i}}} \left[1 - A_{0,i}\, \mathbf{P}_i(t) \cdot \hat{\mathbf{n}} \right] + B \ , \qquad (9.12)$$

where the sum is over the atomic species of the investigated compound, see Fig. 9.6. Moreover, one has to consider that muons captured in different atomic species usually experience different local fields, and that the time evolution of the polarization may depend on the hyperfine interaction between the μ^- magnetic moment and that of the remaining electrons of the muonic atom, and on the interaction of these moments with the surrounding medium. In view of these facts and the large number of parameters involved, the extraction and interpretation of the time evolution of the μ^- spin polarization can become complicated.

---

[15] Information about the muon hyperfine interaction and the nuclear magnetic field can be obtained in μ^-SR experiments with nonzero spin nuclei (Imazato et al. 1984; Brewer et al. 2005; Sugiyama et al. 2020b).

[16] Note that, strictly speaking, the negative muon lifetimes in the isotopes of a given element are slightly different due to the different nuclear radii and hence nuclear capture probabilities (Suzuki et al. 1987). This can obscure the search for subtle effects in the μ^-SR signal.

[17] In Eq. 9.12 there is a minus sign in front of the asymmetry term, because in the μ^- decay the electron is emitted preferentially antiparallel to its spin. On the other hand, the spin of μ^- from the π^- decay is parallel to its momentum.

Fig. 9.6 Backward (with respect to the initial negative muon polarization) histogram recorded in a transverse field of 5 mT in MgH$_2$. The red dots are the experimental data, the green line is the fit using Eq. 9.12 with five contributions represented by the blue lines. Modified from Sugiyama et al. (2018), under CC-BY-4.0 license. © The Authors

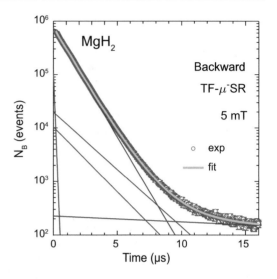

Example μ^-SR in MgH$_2$ As an example of an investigation, where μ^-SR presents some advantages over μ^+SR, we briefly discuss measurements performed on MgH$_2$ with the original aim to determine the parameters characterizing the hydrogen diffusion and the hydrogen desorption temperature in this compound (Sugiyama et al. 2018). In normal μSR such a determination is not trivial, since a motional narrowing of the field observed by the muon, see Sect. 4.3.1.1, may be due either to hydrogen diffusion or diffusion of the muon itself (or from both).[18] μ^-SR does not suffer from this ambiguity so that the signal in MgH$_2$ arising from the negative muons captured by the magnesium atoms is not affected by μ^- diffusion.[19] An important fact is that about 90% of the Mg nuclei (^{24}Mg isotopes) have no spin, so the negative muons captured by the Mg ions sense a local field mainly caused by the surrounding hydrogen spins and are sensitive to possible hydrogen diffusion.[20] The negative muons captured by the 10% abundant ^{25}Mg ions with spin 5/2 precess at a completely different frequency due to the strong contact hyperfine interaction with the ^{25}Mg nucleus.

In the experiment, 50 MeV/c negative muons have been implanted in the sample. Figure 9.6 shows a fitted histogram recorded in a 5 mT TF. To account for the different muonic atoms that can be formed, five different μ^- lifetimes must be

---

[18] As a lighter hydrogen isotope, μ^+ is expected to diffuse faster than hydrogen.

[19] In MgH$_2$ the muons captured by H very quickly transfer to Mg. In molecular compounds, μ^- transfer among the constituent atoms can affect the observed populations and atomic capture rates. μ^- transfer from neutral μH to elements of higher Z is an important process in molecules containing hydrogen atoms (Ponomarev 1973).

[20] Note that the muonic Mg atom, with μ^- in the $1s$ orbit very close to the Mg nucleus ($Z = 12$), compensates a positive charge and represents a Na site ($Z = 11$).

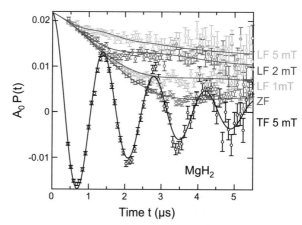

Time t (µs)

introduced: for the ^{24}Mg ions $\tau_{\mu,\text{Mg}} = 1.067(2)\,\mu\text{s}$, for the ^{12}C and ^{16}O ions contained in the sample holder $\tau_{\mu,\text{C}} = 2.0263(15)\,\mu\text{s}$ and $\tau_{\mu,\text{O}} = 1.795(2)\,\mu\text{s}$, for the lead of the beam collimator $\tau_{\mu,\text{Pb}} = 0.0754(10)\,\mu\text{s}$, and finally a very long lifetime of the order of 20 µs for an signal of unknown origin (possibly reflecting a time dependent background).

Figure 9.7 shows ZF, TF and LF asymmetries. Note the very small initial asymmetry of about 0.02. The ZF spectra and the partially decoupled LF spectra indicate a Kubo-Toyabe dynamic relaxation with a slow fluctuation rate, see Eq. 4.62. The measured field width is consistent with dipole calculations and shows that the random magnetic field at the muonic magnesium site is due to the nuclear magnetic fields of the hydrogens in MgH$_2$. The dynamics is best ascribed to thermal vibration and H diffusion by hopping, but, due to the very long measurement time, no temperature dependence of the diffusion parameters could be determined.[21]

The experiment illustrates that μ^-SR provides a local site in addition to the interstitial μ^+ site. If the negative muon is captured by isotopes of light elements with no nuclear spin, such as ^{12}C, ^{16}O, ^{24}Mg, or ^{28}Si, its lifetime is sufficiently long to probe the dynamic behavior of the surrounding atoms.

The immobile character of μ^- bound to a nucleus has also been used to validate ion diffusion rates obtained by μ^+SR (e.g., of Li in battery materials) and to answer the question whether the measured positive muon spin relaxation rate solely reflects ion diffusion or has an unknown contribution from μ^+ diffusion (Sugiyama et al. 2020a; Umegaki et al. 2022).

---

[21] Due to the very low polarization, 200–300 million events (Mev) for TF, and 500–1000 Mev for ZF/LF were necessary to obtain asymmetry spectra as shown in Fig. 9.7. This has to be compared with typical ~10 Mev for μ^+SR.

9.3.2   X-ray triggered μ^-SR

One way to disentangle the μ^-SR contributions from the different elements of a compound is to make use of the characteristic X-rays emitted during the muon cascade to the $1s$ ground state after its capture by an element of the compound. This possibility was considered already in the early days of μ^-SR in Dubna, USSR (Arlt et al. 1973) and later used in studies of semiconductors (Stoykov et al. 2009). In these experiments, negative muons are used to investigate the behavior of acceptor impurities. For example, capture by a Si atom leads to the formation of a muonic atom with an electron shell analogous to that of a conventional Al atom, which acts as an acceptor center.

Triggered μ^-SR has been used to study the binary semiconductor SiC and to enhance the Si contribution. For this, in addition to the μ^-SR spectra, the spectrum of the characteristic X-rays emitted during the muonic atom formation is recorded. The energies of the emitted X-rays characterize the atom that captured the muon, Eq. 9.6, and are practically (within less than picoseconds) coincident with the muon stopping time, recorded by a plastic scintillator. One can tag a muon-electron event with the corresponding transition energy, in this case $K_{\alpha 1,\mu}$ of Si, route the event to a histogram containing the events associated with the specific atom of the compound, and have the contributions from other atoms suppressed (Stoykov et al. 2009). This procedure greatly simplifies the analysis. The disadvantage is a reduction of the event rate by a factor given by the solid angle subtended by the X-ray detector[22] and its efficiency.

9.4   Elemental Analysis

While the use of negative muons for μ^-SR spectroscopy faces important challenges, negative muons are increasingly used for elemental analysis not only in materials science but also in the various disciplines of natural and applied sciences. In fact, negative muons offer a nondestructive method for depth dependent elemental analysis in a wide range of samples including valuable and rare specimens of cultural, historical or archaeological significance.

The possibility to use μ^- for elemental analysis in many fields of science was recognized even before the development of the so-called meson factories and of dedicated centers for applied muon science (Daniel 1969). Numerous experiments have been carried out in the 1970s and 1980s.[23] Studies initially focused on the

---

[22] Semiconductor detectors such as High Purity Germanium (HPGe) detectors are frequently used. Such a detector is placed close to the sample, but outside of the muon beam, and covers at most a solid angle of $0.1 \times 4\pi$ sr.

[23] For some examples of the early phase, see, e.g., (Reidy et al. 1978; Daniel 1984).

possible use of the technique in biology (for a review see Daniel 1987)[24] and in a second step on archaeological objects (Köhler et al. 1981; Daniel et al. 1987).

Despite the initial promising results the technique has been virtually abandoned for several years, probably due to the low μ^- intensities available. Recently it has been revived and several new initiatives have been started at pulsed and continuous muon beam facilities (Ninomiya et al. 2015; Hillier et al. 2018; Kubo 2016; Biswas et al. 2022). New measurements have attracted much attention and expanded the field of applications in archaeology, extraterrestrial geology, battery studies, environmental problems or testing of industrial manufacturing processes.

In analogy to the widely used PIXE technique (Proton or Particle Induced X-ray Emission), this method is commonly called MIXE (Muon Induced X-ray Emission).

9.4.1 Principle

The principle is simple and based on the properties of the muonic atoms discussed so far. As shown in Sect. 9.2, implanted negative muons have similar stopping profiles as positive muons with the characteristic Bragg peak. They stop in the sample at depths (known with $\sim$10% uncertainty) depending on their energy and sample composition, see Sect. 2.1.2 and Fig. 8.1. Here, they form a muonic atom and a series of characteristic X-rays, signaling the presence of a specific element in the specimen, are emitted during the cascade of the muons to their $1s$ ground state, see Eq. 9.6 and Fig. 9.3. In general, only the prompt X-rays (coinciding with the implantation time given by the muon counter in a continuous beam) are recorded. This allows to eliminate other delayed X-rays that may arise from the relaxation of the excited nucleus formed by the nuclear muon capture, see Eq. 9.7.

One of the key features of MIXE is that muons can be implanted relatively deep into the sample. The implantation depth varies from typically about 10 μm in high-Z elements for a muon momentum of 20 MeV/c to several centimeters for 50 MeV/c and typical plastic materials. Another important feature is that the generally high energy of the muonic X-rays, Eq. 9.6, makes them less susceptible to absorption and allows them to escape from the sample and thus to be easily detected by surrounding X-ray detectors. Figure 9.8 shows a setup at PSI for elemental analysis.

The attenuation of X-rays in material depends on absorption and scattering effects such as coherent (Rayleigh) and incoherent (Compton) scattering, photoelectric absorption and pair production. The attenuation of a photon beam is given by the Beer-Lambert-Bouguer law, which, if the attenuation does not vary over the path traversed x, can be written as

$$I(x) = I_0 \, \exp(-\mu x) \ . \tag{9.13}$$

---

[24] Measurements were performed at CERN on animal tissue samples (Daniel et al. 1973) and at Los Alamos on human body parts (Hutson et al. 1976).

Fig. 9.8 View of the
GermanIum Array for
Nondestructive Testing
(GIANT) setup at PSI for
elemental analysis studies.
The muonic X-rays are
detected by 11 HPGe
detectors surrounding the
sample. The apparatus has a
maximum capacity of 22
HPGe detectors. Photo credit
Lars Gerchow, PSI (Gerchow
et al. 2023), under CC-BY-4.0
license. © The Authors

Here, μ is the linear attenuation coefficient,[25] which is material specific and depends on the photon energy, I_0 is the initial photon intensity. Very often, the mass attenuation coefficient μ/ρ is used instead (ρ mass density of the material), since this ratio is rather similar for different materials in the energy range of the X-rays[26] and one writes

$$I(x) = I_0 \, \exp[-(\mu/\rho)\rho x] \ . \tag{9.14}$$

Figure 9.9 displays the mass attenuation coefficient for iron. The electron binding energy of the $1s$ state is 7.112 keV and represents the highest possible energy of a photon emitted during the transition of an electron from a higher level to the $1s$ ground state, for example in an X-ray fluorescence (XRF) measurement. The sharp increase of the X-ray mass attenuation coefficient at the $1s$ binding energy, called K-edge, is clearly visible in Fig. 9.9. It occurs when the energy of an X-ray is just above the ground state binding energy, since a photon with energy above the $1s$-binding energy has a higher probability of being absorbed than a photon with energy just below it.

---

[25] We keep the usual symbol μ for the attenuation coefficient as for the muon because it is always clear from the context which definition applies.

[26] See Hubbell and Seltzer (2004) for tabulated values.

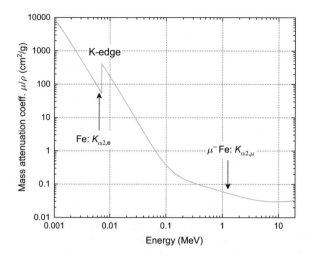

Fig. 9.9 Mass attenuation coefficient for iron in an energy range covering the electronic $K_{\alpha 2,e}$ and the muonic $K_{\alpha 2,\mu}$ transitions (indicated by arrows). Note also the K-edge corresponding to the ionization energy of the electronic $1s$ level

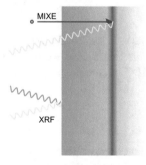

Fig. 9.10 Sketch of the difference between the MIXE technique and the X-ray fluorescence (XRF) technique. In XRF, one observes the emission of characteristic secondary X-rays (yellow) from atoms excited by high-energy X-rays (red). Because the secondary X-rays have limited energy, this elemental analysis technique is limited to surface studies. In the MIXE technique, the characteristic X-rays have a much higher energy. Hence, even if the muons are implanted deep in the material, the probability of observing the characteristic X-rays outside the material is very high. By changing the momentum of the muon beam, one can perform a depth dependent analysis of the elemental composition of the material

Characteristic photons produced by MIXE travel a much longer distance in the material than those produced by X-ray fluorescence or PIXE, see Fig. 9.10. For example, for a photon beam of energy 6.392 keV corresponding to the $K_{\alpha 2,e}$ electronic transition in iron, $\mu \simeq 555.8\,\text{cm}^{-1}$ and $\mu/\rho \simeq 70.71\,\text{cm}^2/\text{g}$. Therefore, a 50% decrease of the photon intensity occurs after a path length of only $\sim 12\,\mu\text{m}$. By comparison, for a photon with energy of the corresponding $K_{\alpha 2,\mu}$ transition

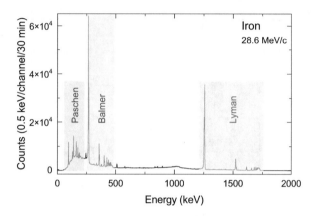

Fig. 9.11 Raw X-ray spectrum of muonic Fe (beam momentum 28.6 MeV/c). Clear lines from the muonic Lyman ($n_f = 1$), Balmer ($n_f = 2$), and Paschen ($n_f = 3$) series are observed. The data were recorded for about 30 minutes at the πE1.1 beamline at PSI with an event rate in the HPGe detector of $\sim$14 kHz. The spectrum shows events recorded within a time window of 50 ns after the muon implantation time. The data are therefore a combination of prompt X-ray events produced during the cascade and prompt and delayed gamma-ray events emitted after the muon nuclear capture

of muonic iron, i.e., 1256 keV,[27] we obtain $\mu/\rho \simeq 5.21 \times 10^{-2}$ cm^2/g and $\mu \simeq 0.41$ cm^{-1}, which means that a 50% drop in photon intensity occurs only after $\sim$17 mm, making the MIXE technique very sensitive.

9.4.2   Typical Spectra

Figure 9.11 shows an X-ray spectrum collected with a single HPGe detector of an iron sample irradiated with a μ^- beam of 28.6 MeV/c momentum. The momentum results in a penetration depth of $\sim$250 μm. The Lyman, Balmer and Paschen muonic series, corresponding to muon cascades from initial muonic levels of higher principal quantum numbers to final levels with $n_f = 1$, $n_f = 2$ and $n_f = 3$, respectively, are clearly observed. This straightforward measurement illustrates well the main advantage of the MIXE technique mentioned above, namely the ability to implant muons deep into the material while still detecting the characteristic X-rays outside the material and thus identifying the element capturing the muon.

A closer look at the lines of the Lyman series, Fig. 9.12a, reveals several features. First, the details of the $2p \rightarrow 1s$ transitions, composed of the $K_{\alpha1, \mu}$ and $K_{\alpha2, \mu}$ lines, can be resolved. The different intensities of these lines (with a ratio close to 2:1) are mainly due to the degeneracies of the levels. In addition to the X-ray lines,

---

[27] This energy is slightly below $207 \times K_{\alpha2,e}$ as it can be seen from Fig. 9.3 where the point for the iron $K_{\alpha2,\mu}$ transition ($Z = 26$) is just below the linear extrapolation, see caption of Fig. 9.3.

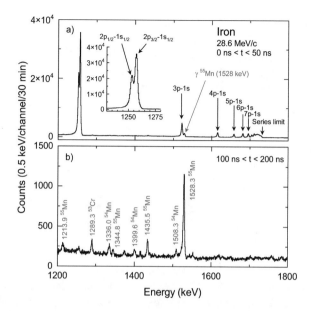

Fig. 9.12 Different time sections used to distinguish between x-rays and gamma rays. Upper panel: Same data as in Fig. 9.11, i.e., the X-ray and gamma-ray events are recorded in a time window of 50 ns after the muon arrival time, with an enlarged view of the Lyman series. Note that the gamma-ray peak at 1528 keV is also visible, see text for details. Bottom panel: Same energy interval with events recorded between 100 and 200 ns after muon arrival. The immediate X-ray events are no longer present, and only the delayed gamma-ray events are visible. Modified from Biswas et al. (2022), under CC-BY-4.0 license. © The Authors

gamma-ray lines due to nuclear deexcitation after muon capture by the nucleus can be identified.

In the energy range shown, the line at about 1528 keV is clearly visible, signaling the release of energy from the $^{55}_{25}\text{Mn}^*$ nucleus created after the μ^- capture into the Fe nucleus and the subsequent neutron emission, see Eq. 9.7 and footnote 13. By performing a cut at later times,[28] Fig. 9.12b, X-ray lines from prompt X-ray events are completely suppressed and only delayed gamma-ray events are observed, resulting from deexcitation processes after the μ^- capture. Thus, the 1528 keV line, which is rather weak in Fig. 9.12a, is now clearly dominant in Fig. 9.12b. Note that the gamma-ray line of Cr shows the emission of protons during the deexcitation process, see footnote 13.

By tracking the intensity of the gamma-ray line as a function of time after muon arrival, the average muon lifetime in a particular element in the sample can be determined. Figure 9.13 shows the intensity of the 1528.3 keV gamma-ray line as a

---

[28] In addition to time cuts, one can perform energy cuts in the gamma-ray spectrum, for example, by selecting a particular isotope line. This increases the intensity of the parent atom or isotope and enhances the specific sensitivity.

Fig. 9.13 Circles: Time evolution of the intensity of the 1528.3 keV gamma-ray line (in 100 ns wide windows) resulting from the nuclear deexcitation after capture of the muon in the Fe nucleus. The line represents a fit of the formula $I(t) = I_0 \exp(-t/\tau_{\mu^-,\mathrm{Fe}})$ to the data, see Eq. 9.11. Modified from Biswas et al. (2022), under CC-BY-4.0 license. © The Authors

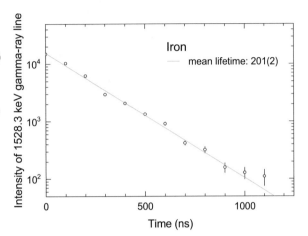

function of time (in 100 ns time windows). The fit gives a mean muon lifetime in Fe of 201(2) ns (Biswas et al. 2022). Note that this line reflects the excitation of a ^{55}Mn nucleus. This nucleus is mainly produced by the nuclear capture in ^{56}Fe and the subsequent emission of a neutron. However, a very small fraction of this nucleus can also be produced by the nuclear capture of a muon by ^{57}Fe and the subsequent emission of two neutrons. Therefore, the mean muon lifetime here is that of natural iron. To obtain the muon lifetime in ^{56}Fe one should use an isotope pure sample.

9.4.3 Depth Dependence

A "sandwich" sample consisting of three layers of Fe/Ti/Cu, each 500 μm thick, illustrates the depth dependent capabilities. By tuning the muon momentum, one can scan the occurrence of the three elements as a function of penetration depth. The top panel of Fig. 9.14 shows Monte Carlo simulations of the muon capture distributions in the three elements as a function of the momenta used in the experiment. The distributions are obtained from the GEANT4 software package (Agostinelli et al. 2003), which calculates the implantation profiles of particles in matter. Since the capture process, which occurs at very low energies ($\sim$ eV), is instantaneous, the curves represent at the same time the capture profiles of the negative muons.[29] Comparing the simulation with the measured intensities of the K_α lines of the three elements (lower panel of Fig. 9.14) a good agreement is found (when comparing the K_α lines the energy dependent efficiency of a HPGe detector has to be taken into account). This example illustrates the ease with which the characteristic X-rays of elements can be detected.

---

[29] Additional materials, as muon counter scintillator and vacuum window, are traversed by the beam prior to the sample.

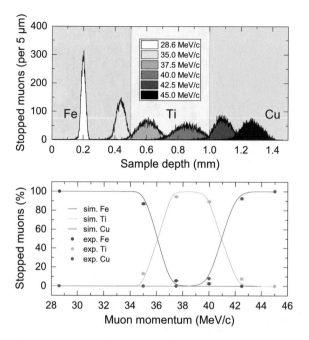

Fig. 9.14 Top panel: Simulation of the capture profiles of μ^- in a Fe/Ti/Cu trilayer (thickness of each layer 500 μm) for different beam momenta (no momentum uncertainty is assumed here). Note that the straggling in the Fe layer increases with increasing penetration, and that the straggling in the Cu layer is smaller compared to that in the Ti layer, which has a lower density. Bottom panel: Comparison between the measured detection probabilities of the three elements of the sample, obtained from the muonic K_α lines (symbols), and those expected from the simulation of the penetration and capture of μ^- in Fe, Ti and Cu (these simulations take into account the 2% momentum bite used for the measurements). Modified from Biswas et al. (2022), under CC-BY-4.0 license. © The Authors

9.4.4 Capture Probability

In the previous examples, the environment of the muon at the end of its trajectory was homogeneous (except at the interfaces of the layers) and only one type of atom could capture the muon. The situation is different when alloys and compounds are studied, or when a quantitative elemental analysis needs to be performed. In this case the muon can be captured by different atomic species with different capture probabilities.

A quantitative understanding of the capture probabilities is important for the application of the MIXE technique in elemental analysis. In the presence of different candidate capturing atoms, the integrated intensities of the $K_{\alpha,\mu}$ lines of the different elements can be used as a first qualitative measure of the elemental composition of the sample. However, to obtain atomic abundances, these intensities must be corrected for the respective capture probabilities per atom.

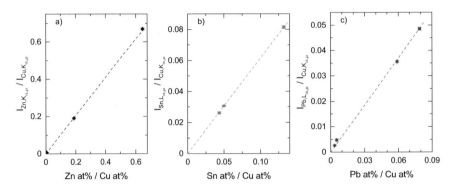

Fig. 9.15 MIXE measurements of alloy samples containing different elements, see text for details. The ratios of the muonic X-ray intensities are plotted as a function of the concentration ratios, see Eq. 9.15. The $L_{\alpha,\mu}$ lines are used for Sn and Pb and the $K_{\alpha,\mu}$ lines for Cu. Note that no correction for the energy dependent detector efficiency has been made. The observed linearity also indicates that the absorption effects are negligible in this case. Modified from Biswas et al. (2023), under CC-BY-4.0 license. © The Authors

Early experiments indicated that the capture probability $P(Z)$ depends mainly on the atomic number Z and only to a small extent on the valence of the element (von Egidy et al. 1984). For a binary compound $Z_k Z'_{k'}$, formed by the elements Z and Z' with concentrations k and k', respectively, the ratio of the measured muonic line intensities $R(Z, Z', k, k')$ (after correction for the energy dependent detector efficiency and, ideally, absorption) can be factorized as

$$R(Z, Z', k, k') = A(Z, Z')\frac{k}{k'} \ , \tag{9.15}$$

where $A(Z, Z')$ is the ratio between capture in atom Z and capture in atom Z'.

Figure 9.15 shows MIXE measurements of different alloy samples containing a large amount of Cu (at% varying from ∼60 to ∼90%), but also containing Zn (∼0 to ∼39%), Sn (∼0 to ∼11)%, Pb (∼0 to ∼7%), etc. (for details about the exact compositions, see Biswas et al. 2023).

The measurements clearly show that the ratios of the muonic X-ray lines are proportional to the ratio of the concentrations. The observed slopes are related to the capture ratios $A(Z, Z')$, i.e., in this case $A(Zn,Cu)$, $A(Sn,Cu)$ and $A(Pb,Cu)$. The remarkable linearity indicates that the capture ratios are independent of the concentrations.[30] It also shows that Eq. 9.15 is not only restricted to binary alloys, but can be extended to multielement alloys (Bergmann et al. 1979).

---

[30] Note that the slopes measured here cannot be directly related to $A(Z, Z')$ because they sometimes reflect the ratios between the L_α and K_α lines. This was done to obtain ratios of lines of similar energies. Also no correction was made for the energy dependent detector efficiency. This omission simply changes the value of the slope, without affecting the observed linearity.

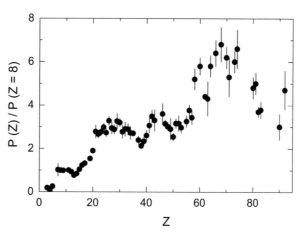

Fig. 9.16 Capture probabilities $P(Z)$ determined for a series of elements. The values are normalized to the probability of capture by oxygen, i.e., $P(Z = 8)$. Modified from von Egidy et al. (1984), © American Physical Society. Reproduced with permission. All rights reserved

Remarkably, it turns out that $A(Z, Z')$ can be satisfactorily expressed by the ratio of the capture probabilities of the two elements Z and Z' (von Egidy & Hartmann 1982)

$$A(Z, Z') = \frac{P(Z)}{P(Z')} \ .$$
(9.16)

This is quite important, as it shows that knowing the generic $P(Z)$ and $P(Z')$ on the one hand, and the measured ratio $R(Z, Z', k, k')$ on the other hand, allows one to directly measure the concentration ratio.[31]

Figure 9.16 shows the capture probabilities $P(Z)$ for different elements (von Egidy et al. 1984). $P(Z)$ grossly depends on Z but clearly does not increase smoothly as a function of Z, since the electronic structure and chemical bonds of the capturing atom have to be taken into account.

9.4.5 Determining the Isotopic Ratio

The nuclear radius of isotopes of a well-defined element follows an $A^{1/3}$ dependence, see Sect. 9.2.1. As a consequence, a variation of the muonic X-ray energy (isotope shift) is expected. This shift is more important for the Lyman series, where the size of the nucleus plays an important role, and for large-Z elements. In addition, the electromagnetic field of the muon distorts the nucleus, especially when the muon is in the $1s$ ground state (Cole 1969). This effect, known as "nuclear polarization",

---

[31] In detail, other factors, such as the valence of the elements, can slightly affect the capture probabilities.

Fig. 9.17 Spectrum showing the $K_{\alpha,1}$ and $K_{\alpha,2}$ lines for a copper sample. The Gaussian curves indicate the ^{63}Cu and ^{65}Cu contributions. Modified from Gerchow et al. (2023), under CC-BY-4.0 license. © The Authors

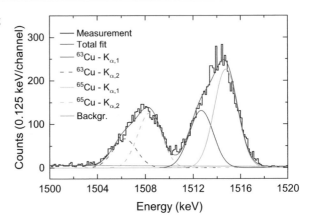

leads to a shift of the binding energy, which can slightly vary for different isotopes (Anderson et al. 1969).

Overall, these shifts open the possibility of determining isotopic ratios by MIXE, at least for elements with atomic number $Z > 20$. This represents an additional advantage over the widely used nondestructive XRF technique, which cannot identify isotopes. Some other techniques provide highly accurate isotopic ratios, but they are all destructive methods.

As an example, Fig. 9.17 displays measurements on a natural Cu sample. Although the data have low statistics, one can clearly identify the $K_{\alpha,1}$ and $K_{\alpha,2}$ lines for the two stable isotopes ^{63}Cu and ^{65}Cu. The isotope ratio obtained in this short measurement, ^{65}Cu/^{63}Cu $= 0.41(3)$, agrees with the expected value of 0.446.

The determination of isotopic ratios has been shown to play an interesting role in the field of archaeology. Thus, by studying the isotopic ratios of an excavated artifact, its provenance (i.e., the ore) can be determined. This is the case for artifacts containing the element Pb, which consists of the four stable isotopes ^{204,206,207,208}Pb (Villa 2009).

9.4.6   Examples

We give here some recent examples of scientific results in different fields that have been obtained by using, among others, the technique of elemental analysis with negative muons.

Archeology   An interesting result is provided by the study of the quality of ancient Roman silver coinage, which can be seen as an indicator of the economic health of a given period. A simple and clear example is provided by the study of a "denarius" of Julia Domna (211–217 AD) (Hampshire et al. 2019). Figure 9.18 displays MIXE measurements performed at different muon momenta. The thickness and composition of the Ag-enriched surface layer, as well as the composition of the

Fig. 9.18 Atomic composition of silver and copper as a function of depth determined in a Julia Domna denarius coin. The lines are exponential fits to the data. The composition inside of the coin reaches 49% silver and 51% copper, whereas a distinct Ag-enriched surface layer is detected. Modified from Hampshire et al. (2019), under CC-BY-4.0 license. © The Authors

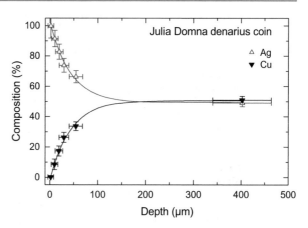

bulk of the coin are obtained. Below a surface layer of about 200 μm, MIXE reveals a composition of the bulk of the sample of about 50/50 between copper and silver.

Extraterrestrial Geology Another recent example is the elemental abundance investigation of samples from the carbonaceous asteroid Ryugu collected by the Hayabusa2 spacecraft (Nakamura et al. 2022). MIXE experiments have been performed to determine the major chemical elements in some Ryugu samples. Carbon, nitrogen, oxygen, sodium, magnesium, silicon, sulfur, iron, and nickel were detected. With the exception of oxygen, their abundances are similar to those observed in a CI chondrite.[32] However, Ryugu is oxygen depleted. In addition, by determining the abundance ratio N/C, it has been possible to estimate that the parent body of Ryugu was formed at a heliocentric distance similar to that found for CI chondrites.

Another new result on meteorites has been obtained from the study of an arrowhead from the Late Bronze Age settlement of Mörigen, Switzerland (Hofmann et al. 2023). Before the smelting[33] of iron from ores began, metallic iron was only available in the form of meteoric iron. This arrowhead is the only artifact from meteoric iron found in archaeological collections of objects from this region. For this object, the question was whether it came from a fragment of the nearby Twannberg meteorite. MIXE measurements performed with muons stopping below the oxidized crust reveal a nickel-rich composition consistent with other iron meteorites but incompatible with the Twannberg meteorite, see Fig. 9.19. It is assumed that the meteorite iron used to produce the arrowhead came from another (Kaalijärv) meteorite (Estonia, about 1500 BC). This indicates that iron meteorites were traded and transported as early as 800 BC or even earlier.

---

[32] CI chondrites are a rare type of stony meteorites.

[33] Smelting is the process of extracting a metal from an ore by the application of heat.

Fig. 9.19 Nickel and cobalt weight concentrations as a function of depth (muon momentum) for the Mörigen arrowhead, compared with unoxidized and oxidized Twannberg meteorite samples. Cobalt shows very similar values in all samples and depths. Modified from Hofmann et al. (2023), © Elsevier. Reproduced with permission. All rights reserved

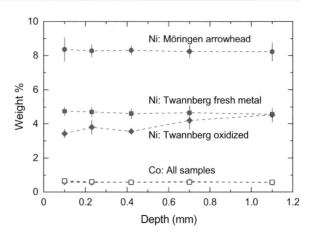

The study of battery materials using the MIXE technique, see for example (Umegaki et al. 2020), opens the possibility to perform operando[34] experiments to detect changes inside batteries during cycling. The use of a continuous muon beam is essential to ensure a short measurement time. Measurements at PSI have initially focused on the investigation of candidate cathode materials for Li-ion batteries. An important goal was to obtain the correct Ni/Mn/Co ratios deep inside a specific cathode (NMC811) and the Ni, Co and Mn depth profiles of a graphite-NMC811 cell were measured. An additional objective was to test the sensitivity of MIXE to detect small changes in the concentration of transition metals in batteries upon aging. In particular, MIXE has been used to detect the dissolution of transition metals from the cathode to the separator (and later to the anode) and to compare the elemental composition of a freshly prepared cell and a cycled cell. Figure 9.20, showing the elemental ratios, strongly suggests a selective loss of nickel towards the interface during cycling.

Another approach would be to determine the lithium content at different depths and to monitor operando the degradation of the battery. However, the detection and quantification of Li within a battery is still challenging because: (i) lithium typically accounts for less than 5 wt% of a Li-ion battery; this means that the battery is mainly composed of elements with a much higher muon capture probability than Li, and (ii) the low energy of the muonic lithium K lines leads to a low escape depth. Therefore, the current route is to test whether MIXE can determine the different oxidation states of transition metals found in battery materials and, as a proxy, to estimate the lithiation state at a given depth in the electrode. This seems possible since a change in the muonic X-ray lines as a function of the oxidation state is expected (Ninomiya et al. 2022).

---

[34] One speaks of operando spectroscopy when the catalyst characterization is performed under reaction conditions and is combined with measurements of the catalytic activity.

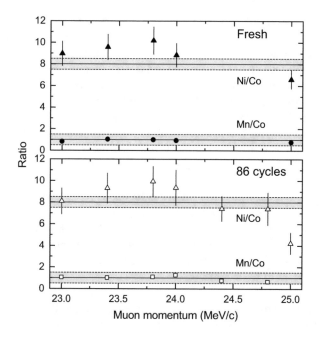

Fig. 9.20 Measurements of the elemental composition of the cathode of a Li-ion battery. By increasing their momenta, muons are implanted deeper into the cathode toward the battery separator. Top panel: Elemental ratios of Ni/Mn/Co determined from a MIXE depth profile across a freshly assembled pouch cell. The bottom panel exhibits the elemental ratios of Ni/Mn/Co of the same cell in the discharged state after 86 cycles. The gray regions indicate the ratios expected from the battery fabrication process. Courtesy of Sayani Biswas, Paul Scherrer Institute. All rights reserved

9.4.7   Characteristics and Comparison with Other Techniques

Specific properties of the MIXE technique favorably compare with other elemental analysis methods.

Depth Resolution   As already mentioned depth dependent measurements can be performed by tuning the muon momentum. This can be done in a depth interval and with a resolution discussed in Sect. 2.2.1. Muon implantation depths can extend to cm, see for example Fig. 2.7.[35] This is a fundamental advantage over techniques such as X-ray fluorescence and X-ray Photoelectron Spectroscopy (XPS), which use X-rays as the primary excitation. Also techniques using other particles for primary excitation, such as electrons for Auger Electron Spectroscopy (AES) or for Scanning

---

[35] Note, however, that the muonic X-rays also require sufficient energy to escape the material or specimen under investigation. For light elements, this limits the accessible depth range. For example, taking a lithium-ion battery study (Umegaki et al. 2020), the $K_{\alpha,\mu}$ line of lithium ($Z = 3$) at about 18 keV has a 50% probability to traverse a 50 μm aluminum case.

Electron Microscope with Energy Dispersive Spectroscopy (SEM/EDS), or protons for Proton-Induced X-ray Emission (PIXE), or techniques such as Nuclear Reaction Analysis with Particle-Induced Gamma-ray Emission (NRA/PIGE) or Rutherford Back Scattering (RBS) have a limited maximum depth range extending from a few to about 10 µm. Techniques using neutrons for primary excitation, such as Prompt Gamma-ray Neutron Activation Analysis (PGAA) or Neutron Activation Analysis (NAA), are sensitive to depths up to several centimeters. However, these techniques face the problem of neutron absorption, which is strongly dependent on the isotopic composition of the sample. In addition, the NAA technique requires that the radioactive isotopes produced have a short lifetime.

Nondestructive Technique The MIXE technique is nondestructive.[36] This is not the case with some of the techniques already mentioned, such as AES, SEM/EDS, or Secondary Ion Mass Spectroscopy (SIMS). Techniques such as Inductive Coupled Plasma Atomic Emission Spectroscopy (ICP-AES) and Inductive Coupled Plasma Mass Spectroscopy (AES-MS), which have extremely good resolution, completely destroy (burn) part of the sample.

Sensitivity and Resolution The MIXE technique is sensitive to all elements and in this sense is truly multielemental. Particularly important is the possibility to detect muonic X-rays of key light elements of organic compounds such as H, C, N, O, due to their energy in the few to several tens of keV.[37] However, the characteristic energies and the lower capture probability of light elements (note that the capture probability varies by a factor 50 between very light and heavy elements, see Fig. 9.16), may represent a limiting factor. The lateral dimension of a negative muon beam (typically in the cm range) is rather large compared to a photon, electron, or proton beam. Therefore, the use of a collimator (although reducing the beam intensity) or of a suitable position sensitive muon start detector (for example a drift chamber or a pixel silicon detector) should possibly be considered. The element detection limit with MIXE is better than 0.1%, which is usually sufficient for many applications, but insufficient for elemental trace analysis.

Absence of Self-induced Effects The MIXE technique is free from self-induced effects such as enhancement or absorption of the emitted characteristic energy. Due to the very low number of muonic atoms in a sample at any given time (only one for a continuous beam and $\sim 10^4$–10^5 at most for a pulsed beam), there is no absorption of

---

[36] A few atoms of the sample will undergo transmutation when the muon is captured by a nucleus, Eq. 9.7. However, this effect is completely negligible. Assuming a typical muon rate of 20 kHz, a transmutation probability of 100% and a measurement time of 30 minutes, less than 2×10^7 transmutations take place. This number must be compared to the Avogadro number, which gives the typical number of atoms or molecules in a sample.

[37] The $2p$-$1s$ transition energies increase from 1.9 keV for muonic hydrogen to 133.5 keV for muonic oxygen (Engfer et al. 1974). This should be compared with the energies of the corresponding electronic transitions with K-edges ranging from 13.6 to 532 eV for XRF investigations.

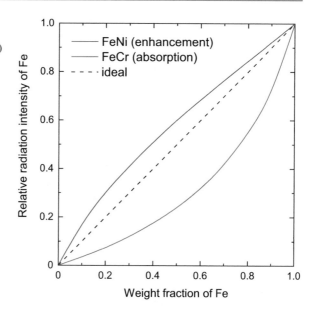

Fig. 9.21 Absorption and enhancement effects in an XRF measurement of (Fe,Cr) and (Fe,Ni) alloys, see text for details. Such effects are not present in the MIXE technique. Modified from Sitko and Zawisza (2012), under CC-BY-3.0 license. © The Authors

the radiation emitted during the muon cascade by another muonic atom. This is not the case with XRF, a powerful technique for determining the elemental composition.

In XRF, when studying an alloy, e.g., $A_{k(A)}B_{k(B)}$, an increase in the intensity of the X-rays associated with the element A can be observed due to the additional excitation from the fluorescent radiation produced by the B elements present in the sample and emitting X-rays with energies higher than the absorption edge of A. In this case, a higher concentration value than the nominal one is extracted. Conversely, the absorption effect is the loss of fluorescence intensity due to partial absorption by an element with a lower absorption edge energy. For example, when determining the concentration of iron in (Fe,Cr) and (Fe,Ni) alloys, absorption of the Fe fluorescence radiation occurs in the (Fe,Cr) alloys because $Z(\mathrm{Cr}) < Z(\mathrm{Fe})$, while enhancement of the line occurs in the (Fe,Ni) alloys because $Z(\mathrm{Ni}) > Z(\mathrm{Fe})$, see Fig. 9.21. Correction for these effects must therefore be considered in XRF.

Continuous vs Pulsed Beam Since the HPGe X-ray detectors have a slow response (much longer than the typical time width of a muon pulse from a pulsed machine), MIXE can only reach its full capability when used with a continuous beam. To avoid distortion of the spectrum, only events with a single hit by detector should be retained. With a pulsed beam, the probability of an HPGe detector being hit by X-rays twice or more is not negligible. This results in a limit of 50 events/s per detector in a beam with a repetition rate of e.g. 50 Hz. In contrast, for a continuous beam, the limit is several thousand events/s per detector. Note, for example, that the spectra in Fig. 9.12 have been obtained in 30 minutes.

In summary, the most interesting features of the MIXE technique are its ability to analyze deep inside the sample or device, its nondestructive nature and its sensitivity to light elements. These properties are very valuable when studying rare and/or valuable samples or specimens (for example, archaeological, cultural heritage objects, meteorites (Terada et al. 2017)) or when studying bulky devices or their components (for example, the cathode material of lithium-ion batteries (Umegaki et al. 2020)).

Exercise

9.1. Hyperfine splitting in muonic hydrogen
Explain the difference in magnitude of the contact hyperfine splitting in conventional hydrogen and in muonic hydrogen.

References

Adkins, G. S. (2008). *American Journal of Physics, 76*, 579.
Agostinelli, S., Allison, J., Amako, K. et al. (2003). *Nuclear Instruments and Methods in Physics Research Section A: Accelerators, Spectrometers, Detectors and Associated Equipment, 506*, 250–303.
Anderson, H. L., Hargrove, C. K., Hincks, E. P. et al. (1969). *Physical Review Letters, 22*, 221.
Arlt, R., Evseev, V. S., Ortlepp, H.-G. et al. (1973). JINR Dubna P15-7202 preprint.
Barrett, R. C. (1974). *Reports on Progress in Physics, 37*, 1.
Bearden J. A. (1967). *Reviews of Modern Physics, 39*, 78.
Bergmann, R., Daniel, H., von Egidy, T. et al. (1979). *Physical Review A, 20*, 633.
Biswas, S., Gerchow, L., Luetkens, H. et al. (2022). *Applied Sciences, 12*, 2541.
Biswas, S., Megatli-Niebel, I., Raselli, L., et al. (2023). *Heritage Science, 11*, 43.
Bohr, N. (1913a). *Philosophical Magazine, 26*, 476.
Bohr, N. (1913b). *Philosophical Magazine, 26*, 1.
Brewer, J. H., Ghandi, K., Froese, A. M., et al. (2005). *Physical Review C, 71*, 058501.
Cole, R. K. (1969). *Physical Review, 177*, 164.
Daniel, H. (1969). *Nuclearmedizin, 8*, 311.
Daniel, H. (1984). *Nuclear Instruments and Methods in Physics Research - Section B, 3*, 65.
Daniel, H. (1987). *Biological Trace Element Research, 13*, 301.
Daniel, H., Hartmann, F. J., Köhler, E., et al. (1987). *Archaeometry, 29*, 110.
Daniel, H., Pfeiffer, H. J., & Springer, K. (1973). *Biomedizinische Technik*, 18, 222.
Devons, S., Duerdoth, I. (1969). In M. Baranger, E. Vogt (Eds.), *Advances in Nuclear Physics: Vol. 2*, 295. Springer. ISBN: 978-1468483437.
Engfer, R., Schneuwly, H., Vuilleumier, J., et al. (1974). In *Atomic Data and Nuclear Data Tables* (Vol. 14), 509. Nuclear charge and moment distributions.
Evseev, V. S. (1975). In V. W. Hughes & C. S. Wu (Eds.), *Muon Physics, Vol. III Chemistry and Solids*. Academic Press, 235, ISBN: 978-0123606037.
Fricke, G., Bernhardt, C., Heilig, K., et al. (1995). *Atomic Data and Nuclear Data Tables, 60*, 177.
Gerchow, L., Biswas, S., Janka, G., et al. (2023). *Review of Scientific Instruments*, 045106.
Grossheim, A., Bayes, R., Bueno, J. F., et al. (2009). *Physical Review D, 80*, 052012.
Hampshire, B. V., Butcher, K., Ishida, K., et al. (2019). *Heritage, 2*, 400.
Hillier, A., Ishida, K., Seller, P., et al. (2018). *JPS Conference Proceedings, 21*, 011042.
Hofmann, B. A., Bolliger, S., Biswas, S., et al. (2023). *Journal of Archaeological Science, 157*, 105827.
Hubbell, J. H., & Seltzer, S. M. (2004). *Tables of X-ray mass attenuation coefficients and mass energy-absorption coefficients (version 1.4)*. National Institute of Standards and Technology.

Hutson, R. L., Reidy, J. J., Springer, K., et al. (1976). *Radiology, 120*, 193.

Imazato, J., Nagamine, K., Yamazaki, T., et al. (1984). *Physical Review Letters, 53*, 1849.

Karr, J.-P., & Marchand, D. (2019). *Nature, 575*, 61.

Köhler, E., Bergmann, R., Daniel, H., et al. (1981). *Nuclear Instruments and Methods, 187*, 563.

Kubo, M. K. (2016). *Journal of the Physical Society of Japan, 85*, 091015.

Measday, D. (2001). *Physics Reports, 354*, 243.

Moseley, H. G. J. (1913). *The London, Edinburgh, and Dublin Philosophical Magazine and Journal of Science, 26*, 1024.

Mukhopadhyay, N. C. (1977). *Physics Reports, 30*, 1.

Musser, J. R., Bayes, R., Davydov, Y. I., et al. (2005). *Physical Review Letters, 94*, 101805.

Nagamine, K. (2003). *Introductory Muon Science*. Cambridge University Press. ISBN: 978-0521593793.

Nakamura, T., Matsumoto, M., Amano, K., et al. (2022). *Science, 379*, eabn8671.

Ninomiya, K., Kajino, M., Nambu, A., et al. (2022). *Bulletin of the Chemical Society of Japan, 95*, 1769–1774.

Ninomiya, K., Kubo, M. K., Nagatomo, T., et al. (2015). *Analytical Chemistry, 87*, 4597.

Pifer, A., Bowen, T., & Kendall, K. (1976). *Nuclear Instruments and Methods, 135*, 39.

Pohl, R., Antognini, A., Nez, F., et al. (2010). *Nature, 466*, 213.

Ponomarev, L. I. (1973). *Annual Review of Nuclear Science, 23*, 395.

Reidy, J. J., Hutson, R. L., Daniel, H., et al. (1978). *Analytical Chemistry, 50*, 40.

Sitko, R., & Zawisza, B. (2012). In S. K. Sharma (Ed.), *X-ray spectroscopy*. IntechOpen, chap. 8.

Stoykov, A., Herlach, D., Scheuermann, R., et al. (2009). *Physica B: Condensed Matter, 404*, 824.

Sugiyama, J., Forslund, O. K., Nocerino, E., et al. (2020a). *Physical Review Research, 2*, 033161.

Sugiyama, J., Umegaki, I., Nozaki, H., et al. (2018). *Physical Review Letters, 121*, 087202.

Sugiyama, J., Umegaki, I., Takeshita, S., et al. (2020b). *Physical Review B, 102*, 144431.

Suzuki, T., Measday, D. F., & Roalsvig, J. P. (1987). *Physical Review C, 35*, 2212.

Terada, K., Sato, A., Ninomiya, K., et al. (2017). *Scientific Reports, 7*, 15478.

Umegaki, I., Higuchi, Y., Kondo, Y., et al. (2020). *Analytical Chemistry, 92*, 8194.

Umegaki, I., Ohishi, K., Nakano, T., et al. (2022). *Journal of Physical Chemistry C, 126*, 10506.

van Dyck, O. B., Hoffman, E. W., Macek, R. J. et al. (1979). *IEEE Transactions on Nuclear Science, 26*, 3197.

Villa, I. M. (2009). *Archaeological and Anthropological Sciences, 1*, 149.

von Egidy, T., & Hartmann F. J. (1982). *Physical Review A, 26*, 2355.

von Egidy, T., Jakubassa-Amundsen, D. H., & Hartmann F. J. (1984). *Physical Review A, 29*, 455.

Winston, R. (1963). *Physical Review, 129*, 2766.

Yamazaki, T., Nagamine, K., Nagamiya, S., et al. (1975). *Physica Scripta, 11*, 133.

Zaretsky, D., & Novikov, V. (1960). *Nuclear Physics, 14*, 540.

Particle Physics Aspects

10

As an elementary particle and a fundamental component of the Standard Model[1] of particle physics, the muon and the precise determination of its properties (allowed and forbidden decay modes, interactions, magnetic moment) play an important role in particle physics, in particular as a playground to unveil "new" physics beyond the Standard Model (Kuno & Okada 2001; Roberts 2007; Gorringe & Hertzog 2015). In this chapter we outline the theoretical treatment of the muon decay to provide an example of a practical application of quantum field theory and to show how the understanding of the weak interaction has evolved. In addition, we present some experiments that underline the importance of the muon in particle physics. We hope that this will also be of interest to researchers in condensed matter and materials science.[2]

10.1 Muon Decay and Lepton Numbers

We recall here the main decay channels. For the other muon properties we refer to Sect. 1.5 and to the Table on Page xviii

$$\mu^+ \rightarrow e^+ + v_e + \bar{v}_\mu$$

$$\mu^- \rightarrow e^- + \bar{v}_e + v_\mu \ . \tag{10.1}$$

The decay is a prominent example of the separate conservation of the muon and electron lepton numbers. Table 10.1 summarizes their values for the three

---

[1] For a comprehensive description of the Standard Model see, e.g., Thomson (2013).

[2] Note that, as in the related Appendix G we will use natural units, as is common practice in particle physics, where $c = 1$, $\hbar = 1$, and $\varepsilon_0 = 1$. Exercise 10.1: what does this mean for the elementary charge?

© Springer Nature Switzerland AG 2024
A. Amato, E. Morenzoni, *Introduction to Muon Spin Spectroscopy*,
Lecture Notes in Physics 961, https://doi.org/10.1007/978-3-031-44959-8_10

Table 10.1 Lepton numbers
of the three lepton families

	L_μ	L_e	L_τ
μ^-, ν_μ^-	+1	0	0
$\mu^+, \bar{\nu}_\mu^-$	−1	0	0
e^-, ν_e^-	0	+1	0
$e^+, \bar{\nu}_e^-$	0	−1	0
τ^-, ν_τ^-	0	0	+1
$\tau^+, \bar{\nu}_{\tau^-}$	0	0	−1

lepton families. The lepton numbers are additive[3] quantum numbers: their sum is conserved in the interaction.

Evidence for a separate conservation of the lepton family numbers (or lepton flavors) comes from the search for and nonobservance of the following decay modes, for which increasingly tighter upper bounds have been established over the years:

$$\mu^+ \rightarrow e^+ + \gamma \qquad \text{Branching ratio} < 4.2 \times 10^{-13}$$

From the MEG II experiment at PSI (Papa 2018)

$$\mu^- \rightarrow e^- + e^+ + e^- \quad \text{Branching ratio} < 1.0 \times 10^{-12}$$

From the SINDRUM experiment at PSI (Bellgardt et al. 1988)

The experimental evidence from the search for violation of the lepton flavor conservation and the standard theory of the electroweak interaction are consistent with the separate conservation of the three lepton numbers. However, experiments with different sources of neutrinos (accelerator, reactor, or atmospheric of solar origin) have shown that a neutrino beam of a given flavor, after propagating over a certain distance, changes its identity and no longer contains only neutrinos of the original flavor (neutrino oscillations or neutrino mixing).[4]

This implies that neutrinos have a nonzero mass, and indicates that the conservation of the lepton family is only approximate. The (classical) Standard Model of particle physics requires neutrinos to be massless, and has to be modified ad hoc to include the neutrino masses.

---

[3] The additive conservation of lepton numbers forbids, for example, the process $\mu^+ + e^- \rightarrow \mu^- + e^+$, whereas a multiplicative conservation rule would allow it.

[4] T. Kajita and A. B. McDonald, from the Super-Kamiokande Collaboration, Japan (Fukuda et al. 1998) and the Sudbury Neutrino Observatory Collaboration, Canada (Ahmad et al. 2001), respectively, received the 2015 Nobel Prize in Physics for the discovery of neutrino oscillations.

10.2 Theory of the Muon Decay

The understanding of the weak interaction responsible for the muon decay has evolved over several decades (Griffiths 2004; Grotz & Klapdor 2018; Okun 2014; Thomson 2013). Important milestones have been the original Fermi theory, then the so-called phenomenological V-A theory, and later the unified theory of the weak and electromagnetic interactions formulated by Glashow-Weinberg-Salam[5] (Weinberg 1980). These are all examples of quantum field theories.

Understanding the theory of the muon decay requires a basic knowledge of the Dirac equation for free particles and of the field quantization. The most important properties are briefly summarized in the Appendix G. For more details we refer to the many textbooks on quantum mechanics and quantum field theory (Bjorken & Drell 1965; Peskin & Schroeder 1995; Lancaster & Blundell 2014; Coleman 2018).

The most general Dirac field operator for free spin $1/2$ particles is given by Eq. G.22 in Appendix G

$$\Psi(x^\beta) = \sum_{s=\pm 1/2} \int \frac{d\mathbf{p}}{(2\pi)^3 2E} \left[b_{\mathbf{p},s} u_{\mathbf{p},s} e^{-ip_\beta x^\beta} + d^\dagger_{\mathbf{p},s} v_{\mathbf{p},s} e^{+ip_\beta x^\beta} \right] . \tag{10.2}$$

The operators $b^\dagger_{\mathbf{p},s}$ and $b_{\mathbf{p},s}$ respectively create and annihilate a spin $1/2$ particle (for example μ^-, e^-, ν_e, or ν_μ) with momentum $\mathbf{p}$ and spin s, $d^\dagger_{\mathbf{p},s}$ and $d_{\mathbf{p},s}$ are the creation and annihilation operators for the corresponding antiparticle (for example μ^+, e^+, $\bar{\nu}_e$, or $\bar{\nu}_\mu$).

To better understand the structure and ingredients of the quantum field theory describing the muon decay, it is instructive to first look at the quantum electrodynamics (QED), one of the most successful quantum field theories. QED is a relativistic theory that describes with great precision all phenomena involving electrically charged spin $1/2$ particles (in general electrons and muons) interacting via the exchange of photons. It represents the quantum counterpart of the classical electromagnetic field theory, where the interaction between currents and the electromagnetic field is described by the Lagrangian density term

$$\mathcal{L}_{\text{int}} = -j_\alpha A^\alpha \quad \text{where} \quad j^\alpha = \begin{pmatrix} \rho \\ \mathbf{j} \end{pmatrix} \quad \text{and} \quad A^\alpha = \begin{pmatrix} \Phi \\ \mathbf{A} \end{pmatrix} . \tag{10.3}$$

Here j^α is the 4-vector current with components charge density ρ, and the conventional current density $\mathbf{j}$. A^α is the electromagnetic 4-vector potential with time-like component Φ, the electric potential, and space-like component $\mathbf{A}$ the magnetic vector potential.

---

[5] For the development of this theory they received the 1979 Nobel Prize in Physics.

The quantization of the fields leads, for the interaction term between fermions of charge q and photons, to the Lagrangian density

$$\mathcal{L}_{\text{QED}} = -j_\alpha A^\alpha = -q\overline{\Psi}\gamma_\alpha\Psi A^\alpha \ , \tag{10.4}$$

where[6]

$$j_\alpha = q\overline{\Psi}\gamma_\alpha\Psi \tag{10.5}$$

is the current density associated with the field Ψ, A^α is the photon field operator, which can also be expressed in term of photon creation and annihilation operators, and $q = -e$ for e^- or μ^-.[7]

In 1934, before the discovery of the parity violation in the β-decay, (Fermi 1934) proposed, in analogy to the interaction term in electromagnetism, see Eqs. 10.4 and 10.5, that the weak interaction responsible for the β-decay ($n \rightarrow p + e^- + \bar{\nu}_e$) can be described by a product of two currents, one $j_\alpha^{n\rightarrow p}$ dealing with the transition of a neutron into a proton and the other $j_{\nu\rightarrow e}^{\alpha}$[8] representing the production of an electron and a neutrino

$$\mathcal{L}_{\text{W}_{\text{Fermi}}} = -\frac{G_{\text{F}}}{\sqrt{2}} j_\alpha^{n\rightarrow p}(x^\beta) j^{\nu_e \rightarrow e,\alpha}(x^\beta)$$

$$= -G_{\text{F}}\left[\overline{\Psi}_p(x^\beta)\gamma_\alpha\Psi_n(x^\beta)\right]\left[\overline{\Psi}_e(x^\beta)\gamma^\alpha\Psi_{\nu_e}(x^\beta)\right] . \tag{10.6}$$

G_{F} is a constant, now named Fermi constant,[9] which determines the coupling strength. Note that all fields are taken at the same space-time point, making the interaction a so-called contact interaction. In principle, Fermi theory can also be applied to the muon decay. In this case, the Lagrangian density is

$$\mathcal{L}_{\text{W}_{\text{Fermi}}} = -\frac{G_{\text{F}}}{\sqrt{2}} j_\alpha^{\mu\rightarrow\nu_\mu}(x^\beta) j^{\nu_e\rightarrow e,\alpha}(x^\beta)$$

$$= -G_{\text{F}}\left[\overline{\Psi}_{\nu_\mu}(x^\beta)\gamma_\alpha\Psi_\mu(x^\beta)\right]\left[\overline{\Psi}_e(x^\beta)\gamma^\alpha\Psi_{\nu_e}(x^\beta)\right] . \tag{10.7}$$

$\mathcal{L}_{\text{W-Fermi}}$ as a product of two polar vectors (VV) is Lorentz invariant. It is also parity invariant. Therefore, it cannot be the complete description of the weak interaction, which later (1956–1957) was found to violate parity conservation first in studies of the K^+ decay (Lee & Yang 1956) and then directly confirmed in the ^{60}Co decay (Wu

---

[6] Note that in the interaction term photons can be created or destroyed individually but that the spin 1/2 particles must be created and destroyed together with antiparticles.

[7] Practically, as it will become clear in the following, the sign of the coupling does not matter.

[8] Both currents are examples of so-called "charged currents", meaning that the charge of the initial state (n, ν_e) is not the same as that of the final state (p, e^-).

[9] The original Fermi's definition of G_{F} included the factor $1/\sqrt{2}$.

et al. 1957). In general, there are only five possible combinations of two spinors and γ-matrices that form Lorentz invariant currents (so-called "bilinear covariants", see also Appendix G.3). So in principle, the Lagrangian density could also contain other terms formed by these covariants. If we want to stay with current-current couplings, the possible interaction terms are restricted to linear combinations of the products SS, VV, TT, AA, PS, and VA. Of these products PS and VA are pseudoscalar and thus do not conserve parity, see Eq. G.31.

A successful generalization of the Fermi current-current interaction is the so-called $V - A$ interaction. Here, in addition to the vector current, an axial current is introduced, so that the Lagrangian contains scalar terms (such as VV or AA) as well as pseudoscalar terms (e.g., VA) (Feynman & Gell-Mann 1958; Sudarshan & Marshak 1958). In its general form, $V - A$ describes a wealth of weak processes and can also be extended to treat simultaneously the weak interaction processes of all three lepton families.

We give here only the part of the Lagrangian density, which is responsible for the μ^- decay[10]

$$\mathcal{L}_{\text{W V-A}} = -\frac{G_F}{\sqrt{2}} \left[\overline{\Psi}_{\nu_\mu}(x^\beta)\gamma_\alpha(1 - \gamma_5)\Psi_\mu(x^\beta) \right] \left[\overline{\Psi}_e(x^\beta)\gamma^\alpha(1 - \gamma_5)\Psi_{\nu_e}(x^\beta) \right] ,$$

(10.8)

where the Dirac fields of the various particles involved in the decay have the form, see Eq. G.22

$$\Psi(x^\beta) = \sum_{s=\pm 1/2} \int \frac{d\mathbf{p}}{(2\pi)^3 2E} \left[b_{\mathbf{p},s} u_{\mathbf{p},s} e^{-ip_\beta x^\beta} + d_{\mathbf{p},s}^\dagger v_{\mathbf{p},s} e^{+ip_\beta x^\beta} \right] .$$

(10.9)

10.3 Calculation of the Muon Decay

The key expression is the differential of the decay rate probability, which has been derived for the muon at rest in the Appendix, see Eq. G.37,

$$d\Gamma_{i \to f} = \frac{(2\pi)^4}{2m_\mu} |\mathcal{M}_{if}|^2 \delta^4(p_0^\alpha - p_1^\alpha - p_2^\alpha - p_3^\alpha) \frac{d\mathbf{p}_1}{(2\pi)^3 2E_1} \frac{d\mathbf{p}_2}{(2\pi)^3 2E_2} \frac{d\mathbf{p}_3}{(2\pi)^3 2E_3} ,$$

(10.10)

---

[10] To simplify the notation, in this chapter, instead of $\mathbf{1}_4$ we simply write 1 for the 4×4 unit matrix (e.g. in Eqs. 10.8 and 10.14), see also Appendix G.

where δ^4 is the four-dimensional δ-function for $\alpha = 0, 1, 2, 3$ and

p_0^α is the μ^- 4-momentum
p_1^α is the $\bar{v}_e$ 4-momentum
p_2^α is the v_μ 4-momentum
p_3^α is the e^- 4-momentum .

The quantity of interest is the matrix element $\mathcal{M}_{if}$, which is directly related to the scattering operator S (also called S-matrix). The matrix element $S_{if} \equiv \langle f | S | i \rangle$ gives the transition amplitude between two asymptotic free states $|i\rangle$ and $|f\rangle$ undergoing an interaction.

It is useful to decompose the scattering matrix into a trivial unity component (corresponding to no interaction or to the unscattered part) and a component describing only the effect of the interaction or the scattered part

$$\langle f | S | i \rangle = \langle f | i \rangle + \langle f | S - 1 | i \rangle$$

$$\langle f | S - 1 | i \rangle = (2\pi)^4 \delta^4 \left(\sum_i p_i^\alpha - \sum_f p_f^\alpha \right) \mathcal{M}_{if}$$

$$= S_{if} \quad \text{for} \quad i \neq f \ . \tag{10.11}$$

p_i^α and p_f^α are the 4-momentum vectors of all particles involved and the δ-function takes into account the conservation of energy and momentum. In time dependent perturbation theory, S_{if} is given by an expression containing the interaction responsible for the transition, which is in our case $-\int d^4x \, \mathcal{L}_{\text{W-A}}$. For the calculation we only have to consider the first order term of the infinite series of S_{if}, for $i \neq f$. Then the transition amplitude is given by the matrix element of the interaction Lagrangian multiplied by i [11]

$$S_{if} = i \langle f | \int d^4x \, \mathcal{L}_{\text{W-A}} | i \rangle \ . \tag{10.12}$$

The initial and final states can be constructed with the creation operators appropriate for each particle, which satisfy the anticommutation relations G.25. We have for the initial state $|i\rangle = |\mu^-(p_0, s_0)\rangle = b_{\mathbf{p}_0, s_0}^\dagger |0\rangle$. The final state $|f\rangle = |\bar{v}_e(p_1, s_1), v_\mu(p_2, s_2), e^-(p_3, s_3)\rangle = |\bar{v}_e(p_1, s_1)\rangle |v_\mu(p_2, s_2)\rangle |e^-(p_3, s_3)\rangle$ can be constructed in a similar way. To evaluate the transition amplitude 10.12 we have to

---

[11] We drop, where clear, the index of the 4-vectors components x^β and p^β and indicate the scalar product with a dot $p \cdot x = p_\beta x^\beta$.

know how the Dirac field operators act on the initial and final single-particle states. We obtain for example

$$\Psi_\mu(x)|\mu^-(p_0, s_0)\rangle = \sum_{s=\pm 1/2} \int \frac{\mathbf{dp}}{(2\pi)^3 2E}$$

$$\times \left[b_{\mathbf{p},s} u_{\mathbf{p},s}\, e^{-ip\cdot x} + d^\dagger_{\mathbf{p},s} v_{\mathbf{p},s}\, e^{+ip\cdot x} \right] b^\dagger_{\mathbf{p_0},s_0} |0\rangle$$

$$= \sum_{s=\pm 1/2} \int \frac{\mathbf{dp}}{(2\pi)^3 2E}\, b_{\mathbf{p},s} b^\dagger_{\mathbf{p_0},s_0} u_{\mathbf{p},s}\, e^{-ip\cdot x} |0\rangle$$

$$= e^{-ip_0\cdot x} u_{\mathbf{p_0},s_0}|0\rangle \ , \tag{10.13}$$

where we made use of the anticommutation properties of the operators, Eq. G.25. Similarly for the final states, for instance $\langle e^-(p_3, s_3)|\, \overline{\Psi}_e(x) = \bar{u}_{\mathbf{p_3},s_3} e^{ip_3\cdot x}\, \langle 0|$.

With this, Eq. 10.8 becomes

$$S_{if} = -i\frac{G_F}{\sqrt{2}} \int d^4x \left[\bar{u}_{\mathbf{p_2},s_2}\gamma_\alpha(1-\gamma_5)u_{\mathbf{p_0},s_0} \bar{u}_{\mathbf{p_3},s_3}\gamma^\alpha(1-\gamma_5)v_{\mathbf{p_1},s_1} \right]$$

$$\times\, e^{-i(p_0-p_1-p_2-p_3)\cdot x} \ . \tag{10.14}$$

Integrating over the space-time 4-vector x the phase terms yield a δ^4-function

$$S_{if} = -i\frac{G_F}{\sqrt{2}}(2\pi)^4\delta^4(p_0 - p_1 - p_2 - p_3)$$

$$\times \left[\bar{u}_{\mathbf{p_2},s_2}\gamma_\alpha(1-\gamma_5)u_{\mathbf{p_0},s_0} \bar{u}_{\mathbf{p_3},s_3}\gamma^\alpha(1-\gamma_5)v_{\mathbf{p_1},s_1} \right] \ . \tag{10.15}$$

Comparing this expression with 10.11, we obtain finally an explicit algebraic expression for the matrix element of the differential decay rate probability

$$\mathcal{M}_{if} = -i\frac{G_F}{\sqrt{2}} \left[\bar{u}_{\mathbf{p_2},s_2}\gamma_\alpha(1-\gamma_5)u_{\mathbf{p_0},s_0} \right]\left[\bar{u}_{\mathbf{p_3},s_3}\gamma^\alpha(1-\gamma_5)v_{\mathbf{p_1},s_1} \right] \ . \tag{10.16}$$

In the following we outline the main steps to evaluate the matrix element and perform the integration over the phase space to obtain the muon lifetime $\tau_\mu = \Gamma^{-1}_{i\to f}$ and the energy and angular distributions of the decay electron. Similar expressions are encountered in the calculation of scattering or particle creation and annihilation processes. We do not give all the calculation steps, which require cumbersome but in principle not difficult handling of γ-matrix multiplications and kinematic integrals, and refer to specialized books for more details (Griffiths 2004; Okun 2014; Grotz & Klapdor 2018; Thomson 2013).

Squaring 10.16

$$|\mathcal{M}_{if}|^2 = \frac{G_F^2}{2} \left[\bar{u}_{\mathbf{p_2},s_2} \gamma_\alpha (1 - \gamma_5) u_{\mathbf{p_0},s_0} \right] \left[\bar{u}_{\mathbf{p_3},s_3} \gamma^\alpha (1 - \gamma_5) v_{\mathbf{p_1},s_1} \right] \times$$
$$\left[\bar{u}_{\mathbf{p_2},s_2} \gamma_\beta (1 - \gamma_5) u_{\mathbf{p_0},s_0} \right]^* \left[\bar{u}_{\mathbf{p_3},s_3} \gamma^\beta (1 - \gamma_5) v_{\mathbf{p_1},s_1} \right]^* . \qquad (10.17)$$

This means that we have to evaluate products such as[12]

$$R = [\bar{u}_a A u_b][\bar{u}_a B u_b]^* \quad \text{or} \quad Q = [\bar{u}_c A v_d][\bar{u}_c B v_d]^* , \qquad (10.18)$$

where A and B are 4×4 matrices and the a, b, c and d indices denote the momentum and helicity (or spin) states of Eq. 10.17. We look at the first term of Eq. 10.18 closely (one can proceed similarly with the Q term). For the complex conjugate it holds[13]

$$[\bar{u}_a B u_b]^* = [\bar{u}_a B u_b]^\dagger = [u_a^\dagger \gamma^0 B u_b]^\dagger =$$
$$= [u_b^\dagger B^\dagger \gamma^{0\dagger} u_a] = [u_b^\dagger \gamma^0 \gamma^0 B^\dagger \gamma^{0\dagger} u_a] = [\bar{u}_b \overline{B} u_a] , \qquad (10.19)$$

where for the matrix we use the notation $\overline{B} = \gamma^0 B^\dagger \gamma^0$.[14] So we have

$$R = [\bar{u}_a A u_b][\bar{u}_a B u_b]^* = [\bar{u}_a A u_b][\bar{u}_b \overline{B} u_a] . \qquad (10.20)$$

Since the polarization of the decaying muon is not fixed and that of the final particles is not measured, we have to average over the two muon spin states and sum over the spin states of the final particles, i.e., we have to perform summations over spins of the form

$$\sum_{s_b} R = \bar{u}_a A \left\{ \sum_{s_b} u_{\mathbf{p_b},s_b} \bar{u}_{\mathbf{p_b},s_b} \right\} \overline{B} u_a = \bar{u}_a A \{ \not{p}_b + m_b \} \overline{B} u_a = \bar{u}_a C u_a . $$
$$(10.21)$$

Here the sum in the curly brackets is the completeness relation for the 4-spinors u, see Eq. G.20, and we have introduced the temporary matrix $C \equiv A \{ \not{p}_b + m_b \} \overline{B}$. To show that one can use the same procedure to perform the sum over the spin of the

---

[12] For example, this is the structure of the first and third terms in Eq. 10.17 and of the second and fourth terms, respectively. Note that the terms in the square brackets are simply complex variables or 1×1 matrices, so that their complex conjugate is the same as the Hermitian conjugate.

[13] Using the properties $(\gamma^0)^2 = \mathbb{1}_4$ and $\gamma^{0\dagger} = \gamma^0$, see Appendix G.2.1.

[14] Remind that for a spinor $\bar{u}$ means $u^\dagger \gamma^0$.

other particle a, we write the matrix multiplication explicitly

$$
\sum_{s_a} \sum_{s_b} R = \sum_{s_a} \bar{u}_a C u_a = \sum_{s_a} \sum_{i,j=1}^{4} \bar{u}_{\mathbf{p}_a,s_a,i} C_{ij} u_{\mathbf{p}_a,s_a,j}
$$

$$
= \sum_{i,j=1}^{4} C_{ij} \sum_{s_a} u_{\mathbf{p}_a,s_a,j} \bar{u}_{\mathbf{p}_a,s_a,i}
$$

$$
= \sum_{i,j=1}^{4} C_{ij} (\not{p}_a + m_a)_{ji} = \mathrm{Tr}[C(\not{p}_a + m_a)]
$$

$$
= \mathrm{Tr}\left[A(\not{p}_b + m_b)\bar{B}(\not{p}_a + m_a)\right] \tag{10.22}
$$

This expression is very useful, because there are no spinors left. Its calculation involves only products of γ-matrices and their traces and the 4-momenta of the particles. If instead of a u a v spinor appears in Eq. 10.18, the corresponding mass of the particle has a minus sign. This is a consequence of the completeness relation for v, Eq. G.20. In the case of the μ^- decay, Eq. 10.17, applying this procedure twice, we find

$$
\frac{1}{2} \sum_{\text{All spins}} |M_{if}|^2 = \frac{G_F^2}{4} \mathrm{Tr}\left[\gamma^\alpha (1 - \gamma^5)(\not{p}_{\mu^-} + m_\mu)\gamma^\beta (1 - \gamma^5)(\not{p}_{\nu_\mu})\right] \times
$$

$$
\times \mathrm{Tr}\left[\gamma_\alpha (1 - \gamma^5)(\not{p}_{\bar{\nu}_e})\gamma_\beta (1 - \gamma^5)(\not{p}_{e^-} + m_e)\right] , \tag{10.23}
$$

where the factor $1/2$ takes into account that we average over the muon spin and where we have indexed the momenta to explicitly identify the particles involved in the decay. The traces of products of γ-matrices can be evaluated using Eq. G.15 so that the average matrix element squared (obtained after summing over all final spin states and averaging the initial muon spin states) can be expressed in terms of the 4-momenta

$$
\langle |M_{if}|^2 \rangle = \frac{1}{2} \sum_{\text{All spins}} |M_{if}|^2 = 64\, G_F^2\, (p_{\mu^-} \cdot p_{\bar{\nu}_e})(p_{\nu_\mu} \cdot p_{e^-}) . \tag{10.24}
$$

This is the expression to insert into Eq. 10.10 yielding

$$
d\Gamma_{i \to f} = 64\, G_F^2\, (p_{\mu^-} \cdot p_{\bar{\nu}_e})(p_{\nu_\mu} \cdot p_{e^-}) \frac{(2\pi)^4}{2m_\mu}
$$

$$
\times \delta^4(p_{\mu^-} - p_{\bar{\nu}_e} - p_{\nu_\mu} - p_{e^-}) \frac{d\mathbf{p}_{\bar{\nu}_e}}{(2\pi)^3 2E_{\bar{\nu}_e}} \frac{d\mathbf{p}_{\nu_\mu}}{(2\pi)^3 2E_{\nu_\mu}} \frac{d\mathbf{p}_{e^-}}{(2\pi)^3 2E_{e^-}} . \tag{10.25}
$$

10.3.1 Energy Distribution of the Decay Electron

From Eq. 10.25 we can derive the differential rates discussed in Sect. 1.6. To obtain
the energy distribution, Eq. 1.14, of the emitted electron (or positron in the case of
the μ^+ decay), we have to express the differential 10.25 as a function of E_{e^-}, i.e.,
we have to integrate over the phase space of ν_μ and $\bar\nu_e$. With $q \equiv p_{\mu^-} - p_{e^-}$,
following integral involving the neutrino momenta must be evaluated

$$I_{\alpha\beta} = \int p_{\bar\nu_e,\,\alpha}\, p_{\nu_\mu,\,\beta} \left[(2\pi)^4 \delta^4(q - p_{\bar\nu_e} - p_{\nu_\mu}) \frac{d\mathbf{p}_{\bar\nu_e}}{(2\pi)^3 2E_{\bar\nu_e}} \frac{d\mathbf{p}_{\nu_\mu}}{(2\pi)^3 2E_{\nu_\mu}} \right] .$$

$$(10.26)$$

This integral can be calculated by exploiting the fact that the term in square brackets
is the Lorentz invariant phase space for a decay of a particle of momentum q into
two particles, yielding[15]

$$I_{\alpha\beta} = \frac{1}{96\pi} \left[q^2 g_{\alpha\beta} + 2q_\alpha q_\beta \right] .$$

$$(10.27)$$

With this expression, the differential emission probability of an electron of energy
E_{e^-} becomes

$$d\Gamma_{i\to f} = \frac{G_F^2}{3\pi m_\mu} p_{\mu^-}^\alpha p_{e^-}^\beta \left[q^2 g_{\alpha\beta} + 2q_\alpha q_\beta \right] \frac{d\mathbf{p}_{e^-}}{(2\pi)^3 2E_{e^-}} .$$

$$(10.28)$$

Using momentum and energy conservation

$$\begin{aligned}
p_{\mu^-}^\alpha p_{e^-}^\beta [q^2 g_{\alpha\beta} + 2q_\alpha q_\beta] &= q^2(p_{\mu^-} \cdot p_{e^-}) + 2(p_{\mu^-} \cdot q)(p_{e^-} \cdot q) = \\
&= (m_\mu^2 + m_e^2 - 2p_{\mu^-} \cdot p_{e^-})(p_{\mu^-} \cdot p_{e^-}) \\
&\quad + 2(m_\mu^2 - p_{\mu^-} \cdot p_{e^-})(-m_e^2 + p_{\mu^-} \cdot p_{e^-}) = \\
&= 3(m_\mu^2 + m_e^2)E_e m_\mu - 4E_{e^-}^2 m_\mu^2 - m_\mu^2 m_e^2 ,
\end{aligned}$$

$$(10.29)$$

and noting that, with $|\mathbf{p}_{e^-}| d|\mathbf{p}_{e^-}| = E_{e^-} dE_{e^-}$, the volume element can be written
in polar coordinates as

$$\mathbf{dp}_{e^-} = |\mathbf{p}_{e^-}|^2 d|\mathbf{p}_{e^-}| d\Omega = |\mathbf{p}_{e^-}| E_{e^-} dE_{e^-} d\Omega = \sqrt{E_{e^-}^2 - m_e^2}\, E_{e^-} dE_{e^-} d\Omega .$$

---

[15] More details about the individual steps can be found in, e.g., Okun (2014), Griffiths (2004).

Performing the integration over the solid angle we obtain the result

$$d\Gamma_{i\to f} = \frac{G_F^2}{12\pi^3 m_\mu}\sqrt{E_{e^-}^2 - m_e^2}\left[3(m_\mu^2 + m_e^2)E_{e^-}m_\mu - 4E_{e^-}^2 m_\mu^2 - m_\mu^2 m_e^2\right]dE_{e^-} .$$

(10.30)

By neglecting the electron mass, this expression simplifies to

$$d\Gamma_{i\to f} = \frac{G_F^2}{12\pi^3}m_\mu E_{e^-}^2\left[3m_\mu - 4E_{e^-}\right]dE_{e^-} .$$

(10.31)

By integrating over the kinematically allowed range of electron energies $0 \leq E_{e^-} \leq m_\mu/2$ we obtain the total decay rate probability (or inverse of the lifetime τ_μ)

$$\Gamma_{i\to f} = \frac{1}{\tau_\mu} = \frac{G_F^2 m_\mu^5}{192\pi^3} .$$

(10.32)

With the reduced energy $\varepsilon \equiv 2E_{e^-}/m_\mu$ and Eq. 10.32 it is clear that Eq. 10.31 is the probability per unit time for the emission of an electron (or for μ^+ positron) of energy E_{e^-} given in Sect. 1.6, Eq. 1.10. See Exercise 10.2 for an hand-waving justification of the dependence of the decay rate probability on G_F and m_μ.

10.3.2 Decay of a Polarized Muon

From the above result, Eq. 10.27, one can work out the angular and energy double differential decay probability, i.e., Eq. 1.10 discussed in Sect. 1.6, which is the basic expression for the μSR applications.

The muon is at rest with its spin pointing along the unit vector $\hat{\mathbf{I}}$. Relativistically, the spin of a particle is described by a 4-vector $\hat{I}^\alpha$, with the properties $\hat{I}^\alpha \hat{I}_\alpha = -1$, $\hat{I}^\alpha p_\alpha = 0$. In the rest frame the spin 4-vector is[16] $\hat{I}^\alpha = \begin{pmatrix} 0 \\ \hat{\mathbf{I}} \end{pmatrix}$.

In a reference frame, where the muon has the 4-momentum p_{μ^-}, the spin 4-vector becomes

$$\hat{I}^\alpha = \begin{pmatrix} \frac{\hat{\mathbf{I}}\cdot\mathbf{p}}{m_\mu} \\ \hat{\mathbf{I}} + \frac{\mathbf{p}(\hat{\mathbf{I}}\cdot\mathbf{p})}{m_\mu(E_\mu+m_\mu)} \end{pmatrix} .$$

(10.33)

---

[16] Note that $\hat{\mathbf{I}}$ is actually the muon spin $\mathbf{I}_\mu$ that we have used throughout the book, normalized to unit.

We can now proceed in the same way as in the previous section. Only now, since we are not averaging over the muon spin, we have to replace the spin completeness relation $\sum_{s=\pm 1/2} u_{\mathbf{p},s} \bar{u}_{\mathbf{p},s} = (p_{\mu^-})_\alpha \gamma^\alpha + m_\mu = \not{p}_{\mu^-} + m_\mu$, Eq. G.20, with the term corresponding to a well-defined muon spin state

$$u_{\mathbf{p},\hat{i}} \bar{u}_{\mathbf{p},\hat{i}} = \frac{1}{2}\left[(\not{p}_{\mu^-} + m_\mu)(1 + \gamma_5 \hat{I}_\nu \gamma^\nu)\right] \ . \tag{10.34}$$

In this case, the trace in Eq. 10.23 modifies to

$$\mathrm{Tr}[\cdots (1-\gamma^5)\left[(\not{p}_{\mu^-} + m_\mu)(1 + \gamma_5 \hat{I}_\nu \gamma^\nu)\right]\gamma^\beta (1-\gamma^5) \cdots =$$

$$= \mathrm{Tr}[\cdots (1-\gamma^5)\left[(\not{p}_{\mu^-} + m_\mu \gamma_5 \hat{I}_\nu \gamma^\nu + m_\mu - \gamma_5 \not{p}_{\mu^-} \hat{I}_\nu \gamma^\nu)\right](1+\gamma^5)\gamma^\beta \cdots =$$

$$= \mathrm{Tr}[\cdots (1-\gamma^5)(\not{p}_{\mu^-} - m_\mu \hat{I}_\nu \gamma^\nu)\gamma^\beta (1-\gamma^5) \cdots \ , \tag{10.35}$$

where we have used $\{\gamma^\nu, \gamma^5\} = 0$ and $(\gamma^5)^2 = \mathbb{1}_4$.

Comparison of this intermediate result with Eq. 10.23 shows that for the calculation of the polarized muon decay we have to replace p_{μ^-} by $(p_{\mu^-} - m_\mu \hat{I})$ in the expressions of Sect. 10.3.1 involving the muon momentum, specifically in Eq. 10.29.

Neglecting the electron mass, the expression

$$p_{\mu^-}^\alpha p_{e^-}^\beta [q^2 g_{\alpha\beta} + 2q_\alpha q_\beta] = q^2(p_{\mu^-} \cdot p_{e^-}) + 2(p_{\mu^-} \cdot q)(p_{e^-} \cdot q) \tag{10.36}$$

modifies with the substitution to

$$q^2(p_{\mu^-} \cdot p_{e^-}) + 2(p_{\mu^-} \cdot q)(p_{e^-} \cdot q) - m_\mu \left[q^2(\hat{I} \cdot p_{e^-}) + 2(\hat{I} \cdot q)(p_{e^-} \cdot q)\right] =$$

$$= (p_{\mu^-} \cdot p_{e^-})[(p_{\mu^-} - p_{e^-})^2 + 2(p_{\mu^-}^2 - (p_{\mu^-} \cdot p_{e^-})] -$$

$$- m_\mu (\hat{I} \cdot p_{e^-})[(p_{\mu^-} - p_{e^-})^2 - 2(p_{\mu^-} \cdot p_{e^-})] \ , \tag{10.37}$$

where we used the fact that $(q \cdot \hat{I}) = ((p_{\mu^-} - p_{e^-}) \cdot \hat{I}) = -(p_{e^-} \cdot \hat{I})$, since $(p_{\mu^-} \cdot \hat{I}) = 0$, and that $p_{e^-}^2 = 0$, since we neglect the electron mass. Furthermore, if we take into consideration that in the muon rest frame

$$\hat{I}^\alpha = \begin{pmatrix} 0 \\ \hat{\mathbf{I}} \end{pmatrix} \ ,$$

then $(p_{e^-} \cdot \hat{I}) = -E_{e^-}(\hat{\mathbf{n}} \cdot \hat{\mathbf{I}})$, where $\hat{\mathbf{n}} = \mathbf{p}_{e^-}/E_{e^-}$ is a unit vector in the direction of emission of the decay electron. Inserting this result into Eq. 10.28 and using the kinematic relations to express the 4-momenta in terms of E_{e^-} and m_μ, we obtain

$$d\Gamma_{i \to f} = \frac{G_F^2}{48\pi^4} m_\mu E_{e^-}^2 \left[(3m_\mu - 4E_{e^-}) + \hat{\mathbf{I}} \cdot \hat{\mathbf{n}}(m_\mu - 4E_{e^-})\right] dE_{e^-} d\Omega \ . \tag{10.38}$$

For the decay of a positive muon, the sign of the term $\hat{\mathbf{I}} \cdot \hat{\mathbf{n}}$ has to be reversed. Finally, with $\varepsilon = E_{e^-}/E_{e,\max} = 2E_{e^-}/m_\mu$ this solution can be written as

$$d\Gamma = W(\varepsilon, \theta)\, d\varepsilon\, d\Omega = \frac{1}{4\pi\tau_\mu} 2\varepsilon^2 [(3 - 2\varepsilon) \pm \hat{\mathbf{I}} \cdot \hat{\mathbf{n}}(2\varepsilon - 1)]\, d\varepsilon\, d\Omega \,, \qquad (10.39)$$

where the positive sign stands for μ^+, see Eq. 1.12, and the negative sign for μ^-.

10.3.3 Decay via Intermediate Vector Boson Exchange

The original Fermi and the $V - A$ theories are based on a point-like four-fermion interaction and are generally referred to as Fermi contact interaction theory. Such a contact interaction is now understood as an effective low-energy limit of the Standard Model, where the weak force is mediated by the exchange of the so-called weak or intermediate vector bosons W^+, W^-, and Z^0 (Thomson 2013). The Fermi constant is related to the fundamental parameters of the Standard Model by the equation

$$\frac{G_F}{\sqrt{2}} = \frac{g^2}{8m_W^2}\left(1 + \sum_i r_i\right) \,. \qquad (10.40)$$

The term $\sum_i r_i$ incorporates the higher-order corrections of the electroweak interaction, g is the weak coupling and $m_W c^2 = 80.377 \pm 0.012\,\text{GeV}$ the W-boson mass (Workman et al. 2022).

The effective four-fermion theory works well at low momenta; for momentum transfers much smaller than m_W, see Fig. 10.1, the Standard Model weak interaction Lagrangian reduces to the $V - A$ Lagrangian. At the lowest-order interaction level of the Standard Model (so-called tree level) $G_F\sqrt{2} = g^2/(8m_W^2)$.

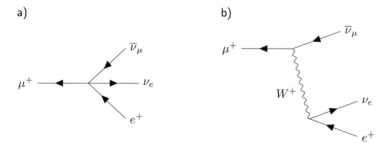

Fig. 10.1 Feynman diagrams for the positive muon decay. (**a**) Fermi (and $V - A$) contact interaction theory (**b**) Standard Model diagram for the W boson mediated weak interaction between the leptonic currents. See footnote 19, Appendix G for the meaning of the arrows

We can empirically understand that the contact interaction is a good approximation from Heisenberg uncertainty relation $\Delta E \Delta t \sim \hbar$. Taking $\Delta E = m_W c^2$, we estimate an upper range of the interaction as $c \Delta t = \hbar / m_W c^2 = 2.5 \cdot 10^{-3}$ fm, which is an extremely short distance considering that the proton radius is ~ 1 fm.

Specifically, a virtual W^+ is exchanged in the μ^+ decay. Figure 10.1 shows the Feynman diagram of the Fermi theory and that of the Standard Model (leading order) (Feynman 1949).[17]

10.4   Muon Lifetime and Determination of the Fermi Constant

It is common to use the Fermi coupling constant G_F to characterize the strength of the weak interaction since, as we will see below, this quantity can be precisely determined from a measurement of the muon lifetime.[18] G_F is a fundamental parameter and plays a key role in precision tests of the Standard Model of electroweak interactions. For a most precise determination one has to go beyond Eq. 10.31, which has been obtained to the lowest order assuming massless final particles. Higher-order QED corrections have to be included in the V-A Fermi theory[19] and the correction for the nonzero electron mass on the phase space taken into account (Workman et al. 2022).

In this case G_F can be written as (van Ritbergen & Stuart 1999, 2000)

$$G_F = \sqrt{\frac{192\pi^3}{\tau_\mu m_\mu^5} \frac{1}{1 + \Delta q}} \quad , \tag{10.41}$$

where Δq encapsulates the theoretical corrections, which have been calculated to sub-ppm precision, and τ_μ and m_μ are measured quantities.[20]

With well-known values of the muon mass and of the theoretical corrections, a precise measurement of the muon lifetime provides an accurate determination of the Fermi constant.[21,22]

---

[17] A Feynman diagram is a graphical representation of a perturbative contribution to the transition amplitude, see for example Eq. 10.16. For any given diagram from a Lagrangian of a quantum field theory, the Feynman rules provide prescriptions for calculating the amplitude.

[18] The actual coupling strength is given by g in Eq. 10.40.

[19] The effect of the W boson exchange, Fig. 10.1b, does not follow from the Fermi theory. It is small, $(3m_\mu^2/5m_W^2)$, and is incorporated in the corrections (van Ritbergen & Stuart 2000).

[20] Additional physically observable quantities, which enter into the calculations of the Fermi constant, are α and m_e.

[21] The positive muon lifetime measurements yield by far the best determination of G_F.

[22] Strictly speaking, the measurement determines the specific coupling of the muon decay $G_{F,\mu}$ which could be compared to other G_F determinations (e.g., from decays of other leptons) as a test of the Standard Model. It is assumed that G_F is universal for the weak interaction.

Several measurements have been performed in the last 40 years. We briefly describe here the recent experiment of the MuLan collaboration, which measured the muon lifetime with an uncertainty of 1 ppm, allowing the determination of G_F to an accuracy of 0.5 ppm (Webber et al. 2011; Tishchenko et al. 2013; Carey et al. 2021).

The experiment was conducted at PSI using the nearly 100% longitudinally polarized surface μ^+ beam from the πE3 beam line. The approach involves the generation of a time-structured muon beam to first prepare a source of muons and then measure the decay positrons. The time structure is imposed on the continuous beam by a fast-switching electrostatic kicker, which is operated at a repetition frequency of about 37 kHz with a beam-on window of 5 μs and a beam-off window of 22 μs. With the kicker high voltage off, the muons are transported straight along the beamline axis to a solid target, where they decay. The outgoing positrons are detected by a symmetric, highly segmented, large-acceptance scintillator array, Fig. 10.2.

In the ideal situation with a perfectly unpolarized muon source, the muon lifetime is obtained directly from a fit of the positron spectra with an exponential function (no μSR signal). It is clear that any degree of polarization modifies the ideal

Fig. 10.2 MuLan setup at the πE3 beamline at PSI (Tishchenko et al. 2013). The picture shows the last focusing quadrupoles of the beamline and the large ($\sim 3\pi$ sr) solid angle detector array for the high precision measurement of the muon lifetime and for the determination of the Fermi constant. Courtesy B. Lauss, Paul Scherrer Institute. All rights reserved

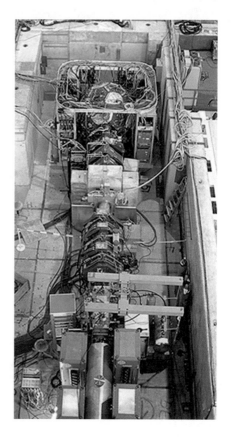

pure lifetime-determined exponential decay and is a source of error if not properly accounted for.

Data have been taken with two different targets corresponding to two strategies for suppressing the μSR signal. The first target was a ferromagnetic alloy (Arnokrome-III), where the large internal field rapidly depolarizes ($\lesssim 50$ ns) the diamagnetic muons. The second target was a quartz crystal (SiO_2). In this case, a small external transverse field has been applied (13 mT) to rapidly dephase the dominant fraction population of paramagnetic muonium atoms. Dephasing during the accumulation period of 5 μs reduces the transverse polarization of the muon ensemble by roughly a factor of about 1000 for the large paramagnetic muonium fraction and 25 for the small (6–7%) "bare" μ^+ population.

Since the μSR signals are of equal magnitude but opposite sign in geometrically opposite detector segments, summed segment-pair time histograms have been used for the analysis, so that residual effects of the μSR signals are largely eliminated.[23]

Before fitting the time histograms to extract the muon lifetime, several corrections have been applied to account for time dependent distortions in the spectra due to pileup of decaying positrons and variations in detector gain. Detailed fitting procedures and consistency checks have been employed to correct for distortions due to residual μSR effects.

The combined weighted average of the two target measurements (about 10^{12} events) yields

$$\tau_\mu = 2196980.3(2.1)(0.7)\,\text{ps} \ , \tag{10.42}$$

where the first term in parenthesis is the statistical and the second the systematic error, corresponding to a total uncertainty of 1 ppm.

From Eq. 10.41 using the CODATA 2010 value for the muon mass $m_\mu = 105.6583715(35)$ MeV (Mohr et al. 2012) and the calculated theoretical corrections contributing to Δq, the MuLan experiment (Tishchenko et al. 2013) yields the most accurate value of the Fermi constant[24,25]

$$G_F = 1.1663787(6) \times 10^{-5}\,\text{GeV}^{-2} \ . \tag{10.43}$$

The 0.5 ppm error is dominated by the 1.0 ppm uncertainty of the muon lifetime measurement with additional contributions of 0.08 ppm from the muon mass value and 0.14 ppm from the theoretical corrections.

---

[23] The nearly 4π isotropic positron detector is also an important element in minimizing μ^+ spin precession and depolarization effects.

[24] The determination of G_F in units of GeV^{-2} from the measurement of τ_μ in ps requires a unit conversion via $\hbar$.

[25] Remember that in this chapter we use natural units. To explicitly include $\hbar$ and c in this expression means to replace G_F with $G_F/(\hbar c)^3$.

10.5 Muon Magnetic Anomaly

The relation between the values of the spin and of the magnetic dipole moment, Eq. 1.20 discussed in Sect. 1.7.1, can be written as

$$\mu_\mu = \gamma_\mu \hbar I_\mu = g_\mu \frac{e}{2m_\mu} I_\mu = 2(1 + a_\mu) \frac{e}{2m_\mu} I_\mu \ , \tag{10.44}$$

where

$$a_\mu = \frac{g_\mu - 2}{2} \ , \tag{10.45}$$

called the muon magnetic anomaly, expresses the deviation from $g_\mu = 2$, the value predicted by the Dirac theory for a pointlike spin $1/2$ particle or antiparticle. Quantum electrodynamics effects, as well as corrections based on the weak and strong interactions lead to a g-factor slightly larger than two. Very precise measurements of g_μ and the comparison with accurate theoretical predictions are a very important test of the Standard Model of particle physics.

A more than two decades standing discrepancy between experiment and theory of the muon magnetic moment anomaly has been giving a possible hint to new physics and has motivated and continues to motivate improved experiments and calculations.

10.5.1 Experiment

The first value of $g_\mu = 2$ to an accuracy of $\sim 10\%$ was obtained in the pioneering experiment of Garwin et al. (1957), who observed the parity violation in the π and μ decays and obtained a first μSR spectrum, see Sect. 1.2. Later, several experiments have been performed to measure g_μ; for a review see Jegerlehner (2017), Roberts (2019).

The muon magnetic anomaly a_μ has been measured with increasing precision by a series of experiments, first at CERN,[26] more recently at Brookhaven National Laboratory (BNL), USA (Bennett et al. 2006) and at Fermilab (FNAL), USA (Abi et al. 2021; Aguillard et al. 2023). The experiment and the data analysis at Fermilab, where the BNL setup has been reassembled and substantially improved, are ongoing at the time of this writing.

The measurement is based on the fact that the magnetic anomaly is proportional to the difference between the Larmor precession and the cyclotron frequency. Following the technique developed in the third CERN experiment (Bailey et al. 1979), which achieved a precision of 7.3 ppm, the muons are stored in a uniform-field storage ring, where electric quadrupoles ensure their vertical confinement.

---

[26] At CERN three experimental enterprises have been undertaken between 1962 and 1979.

For a muon of charge $q = \pm e$ moving in a magnetic field perpendicular to its momentum, the momentum and the spin rotate with the cyclotron and Larmor angular frequencies, respectively

$$\boldsymbol{\omega}_c = -\frac{q\mathbf{B}}{\gamma m_\mu}$$

$$\boldsymbol{\omega}_\mu = -\frac{qg_\mu\mathbf{B}}{2m_\mu} - \frac{q\mathbf{B}}{\gamma m_\mu}(1-\gamma) \ . \tag{10.46}$$

The difference between the two frequencies is the spin precession frequency relative to the momentum[27]

$$\boldsymbol{\omega}_a \equiv \boldsymbol{\omega}_c - \boldsymbol{\omega}_\mu = -\frac{(g_\mu - 2)}{2}\frac{q\mathbf{B}}{m_\mu} = -a_\mu \frac{q\mathbf{B}}{m_\mu} \ . \tag{10.47}$$

Because of a relativistic effect, the electric field of the quadrupole appears in the rest frame of the muon as a magnetic field that affects the spin precession so that, in the presence of $\mathbf{E}$ and $\mathbf{B}$ fields and with the velocity $\boldsymbol{\beta}$ perpendicular to both, the expression 10.47 becomes

$$\boldsymbol{\omega}_a = -\frac{q}{m_\mu}\left[a_\mu\mathbf{B} - \left(a_\mu - \frac{1}{\gamma^2 - 1}\right)\frac{\boldsymbol{\beta}\times\mathbf{E}}{c}\right] \ . \tag{10.48}$$

The coefficient of the electric field term vanishes at a muon momentum of 3.094 GeV/c, corresponding to a Lorentz factor $\gamma_m = \sqrt{1 + (a_\mu)^{-1}} \cong 29.3$ (so-called magic momentum). For the measurement, this value is selected for the central momentum, so that only the magnetic field determines ω_a and Eq. 10.47 can be used.

In the experiment,[28] bunches (about 50 ns wide) of longitudinally polarized muons with the magic momentum are injected and stored in a storage ring (radius 7.1 m, magnetic field 1.45 T), Figs. 10.3, 10.4, and 10.6. The muons have a dilated lifetime of $\gamma_m \tau_\mu = 64.4 \ \mu$s. Each injection (or filling of the storage ring) is followed by a measurement period of typically 700 μs (corresponding to about 10 lifetimes).

As the muons circulate in the ring, the magnetic field rotates their spins in the horizontal plane with respect to the momentum vector with frequency ω_a, Fig. 10.3. The decay electrons or positrons are observed by a set of 24 detectors placed symmetrically around the storage region. The detectors, plastic scintillator-based calorimeters (Pb and scintillating fibers), are read out by photomultiplier tubes. The number of detected electrons/positrons is modulated by the spin precession

---

[27] Note that this expression is also valid nonrelativistically.

[28] We describe here the BNL experiment, noting that the FNAL experiment uses the same concept and apparatus (with improvements).

Fig. 10.3 Principle of the $g - 2$ measurement. Since g_μ is (slightly) greater than 2, the spin vector (blue arrow) advances the momentum vector (black arrow) with a precession frequency proportional to a_μ

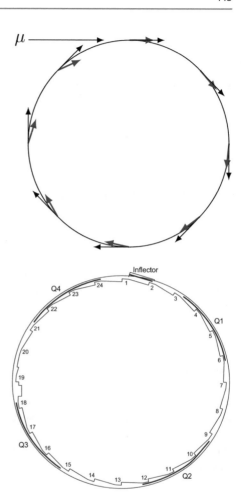

Fig. 10.4 Layout of the storage ring (radius 7.1 m, magnetic field 1.45 T, muon storage region diameter 90 mm) of the BNL $g - 2$ experiment. The 24 numbers represent the locations of the decay electron/positron detectors. Q1 to Q4 are the four electric quadrupole sections for vertical focussing of the beam in the storage ring. Modified from Bennett et al. (2006), © American Physical Society. Reproduced with permission. All rights reserved

frequency, Eq. 10.47, producing a spectrum similar to a μSR spectrum with precession frequency ω_a but dilated lifetime[29],[30]

$$N_e = N_0 \, e^{-\frac{t}{\gamma_m \tau_\mu}} \left[1 \pm A \cos\left(\omega_a t + \phi\right)\right] + B_g \ . \tag{10.49}$$

The value of ω_a is obtained from a least-squares fit to these data, Fig. 10.5, after accounting for many additional small effects (Bennett et al. 2007).

---

[29] This functional form applies to the histogram from one detector for one muon injection but also to the sum of these histograms (Bennett et al. 2007).

[30] Measurements have been performed with both μ^- ($-$ sign) and μ^+ ($+$ sign).

Fig. 10.5 BNL $g-2$
experiment. Decay electron
counts versus time for
3.6×10^9 muon decays in a
data taking period with μ^-.
The data is wrapped around
modulo $100\,\mu s$. The
anomalous spin precession
frequency ω_a with period
$T_a = 2\pi/\omega_a \cong 4.36\,\mu s$ is
clearly visible. Modified from
Bennett et al. (2006),
© American Physical Society.
Reproduced with permission.
All rights reserved

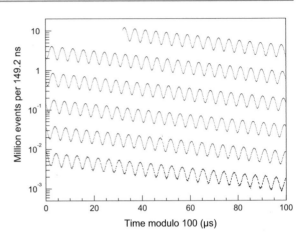

The other quantity to be determined with high precision is the magnetic field,
which is proportional to the free proton spin precession frequency ω_p (μ_p proton
magnetic moment)

$$\hbar\omega_p = 2\mu_p B \ , \tag{10.50}$$

and can be measured by NMR. Great care has been taken to precisely determine the
average magnetic field[31] seen by the stored muons, which is the relevant quantity
entering Eq. 10.47. A trolley inside the vacuum chamber carrying 17 NMR probes
maps the magnetic field by traveling around the ring during the beam-off periods.
Additionally, an array of 378 fixed NMR probes around the ring continuously
monitors the magnetic field during data collection. The average proton frequency ω_p
is determined with a total error of 0.17–0.4 ppm (depending on the data acquisition
run). Using $\mu_\mu = (a_\mu + 1)\frac{e\hbar}{2\mu_\mu}$, Eq. 10.47 can be expressed as a function of the ratio
of the two frequencies ω_a/ω_p measured in the experiment and the ratio of the muon
to proton magnetic moment ratio μ_μ/μ_p, which has been precisely determined from
measurements of the muonium hyperfine levels (Liu et al. 1999)

$$a_\mu = \frac{\dfrac{\omega_a}{\omega_p}}{\dfrac{\mu_\mu}{\mu_p} - \dfrac{\omega_a}{\omega_p}} \ . \tag{10.51}$$

---

[31] Through a series of passive shimming adjustments, the azimuthally averaged magnetic field in
the storage volume has a uniformity of approximately 1 ppm during data collection.

Fig. 10.6 $g - 2$ setup at Fermilab with the muon storage ring previously used for the BNL measurements. Photo credit Reidar Hahn, Fermi National Accelerator Laboratory. Reproduced with permission. All rights reserved

By combining the results from different runs and averaging the results with μ^- and μ^+ beams (i.e., assuming CPT symmetry) a value of a_μ with a total uncertainty of 0.54 ppm[32] was finally achieved in the BNL experiment (Bennett et al. 2006)

$$a_\mu^{\text{exp BNL}} = 116\,592\,080(63) \times 10^{-11}(0.54 \text{ ppm}) \ . \tag{10.52}$$

In the summer of 2013, the storage ring with very highly uniform magnetic field used for the BNL measurements and other parts of the apparatus has been moved from BNL to Fermilab for a new measurement, following the same principle, Fig. 10.6.

The main limitation of the BNL experiment was the statistical uncertainty. The Fermilab experiment plans to collect approximately 20 times the statistics of the BNL measurement, thus reducing the statistical uncertainty to 0.10 ppm. With a number of improvements (e.g., new highly segmented electromagnetic calorimeters for a better determination of ω_a, improved uniformity and monitoring of the magnetic field) the new experiment aims at an accuracy of 0.140 ppm thus improving the precision by a factor of four (Driutti et al. 2019).

Intermediate results from measurements with positive muons at Fermilab have been published in 2021 (Abi et al. 2021) and very recently (Aguillard et al. 2023). The value

$$a_\mu^{\text{exp FNAL}} = 116\,592\,055(24) \times 10^{-11}(0.20 \text{ ppm}) \tag{10.53}$$

---

[32] The total uncertainty consists of a 0.46 ppm statistical uncertainty and a 0.28 ppm systematic uncertainty, combined in quadrature.

represents a more than a factor of two improvement in accuracy and is in good agreement with the previous BNL result. Combining the BNL and FNAL results gives an experimental average

$$a_\mu^{\text{exp FNAL}} = 116\ 592\ 059(22) \times 10^{-11}(0.19\ \text{ppm}) \quad . \tag{10.54}$$

An alternative new experiment is in preparation in Japan at the J-PARC accelerator complex (Abe et al. 2019). It makes use of a low momentum muon beam (300 MeV/c) stored into a small (66 cm diameter), highly uniform storage magnet without electrostatic focusing. The combination of measurements with a different apparatus and experimental approach has been expected to provide deeper insight into the physics of the muon magnetic anomaly and to contribute to the resolution of the long-standing discrepancy between calculated and measured value.

10.5.2 Theory

The muon, like the other leptons, has no internal structure. The value of the magnetic dipole moment comes from the so-called radiative corrections, i.e., from virtual particles that couple to the muon. In principle, these radiative corrections are not limited to those of the Standard Model. All virtual particles existing in nature (even unknown ones) that can couple to the lepton, or to the photon via vacuum polarization loops, can contribute.[33] Therefore, any deviation from predictions of the Standard Model may point to fundamental new physics.[34]

The Standard Model value of a_μ has contributions from QED (with loops involving leptons, e, μ and τ, and photons), strong interaction (with hadrons in vacuum polarization loops), and weak interaction (with loops involving the W, Z, and Higgs bosons), Fig. 10.7.

Therefore one writes

$$a_\mu^{\text{SM}} = a_\mu^{\text{QED}} + a_\mu^{\text{hadr.}} + a_\mu^{\text{weak}} \quad . \tag{10.55}$$

Figure 10.7b shows the diagram of the dominant contribution from QED called Schwinger term (Schwinger 1948). It involves a virtual photon that is emitted and absorbed by the muon before and after its interaction with the magnetic field. It is by far the largest correction, amounting to $\alpha/2\pi \approx 1/860$. The total QED contribution is well understood and accounts for more than 99.99% of the anomaly.

---

[33] Figure 10.7c is an example of a process with a hadronic loop.

[34] The same is true for the other leptons. The electron magnetic anomaly a_e has been measured to an accuracy of 0.24 ppb (Hanneke et al. 2008). For a theoretical comparison, this requires an extension of the QED calculations to the tenth order (5 loop graphs) for an uncertainty of 0.62 ppb (Aoyama et al. 2018). For the muon, the contribution from coupling to heavy particles scales, relative to the electron, as $(m_\mu/m_e)^2 \sim 42,800$. This makes the muon more sensitive to potential new particles and forces than the electron.

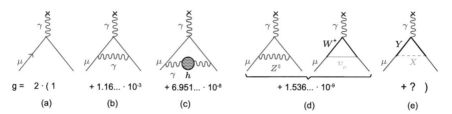

$$g = 2 \cdot (\,1 \qquad + 1.16... \cdot 10^{-3} \qquad + 6.951... \cdot 10^{-8} \qquad + 1.536... \cdot 10^{-9} \qquad + ?\;)$$

(a)   (b)   (c)   (d)   (e)

Fig. 10.7 (a) Examples of Feynman graphs of the contributions to the muon magnetic anomaly a_μ with the relative strengths of the different forces (status of theory in 2019). (b) QED (Schwinger term), (c) strong interaction (with hadrons in vacuum polarization loops), (d) electroweak forces, (e) putative contribution from new physics with a new undiscovered exchange particle X beyond the Standard Model. The external magnetic field is represented by a cross and a photon line coming in from the top. Modified from Roberts (2019), under CC-BY-4.0 license. ©The Authors

The calculations of the different contributions to Eq. 10.55 are being continuously improved. The status in 2020 of the Standard Model calculation of a_μ^{SM} has been discussed in detail and summarized in a review paper (Aoyama et al. 2020), where recommended values are given. Note the extraordinary accuracy of the QED contribution

$$a_\mu^{\text{QED}} = 116\,584\,718.931(104) \times 10^{-11} \; . \qquad (10.56)$$

The hadronic contributions are the most difficult to calculate. They are responsible for most of the theoretical uncertainty, which is also sometimes difficult to assess (Jegerlehner 2018; Aoyama et al. 2020). They are dominated by processes like the one shown in Fig. 10.7c (hadronic vacuum polarization loop) and their estimation relies on experimental data of electron-positron annihilation cross sections into hadrons. Aoyama et al. (2020) recommended a total value for the hadronic contribution $a_\mu^{\text{hadr.}} = 6937(44) \times 10^{-11}$. With the electroweak term $a_\mu^{\text{weak}} = 153.6(1.0) \times 10^{-11}$ the theoretical value of a_μ becomes

$$a_\mu^{\text{SM}} = 116\,591\,810(43) \times 10^{-11} \; . \qquad (10.57)$$

With these values, the difference between measurement and the theory amounts to

$$a_\mu^{\text{exp}} - a_\mu^{\text{SM}} = 249(48) \times 10^{-11} \; , \qquad (10.58)$$

which is a statistically significant 5σ discrepancy and a promising indication of new physics.

More recently, a new calculation of the leading order hadronic contribution has been published, which does not rely on experimental cross sections and instead uses ab initio quantum chromodynamics[35] (Borsanyi et al. 2021; Lehner 2022). The calculation gives a new value of the dominant leading order of the hadronic vacuum polarization, which increases the theoretical value by 144×10^{-11} and its error to 58×10^{-11}.

This result alone, would reduce the long-standing discrepancy between theory and experiment from $5\,\sigma$ to $1.6\,\sigma$, and would shake the hopes for new physics.

Obviously, at the present stage of research no definite conclusion can be drawn. To understand the conflicting results and decide whether the $g - 2$ measurement is giving a glimpse of new physics or not, further theoretical and experimental investigations are under way.

Exercises

10.1. Elementary charge value
Determine the value of the elementary charge in natural units.

10.2. Hand-waving arguments for the $\Gamma_{i \to f}$ expression, Eq. 10.32
Use simple arguments to justify the functional dependence of $\Gamma_{i \to f}$ on the Fermi constant and the muon mass.

References

Abe, M., Bae, S., Beer, G., et al. (2019). *Progress of Theoretical and Experimental Physics, 2019*, 053C02.

Abi, B., et al. (2021). *Physical Review Letters, 126*, 141801. Muon $g - 2$ Collaboration.

Aguillard, D. P., et al. (2023). arXiv:2308.06230. Muon $g - 2$ Collaboration.

Ahmad, Q. R., et al. (2001). *Physical Review Letters, 87*, 071301. SNO Collaboration.

Aoyama, T., Asmussen, N., Benayoun, M., et al. (2020). *Physics Reports, 887*, 1.

Aoyama, T., Kinoshita, T., & Nio, M. (2018). *Physical Review D, 97*, 036001.

Bailey, J., Borer, K., Combley, F., et al. (1979). *Nuclear Physics B, 150*, 1.

Bellgardt, U., Otter, G., Eichler, R., et al. (1988). *Nuclear Physics B, 299*, 1. SINDRUM Collaboration.

Bennett, G. W., et al. (2006). *Physical Review D, 73*, 072003. Muon $g - 2$ Collaboration.

Bennett, G. W., et al. (2007). *Nuclear Instruments and Methods in Physics Research - Section A, 579*, 1096. Muon $g - 2$ Collaboration.

Bjorken, J. D., & Drell, S. D. (1965). *Relativistic Quantum Fields*. McGraw-Hill. ISBN: 978-0070054943.

Borsanyi, S., Fodor, Z., Guenther, J. N., et al. (2021). *Nature, 593*, 51.

Carey, R., Gorringe, T., & Hertzog, D. (2021). *SciPost Physics Proceedings, 5*, 016.

Coleman, S. (2018). *Lectures of Sidney Coleman on Quantum Field Theory*. World Scientific Publishing. ISBN: 981-4635-51-0.

Driutti, A., et al. (2019). *SciPost Physics Proceedings, 1*, 033. $g - 2$ Fermilab Collaboration.

Fermi, E. (1934). *Zeitschrift für Physik, 88*, 161.

---

[35] Quantum chromodynamics (QCD) is the quantum field theory of the strong interaction between quarks mediated by the gluons (Thomson 2013). It is an important part of the Standard Model.

Feynman, R. P. (1949). *Physical Review, 76*, 769.

Feynman, R. P., & Gell-Mann, M. (1958). *Physical Review, 109*, 193.

Fukuda, Y., et al. (1998). *Physical Review Letters, 81*, 1566. Super-Kamiokande Collaboration.

Garwin, R. L., Lederman, L. M., & Weinrich, M. (1957). *Physical Review, 105*, 1415.

Gorringe, T., & Hertzog, D. (2015). *Progress in Particle and Nuclear Physics, 84*, 73.

Griffiths, D. (2004). *Introduction to Elementary Particles*. Wiley-VCH Verlag. ISBN: 978-3527406012.

Grotz, K., & Klapdor, H. V. (2018). *The Weak Interaction in Nuclear, Particle and Astrophysics*. CRC Press. ISBN: 978-0852743126.

Hanneke, D., Fogwell, S., & Gabrielse, G. (2008). *Physical Review Letters, 100*, 120801.

Jegerlehner, F. (2017). *The Anomalous Magnetic Moment of the Muon*. Springer tracts in modern physics (2nd ed.). Springer. ISBN: 978-3319635750.

Jegerlehner, F. (2018). *EPJ Web of Conferences, 166*, 00022.

Kuno, Y., & Okada, Y. (2001). *Reviews of Modern Physics, 73*, 151.

Lancaster, T., & Blundell, S. J. (2014). *Quantum Field Theory for the Gifted Amateur*. Oxford University Press. ISBN: 978-0199699339.

Lee, T. D., & Yang, C. N. (1956). *Physical Review, 104*, 254.

Lehner, C. (2022). *Nature Reviews Physics, 4*, 14.

Liu, W., Boshier, M. G., Dhawan, S., et al. (1999). *Physical Review Letters, 82*, 711.

Mohr, P. J., Taylor, B. N., & Newell, D. B. (2012). *Reviews of Modern Physics, 84*, 1527.

Okun, L. B. (2014). *Leptons and Quarks*. World Scientific Publishing. ISBN: 978-9814603003.

Papa, A. (2018). *EPJ Web of Conferences, 179*, 01018.

Peskin, M. E., & Schroeder, D. V. (1995). *An Introduction to Quantum Field Theory*. Perseus Books Publishing. ISBN: 978-0367320560.

Roberts, B. L. (2007). *Journal of the Physical Society of Japan, 76*, 111009.

Roberts, B. L. (2019). *SciPost Physics Proceedings, 1*, 032.

Schwinger, J. (1948). *Physical Review, 73*, 416.

Sudarshan, E. C. G., & Marshak, R. E. (1958). *Physical Review, 109*, 1860.

Thomson, M. (2013). *Modern Particle Physics*. Cambridge University Press. ISBN: 978-1107034266.

Tishchenko, V., et al. (2013). *Physical Review D, 87*, 052003. MuLan Collaboration.

van Ritbergen, T., & Stuart, R. G. (1999). *Physical Review Letters, 82*, 488.

van Ritbergen, T., & Stuart, R. G. (2000). *Nuclear Physics B, 564*, 343.

Webber, D. M., et al. (2011). *Physical Review Letters, 106*, 041803. MuLan Collaboration.

Weinberg, S. (1980). *Reviews of Modern Physics, 52*, 515; Salam, A., *Reviews of Modern Physics, 52*(3), 525; Glashow, S. L., *Reviews of Modern Physics, 52*(3), 539.

Workman, R. L., et al. (2022). *Progress of Theoretical and Experimental Physics, 2022*, 083C01. Particle Data Group.

Wu, C. S., Ambler, E., Hayward, R. W., et al. (1957). *Physical Review, 105*, 1413.

Conclusions and Outlook 11

The μSR method has established itself as a reliable, useful and versatile technique for obtaining unique, novel and relevant local information on a wide range of physical systems. In addition to standard macroscopic techniques such as magnetometry and transport measurements, it complements measurements with other particle probes such as neutrons, NMR (see Fig. 5.20), but also photons.

Very useful has proved the availability of dedicated instruments open to use by outside scientists and which are continuously improved. The possibility to perform measurements with other particle probes (mainly neutrons and photons) at the same site, as is now the case in some laboratories around the world, has also contributed to the use of the μSR technique. Its applications in the study of condensed matter, materials and applied sciences are growing steadily. The relevance of the technique is also underscored by new initiatives and plans to build μSR facilities, which are at various stages of development in China (Bao et al. 2023), South Korea (Pak et al. 2021), and the United States.

What does the future of μSR hold for the physical sciences? We can make some conjectures or even wish lists, not excluding the possibility that the next breakthrough will surprise us. On general grounds, the versatility of the method (the muon can be implanted in any material), its uniqueness as a local magnetic probe, and the lessons learned from its application to the study of materials and objects that have appeared in the last few decades (e.g., unconventional superconductors, topological materials, skyrmions, and heterostructures) strengthen our conviction that μSR will continue to occupy a valuable place as an experimental tool for the study of physical structures, while evolving in new directions.

The evolution and progress of science is always driven by advances in experimental techniques and theories. Let us first consider the experimental part, or the method itself. Here we can identify three main components that determine the feasibility of an experiment and its scientific caliber, and on which advances have a direct impact. The first is the quality of the beam (intensity, polarization, beam spot, energy range, possibly the temporal structure for pulsed beams). The

© Springer Nature Switzerland AG 2024
A. Amato, E. Morenzoni, *Introduction to Muon Spin Spectroscopy*,
Lecture Notes in Physics 961, https://doi.org/10.1007/978-3-031-44959-8_11

second is the sample environment (with tunable physical parameters such as external fields, temperature, pressure, synchronization with external stimuli such as lasers, photons, electric fields). The third component is the spectrometer (with parameters such as time resolution, solid angle coverage, robustness and reliability in adverse conditions, e.g., in a high magnetic field).

The components of a μSR setup are closely interrelated, so that advances in one component enable advances in another. For example, the development of compact solid-state detectors with excellent time resolution and insensitivity to large magnetic fields has facilitated the construction of high-field instruments. In addition, these characteristics have proven useful for all spectrometers, making solid state detectors the standard component of a detector system.

In the last decade, great progress has been made in extending the range of parameters (such as magnetic field, pressure, and temperature) that are often critical in tuning the phases and driving phase transitions of physical systems. Also important is the simultaneous extension of parameters, e.g., low temperature and high magnetic field or high pressure. It is now possible to perform experiments in a 10 T field at temperatures as low as 10 mK, which is a good field/temperature ratio. For hydrostatic pressure, the current limits are about 3 GPa and 0.3 K.

Developments are underway to further extend these limits, and we can expect interesting advances. However, we must be aware that some limitations are inherent in the current implementation of the μSR technique and pose challenging obstacles to significant advance. For example, increasing the maximum applied magnetic field reduces the radius of curvature of the decay positrons and places spatial limits on their detection. In the case of pressure measurements, the maximum available muon energy limits the wall thickness of the pressure cell that can be penetrated and thus the maximum hydrostatic pressure that can be applied to the sample.[1]

Regarding polarized muon beams, their quality is essentially determined by polarization, intensity (see Exercise 8.1), and by the beam size at the sample position. We believe that especially a reduction of the latter offers a great potential for a new generation of μSR experiments. Currently, the best quality beams available are the surface beam for measurements on bulk samples and the moderated surface muon beam for depth dependent studies of nanometer thin films and heterostructures, see Chaps. 1 and 8. Their size at the sample position are a few mm^2 and $\sim$1–2 cm^2, respectively, and this can be a limiting factor for some topical experiments, since often samples of novel materials or structures are only available in very small sizes. Larger samples (e.g., of crystals or of complex heterostructures and devices) can only be grown or fabricated, if at all, by relaxing quality parameters

---

[1] An interesting promising development in this respect is the application of uniaxial stress. In contrast to hydrostatically pressurized samples, which require muons that are sufficiently energetic (typically $\sim$50 MeV decay muons) to penetrate the walls of the pressure cell, uniaxial pressure devices can be built in which the sample is directly exposed to the beam. Instead of high-energy muons, surface muons can be used. This offers the various advantages of better beam quality, lower energy and a smaller beam spot, allowing the use of thinner and smaller plate-like samples. A convincing example of the potential of this approach is presented in Sect. 6.6.

(density of defects, twin planes, phase purity,... for bulk samples; surface or interface roughness,... for heterostructures).

A significant reduction of the muon beam spot, while maintaining reasonable intensities, cannot be achieved within the scheme of a typical beamline equipped with the standard focusing and deflection elements described in Sects. 1.8.3 and 1.8.4. The initial phase space is too large for this. The goal of reducing the sample size could be achieved if the position of the muon decay were known. The basic principle on which the acquisition of a μSR spectrum is normally based has remained virtually unchanged since the early days of the technique; namely, the time difference between the stopping time of the muon in the sample and the emission time of its decay positron is measured and the event is accumulated in a histogram. As shown in Chap. 3, in a continuous beam this limits the acceptable rate of incoming muons. Knowing the decay position of the muon in the sample (μ^+-e^+ vertex) would represent a quantum leap for the μSR technique. It would ease the sample size limitations and allow an optimum use of the available beam intensity. Vertex reconstruction is routinely used in particle physics experiments, where layers of position sensitive detectors determine the parameters of the helical trajectory of a charged particle in a uniform magnetic field, from which the bend radius and vertex position can be reconstructed. Solid state detectors, which were identified early on as good candidates for position sensitive spectrometers for μSR spectroscopy (Shiroka et al. 2006), have achieved resolutions better than 20 μm (Frühwirth and Strandlie 2021). The requirements of a μSR experiment differ in some respects from those of particle physics experiments; for example, the energies of the decay positrons are rather low, and the magnetic field strength of the spectrometer is a tunable physical parameter and not fixed as in particle physics. However, continued advances in solid-state detector technology promise their use in position resolved μSR in the not too distant future.

In general new major developments represent a significant advance for the μSR method and can open up new fields. This was already the case in the early days of μSR where the development of the surface beam virtually initiated the systematic use of the μSR method in physics, chemistry, and materials science. More recently, the development of tunable low-energy beams has enabled depth dependent investigations in the nanometer range and extended the use of μSR to the study of thin films, heterostructures and near-surface regions. In this category an innovative tool on the wish list of applied muon science is a muon (re)accelerator to be used for the realization of a transmission muon microscope. The idea is to reaccelerate very slow muons (e.g., thermal muons generated by resonant ionization of muonium or to-eV-moderated muons) to 10–30 MeV. This would extend the observable sample thickness from the sub-μm—μm range for a transmission electron microscope to 10–100 μm with spatial resolution in the $\sim$ μm range (microbeam).[2] The recent acceleration of degraded Mu$^-$ to 89 keV using a radio-

---

[2] A microbeam would also allow the use of anvil cells for pressure studies, thus increasing the range of pressures applied and overcoming the limitations of the pressure cells mentioned above.

frequency accelerator is a first small step towards this ambitious goal (Bae et al. 2018).

On the theoretical side, the use of DFT calculations to analyze μSR experiments is still in its infancy, but rapidly progressing, see examples in Chaps. 4, 5, 6, and 7. DFT has already convincingly shown that it can be used to answer two fundamental questions that arise in some μSR experiments, namely the lattice position of the muon site and the degree of perturbation of its local environment. In the future we can expect a more extensive use of DFT calculations, especially in the study of strongly correlated systems on the border of phase transitions.

We can also foresee new avenues of development in the field of data analysis making use of the rapidly expanding tools provided by artificial intelligence and possibly quantum computing. Currently, the main method of analyzing μSR spectra is to choose one or more parameter functions that best fits the data, e.g., by minimizing the sum of the squares of the differences between model and data (mostly asymmetry data) or by maximum likelihood estimation. Powerful, versatile, and portable programs have been developed for this purpose, offering a wide variety of fitting functions (predefined, see Chap. 4, or user-defined) with flexible parameter constraints, data grouping, global fitting options,...; some are available to all μSR users (Suter and Wojek 2012; Pratt 2000). Choosing the appropriate fitting function requires some knowledge of the underlying physics of the system under study. A different approach can be provided by machine learning (ML) techniques,[3] which operate unbiased and without any prior assumptions about the investigated system. The potential of these techniques to analyze μSR asymmetry data has recently been examined (Tula et al. 2021). It is found that ML can reveal correlations within the data and may serve as an unbiased method for detecting phase transitions. This can be a strong point when the changes in the μSR spectrum related to the phase transition are very subtle and no theoretical prediction is available, as is the case for the onset of a TRSB superconducting phase, see Sect. 6.6.

The semiclassical mean-field approach, where the muon spin polarization evolves in a magnetic field of given field distribution, is still widely used and gives reliable results in the static case or when the field dynamics is not too complex. A more accurate description of the system dynamics can be obtained by a fully quantum-mechanical solution of the problem if the Hamiltonian describing the interaction of the muon spin with its surrounding is known and the evolution of the polarization in time is obtained by solving the corresponding Schrödinger equation. While in the past the available computational power was a limitation, the ever-increasing computer power makes such types of calculations more accessible.

By solving a Hamiltonian of dimensionality 2048,[4] the time evolution of the muon spin coupled to the surrounding spin $1/2$ fluorine nuclei in NaF and CaF$_2$ was

---

[3] Machine learning is the design and implementation of computer software that can learn on its own, i.e., without direct instructions. It can be used by a computer system to improve algorithms.

[4] This corresponds in NaF to a 9-spin system of 1 muon, 1 nearest-neighbor Na ($I = 3/2$) and F, and 4 next-nearest-neighbor F spins.

calculated and found to well reproduce the measured ZF polarization (Wilkinson & Blundell 2020).[5] It is reasonable to expect that this analytical approach will become increasingly relevant for a quantitative description of the muon spin relaxation in a wide range of well-characterized materials.

Interesting perspectives open up by considering the quantum mechanical nature of the information obtained in a μSR experiment and viewing the two-state quantum system of the muon spin as the physical realization of a qubit, the building block of a quantum computer.

The Hamiltonian describing the dipole-dipole interaction of N spins contains $O(N^2)$ terms, and the Hilbert-space dimension grows exponentially as 2^N,[6] therefore the number of spins, which can be included in a calculation to-eV-moderated is limited by the resources of a classical (binary) computer.[7] Solving such numerically large problems is considered a natural domain for a quantum computer, which uses resources that scale polynomially with the size of the system under study, rather than exponentially as in the case of classical computing methods.

Recently, a quantum computer approach to analyze the μSR data of CaF_2 (McArdle 2021) has been numerically emulated. In the work a (quantum) algorithm has been constructed, which simulates the evolution of the muon spin polarization on a quantum computer. We refer to the original paper for details, but in essence, the muon and the nuclear spins[8] are each assigned to a qubit, and the dipole-dipole Hamiltonian is mapped to qubit operators. The initial condition, the time evolution and the measurement of the muon spin polarization at different times are expressed in terms of qubit operators. The whole procedure represents the largest simulation of μSR data with a Hilbert space of size 2^{29}. Good agreement is found with the data and the classical analysis of Wilkinson and Blundell (2020). This illustrative example indicates that μSR data might be efficiently analyzed by a quantum computer, and that, on the other hand, a well-characterized muon-spin system might be used to emulate the operations of a quantum computer.

In summary, we believe that we are correct in assuming that that the demand for μSR as a sensitive, efficient and powerful tool for experimental science and its use in a wide variety of fields will not only continue, but will develop significantly in new directions.

---

[5] The muon spin oscillation between a pure and a mixed state can be interpreted as a forward and backward exchange of quantum information between the muon and the fluorine subsystem, and the muon spin relaxation can be interpreted as the dissipation of quantum information carried by the muon spin qubit through decoherence.

[6] We consider here spin $1/2$ nuclei.

[7] There are, however, approximate methods that reduce the memory size needed by the numerical calculations at the cost of introducing additional uncertainty in the result (Celio 1986).

[8] The relevant F and Ca isotopes accounting for practically 100% of the abundance, have spin $1/2$ for F, zero for Ca isotopes.

References

Bae, S., Choi, H., Choi, S., et al. (2018). *Physical Review Accelerators and Beams, 21*, 050101.
Bao, Y., Chen, J., Chen, C., et al. (2023). *Journal of Physics: Conference Series, 2462*, 012034.
Celio, M. (1986). *Physical Review Letters, 56*, 2720.
Frühwirth, R., & Strandlie, A. (2021). *Pattern Recognition, Tracking and Vertex Reconstruction in Particle Detectors*. Springer. ISBN: 978-3-030-65771-0.
McArdle, S. (2021). *PRX Quantum, 2*, 020349.
Pak, K., Park, J., Jeong, J. Y., et al. (2021). *Nuclear Engineering and Technology, 53*, 3344–3351.
Pratt, F. (2000). *Physica B: Condensed Matter, 289–290*, 710.
Shiroka, T., Renzi, R. D., Bucci, C., et al. (2006). *Physica B: Condensed Matter, 374–375*, 494. 10th International Conference on Muon Spin Rotation, Relaxation and Resonance.
Suter, A., & Wojek, B. (2012). *Physics Procedia, 30*, 69. 12th International Conference on Muon Spin Rotation, Relaxation and Resonance (µSR2011).
Tula, T., Möller, G., Quintanilla, J., et al. (2021). *Journal of Physics: Condensed Matter, 33*, 194002.
Wilkinson, J. M., & Blundell, S. J. (2020). *Physical Review Letters, 125*, 087201.

Magnetic Moment and Spin

A

A.1 Magnetic Moment and Angular Momentum

Classically, an electric current I forming a small loop creates a magnetic dipole moment

$$d\boldsymbol{\mu} = I \, \mathbf{dS} \ , \tag{A.1}$$

with units $A \, m^2$. The area enclosed by the loop is $|\mathbf{dS}|$, the vector is normal to the loop and points in the direction defined by the current, following the right-hand rule. For a finite area, the magnetic moment is $\boldsymbol{\mu} = \int d\boldsymbol{\mu} = \int I \, \mathbf{dS}$ (Fig. A.1).

To demonstrate the relationship between magnetic moment and angular momentum consider a particle with mass m, charge value e moving with momentum $p = mv$ on a circular orbit of radius r.

Taking into account that the value of the angular momentum $\mathbf{L} = \mathbf{r} \times \mathbf{p}$ is $L = mvr$, we can write for the associated magnetic moment

$$\mu = IS = \frac{qv}{2\pi r} \pi r^2 = \frac{e}{2m} mvr = \frac{e}{2m} L = \gamma_L L \ , \tag{A.2}$$

where

$$\gamma_L = g_L \frac{e}{2m} \quad \text{with} \quad g_L = 1 \ . \tag{A.3}$$

© Springer Nature Switzerland AG 2024

A. Amato, E. Morenzoni, *Introduction to Muon Spin Spectroscopy*,
Lecture Notes in Physics 961, https://doi.org/10.1007/978-3-031-44959-8

Fig. A.1 A current that loops
around a region creates a
magnetic moment $d\boldsymbol{\mu}$. The
magnetic moment vector
points in the same direction
as **dS**

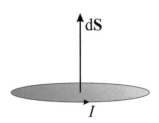

To determine the angular momentum of an electron orbiting around the nucleus in an
atom, we consider the simple Bohr model. The model introduces the quantization
of L in integer units of $\hbar$. Therefore, taking L dimensionless, i.e., replacing the
classical **L** with $\hbar$**L**

$$\mu_L = \frac{e\hbar}{2m}L = \gamma_L \hbar L \ . \tag{A.4}$$

Particles can also possess an intrinsic angular momentum, spin, which is a quantum
mechanical property. The Dirac theory predicts for a spin 1/2 particle like the
electron or the muon a g-factor equal to 2, which gives for example for the intrinsic
magnetic moment associated with the muon spin I_μ the following expression

$$\mu_\mu = \gamma_\mu \hbar I_\mu = g_\mu \frac{e\hbar}{2m_\mu} I_\mu \quad \text{where} \quad \gamma_\mu = g_\mu \frac{e}{2m_\mu} = \frac{e}{m_\mu} \ . \tag{A.5}$$

A.2   Spin Angular Momentum

A.2.1   Spin Operators

We summarize some relations. Commutation relations of the Cartesian coordinates
of the spin operator $\mathbf{S}$[1]

$$[S_x, S_y] = i S_z$$
$$[S_y, S_z] = i S_x$$
$$[S_z, S_x] = i S_y \ . \tag{A.6}$$

Square of the spin magnitude

$$S^2 = S_x^2 + S_y^2 + S_z^2 \ . \tag{A.7}$$

---

[1] Since the spin is a real quantity, the S_i, with $i = x, y, z$, are Hermitian operators, i.e., $S_i = S_i^\dagger$.

It holds

$$[S^2, S_x] = [S^2, S_y] = [S^2, S_z] = 0 \ , \tag{A.8}$$

meaning that the spin magnitude and a Cartesian component (generally taken as the z component) can be measured simultaneously.

$$S\pm = S_x \pm i S_y \tag{A.9}$$

are the raising and lowering operators with properties

$$(S_\pm)^\dagger = S_\mp$$

$$S_+ S_- = S^2 - S_z^2 = S_- S_+$$

$$[S_+, S_z] = -S_+$$

$$[S_-, S_z] = +S_- \ . \tag{A.10}$$

A.2.2   Spin 1/2 States and Pauli Matrices

As a spin $1/2$ particle the muon has two independent spin eigenstates, spin up and spin down

$$|\uparrow_\mu\rangle = |I_\mu = \frac{1}{2}, m_{I_\mu} = \frac{1}{2}\rangle$$

$$|\downarrow_\mu\rangle = |I_\mu = \frac{1}{2}, m_{I_\mu} = -\frac{1}{2}\rangle \ . \tag{A.11}$$

A general spin state can be represented as a linear combination of $|\uparrow_\mu\rangle$ and $|\downarrow_\mu\rangle$ and can also be written as a two-dimensional complex vector called a spinor

$$|\chi\rangle = c_\uparrow |\uparrow_\mu\rangle + c_\downarrow |\downarrow_\mu\rangle \tag{A.12}$$

$$|\chi\rangle = \begin{pmatrix} c_\uparrow \\ c_\downarrow \end{pmatrix} \ . \tag{A.13}$$

The corresponding Hermitian conjugate state is

$$|\chi\rangle^\dagger = \langle\chi| = \begin{pmatrix} c_\uparrow^* & c_\downarrow^* \end{pmatrix} \ , \tag{A.14}$$

and the normalization condition

$$\langle\chi|\chi\rangle = \begin{pmatrix} c_\uparrow^*, c_\downarrow^* \end{pmatrix} \begin{pmatrix} c_\uparrow \\ c_\downarrow \end{pmatrix} = |c_\uparrow|^2 + |c_\downarrow|^2 = 1 \ . \tag{A.15}$$

An arbitrary spin $1/2$ operator A acting on a spinor can be represented as a 2×2 matrix

$$A \mid \chi) = \begin{pmatrix} A_{11} & A_{12} \\ A_{21} & A_{22} \end{pmatrix} \begin{pmatrix} c_\uparrow \\ c_\downarrow \end{pmatrix}$$

$$A^\dagger = \begin{pmatrix} A_{11}^* & A_{21}^* \\ A_{12}^* & A_{22}^* \end{pmatrix} . \tag{A.16}$$

The dimensionless 2×2 Pauli matrices σ_i, $i = x, y, z$, sometimes written in vector form $\boldsymbol{\sigma} = (\sigma_x, \sigma_y, \sigma_z)$, are generally used to represent the spin $1/2$ operator

$$\sigma_x \equiv \begin{pmatrix} 0 & 1 \\ 1 & 0 \end{pmatrix}$$

$$\sigma_y \equiv \begin{pmatrix} 0 & -i \\ i & 0 \end{pmatrix}$$

$$\sigma_z \equiv \begin{pmatrix} 1 & 0 \\ 0 & -1 \end{pmatrix} \tag{A.17}$$

$$\mathbf{I}_\mu = \frac{1}{2}\boldsymbol{\sigma} . \tag{A.18}$$

They satisfy the commutation relations

$$[\sigma_x, \sigma_y] = 2i\sigma_z$$

$$[\sigma_y, \sigma_z] = 2i\sigma_x$$

$$[\sigma_z, \sigma_x] = 2i\sigma_y . \tag{A.19}$$

Furthermore

$$\sigma_x^2 = \sigma_y^2 = \sigma_z^2 = \mathbb{1}_2$$

$$\sigma_x\sigma_y = -\sigma_y\sigma_x = i\,\sigma_z$$

$$\mathrm{Tr}\left(\sigma_i\right) = 0, \quad i = x, y, z , \tag{A.20}$$

where $\mathbb{1}_2$ is the 2×2 unit matrix and Tr the trace operator $\mathrm{Tr}\,\sigma_i \equiv \sum\limits_{n=1}^{3} \sigma_{i,nn}$. Finally, we write the spinor of a spin pointing in an arbitrary direction given by the polar and azymuthal angles θ and ϕ: $\hat{\mathbf{n}} = (n_x, n_y, n_z) = (\sin\theta\cos\phi,\ \sin\theta\sin\phi,\ \cos\theta)$. The spinor must be an eigenstate of $\hat{\mathbf{n}} \cdot \boldsymbol{\sigma}$ with eigenvalue unity

$$\begin{pmatrix} n_z & n_x - in_y \\ n_x + in_y & -n_z \end{pmatrix} \begin{pmatrix} c_\uparrow \\ c_\downarrow \end{pmatrix} = \begin{pmatrix} c_\uparrow \\ c_\downarrow \end{pmatrix} . \tag{A.21}$$

From this, we find that $c_+/c_- = (n_x - in_y)/(1 - n_z) = e^{-i\phi}\cot(\theta/2)$. Noticing the normalization of the state, we obtain (up to an arbitrary phase)

$$\begin{pmatrix} c_\uparrow \\ c_\downarrow \end{pmatrix} = \begin{pmatrix} e^{-i\phi/2}\cos(\theta/2) \\ e^{+i\phi/2}\sin(\theta/2) \end{pmatrix} . \tag{A.22}$$

Magnetic Multipoles

B

We derive here the multipole solutions of the magnetostatic Maxwell equations, which determine the magnetic field between the pole shoes of the magnetic elements used to transport and focus charged particle beams. There are no electric currents in the space under consideration. Therefore, we can write

$$\nabla \times \mathbf{B} = 0$$
$$\nabla \cdot \mathbf{B} = 0 \ , \tag{B.1}$$

and relate $\mathbf{B}$ to the vector potential $\mathbf{A}$ and the scalar potential Φ

$$\mathbf{B} = \nabla \times \mathbf{A}$$
$$\mathbf{B} = -\nabla \Phi \ . \tag{B.2}$$

In a long cylinder approximation, the component of the field along the beam vanishes,[1] i.e., $B_z = 0$ and $\mathbf{A} = (0, 0, A)$. Therefore

$$B_x = \frac{\partial A}{\partial y} = -\frac{\partial \Phi}{\partial x}$$
$$B_y = -\frac{\partial A}{\partial x} = -\frac{\partial \Phi}{\partial y} \ . \tag{B.3}$$

Introducing the complex notation, one can show that a magnetic field $\mathbf{B} = (B_x, B_y, B_z)$ with B_z constant (in our case equal to zero) and B_x, B_y given by

$$B_y + i \, B_x = C_n (x + i \, y)^{n-1} \ , \tag{B.4}$$

---

[1] Ideally dipole and quadrupole elements produce only a field perpendicular to the beam direction.

© Springer Nature Switzerland AG 2024
A. Amato, E. Morenzoni, *Introduction to Muon Spin Spectroscopy*,
Lecture Notes in Physics 961, https://doi.org/10.1007/978-3-031-44959-8

where C_n is a complex constant, satisfies Eq. B.1 (Wolski 2011; Chao & Tigner 1999).[2]

Fields of the form of Eq. B.4 are known as multipole fields, with the index n indicating the order of the multipole. For a dipole field $n = 1$, for a quadrupole field $n = 2$, for a sextupole field $n = 3$ and so on.

Using the principle of superposition, a general magnetic field can be represented by a sum of multipole fields

$$B_y + i\, B_x = \sum_{n=1}^{\infty} C_n (x + i\, y)^{n-1} \; . \tag{B.5}$$

The coefficients C_n characterize the strength and orientation of each multipole component. In polar coordinates, $x = r \cos\theta$ and $y = r \sin\theta$,

$$B_y + i\, B_x = \sum_{n=1}^{\infty} C_n\, r^{n-1} e^{i(n-1)\theta} \; . \tag{B.6}$$

The strength of the field in a "pure" multipole[3] of the order n varies as r^{n-1} with distance from the magnetic axis. Expressing the field in polar components, $B_x = B_r \cos\theta - B_\theta \sin\theta$ and $B_y = B_r \sin\theta + B_\theta \cos\theta$ we can write

$$B_\theta + i\, B_r = \sum_{n=1}^{\infty} C_n\, r^{n-1} e^{in\theta} \; . \tag{B.7}$$

Therefore, for a pure multipole of order n, rotation of the magnet by π/n around the z-axis simply changes the sign of the field as shown in Fig. 1.32 ($n = 1$) and Fig. 1.34 ($n = 2$).

The constants C_n have units that depend on the order of the multipole. The unit of C_1 (dipole) is Tesla; for a quadrupole, the unit of C_2 is $\mathrm{T\,m^{-1}}$; for a sextupole, the unit of C_3 is $\mathrm{T\,m^{-2}}$, and so on. By introducing a reference field B_0 and a reference radius R_0 the multipole components can be expressed in dimensionless units. The parameters B_0 and R_0 can be chosen arbitrarily, but must be specified to correctly interpret the coefficients C_n. Generally, B_0 is taken to be the field at the pole center and R_0 the radial distance of the pole center.

By writing the constant C_n as

$$C_n = \frac{B_0}{R_0^{n-1}} (b_n + i\, a_n) \; , \tag{B.8}$$

---

[2] This can be verified by applying the differential operator $\partial/\partial x + i\partial/\partial y$ to each side of the Eq. B.4.

[3] A pure multipole represents the case where only one C_n coefficient is different from zero and all other are equal to zero.

Eq. B.5 can be expressed as

$$B_y + i \, B_x = B_0 \sum_{n=1}^{\infty} (b_n + i \, a_n) \left(\frac{x + i \, y}{R_0} \right)^{n-1} \, . \qquad (B.9)$$

By convention, a pure multipole is called a normal multipole if $b_n \neq 0$ and $a_n = 0$, whereas it is called a skew multipole if $a_n \neq 0$ and $b_n = 0$. These coefficient are related to the derivatives of the field

$$\frac{\partial^{n-1} B_y}{\partial x^{n-1}} = (n-1)! \frac{B_0}{R_0^{n-1}} b_n \quad \text{and}$$

$$\frac{\partial^{n-1} B_x}{\partial y^{n-1}} = (n-1)! \frac{B_0}{R_0^{n-1}} a_n \, . \qquad (B.10)$$

Figure B.1 shows examples of normal and skew multipole fields.

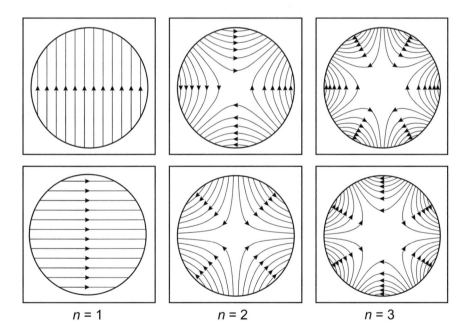

$n = 1$ $\qquad\qquad$ $n = 2$ $\qquad\qquad$ $n = 3$

Fig. B.1 Pure multipole fields. Left: dipole. Middle: quadrupole. Right: sextupole. Fields on the top row are normal (b_n positive, $a_n = 0$); those on the bottom row are skew (a_n positive, $b_n = 0$). Modified from Wolski (2011). Attribution 4.0 International license

Derivation of the TF Abragam Formula

We detail here the derivation of the so-called Abragam formula in transverse field, Eq. 4.70 (Anderson 1954; Abragam 1961). With $\omega_\mu = \gamma_\mu B_{\text{ext}}$, introducing the phase variable $\Phi(t) = \int_0^t \gamma_\mu \delta B_{\mu,z}(t')dt'$, Eq. 4.67 becomes (where $\Re$ denotes the real part)[1]

$$P_{x,\,\text{dyn}}^{\text{TF-G}}(t) = \Re\left\{ e^{i\omega_\mu t} \langle e^{i\Phi(t)} \rangle \right\} \quad . \tag{C.1}$$

Since the random function $\delta B(t')$ is Gaussian, the phase $\Phi(t)$ is also Gaussian distributed, so that the statistical average can be calculated

$$\langle e^{i\Phi(t)} \rangle = \frac{1}{\sqrt{2\pi \langle \Phi^2 \rangle}} \int_{-\infty}^{\infty} e^{-\frac{\Phi^2}{2\langle \Phi^2 \rangle}} e^{i\Phi} d\Phi = e^{-\frac{\langle \Phi^2 \rangle}{2}} \quad . \tag{C.2}$$

This reduces the calculation to computing $\langle \Phi^2 \rangle$, which can be expressed in terms of the autocorrelation function of the fluctuating field $\langle \delta B(t') \delta B(t'') \rangle$

$$\langle \Phi^2 \rangle = \left\langle \left[\int_0^t \gamma_\mu \, \delta B(t') \, dt' \right]^2 \right\rangle = \left\langle \int_0^t \int_0^t \gamma_\mu \, \delta B(t') \, \gamma_\mu \, \delta B(t'') \, dt' dt'' \right\rangle =$$

$$= \gamma_\mu^2 \int_0^t \int_0^t \langle \delta B(t') \, \delta B(t'') \rangle \, dt' dt'' \quad . \tag{C.3}$$

Taking into account that the fluctuating process we are considering is i) random and stationary ii) the autocorrelation function is even and iii) depends only on the time difference $t' - t''$ we have $\langle \delta B(t') \, \delta B(t'') \rangle = \langle \delta B^2 \rangle \, C(t' - t'')$.

---

[1] We temporarily drop the index μ, z in $\delta B_{\mu,z}$.

© Springer Nature Switzerland AG 2024
A. Amato, E. Morenzoni, *Introduction to Muon Spin Spectroscopy*,
Lecture Notes in Physics 961, https://doi.org/10.1007/978-3-031-44959-8

$\langle \delta B^2 \rangle$ is independent of time and has the same value as that of the static field distribution. This corresponds to the assumption that the instantaneous (i.e., snapshot) local field distribution is the static one (as in the strong collision approximation). Equation C.3 is of the form $\int_0^t \int_0^t C(t' - t'') dt' dt''$. One can show that

$$\int_0^t \int_0^t C(t' - t'') dt' dt'' = \int_0^t \int_0^{t'} C(t' - t'') dt' dt'' + \int_0^t \int_{t'}^t C(t' - t'') dt' dt'' =$$

$$= 2 \int_0^t \int_0^{t'} C(t' - t'') dt' dt'' = 2 \int_0^t (t - \tau) C(\tau) d\tau .$$

$$(C.4)$$

The first part of C.4 can be proved by appropriate variable substitutions. The second part by introducing the auxiliary function B such that $\int_0^{t'} C(t' - t'') dt'' = B(t') = \int_0^{t'} C(\tau) d\tau$ (by substituting here $t' - t'' = \tau$).

The identity $\int_0^t B(t') dt' = \int_0^t (t - \tau) C(\tau) d\tau$, can be verified by integration by parts

$$\int_0^t \int_0^{t'} C(t' - t'') dt' dt'' = \int_0^t B(t') dt' = t' B(t') \Big|_0^t - \int_0^t t' B'(t') dt' =$$

$$= t B(t) - \int_0^t t' B'(t') dt'$$

$$= t \int_0^t C(\tau) d\tau - \int_0^t \tau C(\tau) d\tau$$

$$= \int_0^t (t - \tau) C(\tau) d\tau .$$

For a Gaussian-Markovian process, the field autocorrelation function decays exponentially with correlation time τ_c, see Eq. 4.69 (Doob 1942), so that with the Gaussian relaxation rate $\sigma^2 = \gamma_\mu^2 \langle \delta B_{\mu,z}^2 \rangle$

$$\gamma_\mu^2 \langle \delta B_{\mu,z}(0) \delta B_{\mu,z}(\tau) \rangle = \sigma^2 e^{-\frac{\tau}{\tau_c}} .$$

$$(C.5)$$

Inserting this expression in the above equations, performing the integration over τ, and with $\nu = 1/\tau_c$ we obtain the Abragam equation in a transverse field, Eq. 4.70

$$P_{z,\,\text{dyn}}^{\text{TF-G}}(t, \sigma, \nu, \omega_\mu) = \exp\left[-\frac{\sigma^2}{\nu^2} \left[\exp(-\nu t) - 1 + \nu t \right] \right] \cos(\gamma_\mu B_{\text{ext}} t) .$$

Demagnetizing Field

D

Consider a ferromagnetic system with a magnetization $\mathbf{M}$. From

$$\mathbf{B} = \mu_0 \left(\mathbf{M} + \mathbf{H} \right) \tag{D.1}$$

and

$$\nabla \cdot \mathbf{B} = 0 \ , \tag{D.2}$$

when the magnetization encounters the surface of the sample, we have[1]

$$\nabla \cdot \mathbf{H} = -\nabla \cdot \mathbf{M} \ . \tag{D.3}$$

That is, unlike $\mathbf{B}$, $\mathbf{H}$ is not divergence free and behaves as if it originated from an effective magnetic charge density at the surface of the sample

$$\rho_{\mathbf{M}} = -\nabla \cdot \mathbf{M} \ . \tag{D.4}$$

If there are no free currents, the field $\mathbf{H}$ is related to a magnetic potential Φ

$$\mathbf{H} = -\nabla \Phi \ . \tag{D.5}$$

The above Equations give a Poisson Equation for the magnetic potential

$$\nabla^2 \Phi = -\rho_{\mathbf{M}} \ . \tag{D.6}$$

---

[1] Outside the sample $\nabla \cdot \mathbf{H} = 0$.

© Springer Nature Switzerland AG 2024
A. Amato, E. Morenzoni, *Introduction to Muon Spin Spectroscopy*,
Lecture Notes in Physics 961, https://doi.org/10.1007/978-3-031-44959-8

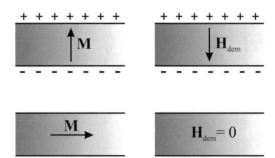

Fig. D.1 Ferromagnetic infinite flat plate (cross section). The upper row shows the situation when the magnetization is perpendicular to the plane of the plate. There is a positive (negative) divergence of **M** at the top (bottom) surface. The field **H**$_{\text{dem}}$ has opposite divergence and direction. This can be seen as positive and negative poles on the surfaces as shown. If the magnetization is along the plane of the plate (bottom row), the only divergence is at the ends (supposedly at infinite distance). There is no divergence for the magnetization and no demagnetization field inside the sample

Solving this equation one obtains the so-called demagnetizing field inside the sample, $\mathbf{H}_{\text{dem}} = \mathbf{B}_{\text{dem}}/\mu_0$, Fig. D.1. In general, the demagnetizing field depends strongly on the shape of the sample and is a function of the magnetization inside the sample. For ellipsoidal bodies the field is uniform and we can write

$$\mathbf{B}_{\text{dem}} = -\tilde{N}\,\mu_0\mathbf{M}\ ,\tag{D.7}$$

where $\tilde{N}$ is the demagnetizing tensor. If **M** is defined along the principal direction of the ellipsoid, the problem can be diagonalized and we obtain

$$\tilde{N} = \begin{pmatrix} N_x & 0 & 0 \\ 0 & N_y & 0 \\ 0 & 0 & N_z \end{pmatrix}\tag{D.8}$$

with unit trace

$$\text{Tr}(\tilde{N}) = N_x + N_y + N_z = 1\ .\tag{D.9}$$

In an ellipsoid placed in a uniform external magnetic field applied along one of the principal directions, the internal fields $\mathbf{B}_{\text{int}} = \mathbf{B}_{\text{ext}} + \mathbf{B}_{\text{dem}}$ and $\mathbf{H}_{\text{int}}$ are parallel to the applied field.

Ideal sample shapes have simple demagnetization factors. For a sphere, $N_x = N_y = N_z = 1/3$. For an infinite disk, if the field is applied perpendicular to the surface $N_\perp = 1$, if parallel $N_\| = 0$. For an infinite long cylinder $N_\perp = 1/2$ and $N_\| = 0$. For demagnetizing factors of general ellipsoids see Osborn (1945), Aharoni (1996); for samples of various shapes see Prozorov and Kogan (2018) and references therein, where the case of diamagnetic samples relevant for superconductors is also discussed.

Units of Hyperfine Constants

Depending on the context and the unit system used, different definitions and units of the hyperfine constant(s) can be found in the literature. In Eq. 5.15, we have introduced a (contact) hyperfine constant as the coupling strength of the local spin-spin interaction with energy dimension $A[\text{eV}]$. This quantity (and its tensor generalization) is encountered in muonium spectroscopy, see, e.g., Eqs. 7.5 and 7.7. Hyperfine constants are also met in the context of Knight-shift measurements. The contact Knight shift $K_{\text{F.cont}}$ in a solid can be expressed in terms of $A[\text{eV}]$, since the electron spin density at the muon site and thus the contact field are proportional to $A[\text{eV}]$, see Eq. 5.74

$$K_{\text{F.cont}} = \frac{A[\text{eV}]}{\mu_0 \hbar^2 \gamma_\mu \gamma_e n_e} \chi_{\text{cond}} \quad . \tag{E.1}$$

Also the Knight-shift due to the dipolar and the RKKY-enhanced contact interactions are proportional to the susceptibilities, see Eq. 5.85 or 5.86, so that, generically, one may write

$$K = A[1] \chi \quad . \tag{E.2}$$

These proportionality constants are also called hyperfine coupling constants. For dimensionless susceptibilities $\chi = M/H$, as in the SI and cgs systems,[1,2] they are

---

[1] In the SI system, the magnetization has the unit $J\,T^{-1}\,m^{-3}$, which reduces to $A\,m^{-1}$, as for the field H. In cgs units, the magnetization is in $erg\,G^{-1}\,cm^{-3}$, which reduces to G, as for the applied field in vacuum ($H = 1$ Oe corresponds to $B = 1$ G) and χ is also dimensionless.

[2] Remember that in the basic SI and cgs units: $1\,J = 1\,kg\,m^2\,s^{-2}$, $1\,T = 1\,kg\,s^{-2}\,A^{-1}$, $1\,erg = 1\,g\,cm^2\,s^{-2}$, $1\,G = 1\,g^{1/2}\,cm^{-1/2}\,s^{-1}$.

© Springer Nature Switzerland AG 2024
A. Amato, E. Morenzoni, *Introduction to Muon Spin Spectroscopy*,
Lecture Notes in Physics 961, https://doi.org/10.1007/978-3-031-44959-8

also dimensionless $A[1]$.[3] The dimensionless susceptibility is often named volume susceptibility.

In addition, other units can be found, see examples in Sect. 5.6.3 where the dipolar and the contact RKKY coupling constants are expressed in kG/μ_B.

This unit is obtained by writing the proportionality between K and χ in the cgs-emu system as

$$K = \frac{A[kG/\mu_B]}{N_A \, C_{\mu_B}} \chi \quad , \tag{E.3}$$

where N_A is Avogadro number and C_{μ_B} is a constant with the value 9.274010×10^{-21} erg $G^{-1} \mu_B^{-1}$ (C_{μ_B} is used to transform the unit of the magnetic moment from μ_B to erg G^{-1}). The susceptibility is expressed as molar susceptibility, with units erg G^{-1} mol^{-1} G^{-1}, which is equivalent to cm^3 mol^{-1}. From this, it follows that the hyperfine coupling constant as defined in Eq. E.3 has the dimension of G/μ_B.

The molar susceptibility, instead of erg G^{-1} mol^{-1} G^{-1} (or erg G^{-1} mol^{-1} Oe^{-1}), can be expressed using "emu";[4] in this case the unit is emu mol^{-1} Oe^{-1}.[5] Units in kG/μ_B are also obtained if, in Eq. E.3, N_A is replaced by the density of magnetic ions (or, for the contact term, conduction electrons n_e) and the susceptibility is in emu cm^{-3} Oe^{-1}.

In SI units A can be expressed in T/μ_B, which is related to the dimensionless $A[1]$ of Eq. E.2 by

$$A[1] = \frac{A[T/\mu_B] V_m}{\mu_0 \, C_{\mu_B}} \quad , \tag{E.4}$$

where V_m is the volume per magnetic ion (or conduction electron, n_e^{-3}) and here $C_{\mu_B} = 9.274010 \times 10^{-24}$ J T^{-1} μ_B^{-1}.

To change from dimensionless to T/μ_B, do the following

$$\frac{A[1] \, \mu_0 \left[\frac{N}{A^2}\right] C_{\mu_B} \left[\frac{J}{T \, \mu_B}\right]}{V_m \left[m^3\right]} = A \left[\frac{T}{\mu_B}\right] \quad . \tag{E.5}$$

Another unit encountered is mol/emu (here emu = cm^3).

---

[3] The symbol "1" for the unit indicates a dimensionless quantity.

[4] The emu, a short for electromagnetic unit, is not a unit in the proper sense. Confusingly, emu can be used to express a magnetic moment, in which case 1 emu corresponds to 1 erg G^{-1} but it can also take the unit of a volume, and in this case 1 emu is equal to 1 cm^3. It would be better to avoid its use, but unfortunately many susceptibility data are given in units making use of emu.

[5] Sometimes the units of the molar susceptibility are given as emu mol^{-1}, see Figs. 5.43 and 5.44. In this case 1 emu stays for 1 cm^3. Similarly, the dimensionless volume susceptibility is often given in units of emu cm^{-3}. Here again, 1 emu stays for 1 cm^3.

To change from mol/emu to dimensionless

$$\frac{A\left[\frac{\text{mol}}{\text{emu}}\right]\ V_m\left[\text{cm}^3\right]\ N_A\left[\frac{1}{\text{mol}}\right]}{4\pi} = A[1] \ , \tag{E.6}$$

and from mol/emu to kG/μ_B

$$\frac{A\left[\frac{\text{mol}}{\text{emu}}\right]\ C_{\mu_B}\left[\frac{\text{erg}}{G\,\mu_B}\right]\ N_A\left[\frac{1}{\text{mol}}\right]}{1000\left[\frac{G}{kG}\right]} = A\left[\frac{kG}{\mu_B}\right] \ . \tag{E.7}$$

Density Matrix

F

The density operator or the density matrix[1] provide a simple description of a statistical mixture of quantum mechanical states[2] as we encountered when calculating the time evolution of the muon spin polarization in muonium, Sect. 7.3.

F.1 Pure Quantum Mechanical State

We first consider a pure state $|\psi(t)\rangle$, expressed by an orthonormal basis of states $|\chi_i\rangle$

$$|\psi(t)\rangle = \sum_i a_i(t)\,|\chi_i\rangle \ .$$

The time evolution of the expectation value of an observable expressed by an operator O can be written with the coefficients $a_i(t)$ as

$$\langle O\rangle(t) = \langle\psi(t)\,|\,O\,|\,\psi(t)\rangle$$

$$= \sum_{n,p} a_n^*(t)\,a_p(t)\,O_{np} \ , \tag{F.1}$$

where

$$O_{np} = \langle\chi_n\,|\,O\,|\,\chi_p\rangle \tag{F.2}$$

---

[1] We use the same symbol ρ for both quantities. It should be clear from the context, which is which.
[2] For an in-depth treatment see, for instance, Blum (2012). For quantum mechanics in general see Messiah (1965), Cohen-Tannoudji et al. (2005), Schwabl (2007, 2005).

© Springer Nature Switzerland AG 2024
A. Amato, E. Morenzoni, *Introduction to Muon Spin Spectroscopy*,
Lecture Notes in Physics 961, https://doi.org/10.1007/978-3-031-44959-8

are the matrix elements of the operator O in the $|\chi_i\rangle$ basis. The time evolution of the pure state $|\psi(t)\rangle$ obeys the Schrödinger equation with the Hamiltonian $\mathcal{H}(t)$

$$i\hbar\frac{d}{dt}|\psi(t)\rangle = \mathcal{H}(t)|\psi(t)\rangle \ . \tag{F.3}$$

Differentiating F.1 with respect to time and making use of the Schrödinger equation, we see that the time evolution of the expectation value obeys the following equation[3]

$$i\hbar\frac{d}{dt}\langle O\rangle(t) = \langle[O,\mathcal{H}(t)]\rangle \ . \tag{F.4}$$

From F.1 we note that the time evolution of the expectation value is contained in the quadratic terms $a_n^*(t)a_p(t)$, which are the expectation values of the time dependent operator $|\psi(t)\rangle\langle\psi(t)|$ since

$$\langle\chi_n|\psi(t)\rangle\langle\psi(t)|\chi_p\rangle = a_n^*(t)a_p(t) \ . \tag{F.5}$$

The operator $|\psi(t)\rangle\langle\psi(t)|$ is called density operator $\rho(t)$. In the $|\chi_i\rangle$ basis it is represented by a matrix called density matrix with elements

$$\rho_{pn}(t) = \langle\chi_p|\rho(t)|\chi_n\rangle = a_n^*(t)a_p(t) \ . \tag{F.6}$$

All the physical properties determined by the state $|\psi(t)\rangle$ can be expressed using the density operator. The state normalization can be written as

$$\sum_n |a_n(t)|^2 = 1 = \sum_n \rho_{nn}(t) = \mathrm{Tr}\,\rho(t) \ , \tag{F.7}$$

the time dependence of the expectation value of an observable as

$$\langle O\rangle(t) = \langle\psi(t)|O|\psi(t)\rangle = \sum_{n,p} a_n^*(t)\,a_p(t)\,O_{np} =$$

$$= \sum_{n,p}\rho_{pn}(t)O_{np} = \sum_p\langle\chi_p|\rho(t)O|\chi_p\rangle$$

$$= \mathrm{Tr}\left(\rho(t)O\right) = \mathrm{Tr}\left(O\rho(t)\right) \ , \tag{F.8}$$

---

[3] We consider here a time independent operator.

and finally, using the Schrödinger equation, the time evolution of the density operator can be expressed as[4]

$$i\hbar \frac{d}{dt}\rho(t) = \left[\mathcal{H}(t), \rho(t) \right] . \tag{F.9}$$

Useful properties of the density operator are

$$\rho^{\dagger}(t) = \rho(t) ,$$

i.e., the operator is Hermitian and (for pure states)[5]

$$\rho^2(t) = \rho(t) \tag{F.10}$$

$$\mathrm{Tr}\, \rho^2(t) = 1 .$$

F.2   Mixed Quantum Mechanical State

If the system is a mixture of pure states $|\psi_k\rangle$ with probabilities p_k, then the density matrix at a given time t is simply the weighted sum of the corresponding density matrices

$$\rho(t) = \sum_k p_k |\psi_k(t)\rangle\langle\psi_k| = \sum_k p_k \rho_k(t) . \tag{F.11}$$

Also in a mixed state F.8 and F.9 hold

$$i\hbar \frac{d}{dt}\rho_k(t) = \left[\mathcal{H}(t), \rho_k(t) \right]$$

$$i\hbar \frac{d}{dt}\rho(t) = \left[\mathcal{H}(t), \rho(t) \right]$$

$$\langle O \rangle(t) = \mathrm{Tr}\left(\rho(t)O \right) = \mathrm{Tr}\left(O\rho(t) \right) . \tag{F.12}$$

The expression for evaluating expectation values is the same for pure and mixed states.

---

[4] If the Hamiltonian is time independent $\mathcal{H}(t) = \mathcal{H}_0$ the equation has the solution $\rho(t) = \exp(-i\mathcal{H}_0 t/\hbar)\,\rho(0)\,\exp(i\mathcal{H}_0 t/\hbar)$.
[5] For mixed states $\mathrm{Tr}\,\rho^2(t) < 1$.

F.3   Time Evolution of an Operator

So far, we have used the Schrödinger representation, where the time evolution of the system is contained in the eigenstates. In the Heisenberg representation the time evolution is contained in the operators. It is easy to modify the above expressions in this case. For example, the expectation value at time t of an operator O in the mixed state represented by ρ, Eq. F.8, becomes

$$\langle O \rangle(t) = \mathrm{Tr}\left(\rho(0)O(t)\right) = \mathrm{Tr}\left(O(t)\rho(0)\right) . \tag{F.13}$$

F.4   Density Matrix of a Spin 1/2 Particle

For a spin $1/2$ particle there are two possible states. Therefore, the density matrix is a Hermitian 2×2 matrix. Any generic 2×2 matrix M can be expressed as a linear combination of the unit matrix $\mathbb{1}_2$ and the Pauli matrices $\boldsymbol{\sigma} = \sigma_x, \sigma_y, \sigma_z$, Eq. A.17.

$$
\begin{aligned}
M &= \begin{pmatrix} m_{11} & m_{12} \\ m_{21} & m_{22} \end{pmatrix} \\
&= \frac{m_{11} + m_{22}}{2}\mathbb{1}_2 + \frac{m_{12} + m_{21}}{2}\sigma_x + i\,\frac{m_{12} - m_{21}}{2}\sigma_y + \frac{m_{11} - m_{22}}{2}\sigma_z ,
\end{aligned}
\tag{F.14}
$$

which can be written in the form

$$M = a_0 \mathbb{1}_2 + \mathbf{a} \cdot \boldsymbol{\sigma} . \tag{F.15}$$

In the general case, the coefficients a_0, a_x, a_y and a_z are complex numbers. If M is Hermitian then the a_0, a_x, a_y and a_z coefficients must be real.

Using the properties of the Pauli matrices, Eq. A.20, it is easy to show that

$$a_0 = \frac{1}{2}\,\mathrm{Tr}\left(M\right) \text{ and}$$

$$\mathbf{a} = \frac{1}{2}\,\mathrm{Tr}\left(M\boldsymbol{\sigma}\right) . \tag{F.16}$$

The density matrix ρ of a spin $1/2$ particle can be written in the form of Eq. F.15 and expressed in terms of its polarization. Taking into account the properties of the trace, Eq. F.16, we obtain in this case [6]

$$a_0 = \frac{1}{2}$$

$$\mathbf{a} = \frac{1}{2} \operatorname{Tr}(\rho(0)\boldsymbol{\sigma}) \text{ and}$$

$$\rho = \frac{1}{2}\left(\mathbb{1} + \mathbf{P} \cdot \boldsymbol{\sigma}\right) \ . \tag{F.17}$$

The time dependence of the polarization vector $\mathbf{P}(t)$ has components defined by the expectation value of the corresponding Pauli matrix and can be expressed by a trace involving the density matrix of the system

$$\mathbf{P}(t) = \langle \boldsymbol{\sigma} \rangle(t) = \operatorname{Tr}\left(\rho(0)\boldsymbol{\sigma}(t)\right) \ . \tag{F.18}$$

F.5   Density Matrix of Muonium

The density matrix of muonium at $t = 0$ is the tensor product of the density matrices of muon and electron.

For the muon [7]

$$\rho_\mu(0) = \frac{1}{2}\left[\mathbb{1}_\mu + \mathbf{P}_\mu(0) \cdot \boldsymbol{\sigma}_\mu\right]$$

$$= \frac{1}{2}\mathbb{1}_\mu + \frac{1}{2}P_{\mu,x}(0)\sigma_{\mu,x} + \frac{1}{2}P_{\mu,y}(0)\sigma_{\mu,x} + \frac{1}{2}P_{\mu,z}(0)\sigma_{\mu,z} \ ,$$

and equivalently for the electron

$$\rho_e(0) = \frac{1}{2}\left[\mathbb{1}_e + \mathbf{P}_e(0) \cdot \boldsymbol{\sigma}_e\right] \ . \tag{F.19}$$

The muonium density matrix is then

$$\rho_{\mathrm{Mu}}(0) = \rho_\mu(0) \otimes \rho_e(0) \ , \tag{F.20}$$

---

[6] We use here and in the following the Heisenberg representation.
[7] We use the indices μ and e to distinguish the vectors and matrices associated with the muon and electron, respectively.

where $\rho_{\mathrm{Mu}}(0)$ is a 4×4 matrix and $\rho_\mu(0)$, $\rho_e(0)$ are 2×2 matrices. Taking into account that $\mathbf{P}_e(0) = 0$ we obtain

$$\rho_{\mathrm{Mu}}(0) = \frac{1}{4}\left[\mathbb{1}_\mu \otimes \mathbb{1}_e + \mathbf{P}_\mu(0) \cdot \boldsymbol{\sigma}_\mu \otimes \mathbb{1}_e\right] \quad, \qquad (\text{F.21})$$

which is usually simply written as[8]

$$\rho_{\mathrm{Mu}}(0) = \rho_\mu(0)\rho_e(0) = \frac{1}{4}\left[\mathbb{1} + \mathbf{P}_\mu(0) \cdot \boldsymbol{\sigma}_\mu\right] \quad. \qquad (\text{F.22})$$

---

[8] The tensor product, which appears in the description of complex systems, is often implicitly assumed. For example, consider two noninteracting particles, one described by n_1 basis vectors $|\chi_1\rangle$ (forming the vector space V_1) and the other described by n_2 basis vectors $|\chi_2\rangle$ (forming the vector space V_2) and the operators O_1 and O_2 each acting on its vector space. A combined operator (for instance the sum $O = O_1 + O_2$) acts on the $n_1 n_2$ dimensional vector space $V_1 \otimes V_2$, which is defined by the tensor product of the individual spaces. The combined operator should be written as $O = O_1 \otimes \mathbb{1}_{n2} + \mathbb{1}_{n1} \otimes O_2$. We have already encountered this situation, for example when describing the muonium spin states in Eq. 7.18. The state $|\uparrow_\mu \uparrow_e\rangle$ should be written as $|\uparrow_\mu\rangle \otimes |\uparrow_e\rangle$, the total spin operator $\mathbf{F} = \mathbf{I}_\mu \otimes \mathbb{1}_e + \mathbb{1}_\mu \otimes \mathbf{S}$.

The tensor products involving matrices, as the density matrices, are performed according to the rules of the matrix direct product or Kronecker product

$$\text{If } A = \begin{pmatrix} a_{11} & a_{12} \\ a_{21} & a_{22} \end{pmatrix} \text{ and } B = \begin{pmatrix} b_{11} & b_{12} \\ b_{21} & b_{22} \end{pmatrix},$$

$$\text{then } A \otimes B = \begin{pmatrix} a_{11} B & a_{12} B \\ a_{21} B & a_{22} B \end{pmatrix} = \begin{pmatrix} a_{11}b_{11} & a_{11}b_{12} & a_{12}b_{11} & a_{12}b_{12} \\ a_{11}b_{21} & a_{11}b_{22} & a_{12}b_{21} & a_{12}b_{22} \\ a_{21}b_{11} & a_{21}b_{12} & a_{22}b_{11} & a_{22}b_{12} \\ a_{21}b_{21} & a_{21}b_{22} & a_{22}b_{21} & a_{22}b_{22} \end{pmatrix} \quad.$$

Relativistic Concepts

G

G.1 Useful Relations of Relativistic Quantum Mechanics

We first introduce the relativistic notation used in Chap. 10 and discuss the Dirac equation, which is a relativistic wave equation describing the behavior of a spin 1/2 particle (and antiparticle) with mass m.

A particle with momentum $\mathbf{p}$ and energy E is described by the relativistic four-dimensional contravariant vector[1]

$$p^{\alpha} = \begin{pmatrix} p^0 \\ p^1 \\ p^2 \\ p^3 \end{pmatrix} = \begin{pmatrix} \frac{E}{c} \\ \mathbf{p} \end{pmatrix} . \tag{G.1}$$

With the relativistic metric tensor[2]

$$g^{\alpha\beta} = g_{\alpha\beta} = \begin{pmatrix} 1 & 0 & 0 & 0 \\ 0 & -1 & 0 & 0 \\ 0 & 0 & -1 & 0 \\ 0 & 0 & 0 & -1 \end{pmatrix} , \tag{G.2}$$

---

[1] Note that, unlike three-dimensional vectors, four-dimensional vectors are not in bold, but are indicated by an index.

[2] With this tensor one can pull upper indices down and vice versa.

© Springer Nature Switzerland AG 2024
A. Amato, E. Morenzoni, *Introduction to Muon Spin Spectroscopy*,
Lecture Notes in Physics 961, https://doi.org/10.1007/978-3-031-44959-8

we have for the covariant vector (lower index)[3]

$$p_\alpha = \sum_\beta g_{\alpha\beta} \, p^\beta \equiv g_{\alpha\beta} \, p^\beta$$

$$= \begin{pmatrix} \dfrac{E}{c} \\ -\mathbf{p} \end{pmatrix} . \tag{G.3}$$

In the following, to simplify the notation, a sum over pairs of upper and lower indices is implied as in G.3. The 4-vector product $p_\beta p^\beta = E^2/c^2 - \mathbf{p}^2 = mc^2$ is the relativistic energy-momentum relation of a particle with rest mass m.

The space-time vectors are

$$x^\alpha = \begin{pmatrix} ct \\ \mathbf{x} \end{pmatrix} \quad \text{and} \quad x_\alpha = \begin{pmatrix} ct \\ -\mathbf{x} \end{pmatrix} \tag{G.4}$$

and $x_\beta x^\beta = c^2 t^2 - \mathbf{x}^2$ is the Lorentz invariant length of the spacetime 4-vector, i.e., the same in every inertial system.[4]

The quantum mechanical 4-momentum operator can be obtained by formally substituting

$$p^\alpha \rightarrow i\hbar \frac{\partial}{\partial x_\alpha} = i\hbar \, \partial^\alpha = i\hbar \begin{pmatrix} \dfrac{1}{c} \dfrac{\partial}{\partial t} \\ -\nabla \end{pmatrix} \quad \text{and}$$

$$\tag{G.5}$$

$$p_\alpha \rightarrow i\hbar \frac{\partial}{\partial x^\alpha} = i\hbar \, \partial_\alpha = i\hbar \begin{pmatrix} \dfrac{1}{c} \dfrac{\partial}{\partial t} \\ \nabla \end{pmatrix} .$$

G.2   Dirac Equation

Note that, as in Chap. 10, for the rest of this Appendix we will use natural units where $c = 1$, $\hbar = 1$ and $\varepsilon_0 = 1$, as is common practice in particle physics.

The Dirac equation for a particle of mass m is, in the absence of an external field (Dirac 1928),[5]

$$i\gamma^\alpha \frac{\partial \Psi}{\partial x^\alpha} - m\Psi = (i\gamma^\alpha \partial_\alpha - m)\Psi = 0 \ ,$$

---

[3] Lowering the indices changes the sign of the "spatial" component of the 4-vector.

[4] This is an example of a so-called Lorentz covariant scalar. In general, a physical quantity is said to be Lorentz covariant if it transforms under a given representation of the Lorentz group. Lorentz covariant scalars are Lorentz invariant.

[5] A unit matrix $\mathbb{1}_4$ is implied with the mass term.

where γ^α is a so-called γ-matrix, see below. Formally, Eq. G.6 corresponds to the relativistic energy-momentum relation for the corresponding operators.[6] Ψ (Dirac spinor or bi-spinor or simply spinor) is a vector of four complex functions interpreted as a superposition of spin up particle, spin down particle, and spin up antiparticle, spin down antiparticle.[7] This is in contrast to the Schrödinger equation involving only a single complex wave function.

$$\Psi(x^\alpha) = \begin{pmatrix} \Psi_0(x^\alpha) \\ \Psi_1(x^\alpha) \\ \Psi_2(x^\alpha) \\ \Psi_3(x^\alpha) \end{pmatrix} . \tag{G.6}$$

The γ^α terms, $\alpha = 0, 1, 2, 3$, are the 4×4 γ-matrices, see Sect. G.2.1 for their properties.[8] It is useful to introduce the adjoint spinor

$$\overline{\Psi}(x^\alpha) = \Psi^\dagger(x^\alpha)\gamma^0 , \tag{G.7}$$

where $\Psi^\dagger(x^\alpha) = \begin{pmatrix} \Psi_0^* & \Psi_1^* & \Psi_2^* & \Psi_3^* \end{pmatrix}$ is the Hermitian adjoint (conjugate transpose) of $\Psi(x^\alpha)$ and $\overline{\Psi}(x^\alpha) = \begin{pmatrix} \Psi_0^* & \Psi_1^* & -\Psi_2^* & -\Psi_3^* \end{pmatrix}$.

The adjoint spinor satisfies the equation

$$\overline{\Psi}(i\gamma^\alpha\partial_\alpha + m) = 0 \tag{G.8}$$

with ∂_α acting left.

G.2.1   Properties of the γ-Matrices

The γ-matrices allow to write the Dirac equation and related expressions in a compact form. There are several representations of them. We use here the Dirac representation, where the γ-matrices, written in terms of the Pauli matrices σ_i, $i = 1, 2, 3$, Eq. A.17, and the 2×2 zero and unit matrices $\mathbf{0}_2$ and $\mathbb{1}_2$ are

$$\gamma^0 = \begin{pmatrix} \mathbb{1}_2 & \mathbf{0}_2 \\ \mathbf{0}_2 & -\mathbb{1}_2 \end{pmatrix} \qquad\qquad \gamma^i = \begin{pmatrix} \mathbf{0}_2 & \sigma_i \\ -\sigma_i & \mathbf{0}_2 \end{pmatrix} , \tag{G.9}$$

$$\mathbb{1}_2 = \begin{pmatrix} 1 & 0 \\ 0 & 1 \end{pmatrix} \qquad \text{and} \quad \mathbf{0}_2 = \begin{pmatrix} 0 & 0 \\ 0 & 0 \end{pmatrix}$$

---

[6] Note that $p^\delta p_\delta - m^2 = (\gamma^\alpha p_\alpha - m)(\gamma^\beta p_\beta + m) = 0$; the Dirac equation corresponds to choosing the first term.

[7] The Dirac spinor is not a relativistic 4-vector and does not transform as such.

[8] The γ-matrices are not 4-vectors and are invariant under a Lorentz transformation, see footnote 4.

or with the direct product[9]

$$\gamma^0 \doteq \sigma_3 \otimes \mathbb{1}_2$$
$$\gamma^i = i \sigma_2 \otimes \sigma_i \ .$$
(G.10)

The covariant matrices are given by $\gamma_\alpha = g_{\alpha\beta} \gamma^\alpha$, i.e. $\gamma_0 = \gamma^0$ and $\gamma_i = -\gamma^i$. It holds $\gamma^\alpha g_{\alpha\beta} \gamma^\beta = g_{\alpha\beta} \gamma^\alpha \gamma^\beta$.

A useful auxiliary matrix (representation independent) is

$$\gamma^5 \equiv i \gamma^0 \gamma^1 \gamma^2 \gamma^3 = -i \gamma_0 \gamma_1 \gamma_2 \gamma_3 = \gamma_5 = \begin{pmatrix} 0_2 & \mathbb{1}_2 \\ \mathbb{1}_2 & 0_2 \end{pmatrix} = \sigma_1 \otimes \mathbb{1}_2 \ . \quad \text{(G.11)}$$

The traceless γ-matrices satisfy many identities. We give here only a few:

- Hermitian conjugation
  $\gamma^0 = \gamma^{0\dagger}$ and $\gamma^5 = \gamma^{5\dagger}$ are Hermitian,[10]
  γ^i $(i = 1, 2, 3)$ are anti-Hermitian $\gamma^{i\dagger} = -\gamma^i$.
- Anticommutation

$$\{\gamma^\alpha, \gamma^\beta\} = 2g^{\alpha\beta}$$
$$\{\gamma^\alpha, \gamma^5\} = 0 \ .$$
(G.12)

Useful identities follow from the anticommutation relations, e.g.,

$$(\gamma^0)^2 = \mathbb{1}_4$$
$$(\gamma^\alpha)^2 = -\mathbb{1}_4$$
$$\gamma^\alpha \gamma_\alpha = 4 \mathbb{1}_4 \ .$$
(G.13)

- Unitarity. The γ-matrices are unitary $\gamma^{\alpha\dagger} = (\gamma^\alpha)^{-1}$. For all matrices

$$\gamma^0 \gamma^\alpha \gamma^0 = \gamma^{\alpha\dagger} \ . \quad \text{(G.14)}$$

---

[9] Note that $\sigma_i = \sigma^i$.
[10] $(K_{ij})^\dagger = (K_{ji})^*$.

- Traces[11]

$$\mathrm{Tr}(\gamma^{\alpha}) = \mathrm{Tr}(\gamma^5) = 0$$

$$\mathrm{Tr}(\gamma^{\alpha}\gamma^{\beta}) = 4g^{\alpha\beta}$$

$$\mathrm{Tr}(\gamma^{\alpha}\gamma^{\beta}\gamma^5) = 0_4 \qquad (\mathrm{G.15})$$

$$\mathrm{Tr}(\gamma^{\alpha}\gamma^{\beta}\gamma^{\lambda}\gamma^{\sigma}) = 4(g^{\alpha\beta}g^{\lambda\sigma} - g^{\alpha\lambda}g^{\beta\sigma} + g^{\alpha\sigma}g^{\beta\lambda})$$

$$\mathrm{Tr}(\gamma^5\gamma^{\alpha}\gamma^{\beta}\gamma^{\lambda}\gamma^{\sigma}) = 4i\epsilon^{\alpha\beta\lambda\sigma} \ .$$

G.2.2 Free Particle Solutions of the Dirac Equation

For a particle at rest ($\mathbf{p} = 0$), inserting the spinor

$$\Psi = e^{-iEt}u, \quad \text{with} \quad u = \begin{pmatrix} u_0 \\ u_1 \\ u_2 \\ u_3 \end{pmatrix} \qquad (\mathrm{G.16})$$

in the Dirac equation we obtain four solutions: two with positive energy $+m$ and two with negative energy $-m$.

The corresponding spinors are

$$u_{\uparrow,+} = \begin{pmatrix} 1 \\ 0 \\ 0 \\ 0 \end{pmatrix} \quad u_{+,\downarrow} = \begin{pmatrix} 0 \\ 1 \\ 0 \\ 0 \end{pmatrix} \quad v_{-,\downarrow} = \begin{pmatrix} 0 \\ 0 \\ 1 \\ 0 \end{pmatrix} \quad v_{-,\uparrow} = \begin{pmatrix} 0 \\ 0 \\ 0 \\ 1 \end{pmatrix} ,$$

$$(\mathrm{G.17})$$

with phase e^{-imt} for the u_+ states and e^{+imt} for the v_- states. These spinors are spin eigenstates. Solutions with negative energy and in the spin down state ($\downarrow$) are associated with antiparticles with positive energies and in the spin up state ($\uparrow$). Similarly for the spin up state.[12] From the solution for a particle at rest, the solution for a moving particle with momentum $\mathbf{p}$ can be obtained by Lorentz transformation,

---

[11] $\epsilon^{\alpha\beta\lambda\sigma}$ is a completely antisymmetric tensor of rank four, i.e., antisymmetric in every pair of superscripts, $\epsilon^{0123}=1$; zero if any two indices are equal.

[12] Note that in the spinors the arrow gives the spin value that we associate with the physical states, i.e., for the u_+ states it is that of a particle, for the v_- states it is that of the antiparticle.

or using a plane wave Ansatz of the form $\Psi = e^{-ip_\alpha x^\alpha} u$. The solution reads

$$
u_{\mathbf{p},\uparrow} = \sqrt{E + m} \begin{pmatrix} 1 \\ 0 \\ \dfrac{p_z}{E + m} \\ \dfrac{p_x + ip_y}{E + m} \end{pmatrix}
\qquad
u^{\mathbf{p},\downarrow} = \sqrt{E + m} \begin{pmatrix} 0 \\ 1 \\ \dfrac{p_x - ip_y}{E + m} \\ \dfrac{-p_z}{E + m} \end{pmatrix}
$$

$$
v_{\mathbf{p},\downarrow} = \sqrt{E + m} \begin{pmatrix} \dfrac{p_z}{E + m} \\ \dfrac{p_x + ip_y}{E + m} \\ 1 \\ 0 \end{pmatrix}
\qquad
v_{\mathbf{p},\uparrow} = \sqrt{E + m} \begin{pmatrix} \dfrac{p_x - ip_y}{E + m} \\ \dfrac{-p_z}{E + m} \\ 0 \\ 1 \end{pmatrix} .
$$

$$(G.18)$$

The spinors $u_{\mathbf{p},\uparrow} e^{-ip_\alpha x^\alpha}$ and $u_{\mathbf{p},\downarrow} e^{-ip_\alpha x^\alpha}$ represent propagating plane waves of fermions with momentum $\mathbf{p}$, energy E and spin up ($\uparrow$) or down ($\downarrow$), respectively. The negative energy solutions $v_{\mathbf{p},\uparrow} e^{+ip_\alpha x^\alpha}$ and $v_{\mathbf{p},\downarrow} e^{+ip_\alpha x^\alpha}$ are states with momentum $-\mathbf{p}$, energy $-E$ and spin down and up, respectively. They represent antifermions with momentum $\mathbf{p}$, energy E and spin orientation up $\uparrow$ or down $\downarrow$, respectively.

Note that the spin state, that we use to label the spinors, is that in the rest frame of the particle/antiparticle. In general, the spin operator does not commute with the Dirac Hamiltonian for a free particle/antiparticle. Only in the rest frame or for particles/antiparticles travelling in the $\pm\hat{\mathbf{z}}$ direction (i.e., $p_x = p_y = 0$), the u and v spinors are spin eigenvectors.[13]

In summary, the wave functions for fermions and antifermions are given by the following plane waves (for simplicity, we denote the $\uparrow$ and $\downarrow$ states by $s = \pm 1/2$, respectively)

$$
\begin{aligned}
u_{\mathbf{p},s}\, e^{-ip_\alpha x^\alpha}, &\qquad s = \pm 1/2 \\
v_{\mathbf{p},s}\, e^{ip_\alpha x^\alpha}, &\qquad s = \pm 1/2 .
\end{aligned}
$$

$$(G.19)$$

The spinors in Eq. G.19 are orthogonal to each other but not normalized to one. The normalization of the spinors must be chosen, so that the probability of finding a particle in a volume V is invariant under Lorentz transformation. The total probability is given by $\int u^\dagger u\, dV$, where u is one of the spinors of Eq. G.19.

---

[13] An operator that commutes with the Hamiltonian is the helicity operator, which has eigenvalues ± 1 and gives the projection of the particle spin along the direction of its momentum.

Under a Lorentz transformation the volume is not preserved but scales as the inverse of the Lorentz factor $(E/m)^{-1}$. To ensure a Lorentz invariant total probability, the probability density $u^\dagger u$ must have a normalization proportional to E. Following the usual convention[14] we have chosen for the u and v spinors, Eq. G.19, a normalization such that $u^\dagger_{\mathbf{p},s} u_{\mathbf{p},s} = v^\dagger_{\mathbf{p},s} v_{\mathbf{p},s} = 2E, s = \pm 1/2$. These spinors form a complete vector system.

Given the above normalization and summing over the two spin states, one obtains the following completeness relations for the u and v spinors, which are very useful for the calculation of spin sums in the transition matrix elements, see Sect. 10.3[15]

$$\sum_{s=\pm 1/2} u_{\mathbf{p},s} \bar{u}_{\mathbf{p},s} = p_\alpha \gamma^\alpha + m = \slashed{p} + m$$

$$\sum_{s=\pm 1/2} v_{\mathbf{p},s} \bar{v}_{\mathbf{p},s} = p_\alpha \gamma^\alpha - m = \slashed{p} - m \ . \tag{G.20}$$

The general solution of the Dirac equation for a free particle is a linear superposition of the complete set of plane wave solutions for spin $1/2$ fermions and antifermions

$$\Psi(x^\alpha) = N \sum_{s=\pm 1/2} \int d\mathbf{p} \left[b_{\mathbf{p},s} u_{\mathbf{p},s} e^{-ip_\alpha x^\alpha} + d^*_{\mathbf{p},s} v_{\mathbf{p},s} e^{+ip_\alpha x^\alpha} \right] \ . \tag{G.21}$$

Here b and d^* are arbitrary functions of $\mathbf{p}$, s and N is an appropriate normalization.

G.3   Dirac Field Operators

To describe the muon decay, we use the formalism of quantum field theory (QFT). QFT is here a mathematical framework for describing elementary particles and their interactions. QFT describes the particles as excited states of associated quantum fields, which allows to describe the creation and annihilation of particles and antiparticles in a natural and consistent way. In QFT, the classical fields and quantum mechanical wave functions become field operators.

---

[14] Note that there are other normalization choices for Dirac states in the literature.
[15] With the Feynman notation $\slashed{p} \equiv p_\alpha \gamma^\alpha = p^\alpha \gamma_\alpha = \gamma^\alpha p_\alpha$.

The most general Dirac field operator is obtained by replacing the Fourier coefficients in Eq. G.21 by creation and annihilation operators[16,17,18]

$$\Psi(x^\alpha) = \sum_{s=\pm 1/2} \int \frac{d\mathbf{p}}{(2\pi)^3 2E} \left[b_{\mathbf{p},s} \, u_{\mathbf{p},s} \, e^{-ip_\alpha x^\alpha} + d^\dagger_{\mathbf{p},s} \, v_{\mathbf{p},s} \, e^{+ip_\alpha x^\alpha} \right] \ . \qquad \text{(G.22)}$$

The operators $b^\dagger_{\mathbf{p},s}$ and $b_{\mathbf{p},s}$ create and annihilate a spin $1/2$ particle, respectively (e.g., μ^- or e^-) out of the vacuum[19] with momentum $\mathbf{p}$ and spin s. $d^\dagger_{\mathbf{p},s}$ and $d_{\mathbf{p},s}$ are the creation and annihilation operators for the corresponding antiparticle with the same spin and energy of the particle, but with opposite charge (e.g., μ^+ or e^+).[20] For completeness, we also give the expression for $\overline{\Psi}$, which is obtained from G.22 by applying Hermitian conjugation and right multiplication with γ^0

$$\overline{\Psi}(x^\alpha) = \sum_{s=\pm 1/2} \int \frac{d\mathbf{p}}{(2\pi)^3 2E} \left[b^\dagger_{\mathbf{p},s} \, \overline{u}_{\mathbf{p},s} \, e^{ip_\alpha x^\alpha} + d_{\mathbf{p},s} \, \overline{v}_{\mathbf{p},s} \, e^{-ip_\alpha x^\alpha} \right] \ . \qquad \text{(G.23)}$$

The effect of Ψ and $\overline{\Psi}$ on physical states can be understood as follows: Ψ annihilates an incoming particle (particle in the initial state) or creates an outgoing antiparticle (antiparticle in the final state). On the other hand $\overline{\Psi}$ creates an outgoing particle (particle in the final state) or annihilates an incoming antiparticle (antiparticle in the initial state). As a field with half-integer spin, Ψ has to satisfy the Fermi-Dirac statistics and therefore the equal-time anticommutation relations

$$\begin{aligned} \left\{ \Psi(t, \mathbf{x}), \Psi^\dagger(t, \mathbf{x}') \right\} &= \delta^3(\mathbf{x} - \mathbf{x}') \\ \left\{ \Psi(t, \mathbf{x}), \Psi(t, \mathbf{x}') \right\} &= \left\{ \Psi^\dagger(t, \mathbf{x}), \Psi^\dagger(t, \mathbf{x}') \right\} = 0 \ . \end{aligned} \qquad \text{(G.24)}$$

---

[16] For simplicity, we use the same symbol Ψ for the Dirac field operators as for the Dirac spinors.

[17] Again, different normalization choices are found in the literature. The anticommutation factors in G.25 must be adapted accordingly.

[18] $d\mathbf{p}/(2\pi)^3$ is the number of states per unit volume in $d\mathbf{p}$. The factor $2E$ is a consequence of our normalization choice to make $d\mathbf{p}/[(2\pi)^3 2E]$ Lorentz invariant, see Sect. G.4.

[19] Dirac interpreted the vacuum as the state in which all the negative energy states are filled, so-called Dirac sea. The negative energy solutions are holes in the vacuum and correspond to antiparticle states. The infinite Dirac sea poses several problems; for instance it has infinite energy. In his construction of QED, Feynman used the interpretation of the Swiss physicist Ernst Stückelberg (Stückelberg 1941), where the antiparticles are particles going backward in time. In a Feynman graph, although antiparticles move forward in time, they are usually drawn with an arrow pointing backwards in time to emphasize that they are antiparticles, see Fig. 10.1.

[20] As noted above, $d^\dagger_{\mathbf{p},s}$ can be seen as the annihilation operator of a negative energy state of the Dirac equation with wave function $v_{\mathbf{p},s}$. In the Dirac picture, this hole in the Dirac sea corresponds to the creation of a positive energy antiparticle from the vacuum.

For the creation/annihilation operators, this gives the following anticommutation relations

$$\left\{ b_{\mathbf{p},s} , b^{\dagger}_{\mathbf{p}',s'} \right\} = (2\pi)^3 (2E)\delta(\mathbf{p} - \mathbf{p}')\delta_{s,s'}$$

$$\left\{ b_{\mathbf{p},s} , b_{\mathbf{p}',s'} \right\} = \left\{ b^{\dagger}_{\mathbf{p},s}, b^{\dagger}_{\mathbf{p}',s'} \right\} = 0$$

$$\left\{ d_{\mathbf{p},s} , d^{\dagger}_{\mathbf{p}',s'} \right\} = (2\pi)^3 (2E)\delta(\mathbf{p} - \mathbf{p}')\delta_{s,s'}$$

$$\left\{ d_{\mathbf{p},s} , d_{\mathbf{p}',s'} \right\} = \left\{ d^{\dagger}_{\mathbf{p},s}, d^{\dagger}_{\mathbf{p}',s'} \right\} = 0 \ .$$

(G.25)

What happens when we go from one inertial system to another, or when we do a parity operation?

Relativistic Covariance Consider a Lorentz transformation L that transforms the space-time coordinates from one inertial system to another

$$L :\to x'^{\alpha} = \Lambda^{\alpha}_{\beta} x^{\beta} \ .$$

(G.26)

Relativistic covariance means that the spinor $\Psi(x^{\alpha})$ in the inertial frame transforms into a spinor in the other frame $\Psi'(x'^{\alpha})$ by a linear transformation, so that the Dirac equation has the same form in both systems[21]

$$i\gamma^{\alpha} \frac{\partial \Psi}{\partial x^{\alpha}} - m\Psi = 0$$

$$i\gamma^{\alpha} \frac{\partial \Psi'}{\partial x'^{\alpha}} - m\Psi = 0 \ .$$

(G.27)

Parity Operation The parity operation P changes the sign of the spatial coordinates while keeping the time coordinate unchanged

$$P: \quad x^{\alpha} = \begin{pmatrix} t \\ \mathbf{x} \end{pmatrix} \to \tilde{x}^{\alpha} = \begin{pmatrix} t \\ -\mathbf{x} \end{pmatrix} \ .$$

(G.28)

The invariance of the Dirac equation under parity leads to the following spinor transformation

$$P: \quad \Psi(t, \mathbf{x}) \to \gamma^0 \Psi(t, -\mathbf{x}) \ .$$

(G.29)

The space-time and momentum 4-vectors are example of (polar) vectors. Axial vectors, under the action of the parity transformation, change the sign of the first

---

[21] Note that both the Dirac field operator $\Psi(x^{\alpha})$ and the coordinate x^{α} change under the inertial frame transformation, while the γ-matrices do not, see also footnote 8.

component, while keeping the other three unchanged.

$$
P: \quad A^{\alpha} = \begin{pmatrix} A^0 \\ A^1 \\ A^2 \\ A^3 \end{pmatrix} \rightarrow \quad A^{\alpha} = \begin{pmatrix} -A^0 \\ A^1 \\ A^2 \\ A^3 \end{pmatrix} . \tag{G.30}
$$

The simplest set of covariants that can be made from Dirac spinors and γ-matrices, together with their properties are[22]

$$
\begin{aligned}
\overline{\Psi}\Psi & \quad \text{scalar (S)} \\
\overline{\Psi}\gamma^5\Psi & \quad \text{pseudoscalar (P)} \\
\overline{\Psi}\gamma^{\alpha}\Psi & \quad \text{vector (V)} \\
\overline{\Psi}\gamma^5\gamma^{\alpha}\Psi & \quad \text{axial vector (A)} \\
\overline{\Psi}\gamma_{\beta}\gamma^{\alpha}\Psi & \quad \text{tensor (T)} .
\end{aligned} \tag{G.31}
$$

G.4   Fermi's Golden Rule and Lorentz Invariance

We derive here the relativistic expression for the Fermi's golden rule, which is used in Sect. 10.3 to calculate the muon decay and its properties. In quantum mechanics Fermi's golden rule gives the transition probability per unit time (transition or decay rate) from an initial energy eigenstate to a final state, as a result of a weak perturbation V_{int}. To first-order

$$
\Gamma_{i \rightarrow f} = 2\pi \left| \langle f | V_{\text{int}} | i \rangle \right|^2 \rho(E_f) . \tag{G.32}
$$

The matrix element contains the information about the dynamics of the interaction. The phase space factor $\rho(E_f) = \left| dN_f/dE \right|_{E=E_f}$ is the density of available final states with energy E_f and contains only kinematic information.

The relativistic version of the Fermi's golden rule can be obtained formally from the S-matrix scattering theory see, e.g., Alvarez-Gaumé and Vázquez-Mozo (2011). We give here a simpler derivation for the specific case of a particle decaying into three particles, as in the muon decay.

---

[22] A pseudoscalar changes sign under a parity operation. The helicity is an example of a pseudoscalar. By converse a scalar remains unchanged. The spinors can also belong to different particle species.

First, we note that by explicitly taking the energy conservation ($E_f = E_i$) into account with a δ-function

$$\rho(E_f) = \left| \frac{dN_f}{dE} \right|_{E=E_f} = \int \frac{dN_f}{dE} \delta(E - E_i) dE \ , \tag{G.33}$$

Eq. G.32 becomes

$$\Gamma_{i \to f} = 2\pi \int |V_{if}|^2 \delta(E - E_i) dN_f \ , \tag{G.34}$$

where $V_{if} \equiv \langle f | V_{int} | i \rangle$ and the integration is over final states of any energy.

In the case of a three-body decay ($0 \to 1 + 2 + 3$) for dN_f we only need to consider two particles, since the conservation of momentum fixes the state of the third particle. Nonrelativistically, with the usual normalization of one particle per unit volume, G.34 becomes

$$\Gamma_{i \to f} = 2\pi \int |V_{if}|^2 \delta(E_0 - E_1 - E_2 - E_3) \frac{d\mathbf{p}_1}{(2\pi)^3} \frac{d\mathbf{p}_2}{(2\pi)^3} \ . \tag{G.35}$$

We can explicitly include momentum conservation in this formula by introducing a δ-function and integrating over the momentum of the third particle

$$\Gamma_{i \to f} = (2\pi)^4 \int |V_{if}|^2 \delta(E_0 - E_1 - E_2 - E_3)$$

$$\delta(\mathbf{p}_0 - \mathbf{p}_1 - \mathbf{p}_2 - \mathbf{p}_3) \frac{d\mathbf{p}_1}{(2\pi)^3} \frac{d\mathbf{p}_2}{(2\pi)^3} \frac{d\mathbf{p}_3}{(2\pi)^3} \ . \tag{G.36}$$

In Sect. G.2.2 we have seen that a normalization of one particle per volume of the Dirac solutions, $\int u'^\dagger u' dV = \int v'^\dagger v' dV = 1$ as in the previous equations, is not Lorentz invariant and that a common convention to ensure Lorentz invariance is to normalize to $2E$ particles per unit volume. This corresponds to replacing the spinors u' and v' describing the particles involved in the muon decay by $u = \sqrt{2E}u'$ and $v = \sqrt{2E}v'$, respectively.

Introducing a Lorentz invariant matrix element $\mathcal{M}$ written in terms of the spinors normalized to $2E$ and noting that

$$\mathcal{M}_{i \to f} = \sqrt{2E_0 2E_1 2E_2 2E_3} \, V_{if} \ ,$$

we can now rewrite G.36 and obtain

$$\Gamma_{i \to f} = \frac{(2\pi)^4}{2E_0} \int \left| \mathcal{M}_{if} \right|^2 \delta(E - E_i)$$

$$\delta(\mathbf{p}_0 - \mathbf{p}_1 - \mathbf{p}_2 - \mathbf{p}_3) \frac{d\mathbf{p}_1}{(2\pi)^3 2E_1} \frac{d\mathbf{p}_2}{(2\pi)^3 2E_2} \frac{d\mathbf{p}_3}{(2\pi)^3 2E_3} \quad . \tag{G.37}$$

This expression is actually the same as Eq. G.36, but in this form its properties under Lorentz transformation are more evident. The integral is Lorentz invariant because $\mathcal{M}_{i \to f}$ as well as the phase space are Lorentz invariant.

The transition rate $\Gamma_{i \to f}$, which is inversely proportional to the energy of the decaying particle, has the transformation properties that we expect from the relativistic time dilation. For a decay of a particle at rest $\Gamma_{i \to f}^{-1} = \tau$, where τ is the lifetime of the particle, which is always measured in its rest frame.

References

Abragam, A. (1961). *The Principles of Nuclear Magnetism*. Clarendon Press. ISBN: 978-0198520146.

Aharoni, A. (1996). *Introduction to the Theory of Ferromagnetism*. Oxford University Press. ISBN: 978-0198508090.

Alvarez-Gaumé, L., & Vázquez-Mozo, M. A. (2011). *An Invitation to Quantum Field Theory*. Springer. ISBN: 978-3642237270.

Anderson, P. W. (1954). *Journal of the Physical Society of Japan, 9*, 316.

Blum, K. (2012). *Density Matrix Theory and Applications*. Springer-Verlag. ISBN: 978-3642205606.

Chao, A. W., & Tigner, M. (1999). *Handbook of Accelerator Physics and Engineering* (1st ed.). World Scientific.

Cohen-Tannoudji, C., Laloë, F., & Diu, B. (2005). *Quantum mechanics*. John Wiley & Sons. ISBN: 9780471569527.

Dirac, P. A. M. (1928). *Proceedings of the Royal Society A: Mathematical, Physical and Engineering Sciences, 117*, 610.

Doob, J. L. (1942). *Annals of Mathematics, 43*, 351.

Messiah, A. (1965). *Quantum Mechanics*. North-Holland. ISBN: 978-0720400441.

Osborn, J. A. (1945). *Physical Review, 67*, 351.

Prozorov, R., & Kogan, V. G. (2018). *Physical Review Applied, 10*, 014030.

Schwabl, F. (2005). *Advanced Quantum Mechanics*. Advanced texts in physics. Springer. ISBN: 3-540-28528-8.

Schwabl, F. (2007). *Quantum Mechanics*. Springer. ISBN: 3-540-71933-4.

Stückelberg, E. C. G. (1941). *Helvetica Physica Acta, 14*, 51.

Wolski, A. (2011). arXiv:1103.0713. CERN-2010-004 Report (pp. 1–38).

Solutions of the Exercises

1.1. Cosmic muons and relativity

Classically:

Flight time (assuming a muon velocity of $v \simeq c$)

$$t = \frac{d}{v} = \frac{15000}{3 \times 10^8} \text{ s} = 50 \times 10^{-6} \text{ s}$$

The flight time is equal to n lifetimes

$$n = \frac{t}{\tau_\mu} = \frac{50 \times 10^{-6}}{2.197 \times 10^{-6}} \simeq 22.8$$

Expected number of muons after n lifetimes

$$N_{\text{Earth, class.}} = \frac{N_0}{e^n} \simeq 1.25 \times 10^{-4} \text{ muons}$$

Relativistically:

Kinetic energy $E_{\text{kin}} \simeq 4 \text{ GeV}$

Total muon energy

$$E_\mu = m_\mu c^2 + E_{\text{kin}} = 0.106 \text{ GeV} + 4 \text{ GeV} = \gamma m_\mu c^2$$

Dilation factor

$$\gamma = \frac{E_\mu}{m_\mu c^2} \simeq 38.7$$

Muon lifetime as measured by an earthbound observer

$$\tau_{\mu, \text{Earth}} = \gamma \, \tau_\mu \simeq 85 \times 10^{-6} \text{ s}$$

© Springer Nature Switzerland AG 2024
A. Amato, E. Morenzoni, *Introduction to Muon Spin Spectroscopy*,
Lecture Notes in Physics 961, https://doi.org/10.1007/978-3-031-44959-8

For this observer, the flight time corresponds to n_{Earth} lifetimes

$$n_{\text{Earth}} = \frac{t}{\tau_{\mu,\,\text{Earth}}} = \frac{50 \times 10^{-6}}{85 \times 10^{-6}} \simeq 0.59$$

Number of muons at the Earth level

$$N_{\text{Earth, rel.}} = \frac{N_0}{e^{\,n_{\text{Earth}}}} \simeq 0.55 \times 10^6 \text{ muons} \ .$$

1.2. Threshold proton energy for pion production
We consider the reaction

$$p + p \rightarrow p + p + \pi^+$$

Energy and momentum conservation can be expressed with the four-momentum vector, see Appendix G

$$p^\nu = \begin{pmatrix} p^0 \\ p^1 \\ p^2 \\ p^3 \end{pmatrix} = \begin{pmatrix} \frac{E}{c} \\ \mathbf{p} \end{pmatrix}$$

$$p^\nu_{\text{tot}} = p'^\nu_{\text{tot}} \ ,$$

where p^ν_{tot} is the total four-momentum of the incoming particles and p'^ν_{tot} that of the outgoing particles. The quantity (sum over ν)

$$p^\nu_{\text{tot}} \, p_{\nu\,\text{tot}} = p'^\nu_{\text{tot}} \, p'_{\nu\,\text{tot}}$$

is conserved in the reaction and is a Lorentz invariant, i.e., it is the same in the laboratory and in the center of mass system.
We evaluate $p^\nu_{\text{tot}} \, p_{\nu\,\text{tot}}$ in the laboratory system

$$c^2 p^\nu_{\text{tot}} \, p_{\nu\,\text{tot}} = E^2_{\text{tot}} - p^2_{\text{tot}} c^2 = (E_p + m_p c^2)^2 - (E_p - m_p c^2)^2 \ ,$$

where E_p is the energy of the incoming proton in the laboratory system.
In the center of mass system

$$c^2 p'^\nu_{\text{tot}} \, p'_{\nu\,\text{tot}} = (2 m_p c^2 + m_\pi c^2)^2 \ ,$$

where we have taken into account that at the threshold the total energy in the final state is just the rest mass energy of the particles after the reaction. Equating the two expressions we obtain for the minimum proton energy in the laboratory to create a positive pion

$$E_p^{\text{thr}} = m_p c^2 \left(1 + \frac{m_\pi^2 + 4m_\pi m_p}{2m_p^2} \right) ,$$

which corresponds to a kinetic energy $E_p^{\text{thr}} - m_p c^2 \simeq 289.5$ MeV.

1.3. Pion decay kinematics
Decay

$$\pi^+ \to \mu^+ + \nu_\mu$$

Energy conservation (assuming $m_\nu = 0$)

$$E_\pi = m_\pi c^2 = E_\mu + E_\nu$$

Momentum conservation

$$0 = \mathbf{p}_\mu + \mathbf{p}_\nu$$
$$|\mathbf{p}_\mu| = |\mathbf{p}_\nu| \equiv k$$

Combining both equations

$$m_\pi c^2 = \sqrt{m_\mu^2 c^4 + k^2 c^2} + k c$$

$$m_\pi c^2 - k c = \sqrt{m_\mu^2 c^4 + k^2 c^2}$$

$$m_\pi^2 c^4 - 2k m_\pi c^3 + k^2 c^2 = m_\mu^2 c^4 + k^2 c^2$$

$$2k m_\pi c^3 = m_\pi^2 c^4 - m_\mu^2 c^4$$

$$k = \frac{m_\pi^2 - m_\mu^2}{2m_\pi} c = 29.7921 \text{ MeV}/c$$

The energy of the muon is given by

$$E_\mu = \sqrt{m_\mu^2 c^4 + k^2 c^2} = \sqrt{m_\mu^2 c^4 + \left(\frac{m_\pi^2 - m_\mu^2}{2m_\pi} c\right)^2 c^2}$$

$$E_\mu = \sqrt{\left(\frac{m_\pi^2 + m_\mu^2}{2m_\pi} c^2\right)^2}$$

$$E_\mu = \frac{m_\pi^2 + m_\mu^2}{2m_\pi} c^2 = 109.78 \text{ MeV}$$

The kinetic energy is therefore

$$E_\mu^{\text{kin}} = E_\mu - m_\mu c^2 = \frac{(m_\pi - m_\mu)^2}{2m_\pi} c^2 = 4.12 \text{ MeV} \ .$$

1.4. Muon decay kinematics
Decay

$$\mu^+ \rightarrow e^+ + \bar{\nu}_\mu + \nu_e$$

From energy and momentum conservation (taking into account that the muon is at rest and neglecting the neutrino masses)

$$E_\mu = E_{e^+} + p_{\bar{\nu}_\mu} c + p_{\nu_e} c$$

$$0 = \mathbf{p}_{e^+} + \mathbf{p}_{\bar{\nu}_\mu} + \mathbf{p}_{\nu_e}$$

The positron momentum is at its maximum when both neutrinos are traveling in the same direction, opposite to that of the positron

$$p_{e^+}^{\text{max}} = p_{\bar{\nu}_\mu} + p_{\nu_e}$$

Rewriting the energy conservation

$$m_\mu c^2 = E_{e^+} + p_{e^+}^{\text{max}} c$$

$$m_\mu c^2 = \sqrt{m_{e^+}^2 c^4 + (p_{e^+}^{\text{max}})^2 c^2} + p_{e^+}^{\text{max}} c$$

$$m_\mu c^2 - p_{e^+}^{\text{max}} c = \sqrt{m_{e^+}^2 c^4 + (p_{e^+}^{\text{max}})^2 c^2}$$

$$\left(m_\mu c^2 - p_{e^+}^{\text{max}} c\right)^2 = m_{e^+}^2 c^4 + (p_{e^+}^{\text{max}})^2 c^2$$

$$m_\mu^2 c^4 - 2 m_\mu p_{e+}^{max} c^3 = m_{e+}^2 c^4$$

$$p_{e+}^{max} = \frac{m_\mu^2 - m_{e+}^2}{2m_\mu} c = 52.828 \text{ MeV/c}$$

The maximum energy of the positron is given by

$$E_{e+}^{max} = \sqrt{m_{e+}^2 c^4 + (p_{e+}^{max})^2 c^2}$$

$$= \sqrt{m_{e+}^2 c^4 + \left(\frac{m_\mu^2 - m_{e+}^2}{2m_\mu} c\right)^2 c^2}$$

$$= \sqrt{\left(\frac{m_\mu^2 + m_{e+}^2}{2m_\mu} c^2\right)^2}$$

$$= \frac{m_\mu^2 + m_{e+}^2}{2m_\mu} c^2 = 52.830 \text{ MeV}$$

To determine the mean positron energy, consider the Michel spectrum ($\varepsilon = 2E_{e+}/(m_\mu c^2)$, $E_{e+}^{max} \cong m_\mu c^2/2$)

$$W(\varepsilon) \, d\varepsilon = \frac{1}{\tau_\mu} 2\varepsilon^2(3 - 2\varepsilon)d\varepsilon$$

Average energy

$$\bar\varepsilon = \frac{\int_0^1 \varepsilon W(\varepsilon) \, d\varepsilon}{\int_0^1 W(\varepsilon) \, d\varepsilon} = \frac{7}{10}$$

$$\overline{E}_{e+} \cong \frac{7}{10} \frac{m_\mu c^2}{2} = 36.980 \text{ MeV} .$$

1.5. Pion production in a graphite target

The positive pion rate $N_{\pi+}$ is given by

$$N_{\pi+} = I_p \sigma_{\pi+} N_{at} d ,$$

where I_p is the proton rate impinging on a target of thickness $d = 6$ cm

$$I_p = \frac{2.400 \times 10^{-3} \text{ A}}{1.602 \times 10^{-19} \text{ C}} = 1.5 \times 10^{16} \text{ protons/s}$$

For a typical pion production cross section at $\sim$600 MeV, see Fig. 1.9,

$$\sigma_{\pi^+} \cong 1 \, \text{fm}^2 = 1 \times 10^{-26} \, \text{cm}^2$$

The number of atoms/cm^3 in the target (N_A, Avogadro constant, A_C and ρ_C molar mass and density of graphite, respectively)

$$N_{at} = \frac{N_A}{A_C} \rho_C = \frac{6.022 \times 10^{23} \, \text{mol}^{-1}}{12 \, \text{g mol}^{-1}} 2.26 \, \text{g cm}^{-3} = 1.13 \times 10^{23} \, \text{atoms cm}^{-3}$$

Therefore, the number of pions produced over the entire 4π solid angle is

$$N_{\pi^+} \cong 1 \times 10^{14} \, \pi^+/\text{s}$$

Note that for one proton the probability P_{coll} to have a collision with a nucleon in the target is quite small

$$P_{coll} = \sigma_{\pi^+} N_{at} \, d \cong 0.007 \ .$$

1.6. Momentum of forward and backward decay muons

The solution can be found by transforming muon momentum and energy (from Exercise 1.3) from the pion center of mass system (pion at rest) to the laboratory frame moving with velocity β_π with respect to the pion center of mass system. We use the muon momentum four-vector

in the laboratory frame
$$\begin{pmatrix} \frac{E'_\mu}{c} \\ \mathbf{p}'_\mu \end{pmatrix}$$

and in the pion rest frame
$$\begin{pmatrix} \frac{E_\mu}{c} \\ \mathbf{p}_\mu \end{pmatrix}$$

Assuming that the pion moves in the x direction, the two systems are related by a Lorentz transformation

$$\begin{pmatrix} \frac{E'_\mu}{c} \\ p'_\mu \\ 0 \\ 0 \end{pmatrix} = \begin{pmatrix} \gamma_\pi & \beta_\pi \gamma_\pi & 0 & 0 \\ \beta_\pi \gamma_\pi & \gamma_\pi & 0 & 0 \\ 0 & 0 & 1 & 0 \\ 0 & 0 & 0 & 1 \end{pmatrix} \begin{pmatrix} \frac{E_\mu}{c} \\ p_\mu \\ 0 \\ 0 \end{pmatrix}$$

For forward and backward emitted muons the momentum is p_μ and $-p_\mu$, respectively. The corresponding momenta in the laboratory frame are

$$p'_\mu = \beta_\pi \gamma_\pi \frac{E_\mu}{c} \pm \gamma_\pi p_\mu$$

$$= \frac{p_\pi}{m_\pi c^2} E_\mu \pm \frac{E_\pi}{m_\pi c^2} p_\mu$$

Using the expressions for p_μ and E_μ from the solution of Exercise 1.3 we obtain

$$p'_\mu = \frac{p_\pi (m_\pi^2 + m_\mu^2)c^2 \pm \sqrt{p_\pi^2 c^2 + m_\mu^2 c^4} \, (m_\pi^2 - m_\mu^2)c}{2 m_\pi^2 c^2}$$

From a pion decaying in flight with $p_\pi = 200$ MeV/c we obtain for a forward decay muon $p'_\mu = 209.4$ MeV/c and for a backward decay muon $p'_\mu = 105.25$ MeV/c, see Fig. 1.31.
The decay length of a pion λ_π is determined by its lifetime ($\tau_\pi = 26$ ns)

$$\lambda_\pi = \beta_\pi \, c \, \gamma_\pi \, \tau_\pi \ ,$$

numerically

$$\lambda_\pi \, [\text{m}] = 0.055 \, p_\pi \, [\text{MeV/c}]$$

For 200 MeV/c pions, therefore, one needs a solenoidal decay channel of about 10 m or more in length.

2.1. Energy loss of cosmic muons in the atmosphere

The interaction depth of muons through the atmosphere can be determined in different ways. From Fig. 1.6 we read a value ~ 1000 g cm^{-2}. This corresponds to taking the midpoint between the density at sea level (0.001225 g cm^{-3}) and the one at $h = 15{,}000$ m where the muons are created (about 0.000194 g cm^{-3}) and multiplying it by the distance the muon travels to sea level

$$x = \frac{\rho_1 + \rho_2}{2} h = 1064 \ \text{g cm}^{-2}$$

A better approximation can be obtained by using the so-called "barometric formula", which gives the evolution of pressure and density with altitude. Neglecting the variation of the temperature with altitude
The barometric formula is

$$\rho(h) = \rho_0 \exp\left[-\frac{g_0 \, M \, h}{R^* \, T_b}\right] \ ,$$

where

ρ = mass density (kg m^{-3}) with $\rho_0 = 1.2250$ kg m^{-3}
T_b = standard temperature, 288.15 K
h = altitude above sea level
R^* = universal gas constant for air, 8.3145 N m mol^{-1} K^{-1}
g_0 = gravitational acceleration, 9.80665 m s^{-2}
M = molar mass of Earth's air, 0.02897 kg mol^{-1}

$$x = \int_0^h \rho_0 \exp\left[-\frac{g_0\, M\, h'}{R^*\, T_b}\right] dh'$$

$$= -\frac{\rho_0\, R^*\, T_b}{g_0 M}\left(\exp\left[-\frac{g_0\, M\, h}{R^*\, T_b}\right] - 1\right)$$

$$= 858.7 \;\; \mathrm{g\,cm}^{-2}$$

From Fig. 2.5 we see that the energy loss of a 6 GeV muon is typically 2 MeV cm^2 g^{-1}. Multiplying this number by the interaction length estimated by both methods gives an energy loss of about 2 GeV.

2.2. Energy loss of decay muons
From Eq. 2.17, the energy loss can be written as

$$\frac{dE}{dx} \approx -\frac{b}{\beta^2}$$

We neglect here the weak energy dependence of b. For the muon range we then have

$$R = \int_{E_\mu}^{m_\mu c^2} \frac{1}{dE/dx}\, dE$$

$$= -\frac{1}{b}\int_{E_\mu}^{m_\mu c^2} \beta^2 dE \;\;,$$

with

$$E_\mu = m_\mu \gamma c^2 = \frac{m_\mu c^2}{\sqrt{1 - \beta^2}}$$

$$\beta^2 = 1 - \frac{m_\mu^2 c^4}{E_\mu^2}$$

So

$$R = \frac{1}{b}\int_{m_\mu c^2}^{E_\mu}\left(1 - \frac{m_\mu^2\, c^4}{E^2}\right) dE$$

$$= \frac{1}{b}\left|E + \frac{m_\mu^2\, c^4}{E}\right|_{m_\mu c^2}^{E_\mu}$$

$$= \frac{1}{b}\left(E_\mu + \frac{m_\mu^2 c^4}{E_\mu} - 2m_\mu c^2\right)$$

$$= \frac{1}{b}\frac{(E_\mu - m_\mu c^2)^2}{E_\mu} = \frac{1}{b}\frac{E_{kin}^2}{E_\mu}.$$

2.3. Bragg curve

Using Eq. 2.24 the energy loss, within a classical approximation, can be written as

$$-\frac{d}{dx}(\frac{1}{2}m_\mu \beta^2(x)c^2) = \frac{b}{\beta^2(x)}$$

$$-m_\mu c^2 \beta(x)\frac{d\beta}{dx} = \frac{b}{\beta^2(x)}$$

$$-\beta^3(x)\,d\beta = \frac{b}{m_\mu c^2}dx$$

$$\beta^4(x) - \beta_0^4 = -\frac{4b}{Mc^2}x$$

$$\beta^2(x) = \beta_0^2\sqrt{1 - \frac{4b}{m_\mu c^2 \beta_0^4}x}$$

$$\beta^2(x) = \beta_0^2\sqrt{1 - \frac{x}{R}}$$

With this depth dependence of $\beta(x)$, we can rewrite Eq. 2.24 as

$$-\frac{dE}{dx}(x) = \frac{b}{\beta_0^2\sqrt{1 - \frac{x}{R}}}$$

Qualitative estimate of the
Bragg curve obtained with
the calculated equation

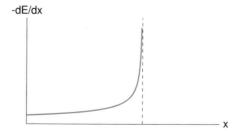

-dE/dx

x

This is of course a rough estimate, since the stopping power does not follow Eq. 2.24 for all velocities, and we have neglected energy straggling and scattering effects. When this is taken into account, the peak (Bragg peak) is less pronounced and its maximum occurs before the full range is reached.

3.1. Initial polarization in NMR

Consider a compound containing an isotope with spin I_N. The z component of the nuclear magnetic moment is

$$\mu_{N,z} = \gamma_N \hbar I_{N,z} \ ,$$

where $I_{N,z}$ can take one of the $2I_N + 1$ values between $-I_N$ and $+I_N$. The corresponding energies in a magnetic field B_z are

$$E_{I_{N,z}} = -\mu_{N,z} B_z = -\gamma_N \hbar \, I_{N,z} B_z$$

Under the influence of the field the spins align producing a macroscopic polarization (and magnetization). A level with energy $E_{I_{N,z}}$ is occupied with a probability given by the Boltzmann distribution

$$\frac{e^{-E_{I_{N,z}}/k_B T}}{\displaystyle\sum_{I_{N,z}=-I_N}^{I_{N,z}=+I_N} e^{-E_{I_{N,z}}/k_B T}}$$

The average spin $\langle I_{N,z}\rangle$ is obtained by summing over all values of $I_{N,z}$, The calculation takes into account that $E_{I_{N,z}}$ is small with respect to the thermal energy. So the exponentials can be expanded. Keeping only the first two terms and remembering that $\sum_{I_{N,z}=-I_N}^{I_{N,z}=+I_N} I_{N,z} = 0$

$$
\langle I_{N,z}\rangle = \frac{\displaystyle\sum_{I_{N,z}=-I_N}^{I_{N,z}=+I_N} I_{N,z} e^{-E_{I_{N,z}}/k_B T}}{\displaystyle\sum_{I_{N,z}=-I_N}^{I_{N,z}=+I_N} e^{-E_{I_{N,z}}/k_B T}} = \frac{\gamma_N \hbar \, I_{N,z} B_z}{k_B T} \frac{\displaystyle\sum_{I_{N,z}=-I_N}^{I_{N,z}=+I_N} I_{N,z}^2}{\displaystyle\sum_{I_{N,z}=-I_N}^{I_{N,z}=+I_N} 1}
$$

$$
= \frac{\gamma_N \hbar I_N (2I_N + 1)(I_N + 1)}{3 k_B T (2I_N + 1)} B_z = \frac{\gamma_N \hbar I_N (I_N + 1)}{3 k_B T} B_z
$$

If we take as an example ^{63}Cu, an isotope very often used for cuprate NMR studies, we have for the initial polarization with $I_{Cu} = 3/2$, $\gamma_{Cu} = 2\pi \times 11.3$ MHz/T in a field of 1 T and at 1 K

$$P(0) = \frac{\langle I_{Cu,z} \rangle}{I_{Cu}} = 4.52 \times 10^{-4} \ ,$$

which has to be compared with $P(0) \cong 1$ for μSR.

3.2. Sensitivity to a local field

In a local magnetic field the muon spin precesses with frequency

$$\nu_\mu = \frac{\gamma_\mu}{2\pi} B_\mu$$

We can estimate the minimum field value that the muon can detect by taking into account that in a typical μSR experiment the time window of the signal extends to about 10 μs. We need to observe about a quarter of the spin precession period to determine the field, i.e.,

$$\nu_{min} = \frac{1}{T} = 0.025 \text{ MHz} \ ,$$

which corresponds to a minimum field of 1.8×10^{-4} T.

In practice, even smaller magnetic fields, e.g. arising from nuclear moments, can be detected. In this case, the cumulative effect of the nuclear magnetic moment surrounding the muon leads to a detectable depolarization from fields of the order of a fraction of 10^{-4} T (fraction of Gauss).

For electronic moments $B_\mu \approx B_{dip} \approx \frac{\mu_0}{4\pi} \frac{\mu_j}{R_j^3} \approx \frac{\mu_j[\mu_B]}{R_j^3[\text{Å}^3]}$[T].

This means that assuming a distance between muon and electronic moment of $R_j \approx$ 2 Å, electronic moments of the order of $10^{-3}\mu_B$ can be detected.

3.3. Asymmetry expression in a real spectrometer

We consider the geometry of Fig. 3.10 with a Backward (B) and Forward (F) detector pair. The positron events in the histograms are given by

$$N_B(t) = N_{0,B} \, e^{-\frac{t}{\tau_\mu}} \left[1 + A_{0,B} \, P(t) \right] + B_{g,B}$$

$$N_F(t) = N_{0,F} \, e^{-\frac{t}{\tau_\mu}} \left[1 - A_{0,F} \, P(t) \right] + B_{g,F}$$

With the definitions $\alpha = N_{0,F}/N_{0,B}$ and $\beta = A_{0,F}/A_{0,B}$ we can write N_F in terms of the B detector parameters

$$N_F(t) - B_{g,F} = \alpha N_{0,B}\, e^{-\frac{t}{\tau_\mu}}\left[1 - \beta A_{0,B}\, P(t)\right]$$

Solving for $A_{0,B}\, P(t)$

$$\alpha\big(N_B(t) - B_{g,B}\big) - \big(N_F(t) - B_{g,F}\big) = A_{0,B}\, P(t)\alpha N_{0,B}(1+\beta)\, e^{-\frac{t}{\tau_\mu}}$$

$$\big(N_F(t) - B_{g,F}\big) + \alpha\beta\big(N_B(t) - B_{g,B}\big) = \alpha N_{0,B}(1+\beta)\, e^{-\frac{t}{\tau_\mu}}$$

$$A_{0,B}\, P(t) = \frac{\alpha\big(N_B(t) - B_{g,B}\big) - \big(N_F(t) - B_{g,F}\big)}{\big(N_F(t) - B_{g,F}\big) + \alpha\beta\big(N_B(t) - B_{g,B}\big)}$$

which can be written in terms of the raw asymmetry

$$A_{raw}(t) = \frac{\big[N_B(t) - B_{g,B}\big] - \big[N_F(t) - B_{g,F}\big]}{\big[N_B(t) - B_{g,B}\big] + \big[N_F(t) - B_{g,F}\big]},$$

$$A_{0,B}\, P(t) = \frac{(\alpha+1)A_{raw}(t) + (\alpha-1)}{(\alpha\beta+1) + (\alpha\beta-1)A_{raw}(t)}$$

or vice versa

$$A_{raw}(t) = \frac{A_{0,B}\, P(t)\big(\alpha\beta+1\big) - (\alpha-1)}{(\alpha+1) - A_{0,B}\, P(t)\big(\alpha\beta-1\big)}.$$

3.4. Evaluation of the parameter β
We consider the two most likely situations

1. The two detectors in the pair have different solid angles.
   Averaging the asymmetry over a finite solid angle gives a correction $\sin(\Delta\theta/2)/(\Delta\theta/2)$ to the ideal value $\bar A = 1/3$, see footnote 9, Chap. 3. Taking typical values of one detector subtending an angle $\Delta\theta = 45°$ and the other $55°$ gives $\beta = 0.987$.
2. The positrons reaching the two detectors must pass through different amounts of material. We assume that for one detector there is no absorption and that for the other one the positrons below an energy ε_{min} remain undetected due to absorption

or scattering. For the second detector Eq. 1.15 modifies to

$$A_{0,\Delta\Omega} = \int_{\varepsilon_{min}}^{1} a(\varepsilon)E(\varepsilon)\, d\varepsilon = \varepsilon^4\Big|_{\varepsilon_{min}}^{1} - \frac{2}{3}\varepsilon^3\Big|_{\varepsilon_{min}}^{1} = \frac{1}{3} - \varepsilon_{min}^4 + \frac{2}{3}\varepsilon_{min}^3 \ .$$

For a cutoff energy $\varepsilon_{min} = 0.2$, this increases the asymmetry by 0.00373, corresponding to $\beta = 0.989$.

3.5. Variation of the asymmetry with the parameter α
We start with Eq. 3.16 with $\beta = 1$

$$A_0\, P(t) = \frac{(\alpha+1)A_{raw}(t) + (\alpha-1)}{(\alpha+1) + (\alpha-1)A_{raw}(t)}$$

Assuming $(\alpha - 1) \ll 1$, we can approximate

$$A_0\, P(t) \cong \left(A_{raw}(t) + \frac{\alpha-1}{\alpha+1}\right)\left(1 - A_{raw}(t)\frac{\alpha-1}{\alpha+1}\right)$$

$$\cong \left(A_{raw}(t) + \frac{\alpha-1}{2}\right)\left(1 - A_{raw}(t)\frac{\alpha-1}{2}\right)$$

$$\Delta\big(A_0\, P(t)\big) \cong \Delta\big(\frac{\alpha-1}{2}\big)\big(1 - A_{raw}(t)^2\big) \cong \frac{\Delta\alpha}{2}$$

Thus, a change in α, which for the L and R detectors can be caused by a lateral shift of the beam, will cause an approximate shift of the asymmetry by half the change in α.

4.1. Time evolution of the polarization
We decompose $\mathbf{P}(t)$ in components parallel and perpendicular to the local field $\mathbf{B}_\mu$, see Fig. 4.1

$$\mathbf{P}(t) = \mathbf{P}_\parallel(t) + \mathbf{P}_\perp(t)$$

From the equation of motion 1.24

$$\frac{d\mathbf{P}(t)}{dt} = \gamma_\mu \left(\mathbf{P}(t) \times \mathbf{B}_\mu\right) \ ,$$

it follows that the parallel component is time independent.

We want to determine $P_z(t)$

$$P_z(t) = \frac{\mathbf{P}(t) \cdot \mathbf{P}(0)}{P(0)} = \frac{1}{P(0)} \left(\mathbf{P}_\parallel^2 + \mathbf{P}_\perp(t) \cdot \mathbf{P}_\perp(0) \right) ,$$

where

$$\mathbf{P}_\parallel = P(0) \cos\theta \, \hat{\mathbf{b}}$$

The perpendicular component of the polarization precesses in the plane perpendicular to $\hat{\mathbf{b}}$ defined by the Cartesian vectors $\hat{\mathbf{x}}_\perp$ and $\hat{\mathbf{y}}_\perp$

$$\mathbf{P}_\perp(t) = \begin{pmatrix} \cos(\omega_\mu t) & \sin(\omega_\mu t) \\ -\sin(\omega_\mu t) & \cos(\omega_\mu t) \end{pmatrix} \mathbf{P}_\perp(0)$$

$$\mathbf{P}_\perp(t) = P_\perp(0) \left[\sin(\omega_\mu t) \hat{\mathbf{x}}_\perp + \cos(\omega_\mu t) \hat{\mathbf{y}}_\perp \right] ,$$

with

$$\mathbf{P}_\perp(t) \cdot \mathbf{P}_\perp(0) = P_\perp^2(0) \cos(\omega_\mu t) = P^2(0) \sin^2\theta \cos(\omega_\mu t)$$

we obtain

$$P_z(t) = P_z(0) \left[\cos^2\theta + \sin^2\theta \, \cos(\omega_\mu t) \right] .$$

4.2. Time evolution of the polarization for planar field isotropy
We use Eq. 4.1, which we have to average over θ ($\phi = 0$)

$$P_z(t) = P_z(0) \frac{1}{2\pi} \int_0^{2\pi} \left[\cos^2\theta + \sin^2\theta \, \cos(\gamma_\mu B_\mu t) \right] d\theta$$

With

$$\int \cos^2\theta \, d\theta = \frac{1}{2}(1 + \frac{1}{2}\sin 2\theta) \qquad \text{and}$$

$$\int \sin^2\theta \, d\theta = \frac{1}{2}(1 - \frac{1}{2}\sin 2\theta) ,$$

we obtain

$$P_z(t) = P_z(0) \left[\frac{1}{2} + \frac{1}{2} \cos(\gamma_\mu B_\mu t) \right]$$

As in the spherical isotropic case leading to Eq. 4.11, this result can be guessed by considering that a planar randomly oriented local field, corresponds to a $1/2$ parallel or antiparallel component to the initial muon spin direction and $1/2$ perpendicular to it.

4.3. Discriminate between single valued field and isotropic distribution when measuring a polarization function given by Eq. 4.11
$P(0) = P_z(0)$ and $P_z(t)$ is the same in both cases

$$P_z(t) = P(0)\left[\frac{1}{3} + \frac{2}{3}\cos(\gamma_\mu B_\mu t)\right]$$

However, $P_y(t)$ is different.
For the single valued field with $\theta = \arccos(1/\sqrt{3}) = 54.736°$ and $\phi = 0°$

$$P_y(t) = P(0)\left[\frac{1}{2}\sin 2\theta \sin\phi[1 - \cos(\gamma_\mu B_\mu t)] + \sin\theta \cos\phi \sin(\gamma_\mu B_\mu t)\right]$$

becomes

$$P_y(t) = 0.40825\, P(0)\sin(\gamma_\mu B_\mu t)$$

For the isotropic case $P_y(t)=0$ when averaging over ϕ and θ.
This is an example where, to obtain the full information, it is necessary to measure not only $P_z(t)$ but also a transverse component of the polarization vector.

4.4. Short time behavior of Gaussian and Lorentzian Kubo-Toyabe functions
Expanding the exponential function we have for ZF

$$P_z^{GKT}(t) = \frac{1}{3} + \frac{2}{3}(1 - \sigma^2 t^2)\exp\left(-\frac{\sigma^2 t^2}{2}\right)$$

$$\simeq \frac{1}{3} + \frac{2}{3}(1 - \sigma^2 t^2)\left(1 - \frac{\sigma^2 t^2}{2} + \frac{\sigma^4 t^4}{8} + \cdots\right)$$

$$= \frac{1}{3} + \frac{2}{3} - \frac{\sigma^2 t^2}{3} - \frac{2\sigma^2 t^2}{3} + \frac{\sigma^4 t^4}{3} + \frac{\sigma^4 t^4}{12} + \cdots$$

$$= 1 - \sigma^2 t^2 + \frac{5\sigma^4 t^4}{12} + \cdots$$

$$\simeq \exp\left(-\sigma^2 t^2\right)$$

For TF

$$P_x^{\text{TF-GKT}}(t) = \exp\left(-\frac{\sigma^2 t^2}{2}\right)$$

$$\simeq 1 - \frac{\sigma^2 t^2}{2} + \frac{\sigma^4 t^4}{8} + \cdots$$

In ZF, the exponent of the Gaussian function is twice as large as in TF, or σ is a factor $\sqrt{2}$ higher. This can be understood qualitatively. In TF it is mainly the field broadening in the z direction that adds to the external field and contributes to the depolarization of the muon ensemble. In ZF, the broadening along the z direction (parallel to the initial spin direction) does not produce depolarization; however, the two transverse components x and y produce spin precession and hence depolarization. This discussion assumes that we can use the same field broadening for ZF and TF. This is, however, not always the case, see Sect. 5.7.1 and Eq. 5.105. For the Lorentz-Kubo-Toyabe function

$$P_z^{\text{LKT}}(t) = \frac{1}{3} + \frac{2}{3}(1 - at)\exp(-at)$$

$$\simeq \frac{1}{3} + \frac{2}{3}(1 - at)\left(1 - at + \frac{a^2 t^2}{2} + \cdots\right)$$

$$= \frac{1}{3} + \frac{2}{3} - \frac{2at}{3} - \frac{2at}{3} + \frac{2a^2 t^2}{3} + \frac{a^2 t^2}{3} + \cdots$$

$$= 1 - \frac{4at}{3} + a^2 t^2 + \cdots$$

$$\simeq \exp\left(-\frac{4at}{3}\right)$$

This is to be compared with

$$P_x^{\text{TF-LKT}}(t) = \exp(-at)$$

$$\simeq \left(1 - at + \frac{a^2 t^2}{2} + \cdots\right)$$

Similarly, the damping is higher in ZF than in TF.

5.1. Non-singularity of the dipolar field term, Eq. 5.7

The field components under consideration are of the form

$$\frac{\mu_0}{4\pi} \sum_{\beta=1,3} (\nabla_\alpha \nabla_\beta - \frac{1}{3}\nabla^2\delta_{\alpha\beta}) \frac{\mu_{\mu,\beta}}{\sqrt{x_1^2 + x_2^2 + x_3^2}} \quad ,$$

where $(\nabla_\alpha \nabla_\beta - \frac{1}{3}\nabla^2\delta_{\alpha\beta})/x$ is a 3×3 matrix.
To show that it is not singular at $x=0$, we integrate it over a sphere S.
In Cartesian coordinates

$$I_{\alpha\beta} = \int_S (\nabla_\alpha \nabla_\beta - \frac{1}{3}\nabla^2\delta_{\alpha\beta}) \frac{1}{\sqrt{x_1^2 + x_2^2 + x_3^2}} dx_1 dx_2 dx_3$$

It is easy to show that the integral is the 3×3 zero matrix $\mathbf{0}_3$. To convince ourselves, it is sufficient to consider one diagonal and one off-diagonal term, since the other terms give the same result, by symmetry.

$$I_{11} = \int_S \left[\frac{\partial^2}{\partial x_1^2} - \frac{1}{3}(\frac{\partial^2}{\partial x_1^2} + \frac{\partial^2}{\partial x_2^2} + \frac{\partial^2}{\partial x_3^2}) \right] \frac{1}{\sqrt{x_1^2 + x_2^2 + x_3^2}} dx_1 dx_2 dx_3$$

$$= \int_S \left[\frac{2}{3}\frac{\partial^2}{\partial x_1^2} - \frac{1}{3}\frac{\partial^2}{\partial x_2^2} - \frac{1}{3}\frac{\partial^2}{\partial x_3^2}) \right] \frac{1}{\sqrt{x_1^2 + x_2^2 + x_3^2}} dx_1 dx_2 dx_3$$

$$I_{12} = \int_S \frac{\partial^2}{\partial x_1 \partial x_2} \frac{1}{\sqrt{x_1^2 + x_2^2 + x_3^2}} dx_1 dx_2 dx_3$$

I_{11} contains terms of the form

$$\int_S \frac{\partial^2}{\partial x_1^2} \frac{1}{\sqrt{x_1^2 + x_2^2 + x_3^2}} dx_1 dx_2 dx_3 =$$

$$= \int_S \left[\frac{3x_1^2}{(x_1^2 + x_2^2 + x_3^2)^{\frac{5}{2}}} - \frac{1}{(x_1^2 + x_2^2 + x_3^2)^{\frac{3}{2}}} \right] dx_1 dx_2 dx_3$$

$$= \int_S \left[\frac{2x_1^2 - x_2^2 - x_3^2}{(x_1^2 + x_2^2 + x_3^2)^{\frac{5}{2}}} \right] dx_1 dx_2 dx_3 = 0$$

For I_{12} we have

$$I_{1,2} = \int_S \frac{\partial^2}{\partial x_1 \partial x_2} \frac{1}{\sqrt{x_1^2 + x_2^2 + x_3^2}} dx_1 dx_2 dx_3$$

$$= \int_S \frac{\partial}{\partial x_1} \frac{-x_2}{(x_1^2 + x_2^2 + x_3^2)^{\frac{3}{2}}} dx_1 dx_2 dx_3$$

$$= \int_S \frac{3 x_1 x_2}{(x_1^2 + x_2^2 + x_3^2)^{\frac{5}{2}}} dx_1 dx_2 dx_3 = 0$$

By symmetry and since all Cartesian directions are equivalent, it follows that $I_{\alpha\beta} = 0$, for α and $\beta = 1, 2, 3$.

5.2. Simple derivation of the contact field
The **B** field inside a magnetized sphere is given by

$$\mathbf{B} = \mu_0(\mathbf{M} + \mathbf{H})$$

In zero external field only the demagnetizing field contributes to **H**

$$\mathbf{H} = \mathbf{H}_{\text{dem}} = -N\mathbf{M} = -\frac{1}{3}\mathbf{M}$$

so that

$$\mathbf{B} = \frac{2}{3}\mu_0\mathbf{M}$$

We consider a sphere of arbitrarily small radius R (and volume V) around the muon and an electron with a density at the muon site of $|\psi_e(0)|^2$. In the sphere, this density produces a magnetization

$$\mathbf{M} = \frac{\mu_e \int_V |\psi_e(r)|^2 d^3r}{V} \xrightarrow{R \to 0} \mu_e |\psi_e(0)|^2$$

Therefore, the contact field sensed by the muon is

$$\mathbf{B}_{\text{cont}} = \frac{2\mu_0}{3} \mu_e \, |\psi_e(0)|^2 \quad .$$

5.3. Dynamical range, μSR fluctuation time window

Lower limit. Very slow fluctuations are observed in the $1/3$ tail of the Kubo-Toyabe function, Eq. 4.62

$$P(t) \sim \exp\left(-\frac{2t}{3\tau_c}\right) = \exp\left(-\frac{t}{T}\right)$$

$$T = \frac{3}{2}\tau_c$$

Over a typical time window of 10 μs a decay due to a relaxation time $T \sim 10 - 100\,\mu$s is observable, which corresponds to $\tau_c \sim 10^{-5}$–10^{-4} s.

Upper limit. Very fast fluctuations lead to an exponential decay of the polarization

$$P(t) \sim \exp\left(-2\sigma^2\tau_c t\right) = \exp\left(-\frac{t}{T}\right)$$

$$T = \frac{1}{2\sigma^2\tau_c}$$

Again, with an observable $T \sim 10$–100 μs and a typical field width in magnetic materials of 1 kG $= \sigma/\gamma_\mu$ we obtain $\tau_c \sim 10^{-11}$–10^{-12} s.

5.4. Error of the first order Knight-shift expression

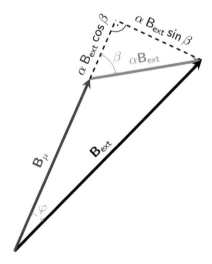

We assume $|\mathbf{B}_\mu - \mathbf{B}_{ext}| = \alpha B_{ext}$ with $\alpha \ll 1$.

We first calculate the exact Knight-shift expression

$$K_{exp} = \frac{|\mathbf{B}_\mu| - |\mathbf{B}_{ext}|}{|\mathbf{B}_{ext}|}$$

From the figure above

$$(B_\mu + \alpha B_{ext} \cos \beta)^2 + (\alpha B_{ext} \sin \beta)^2 = B_{ext}^2$$

$$B_\mu^2 + 2\alpha B_\mu B_{ext} \cos \beta + B_{ext}^2(\alpha^2 - 1) = 0 \; ,$$

solving for B_μ and taking the positive solution

$$B_\mu = -\alpha B_{ext} \cos \beta + \sqrt{B_{ext}^2(1 + \alpha^2 \cos^2\beta - \alpha^2)} \; ,$$

and the terms to order α^2 we obtain

$$\frac{B_\mu}{B_{ext}} = -\alpha \cos \beta + \frac{1}{2}\alpha^2 \sin^2\beta + 1$$

Therefore for the exact expression

$$K_{exp} = \frac{|\mathbf{B}_\mu| - |\mathbf{B}_{ext}|}{|\mathbf{B}_{ext}|} = -\alpha \cos \beta + \frac{1}{2}\alpha^2 \sin^2\beta$$

This must be compared to the approximate expression

$$K'_{exp} = \frac{\mathbf{B}_{ext} \cdot (\mathbf{B}_\mu - \mathbf{B}_{ext})}{B_{ext}^2} = \frac{B_{ext} B_\mu \cos \varphi - B_{ext}^2}{B_{ext}^2}$$

With

$$\cos \varphi = \frac{B_\mu + \alpha B_{ext} \cos \beta}{B_{ext}} \; ,$$

and using the above result for $\frac{B_\mu}{B_{ext}}$ we obtain

$$K'_{exp} = \frac{\mathbf{B}_{ext} \cdot (\mathbf{B}_\mu - \mathbf{B}_{ext})}{B_{ext}^2} = -\alpha \cos \beta + \alpha^2 \sin^2\beta$$

Comparing this result with the exact expression above, we see that the error introduced by using the approximate Knight-shift expression, $\frac{1}{2}\alpha^2 \sin^2 \beta$, is only of the order of α^2 and therefore negligible. It is maximal when $\mathbf{B}_\mu - \mathbf{B}_{ext}$ is perpendicular to $\mathbf{B}_\mu$.

5.5. Hyperfine constants from temperature dependent Knight-shift measurements

The purpose of this exercise is to become familiar with the units kG/μ_B. From Fig. 5.43 we can roughly read

$$K_\mu^a \cong 0.002 \quad \text{for } \chi_a \cong 0.011 \text{ emu/mol}$$

$$K_\mu^b \cong 0.0001 \quad \text{for } \chi_b \cong 0.010 \text{ emu/mol}$$

$$K_\mu^c \cong -0.001 \quad \text{for } \chi_c \cong 0.009 \text{ emu/mol}$$

With

$$K_\mu^a(T) = (\tilde{\mathcal{D}}_{\mathbf{r}_\mu}^{aa} + A_{\text{cont},fd,\mathbf{r}_\mu}) \, \chi_a(T)$$

$$K_\mu^b(T) = (\tilde{\mathcal{D}}_{\mathbf{r}_\mu}^{bb} + A_{\text{cont},fd,\mathbf{r}_\mu}) \, \chi_b(T)$$

$$K_\mu^c(T) = (\tilde{\mathcal{D}}_{\mathbf{r}_\mu}^{cc} + A_{\text{cont},fd,\mathbf{r}_\mu}) \, \chi_c(T)$$

The terms $(\tilde{\mathcal{D}}_{\mathbf{r}_\mu}^{ii} + A_{\text{cont},fd,\mathbf{r}_\mu})$ in kG/μ_B for $i = (a, b, c)$ can be determined using Eq. E.3

$$(\tilde{\mathcal{D}}_{\mathbf{r}_\mu}^{ii} + A_{\text{cont},fd,\mathbf{r}_\mu}) \, [kG/\mu_B] = \frac{K_\mu^i N_A C \mu_B}{\chi_i \, [\text{emu/mol}]} \quad,$$

giving

$$\tilde{\mathcal{D}}_{\mathbf{r}_\mu}^{aa} + A_{\text{cont},fd,\mathbf{r}_\mu} \cong 1 \text{ kG}$$

$$\tilde{\mathcal{D}}_{\mathbf{r}_\mu}^{bb} + A_{\text{cont},fd,\mathbf{r}_\mu} \cong 0.06 \text{ kG}$$

$$\tilde{\mathcal{D}}_{\mathbf{r}_\mu}^{cc} + A_{\text{cont},fd,\mathbf{r}_\mu} \cong -0.62 \text{ kG}$$

Making use of the traceless property of the dipolar tensor as an additional equation, we obtain

$$A_{\text{cont},fd,\mathbf{r}_\mu} \cong 0.15 \text{ kG}$$

$$\tilde{\mathcal{D}}_{\mathbf{r}_\mu}^{aa} \cong 0.85 \text{ kG}$$

$$\tilde{\mathcal{D}}_{\mathbf{r}_\mu}^{bb} \cong -0.09 \text{ kG}$$

$$\tilde{\mathcal{D}}_{\mathbf{r}_\mu}^{cc} \cong -0.77 \text{ kG}$$

These values compare well with those given in the text.

5.6. Total dipolar tensor of a cubic crystal

We recall the general expressions 5.19 and 5.80

$$\tilde{\mathcal{D}}_{\mathbf{r}_\mu}^{\alpha\beta} = \frac{v_0}{\mu_0} \sum_{\text{Lor}} \mathcal{D}_{j,\alpha\beta}$$

$$\mathcal{D}_{j,\alpha\beta}(\mathbf{r}_\mu) = \frac{\mu_0}{4\pi R_j^3} \left(\frac{3R_{j,\alpha} R_{j,\beta}}{R_j^2} - \delta_{\alpha\beta} \right) ,$$

where the jth-moment is at lattice position $\mathbf{r}_j$ and at position $\mathbf{R}_j = \mathbf{r}_j - \mathbf{r}_\mu$ from the muon site.

We take as Cartesian system of coordinates the crystal axis of the cubic crystal and consider the muon site $\mathbf{r}_\mu = (\frac{1}{2}, 0, 0)$, see Fig. 5.46 left panel.

Because of symmetry only the non-diagonal terms of $\tilde{\mathcal{D}}_{(\frac{1}{2},0,0)}$ are different from zero. Also from symmetry considerations $\tilde{\mathcal{D}}_{(\frac{1}{2},0,0)}^{yy} = \tilde{\mathcal{D}}_{(\frac{1}{2},0,0)}^{zz}$. This, together with the traceless property of the dipolar tensor gives

$$\tilde{\mathcal{D}}_{(\frac{1}{2},0,0)}^{xx} = -\frac{1}{2} \tilde{\mathcal{D}}_{(\frac{1}{2},0,0)}^{yy} = -\frac{1}{2} \tilde{\mathcal{D}}_{(\frac{1}{2},0,0)}^{zz}$$

with $\tilde{\mathcal{D}}_{(\frac{1}{2},0,0)}^{xx} \equiv \tilde{\mathcal{D}}$ we obtain the $\tilde{\mathcal{D}}_{(\frac{1}{2},0,0)}$ tensor of Eq. 5.91. Similarly for the two other sites.

5.7. Magic angle for zero dipolar contribution

We recall the expression for the dipolar contribution, Eq. 5.87

$$
\begin{aligned}
B'_{\text{dip},\parallel} = &\tfrac{1}{3}(\tilde{\mathcal{D}}_{\mathbf{r}_\mu}^{xx}\chi_x + \tilde{\mathcal{D}}_{\mathbf{r}_\mu}^{yy}\chi_y + \tilde{\mathcal{D}}_{\mathbf{r}_\mu}^{zz}\chi_z) B_{\text{ext}} \\
&+ \tfrac{2}{3}[\tilde{\mathcal{D}}_{\mathbf{r}_\mu}^{zz}\chi_z - \tfrac{1}{2}(\tilde{\mathcal{D}}_{\mathbf{r}_\mu}^{xx}\chi_x + \tilde{\mathcal{D}}_{\mathbf{r}_\mu}^{yy}\chi_y)] P_2^0(\cos\theta) B_{\text{ext}} \\
&- \tfrac{1}{3}\tilde{\mathcal{D}}_{\mathbf{r}_\mu}^{xz}(\chi_x + \chi_z) P_2^1(\cos\theta)\cos\phi B_{\text{ext}} \\
&- \tfrac{1}{3}\tilde{\mathcal{D}}_{\mathbf{r}_\mu}^{yz}(\chi_y + \chi_z) P_2^1(\cos\theta)\sin\phi B_{\text{ext}} \\
&+ \tfrac{1}{6}(\tilde{\mathcal{D}}_{\mathbf{r}_\mu}^{xx}\chi_x - \tilde{\mathcal{D}}_{\mathbf{r}_\mu}^{yy}\chi_y) P_2^2(\cos\theta)\cos 2\phi B_{\text{ext}} \\
&+ \tfrac{1}{6}\tilde{\mathcal{D}}_{\mathbf{r}_\mu}^{xy}(\chi_x + \chi_y) P_2^2(\cos\theta)\sin 2\phi B_{\text{ext}} ,
\end{aligned}
\tag{5.87}
$$

where the external field is expressed with the polar and azimuthal angles in the crystal frame $\mathbf{B}_{\text{ext}} = B_{\text{ext}}(\sin\theta\cos\phi, \sin\theta\sin\phi, \cos\theta)$.

For a cubic crystal $\chi_x = \chi_y = \chi_z \equiv \chi$. With the dipolar tensors at the muon sites given by Eq. 5.91 only the second term of the above equation is not zero

$$B'_{\text{dip},\parallel} = \frac{2}{3}[\tilde{\mathcal{D}}^{zz}_{\mathbf{r}_\mu}\chi_z - \frac{1}{2}(\tilde{\mathcal{D}}^{xx}_{\mathbf{r}_\mu}\chi_x + \tilde{\mathcal{D}}^{yy}_{\mathbf{r}_\mu}\chi_y)]\, P_2^0(\cos\theta) B_{\text{ext}}$$

Inserting the values of the dipolar constants and susceptibilities we obtain

$$B'_{\text{dip},\parallel,(\frac{1}{2},0,0)} = -\frac{1}{4}\tilde{\mathcal{D}}(3\cos^2\theta - 1)\chi\, B_{\text{ext}}$$

$$B'_{\text{dip},\parallel,(0,\frac{1}{2},0)} = -\frac{1}{4}\tilde{\mathcal{D}}(3\cos^2\theta - 1)\chi\, B_{\text{ext}}$$

$$B'_{\text{dip},\parallel,(0,0,\frac{1}{2})} = \frac{1}{2}\tilde{\mathcal{D}}(3\cos^2\theta - 1)\chi\, B_{\text{ext}}$$

This shows that in the paramagnetic phase, when the external field is directed along one of the diagonals of the cube corresponding to the magic angle $\theta_m = \arccos(1/\sqrt{3}) \approx 54.7°$, the dipolar contribution to the muon local field disappears.

6.1. Superconducting gap

The temperature dependence of the normalized gap function $\Delta(T)/\Delta(0)$ obtained from the weak-coupling self-consistent equation of BCS theory (Tinkham 1996) is shown in the Figure below. Numerical solutions are tabulated in Mühlschlegel (1959) and Johnston (2013).

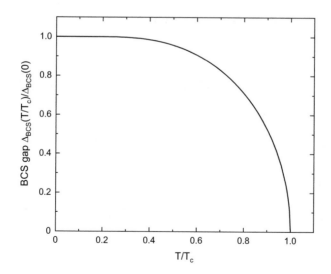

Normalized BCS gap as a function of the normalized temperature T/T_c

To analyze μSR data on superconductors, parametrizations of the gap functions are often used. One of the first parametrizations of the BCS gap function has been given by (Sheahen 1966)

$$\Delta(T) = \Delta(0) \sqrt{\cos\left[\frac{\pi}{2}\left(\frac{T}{T_c}\right)^2\right]}$$

Most commonly used in μSR is the expression given by Carrington & Manzano (2003)

$$\Delta(T) = \Delta(0) \tanh\left[1.82[1.018(T_c/T - 1)]^{0.51}\right]$$

The BCS equation for weak coupling has also been used for nonsymmetric gaps (e.g. for d-wave (Won & Maki 1994)). Also in Gross et al. (1986), besides the s-wave gap on a spherical Fermi surface, several uniaxial gaps with $\hat{\mathbf{l}}$ as local axis of gap symmetry have been considered. The corresponding normalized gap functions can be expressed by a universal function with the parameter $\Delta(0)/k_B T_c$ and the parameter $a'\Delta C/C$, which depends on the geometry of the pairing state and is related to the specific heat jump at T_c

$$\Delta(T) = \Delta(0) \tanh\left[\frac{\pi k_B T_c}{\Delta(0)}\sqrt{a'\frac{\Delta C}{C}(T_c/T - 1)}\right]$$

We summarize here parametrizations for various gaps.

For an isotropic gap

$a' = \frac{2}{3}$, $\frac{\Delta(0)}{k_B T_c} = 1.764$, $\frac{\Delta C}{C} = 1.43$, $a = a'\frac{\Delta C}{C} = 0.953$, $D \equiv \frac{\pi k_B T_c}{\Delta(0)}\sqrt{a} = 1.739$

For an axial p-wave gap (two point nodes) $\Delta_\mathbf{k} = \Delta_0(T)|\hat{\mathbf{k}} \times \hat{\mathbf{l}}|$

$a' = 1$, $\frac{\Delta(0)}{k_B T_c} = 2.03$, $\frac{\Delta C}{C} = 1.19$, $a = a'\frac{\Delta C}{C} = 1.19$, $D \equiv \frac{\pi k_B T_c}{\Delta(0)}\sqrt{a} = 1.688$

For a polar p-wave gap (equatorial line of nodes) $\Delta_\mathbf{k} = \Delta_0(T)\hat{\mathbf{k}} \cdot \hat{\mathbf{l}}$

$a' = 2$, $\frac{\Delta(0)}{k_B T_c} = 2.46$ $\frac{\Delta C}{C} = 0.78$, $a = a'\frac{\Delta C}{C} = 1.56$ $D \equiv \frac{\pi k_B T_c}{\Delta(0)}\sqrt{a} = 1.595$

Other parametrizations for isotropic and nonisotropic gaps can be found in Prozorov & Giannetta (2006)

$$\Delta(T) = \Delta(0) \tanh\left[\frac{\pi k_B T_c}{\Delta(0)}\sqrt{a\,(T_c/T - 1)}\right]$$

For the s-wave $\Delta(T, \phi) = \Delta(T)$

$a = 1,$ $\qquad \frac{\Delta(0)}{k_B T_c} = 1.764,$ $\qquad D \equiv \frac{\pi k_B T_c}{\Delta(0)}\sqrt{a} = 1.781$

For the d-wave $\Delta(T, \phi) = \Delta(T)\cos 2\phi$

$a = \frac{4}{3},$ $\qquad \frac{\Delta(0)}{k_B T_c} = 2.14,$ $\qquad D \equiv \frac{\pi k_B T_c}{\Delta(0)}\sqrt{a} = 1.694$

For the extended d-wave $\Delta(T, \phi) = \Delta(T)\big[B\cos(2\phi) + (1 - B)\cos(6\phi)\big].$[1]
For $B = 1.43$

$a = 0.72,$ $\qquad \frac{\Delta(0)}{k_B T_c} = 1.55,$ $\qquad D \equiv \frac{\pi k_B T_c}{\Delta(0)}\sqrt{a} = 1.72$

For the $(s + g)$-wave $\Delta(T, \theta, \phi) = \frac{\Delta(T)}{2}\big[1 - \sin^4\theta\cos 2\phi\big]$

$a = 2,$ $\qquad \frac{\Delta(0)}{k_B T_c} = 2.77,$ $\qquad D \equiv \frac{\pi k_B T_c}{\Delta(0)}\sqrt{a} = 1.60$

All above expressions are of the form

$$\Delta(T) = \Delta(0) \tanh\left[D\sqrt{(T_c/T - 1)}\right]$$

We remark that D does not vary very much for different gap symmetries.[2] The following Figure shows the relative differences between the numerical solution of the BCS gap equation (Mühlschlegel 1959) and parametrizations for different gap symmetries (Sheahen 1966; Carrington & Manzano 2003; Gross et al. 1986; Prozorov & Giannetta 2006).

---

[1] This expression gives for $B = 1$ the d-wave gap. For our comparison we take the unusual value of $B = 1.43$ found for optimally electron-doped cuprates (Prozorov 2008).

[2] Note that in a fit $\Delta(0)$ and T_c are the fit parameters, although originally D is obtained from a theoretical ratio of these two quantities. This is one of the lacks of self-consistency of the α-model, which is however empirically very successful in extracting gap values, see also remarks in footnote 31, Chap. 6.

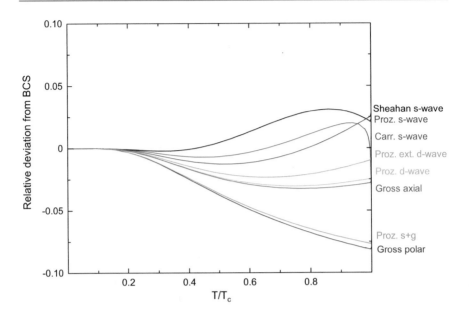

Relative differences between the temperature dependence of the normalized BCS gap and parametrizations of the normalized gap for different pairing state symmetries (Sheahen 1966; Carrington & Manzano 2003; Prozorov & Giannetta 2006; Gross et al. 1986)

Most of the gap curves differ only by ±3% from the BCS dependence, except for the polar gap, which has a maximum deviation of 10% very close to T_c where the gap is small anyway.

This finding supports the use of the α-model also outside the BCS s-wave weak coupling regime.

6.2. London model

From Eqs. 6.57 and 6.49

$$k^2 \geq \frac{16\pi^2}{3d^2} \quad \text{with} \quad \frac{1}{d^2} = \frac{\langle B_z \rangle \sqrt{3}}{2\Phi_0}$$

From Ginzburg-Landau theory (Tinkham 1996)

$$B_{c1} \approx \frac{\Phi_0}{4\pi\lambda^2} \ln\kappa \ ,$$

so that the condition $k^2\lambda^2 \gg 1$ translates into

$$\frac{2\pi}{\sqrt{3}} \frac{\langle B_z \rangle}{B_{c1}} \ln\kappa \gg 1 \ ,$$

which is always satisfied in extreme type-II superconductors, where the London model applies, or when $\langle B_z \rangle \gg B_{c1}$.

7.1. Level crossing in vacuum muonium

From Eq. 7.16, for a crossing between level $|1\rangle$ and $|2\rangle$ we must have

$$\frac{1}{4}\hbar\omega_0 + \frac{1}{2}\hbar(\omega_e - \omega_\mu) = -\frac{1}{4}\hbar\omega_0 + \underbrace{\frac{1}{2}\hbar\left[(\omega_e + \omega_\mu)^2 + \omega_0^2\right]^{1/2}}_{\cong \frac{1}{2}\hbar(\omega_e + \omega_\mu)},$$

from which it follows

$$\frac{1}{2}\omega_0 \cong \omega_\mu = \gamma_\mu B_{cross}$$

$$B_{cross} \cong \frac{\omega_0}{2\gamma_\mu} = \frac{2\pi \times 4463.3 \text{ MHz}}{2 \times 2\pi \times 135.54 \text{ MHz T}^{-1}} \cong 16.46 \text{ T}$$

The existence of a crossing is expected based on the following argument. In the left panel of Fig. 7.2 the eigenstate $|2\rangle$ has a higher energy than $|1\rangle$. However, at very high fields, the most energetic state must be the eigenstate $|2\rangle \rightarrow |\downarrow_\mu\uparrow_e\rangle$, which corresponds to the magnetic moments of muon and electron pointing antiparallel to the external field direction, thus maximizing the Zeeman energy (remember that the spin direction of the electron is opposite to the magnetic moment). Similarly, the lowest energy state is the eigenstate $|4\rangle \rightarrow |\uparrow_\mu\downarrow_e\rangle$, which corresponds to both magnetic moments pointing parallel to the external field.

7.2. Muon spin polarization in muonium in a longitudinal field

The starting point is Eq. 7.29

$$\mathbf{P}(t) = \frac{1}{4}\sum_{k=1}^{4}\langle k|\mathbf{P}(0)\cdot\boldsymbol{\sigma}_\mu|n\rangle\,\langle n|\boldsymbol{\sigma}_\mu|k\rangle\,\exp(i\omega_{nk}t)$$

At $t = 0$ the muon spin is pointing in the direction of the applied field

$$\mathbf{P}(0) = P_z\,\hat{\mathbf{z}}\,,\quad P_z = 1 \text{ and } \mathbf{B}_{ext} = B_{ext}\,\hat{\mathbf{z}}$$

Therefore, $\mathbf{P}(0)\cdot\boldsymbol{\sigma}_\mu = \sigma_{\mu,z}$.[3] and Eq. 7.29 becomes

$$P_z(t) = \frac{1}{4}\sum_{k,n}\langle k|\sigma_{\mu,z}|n\rangle\,\langle n|\sigma_{\mu,z}|k\rangle\,\exp(i\omega_{nk}t)$$

---

[3] If the initial polarization is in the $i = x, y, z$ direction $\mathbf{P}_\mu(0)\cdot\boldsymbol{\sigma}_\mu = \sigma_{\mu,i}$ and $\rho(0) = 1/4(1+\sigma_{\mu,i})$.

Using the eigenstates $|k\rangle$ given in Eq. 7.18 we calculate the matrix elements

$$
\begin{aligned}
P_z(t) &= \frac{1}{4}\Big[\langle 1|\sigma_{\mu,z}|1\rangle^2 + \langle 2|\sigma_{\mu,z}|2\rangle^2 + \langle 3|\sigma_{\mu,z}|3\rangle^2 + \langle 4|\sigma_{\mu,z}|4\rangle^2 \\
&\quad + \langle 2|\sigma_{\mu,z}|4\rangle \langle 4|\sigma_{\mu,z}|2\rangle \exp(-i\omega_{24}\,t) \\
&\quad + \langle 4|\sigma_{\mu,z}|2\rangle \langle 2|\sigma_{\mu,z}|4\rangle \exp(i\omega_{24}\,t)\Big] \\
&= \frac{1}{4}\Big[1^2 + (\sin^2\delta - \cos^2\delta)^2 + (-1)^2 + (\cos^2\delta - \sin^2\delta)^2 \\
&\quad + 4\sin^2\delta\,\cos^2\delta\,\exp(-i\omega_{24}\,t) + 4\sin^2\delta\,\cos^2\delta\,\exp(i\omega_{24}\,t)\Big] \\
&= \frac{1}{4}\Big[2 + 2\left(\cos^2\delta - \sin^2\delta\right)^2 + 4\cos^2\delta\,\sin^2\delta\left(e^{i\omega_{24}t} + e^{-i\omega_{24}t}\right)\Big] \\
&= \frac{1}{2}\Big[1 + \left(\cos^2\delta - \sin^2\delta\right)^2 + 4\cos^2\delta\,\sin^2\delta\,\cos(\omega_{24}\,t)\Big]
\end{aligned}
$$

With $(\cos^2\delta - \sin^2\delta)^2 = (\sin^2\delta - \cos^2\delta)^2 = x_B^2/(1 + x_B^2)$ and $4\sin^2\delta\,\cos^2\delta = 1/(1 + x_B^2)$ we obtain Eq. 7.30

$$
\begin{aligned}
P_z(t) &= \frac{1}{4}\left[2 + \frac{2x_B^2}{1 + x_B^2} + \frac{1}{1 + x_B^2}\cos(\omega_{24}\,t)\right] \\
&= \frac{1}{2}\left[1 + \frac{1}{1 + x_B^2}\left(x_B^2 + \cos(\omega_{24}\,t)\right)\right] \quad .
\end{aligned}
$$

7.3. Time resolution of a spectrometer and detectable frequencies

We consider how the asymmetry signal from a precession at frequency ν

$$
A(t) = A_0 \cos(2\pi \nu t)
$$

is reduced by a finite time resolution.

We assume the normalized time resolution function of the spectrometer to be a Gaussian with standard deviation σ_r

$$
\frac{1}{\sqrt{2\pi}\,\sigma_r}\exp\left(-\frac{t'^2}{2\sigma_r^2}\right)
$$

The observable asymmetry becomes

$$
\begin{aligned}
A_{\mathrm{obs}}(t) &= A_0 \frac{1}{\sqrt{2\pi}\,\sigma_r}\int_{-\infty}^{\infty}\exp\left(-\frac{t'^2}{2\sigma_r^2}\right)\cos[2\pi\nu(t - t')]dt' \\
&= A_0 \frac{1}{\sqrt{2\pi}\,\sigma_r}\left[\int_{-\infty}^{\infty}\exp\left(-\frac{t'^2}{2\sigma_r^2}\right)\cos(2\pi\nu t')dt'\right]\cos(2\pi\nu t)
\end{aligned}
$$

$$= A_0 \exp\left[-2(\pi v \sigma_r)^2\right] \cos(2\pi vt)$$

$$= A(v) \cos(2\pi vt) \ ,$$

where we use the identity

$$\cos[2\pi v(t - t')] = \cos(2\pi vt)\cos(2\pi vt') + \sin(2\pi vt)\sin(2\pi vt') \ ,$$

take into account that the odd sine term does not contribute to the integral, and use the well-known Fourier transform of a Gaussian function (Abramowitz & Stegun 1972).

We see that the finite time resolution reduces the measured asymmetry amplitude by a frequency dependent factor

$$\frac{A(v)}{A(0)} = \exp\left[-2(\pi v \sigma_r)^2\right]$$

From this expression it follows that, as a rule of thumb, to measure a frequency v one needs a time resolution (FWHM) $\Delta t = \sqrt{8 \ln 2}\, \sigma_r = 2.355\, \sigma_r$ of the order of $1/2v$.

7.4. Frequency beating in muonium
From Eq. 7.16

$$v_1 = \frac{1}{4}v_0 + \frac{1}{2}(v_e - v_\mu)$$

$$v_2 = -\frac{1}{4}v_0 + \frac{1}{2}\left[(v_e + v_\mu)^2 + v_0^2\right]^{1/2}$$

$$v_3 = \frac{1}{4}v_0 - \frac{1}{2}(v_e - v_\mu)$$

$$v_4 = -\frac{1}{4}v_0 - \frac{1}{2}\left[(v_e + v_\mu)^2 + v_0^2\right]^{1/2}$$

Hence for the triplet frequencies

$$v_{12} = \frac{1}{2}v_0 + \frac{1}{2}(v_e - v_\mu) - \frac{1}{2}\left[(v_e + v_\mu)^2 + v_0^2\right]^{1/2}$$

$$v_{23} = -\frac{1}{2}v_0 + \frac{1}{2}(v_e - v_\mu) + \frac{1}{2}\left[(v_e + v_\mu)^2 + v_0^2\right]^{1/2} \ ,$$

for the splitting

$$v_{23} - v_{12} = -v_0 + v_0 \sqrt{\left(1 + \frac{(v_e + v_\mu)^2}{v_0^2}\right)}$$

$$\cong \frac{(v_e + v_\mu)^2}{2v_0} \quad ,$$

and finally

$$v_0 \cong \frac{(v_e + v_\mu)^2}{2(v_{23} - v_{12})}$$

This shows that, for low fields, the hyperfine frequency can be obtained from the precession and beating frequencies.

8.1. Figure of merit

The physical information is in the asymmetric part of the spectrum. To be significant, it must be greater than the statistical noise.
Thus, for each channel of a μSR histogram

$$A_0 N_i \gtrsim \sqrt{N_i} \qquad \text{or}$$

$$A_0^2 N_i \gtrsim 1 \quad ,$$

where A_0 is the experimental asymmetry of the spectrometer and N_i is the number of decays in the channel i recorded in a measurement time t and proportional to the beam intensity I.
From this it follows that the product $A_0^2 I$ can be taken as figure of merit of a μSR setup. The expression shows that a gain in asymmetry has a quadratic weight, while an increase in beam intensity has only a linear weight.

9.1. Hyperfine splitting in muonic hydrogen

The hyperfine interaction between a negative muon and a nucleus in a muonic atom is analogous to that in a normal electronic atom.
For the contact hyperfine splitting in conventional hydrogen, Eq. 7.7, we have

$$A_H \propto \mu_e \mu_p |\psi_{1s}(0)|^2 \propto \mu_e \mu_p \frac{1}{a_H^3}$$

Assuming that we can use the same expression for the muonic wave function in the $1s$ state, the contact hyperfine splitting in muonic hydrogen is

$$A_{\mu p} \propto \mu_\mu \mu_p |\psi_{1s,\mu}(0)|^2 \propto \mu_\mu \mu_p \frac{1}{a_{\mu p}^3}$$

Considering that the magnetic moments of electron and muon are inversely proportional and that the orbital radii are proportional to their masses we obtain for the ratio

$$\frac{A_{\mu p}}{A_H} = \left(\frac{m_\mu}{m_e}\right)^2 = 207^2 = 4.28 \times 10^4$$

Experimentally we have, see footnote 10, Chap. 9

$$\frac{A_{\mu p}}{A_H} = \frac{0.183 \text{ eV}}{5.88 \times 10^{-6} \text{ eV}} = 3.11 \times 10^4 \ .$$

10.1. Elementary charge value
In natural units $\hbar = c = 1 = \varepsilon_0$. The value of e can be calculated by taking into account that the fine structure constant α has the same value in the different systems of units.

In S.I.

$$\alpha = \frac{e^2}{4\pi\varepsilon_0\hbar c} \cong \frac{1}{137}$$

So in natural units we have

$$e = \sqrt{4\pi\alpha} \cong 0.30282 \ .$$

10.2. Hand-waving arguments for the $\Gamma_{i \to f}$ expression, Eq. 10.32
The decay rate probability $\Gamma_{i \to f}$ must depend quadratically on the coupling constant G_F because the decay process has been treated in a first-order approximation, which involves squaring the transition amplitude proportional to the coupling constant. The other parameter that determines the physics of the muon decay is the muon mass, since we have neglected the electron mass.
From dimension-based arguments $\Gamma_{i \to f}$ must be $\propto m_\mu^5$, since in natural units $\Gamma_{i \to f}$ as well as the mass have the dimension of an energy, and G_F that of an inverse energy squared. Therefore $\Gamma_{i \to f} \propto G_F^2 m_\mu^5$.

References

Abramowitz, M., & Stegun, I. A. (1972). *Handbook of Mathematical Functions with Formulas, Graphs, and Mathematical Tables*. Wiley. ISBN: 978-0486612720.

Carrington, A., & Manzano, F. (2003). *Physica C: Superconductivity, 385,* 205.

Gross, F., Chandrasekhar, B. S., Einzel, D., et al. (1986). *Zeitschrift für Physik B Condensed Matter, 64,* 175.

Johnston, D. C. (2013). *Superconductor Science and Technology, 26,* 115011.

Mühlschlegel, B. (1959). *Zeitschrift für Physik, 155,* 313.

Prozorov, R. (2008). *Superconductor Science and Technology, 21,* 082003.

Prozorov, R., & Giannetta, R. W. (2006). *Superconductor Science and Technology, 19*, R41;
     Erratum: Prozorov, R. *Superconductor Science and Technology, 21*, 082003.
Sheahen, T. P. (1966). *Physical Review, 149*, 368.
Tinkham, M. (1996). *Introduction to Superconductivity*. Mc Graw-Hill. ISBN: 978-0070648784.
Won, H., & Maki, K. (1994). *Physical Review B, 49*, 1397.

Index

A

Abragam function, 127, 135, 467
Abrikosov, Alexei, 237
AgMn, 191
α-model, 246, 295, 518
Angular momentum, 457
Antiferromagnetism, 89, 96, 178
Apodization, 256
Archeology, 416
(Ga,Mn)As, 386
Asteroid, 417
Asymmetry
    α parameter, 81
    β parameter, 81
    energy dependence, 17
    parameter, 67
Asymmetry signal, *see* μSR signal
Atomic capture, 394–397, 399, 402, 410
    probability, 413–415
Avalanche photodiode (APD), *see* Silicon
        photomultiplier
Avoided level crossing (ACL) technique, 65

B

Background, 67, 72, 74, 77
Backward muon, 30, 395
$BaFe_2As_2$, 166
Bardeen, John, 226
$BaSi_2$, 317
BCS theory, 226
Beam cooling technique, 343
BeAu, 294
Becquerel, Henri (Antoine), 2
Bessel function, 110, 132, 184, 187, 251
β-Sn, 294
β-NMR, 353
Bethe formula, *see* Eenergy loss
$Bi_2Sr_2CaCu_2O_8$, 264
Bogoliubov quasiparticle, 242

C

$Ca_{10}Cr_7O_{28}$, 138
CdS, 333
CdSe, 335
CdTe, 335
CeB_6, 207
$CeCu_5Au$, 102, 185
$CeCu_6$, 102, 203
Central limit theorem, 98
$CeRhIn_5$, 90
$CeRu_2Si_2$, 204
Ce_7Ni_3, 173
Charge density, 57, 340
    enhancement, 57, 151
Chemistry, 165
Chiral superconductor, 292
Clay, Jacob, 4
Clebsch-Gordan coefficient, 15
Clogston-Jaccarino plot, 202, 203, 209, 211
Cloud chamber, 5
Cloud muon, 395, 403
Cockcroft-Walton accelerator, 25
Coherence length, 228, 230, 232
Condensation energy, *see* Superconducting
        condensation energy
Contact magnetic field, 151, 156, 193
Continuous beam, 65, 71, 421
Conventional superconductor, 243
Cooper, Leon, 226
Cooper pair, 226, 242, 289, 374
Correlation time, *see* Fluctuation time
Cosmic rays, 7

© Springer Nature Switzerland AG 2024
A. Amato, E. Morenzoni, *Introduction to Muon Spin Spectroscopy*,
Lecture Notes in Physics 961, https://doi.org/10.1007/978-3-031-44959-8

Printed in the United States
by Baker & Taylor Publisher Services